Under Two Flags

A STORY OF THE HOUSEHOLD
AND THE DESERT

Ouida

'Cœur Vaillant se fait Royaume'

Introduced by

John Sutherland

D1304747

Oxford   New York
OXFORD UNIVERSITY PRESS
1995

Oxford University Press, Walton Street, Oxford OX2 6DP

Oxford New York
Athens Auckland Bangkok Bombay
Calcutta Cape Town Dar es Salaam Delhi
Florence Hong Kong Istanbul Karachi
Kuala Lumpur Madras Madrid Melbourne
Mexico City Nairobi Paris Singapore
Taipei Tokyo Toronto
and associated companies in
Berlin Ibadan

Oxford is a trade mark of Oxford University Press

British Library Cataloguing in Publication Data
Data available

Library of Congress Cataloging in Publication Data
Ouida, 1839–1908.
Under two flags : a story of the household and the desert / Ouida;
introduced by John Sutherland.
p.   cm. — (Oxford popular fiction)
Includes bibliographical references.
1. Man-woman relationships—Algeria—Fiction. 2. France. Armée.
Legion étrangère—Fiction. 3. British—Travel—Algeria—Fiction.
I. Title.   II. Series.
PR4527.U59   1995   823'.8—dc20   94–28425
ISBN 0–19–282328–0

1 3 5 7 9 10 8 6 4 2

Typeset by Best-set Typesetter Ltd., Hong Kong
Printed in Great Britain
by Biddles Ltd Guildford and King's Lynn

OXFORD POPULAR FICTION

General Editor ·Professor David Trotter
Associate Editor Professor John Sutherland
Department of English, University College London

Amongst the many works of fiction that have become bestsellers and have then sunk into oblivion a significant number live on in popular consciousness, achieving almost folkloric status. Such books possess, as George Orwell observed, 'native grace' and have often articulated the collective aspirations and anxieties of their time more directly than so-called serious literature.

The aim of the Oxford Popular Fiction series is to introduce, or reintroduce, some of the most influential literary myth-makers of the last 150 years—bestselling works of British and American fiction that have helped define a new style or genre and that continue to resonate in popular memory. From crime and historical fiction to romance, adventure, and social comedy, the series will build up into a library of books that lie at the heart of British and American popular culture.

CONTENTS

Contents

VOLUME III

INTRODUCTION

The terms 'bestseller' and 'popular fiction' are blunt critical instruments. Standing out from the category of works which have sold in their millions and promptly sunk into oblivion are a few which have achieved folkloric longevity and ubiquity. These popular 'classics' will be familiar even to those who have never read them, or may not even recognize their titles. *Under Two Flags: A Tale of the Household and the Desert* is one such novel. It defined a set of romantic stereotypes and narrative formulae now inseparably associated with the 'Foreign Legion'. They have been reworked for successive generations since Ouida's bestselling romance of 1867. As one of its less creditable achievements *Under Two Flags* must have incited any number of weak-minded young men to enlist in what is, in actuality, a brutally unromantic branch of the French military.

The most influential imitation of *Under Two Flags*—one which has to a large extent obliterated Ouida's archetypal text in the popular mind—is P[ercival] C[hristopher] Wren's *Beau Geste* (1924). 'Major' Wren (as he preferred to be known) was an interesting man and an inferior—but very successful—novelist. On graduating from Oxford in the early 1900s Wren resolved with a young Briton's fearlessness to see all five continents of the world before settling down. During these *Wanderjahren* he served briefly in North Africa as a legionnaire. Details are sparse (one would like to know, for instance, how Wren got out; even today tourists meet luckless Britons in Corsica who despair of having signed on for the minimum engagement of five years—penalties for desertion from the Legion have always been Draconian). Wren was almost certainly inspired to join up by romantic recollections of *Under Two Flags*, which he would have read like every schoolboy of his class.

In fact, although everyone associates *Under Two Flags* with 'the Legion', the regiment is never mentioned by Ouida. It is worth spelling this out in detail, since much misapprehension has gathered round the point. In Vol. 1, Chapter 13 ('In the Café of the Chasseurs') it is clear that the hero, Bertie Cecil, having cast the dice to see whether he will ride with the Arabs or the Franks, enlists in the second regiment of the 'Chasseurs d'Afrique'. The Chasseurs were formed in 1831 to put down the mounted Arab insurgents of Algeria (a colonial possession which France had come by almost accidentally the year before). The Chasseurs were not one of the *régiments étrangères*. Although they initially accepted *indigènes* (Arabs) these were soon separated into the 'Spahi' mounted units. It is clear that Bertie enlists as the Frenchman 'Louis Victor', trading on his flawless knowledge of the language.

It is an important detail. Bertie joins a *corps d'élite*. The Foreign Legion was also founded in 1831, and also designed to assist in pacifying Algeria. But its recruits were the scum of Europe and (most important to Ouida) infantrymen who marched like mules with fifty pounds of equipment on their backs and the private soldier's distinctive white kepi on their heads. It is unthinkable that Ouida's fine-boned hero should mess with such *canaille*. They were, apart from anything else, mercenaries. It is not even clear that Ouida, whose grasp of military matters is often patchy, knew of the existence of the Legion. She absorbed her expertise about France's colonial army from a book which had caused some stir in 1864, when it was translated into English—Antoine Camus's *Zéphyrs, Spahis, Turcos, Tringlos: Les Bohèmes du Drapeau*. Camus does not deal with the Legion.

Wren it was who made the Foreign Legion world famous and glamorous, as a kind of Edwardian superstructure to Ouida's mid-Victorian romantic edifice. His reputation (and with it that of the Legion, where tough but sensitive men go to 'forget') took off with *Beau Geste* (1924). Ouida's Bertie Cecil, it will be remembered, has as one of his nicknames 'Beau Lion'. Astutely, Wren followed up his hit with *Beau Sabreur* (1926; the title is lifted from a dashing character in another of Ouida's novels, *Strathmore*, 1865); *Beau Ideal* (1928) and *The Good Gestes* (1929). All featured depictions of Legion battle, love, and glory and drew substantially on Ouida's original text. Although he dabbled in other popular genres, it was for his stories of the Legion that Wren became a household name in Britain and America. His leap into fame in 1924 was assisted by the cult of desert erotomania, inaugurated by Edith M. Hull's and Rudolph Valentino's *The Sheik* (1921), Sigmund Romberg's operetta *The Desert Song* (1926), and the postwar 'Lawrence of Arabia' mystique.

The basic *donnée* of Wren's string of Geste romances—that of the English aristocrat self-sacrificially joining the French Army to protect the good name of those near to him—derives directly from *Under Two Flags*. In Ouida's novel, the Hon. Bertie Cecil (an 'Hon.' if there ever was one, as Monica Stirling notes) is driven by a near-suicidal *noblesse oblige*. He must protect his younger brother Berkeley (no 'Hon.', he) who has forged a bill. Bertie must also protect the name (even though it is by no stretch a good name) of the woman with whom he has been amusing himself, 'the titled and wedded' Lady Guenevere. Bertie could, of course, exonerate himself with a couple of explanatory sentences: 'Berkeley did it. I was riding with Lady Guenevere in the Park. What was I doing with her?

Mind your own damn business!' Rather than take this dishonourable but rational course, he resolves to seek probable death and certain hardship under the burning sun of Algeria.

In Wren's novel there is an analogous but rather more ingenious dilemma of honour. Michael 'Beau' Geste discovers that his aunt's renowned heirloom, the Blue Water sapphire, is paste. To protect the family's honour (and its evidently rocky finances) he 'steals' the gem, and buries himself in the Foreign Legion. (Beau is accompanied, in an unlikely plot turn borrowed from Kipling's *Soldiers Three*, by his two brothers.) In both Ouida's and Wren's novels, the hero suffers the torments of the damned under a sadistic superior: Colonel (and Marquis) Raoul de Châteauroy in *Under Two Flags*, the *sous officier* Lejaune in *Beau Geste*. In Ouida's novel Châteauroy finally provokes Bertie/Louis into striking him—a death warrant by the stern code of his army. In Wren's novel Lejaune, having manned the battlements of Fort Zinderneuf with the corpses of shot legionnaires, is left alone with Beau. Both expire bloodily. (Wren's extravagantly morbid last act was clearly inspired by the 1914–18 carnage in the trenches, which he witnessed as an over-age and eventually wounded volunteer.)

The Foreign Legion romance might have petered out after Wren's exploitative string of bestsellers but for the geographical accident of Hollywood being in Southern California. Just as Tarzan swung through numberless sequels among the lianas in the conveniently nearby Los Angeles Arboretum, and cowboys shot it out in the chaparral of Griffith Park, so the dunes of Death Valley and the Coachella Dunes in Arizona encouraged early film-makers to numerous makes and remakes of Foreign Legion epics. Ersatz Saharan location shooting, complete with palms and oases, could be had on low budget, for the modest expense of a six-hour journey on good roads, a jerry-built desert fort, and the hire of a brace of camels from the Los Angeles Zoo. There have been three full-dress versions of *Beau Geste* (in 1926, 1936, 1966—the last with a sublimely miscast Telly Savalas playing Lejaune). There have also been three film adaptations of *Under Two Flags*; in 1916, 1922, and 1936. The last (the first talkie version, and a $2-million budget spectacle) is by far the best. It starred the mature Ronald Colman—an expatriate English actor who specialized in Anglo-Saxon sang-froid (he also played on screen Sydney Carton, Raffles, Bulldog Drummond, Rudolf Rassendyll, and—inevitably—Beau Geste).

The film versions of *Under Two Flags* betray Ouida's original novel in two significant details. The first and most irritating is the insertion of the Foreign Legion into the action. This renders Bertie a footslogging infantryman rather than the magnificent cavalryman ('chevalier') of Ouida's superheated imagination. The second departure from the author's original conception is Hollywood's incorrigible over-pneumatizing of Cigarette, so that she can be played by some current sex goddess (in the 1922 version, as performed by Priscilla Dean, she is made half-Arab so that she may lounge voluptuously in oriental silks on a divan). This revision of Ouida's 17-year-old 'unsexed' *gamine* into a busty beauty of mature years flouts the image of Cigarette given the reader in Vol. 2, Chapter 1.

She was very pretty, audaciously pretty, though her skin was burned to a bright sunny brown, and her hair was cut as short as a boy's, and her face had not one regular feature in it. But then—regularity! who wanted it, who would have thought the most pure classic type a change for the better, with those dark, dancing, challenging eyes; with that arch, brilliant, kitten-like face, so sunny, so *mignon*, and those scarlet lips like a bud of camellia that were never so handsome as when a cigarette was between them.

Claudette Colbert, in the 1936 version, comes close at times to capturing Ouida's vision, but at 31 and with her Hollywood star baggage she cannot quite manage the pubescent and guttersnipe nuances.

In addition to the straight adaptations, Hollywood produced a clutch of freer versions of *Under Two Flags*. They include *The Winding Stair* (1925), *The Silent Lover* (1926), *The Foreign Legion* (1928), *Three Men and a Maid* (1929), *Morocco* (1930), and *Women Everywhere* (1930). Among the interesting twists imposed on Ouida's plot was that in *The Foreign Legion* Châteauroy is made Bertie's father. There is, the reader will recall, some warrant in Ouida's novel for this invention. Bertie does indeed have mysterious parentage. Evil tongues (to which Viscount Royallieu has lent too-willing credence) have linked Bertie's dead mother with a dashing cousin, Alan Bertie—'a fearless and chivalrous soldier'. Bertie bears a suspiciously close likeness to this relative, who left England (consumed, presumably, with hopeless love) and died at the head of 'Bertie's Horse, the most famous of all the wild Irregulars of the East'. It would require improbabilities beyond even Ouida's invention to link Alan with the villainous 'Faucon Noir', Châteauroy. None the less, the novel persistently hints at connection. Bertie's birthright name (Royallieu—royal place) is oddly analogous to Châteauroy (house of the king). Nor is any plausible reason ever given for the commandant's visceral hatred for Louis Victor.

Hollywood has shown extraordinary pertinacity in dispatching the most unlikely matinée idols to serve in the Legion and stumble through their parts in the stilted English that is to be taken by the audience as idiomatic French (Ouida's incomparably haughty glosses of service *argot*—lifted in the main from Camus—are one of the many incidental pleasures of *Under Two Flags*). Gary Cooper was drafted in 1936 (*Beau Geste*), George Raft in 1948 (*Outpost in Morocco*), Alan Ladd in 1953 (*Desert Legion*), and Brian Keith in 1958 (*Desert Hell*). Even Gene Hackman—fresh from his New York ethnic triumphs as Popeye Doyle was given a more bizarre French connection in the Legion melodrama *March or Die* (1977), a film whose box office disaster may have deterred film-makers from the dunes of Death Valley for a while.

Some of the most interesting film adaptations over the years have been the burlesques of *Under Two Flags* and its progeny, which make free play with the conventions and stock episodes patented by Ouida and Wren. They include mirage jokes, 'I joined to forget' jokes, brutal commandant jokes, besieged desert fort jokes, 'cafard' jokes, fiendish Arab torture jokes, luscious camp-follower jokes, Casbah jokes. Laurel and Hardy's *Sons of the Desert* (1933), Hope and Crosby's *Road to Morocco* (1942), *Abbot and Costello in the Foreign Legion* (1950), and the 'Carry-on' team's hilarious *Follow that Camel* (1967; it features Phil Silvers as Sergeant Bilko alias Lejaune) are distant, deformed, but highly characteristic offspring of Ouida's original romance.

Unfortunately, Ouida tends to be lumped in the public mind with her imitator, the much inferior Marie Corelli. Louisa Ramé (to use her birth name—whose foreignness, unlike Mary Mackay's, was half genuine) was a novelist who, in her day, had the admiration of Henry James and other respectable judges (as Corelli did not). Ouida deserves rather more of a literary reputation than posterity has allowed her. Born in Bury St Edmunds, Suffolk, in 1839, Louisa was the daughter and only child of Susan Ramé (née Sutton) and her husband Louis. 'Ouida'—the child's mispronunciation of 'Louisa'—was brought up in a bilingual household (under two flags, one might say). Her mother was English, the moderately prosperous daughter of a local wine merchant, and a model of middle-class Victorian maternal rectitude. (Mrs Ramé acted as her daughter's chaperone until her death in 1893—long after her daughter needed any protection from the male sex.) The dominant influence on Ouida's personality, however, was her French father. It was to his flag she cleaved. When, in much later life, Bury St Edmunds offered a monument to her as a famous daughter of the town she disdained 'this tomfoolery in Suffolk'. 'I identify myself with father's French race and blood,' she grandly proclaimed. There was no monument during her lifetime.[1]

---

[1] Biographical details are taken from Monica Stirling, *The Fine and the Wicked: The Life and Times of Ouida* (London, 1958).

Ouida's father, Louis Ramé, is a man of mystery. No biographer has penetrated the obscurities of his life. On the face of it, he was a humble French schoolmaster in a sleepy East Anglian town. It was as a teacher that Louis presumably met and wooed one of his pupils Susan Sutton (he was evidently some twenty years older than she). But in another more glamorous part of his life, Louis was allegedly a secret agent in the pay of the exiled Louis Napoleon. As Monica Stirling puts it: 'Bury St Edmunds seems an incongruous place in which to find a Bonapartist agent—so incongruous that Louis Ramé probably was just this.' Whatever his reasons, Louis made frequent and long trips abroad on unspecified missions. He was also a 'Carbonarist'—a member of the secret society dedicated to the liberation of Italy. (His daughter was, in time, to conceive an even more intense passion for that country than for her 'native' France.)

Ouida evidently had a happy girlhood in Bury St Edmunds—a happiness only upset by her father's mysterious (and as she must have feared dangerous) trips. The Crimean War which coincided with her puberty made an indelible impression—more specifically, the glorious six-hundred charging into the mouths of the Russian cannon exalted her. The English cavalryman was thereafter to be Ouida's highest ideal of manhood. It was important to the girl too that the Crimean War forged what came to be known as the *Entente Cordiale*. England and France, historical enemies, were now allies (united against Prussian militarism). The alliance combined important elements in her own life.

Ironically, it was at this point that the little Anglo-French alliance that was the Ramé family broke apart. Louis seems never to have lived at home again, and only to have visited his family rarely. Ouida, her mother, and her grandmother (together with Ouida's dog Beausire—'beautiful father') moved to London in 1857, where they were apparently supported by Mrs Ramé's savings and remittances from the absent Louis. Ouida never returned to Bury St Edmunds again for as long as she lived. The English provinces were behind her forever.

Many young Victorian ladies dreamed of liberating themselves from the oppression of impoverished gentility by writing a bestselling novel. Ouida, like Charlotte Brontë, was one of the few who succeeded and— even more unusually—she succeeded at her first serious try. One of their neighbours in Hammersmith, where the Ramés first resided, was a brother of the magazine editor, W. Harrison Ainsworth. In his day (now some quarter-century passed) Ainsworth had been a dandy and a bestselling novelist. He was still very vain and time had revealed him to be a second-rate writer. But he had a keen eye for talent—particularly young female talent. It was he, for instance, who 'discovered' Mrs Henry

Wood in 1859 after *East Lynne* had been turned down by other editors and publishers. Ainsworth serialized the story in one of his magazines, and it went on to become a huge bestseller. He recognized a similar talent in 'Ouida' (now her pen-name; she guarded her real identity jealously—perhaps as a sales gimmick, more likely from motives of shyness). No details survive, but it is likely that Louisa flattered and played on the old dandy's susceptibilities. Whatever the inducements Ainsworth in 1859 accepted her first story, *Dashwood's Drag; or the Derby and What Came of it*, a 'fast' story of the turf, and thereafter commissioned a string of similar dashing, horsey, high-life tales. They cannot have earned her more than about £20 apiece, but she turned them out in large quantity and at high speed. There followed the inevitable three-volume novel *Granville de Vigne* in 1861 which Ainsworth again serialized in one of his magazines. He also introduced the young author to his publishers, Tinsley Brothers and Chapman & Hall, who took over the book issue of her novels.

Ouida became a favourite author with military readers who liked her brio, her 'fast' amorality (tame enough by the standards of later decades), her aura of expensive femininity (now enhanced by the services of Worth of Paris), and her frank hero-worship of the British officer class. The more respectable sectors of society chose to be appalled. *Blackwood's Magazine* declared in 1867: 'We do not feel ourselves capable of noticing . . . certain very fine and very nasty books signed with the name of Ouida. It is supposed to be a woman.' Such censure (in which the majority of respectable journals joined) served only to make her novels more interesting to the general reader, bored with the decencies of mid-Victorian domestic fiction and mild Trollopian intrigues in cathedral towns.

Ouida consorted socially with notable military men and adventurers—though always apparently with her mother present as the guarantor of her respectability. Now renamed 'Louise de la Ramée' she was hostess to a salon at the fashionable Langham Hotel where she cultivated men with interesting pasts, permitting them to smoke and talk as freely as if she and her mother were not there. Ouida's set around the time she wrote *Under Two Flags* is vividly described by Monica Stirling (*The Fine and the Wicked*, p. 64):

Her regular guests included: Richard Burton, the explorer, who looked like a handsome Arab and stained his under eyelids with kohl . . . Colonel Meadows Taylor, who served in the Army of the Nizam of Hyderabad, was put in charge of some of the ceded districts of Deccan after the mutiny, and wrote more than half a dozen books, such as *Confessions of a Thug* . . . Major Brackenbury, the *Times'* military correspondent; Colonel Pemberton, another *Times* correspondent, killed in the Franco-Prussian war; Whyte-Melville, who joined the 93rd Highlanders at

seventeen, retired from the Coldstream Guards as a captain ten years later, volunteered for the Crimea, was appointed major of the Turkish irregular cavalry, and
wrote very succesful novels of fashionable and sporting life [he was, rather
unkindly, called 'Ouida in breeches' behind his back] . . . Bierstadt, the American
landscape-painter, and the poet Longfellow; Lieutenant-General Sir Edward
Bruce Hamley, who had served in the Crimea, was later Commandant of the Staff
College, and wrote a score of books on literary as well as military subjects . . .
George Lawrence, author of the bestselling novel *Guy Livingstone*; Hamilton Aidé,
the English poet and novelist, born in Paris of an Armenian father and an English
mother, who studied at Bonn and served in the British army before concentrating
on literature . . . Major-General Breckinridge of the Confederate Army, later Vice-
President of the United States . . . Bulwer Lytton, the Byronic author of *The Last
Days of Pompeii*.

   Under Two Flags (1867) was Ouida's fourth full-length work, and was to
be the biggest seller of her career (it racked up more than sixty editions
in ten years). It is important, however, to scotch the legend propagated by
her biographers that the novel made her rich.[2] Ouida had been writing
stories for *The British Army and Navy Review* since 1864 and from August
1865 she serialized *Under Two Flags* in its pages. For each instalment she
received a measly £6. The journal ceased publication in June 1866 with
the climax of the novel unpublished. Ouida (no less than her tantalized
readers) was eager to find a publisher for the complete text. Bentley
turned it down on the advice of their reader who declared *Under Two
Flags* to be a 'very unmoral book'. Chapman & Hall eventually took it
paying, as Celia Phillips estimates, something under £200 for the entire
copyright. Ouida may have had about as much again from America (where
there were around fifteen pirated editions). In all, she almost certainly
received under £500 for the novel. In 1876 Chapman sold the copyright
to Chatto & Windus who brought out some 700,000 cheap editions of
Under Two Flags during Ouida's lifetime, for which she received nothing.
Nor did she receive a penny from the scores of theatrical adaptations, nor
did her estate profit from the film versions. She was, eventually, to die
destitute while the copyright was still coining money for Chatto &
Windus. The success of the novel did, of course, boost her reputation and
enabled her to make good bargains with subsequent publishers. But none
of her other novels sold as well. She was, in short, robbed.

   Under Two Flags contained what were by now Ouida's trademarks:
beautiful women, reckless guardsmen, superb horseflesh, exotic lo-
cations, affairs of honour, and panting romance. But the formative influence on the plot was the final act of Louis Ramé's strange domestic

---

[2] The publishing history of which follows is taken from Celia Phillips, '*Under Two Flags*:
The Publishing History of a Best-Seller', *Publishing History* (1978), iii. 67–9.

career. In 1863 the schoolmaster-spy left England once more. It seems that Ouida never saw him again. Nor did she or her mother ever discover what had happened to him. The persistent rumour was that he went undercover and died in the Commune disturbances of 1870–2. Four years after his departure, the 29-year-old Ouida dramatized her romantic fantasies about her lost father in *Under Two Flags*. The novel is propelled by a girl's immature wish fulfilments and wild romantic speculations. He had *not* deserted them—there must have been motives of great honour in his going, in his never writing, in his having disappeared from the face of the earth. It was honour which kept him silent. Nothing else would have induced Papa to have left his Louise without so much as a word of farewell or a token of love. (The banal likelihood, which no biographer seems to have investigated, is that Ramé was a bigamist, with another family in France.)

Ouida projected herself doubly into the novel. She was Cigarette, the 'unsexed' camp follower, who (preposterously) wins the Cross of the Legion of Honour for gallantry on the battlefield and who dies hurling herself into the firing squad's fusillade of bullets, shielding the man she loves from certain death. What Ouida intended by this ballistically impossible finale is that when her father read her novel, he should realize that she was still willing to die for him. *Under Two Flags*'s writing exceeds even its own higher pitches of melodrama in describing Cigarette's desperate gallop across the moonlit desert, pardon in hand, to prevent death at dawn for Louis. Her own horse, Etoile-Filante, has collapsed under her; she has borrowed a steed from her enemies, the Arabs, who honour the young girl's noble mission:

borne by the fleetness of the desert-bred beast, she went away through the heavy, bronze-hued dullness of the night. Her brain had no sense, her hands had no feeling, her eyes had no sight; the rushing of waters was loud on her ears, the giddiness of fasting and of fatigue sent the gloom eddying round and round like a whirlpool of shadow. Yet she had remembrance enough left to ride on, and on, and on without once flinching from the agonies that racked her cramped limbs and throbbed in her beating temples; she had remembrance enough to strain her blind eyes toward the east and murmur, in her terror of that white dawn, that must soon break, the only prayer that had been ever uttered by the lips no mother's kiss had ever touched: '*O God! keep the day back!*' (Vol. 3, Chapter 11)

In addition to dramatizing herself as the tomboy Cigarette, Ouida also projected herself into the part of 'Petite Reine', the exquisitely well-behaved, daintily dressed, and stay-at-home little girl who last saw Bertie when she was 8 years old, but who has loved him ever since. The Princess Venetia (as she grows up) will never marry (as Ouida never married) unless, by some miracle, Bertie returns from the dead—which, by the friendly mechanisms of romance, he does. Louis Ramé did not.

The two women who love Bertie are both 'children'. Cigarette is we calculate (by reference to her infantine heroism in the Revolution of 1848) 17 at the time of the main action of the novel. 'Petite Reine'— adding Bertie/Louis's twelve years of service to her childhood age—must be a couple of years older, just 20. Bertie himself, with his incipiently 'silvered' hair, must be something over 40. He could be their father.

Although she was still under 30 at the time she wrote *Under Two Flags*, Ouida was an experienced and highly professional novelist. She had perfected her technique on her chosen literary instrument—the so-called 'Sensation Novel'. *Under Two Flags* features many of the gimmicks of the genre: notably the 'dead but not dead' device. Popular novels of the 1860s and 1870s are replete with characters who 'seem' to have died, but who return under deep disguise into their previous lives. 'Dead but not dead' heroes and heroines proliferate in the novels of Wilkie Collins, Charles Reade, and Mary Braddon (notably, *Lady Audley's Secret*, 1862). The trope is even found in Dickens's *Our Mutual Friend* (1865) where the drowned John Harmon returns as Rokesmith. But Ouida's immediate inspiration for *Under Two Flags* was *East Lynne* (1861). In that novel, the faithless Lady Isabel Vane is supposed to have been killed in a train wreck. But the smashed corpse was not hers. She returns as the French governess 'Madame Vine', and actually takes up a position teaching her own young children who, in the way of sensation fiction, do not recognize their own mother behind the green glasses and lost good looks. One of the children dies, provoking the famous ejaculation 'Dead! dead! and never called me mother!'

Ouida borrows Wood's invention in Bertie's disappearance from the face of the earth (in a train wreck among which are unrecognizable corpses) and his return as 'Louis Victor'. There is also a clear echo of *East Lynne* in Bertie's agonies when the unworthy brother who now has his title, his best friend who has always believed him innocent, and the woman who has always loved him appear (most improbably) in Algeria— and, of course, do not recognize the Hon. Bertie Cecil in the bronzed, aged, and *déclassé* Louis Victor.

Sensation novels were—as *Punch* cartoons never tired of pointing out—typically written for girls. Ouida disdained this market. 'Je n'écris pas pour les femmes. J'écris pour les militaires', she liked to declare. The point is stressed in the dedication of *Under Two Flags* to Colonel George Poulett Cameron (1806–82), an old war-horse and author of *The Romance of Military Life* (1853). It is further stressed in the 'Avis au Lecteur' which announces: 'This story was originally written for a military periodical [i.e. *The British Army and Navy Review*]. It has been fortunate enough to receive much commendation from military men, and for them it is now specially issued in its present form [i.e. as a three-volume novel].' Numerous digressions in the novel certify Ouida's writing for comrades in the officers' mess—for example the long eulogy on Tommy Atkins in Vol. 2, Chapter 3: 'There aren't better stuff to make soldiers out of nowhere than Englishmen, God bless 'em! but they're badgered, they're horribly badgered.'

Ouida, of course, was not principally occupied with discussing the conditions of service for other ranks in the British Army. She left that to the politicians. (Edward Cardwell, the Secretary of War, was in fact bringing in far-reaching reforms of the British Army as she wrote.) Ouida set out in her novel to delineate, with all the garish pigments of romantic fiction, the 'beau idéal' of the officer, 1860s-style. Accurate in its observations of the mannerisms and affectations of the swell-officer class, *Under Two Flags* was also immensely influential in propagating the approved fashions of the day. Young officers must have read the novel not just as a flattering looking-glass, but as a manual in which they might pick up tips as to how they should comport themselves.

For modern readers, the ideal guards' officer of 130 years ago strikes some distinctly odd notes. Particularly, the pose of listlessness seems odd in professional men of action. The roots of this mannerism are interesting to trace. In 1815, after Waterloo, Walter Scott went to Paris. He saw there the victorious young English officers and noted their extraordinary *nonchalance*, how studiously unexcited they were by the greatest triumph ever won by an army. Scott assumed the young officers were imitating the sang-froid of their commander, Wellington. It is as likely that they were influenced by Byron's *Childe Harold*, the first two cantos of which had come out in 1812. The new cult of military *indifférence* attained an acme in Mrs Gore's *Cecil* (1841). Gore's romance glorified the sub-Byronic ideal of the epicene, hyper-cultivated English officer, who for all his 'coxcombry' has an indomitable inner strength which will be called out by moments of supreme crisis. Until that emergency, the hero exists in a condition of latency—or 'ennui'—relieved only by beautiful women, gaming, and peacock-like decoration of his own person. This ideal of the English officer class was given a new, post-Crimean lease of life by the immensely popular hero of G. A. Lawrence's *Guy Livingstone* (1857). It was in the Crimea, of course, that the British officer picked up (from the

Russian foe) the practice of smoking the so-called 'cigarettes'—in the 1860s the bored young man's inevitable appendage.[3] Until war, or some other great trial comes along, Bertie whiles away the time with 'Highland shooting, his Baden gaming, his prize-winning schooner among the RVY Squadron, his September battues, his Pytchley hunting, his pretty expensive Zu-Zus [i.e. mistresses] and other toys, his drag for Epsom and his trap and hack for the Park, his crowd of engagements through the season, and his bevy of fair leaders of the fashion.' And, of course, cigarettes.

A main component in the ideal of the English officer class promoted by Gore, Lawrence, Ouida (and a host of imitators) was that of aristocratic amateurism. When a mess comrade upbraids him for training for horse-races on champagne, burgundy, and night-long assignations with fascinating women, Bertie's friend Rockingham ('the Seraph') answers on his behalf. Bertie he declares 'don't believe in training. Nor do I . . . it's utter bosh. You might as well be in purgatory; besides, it's no more credit to win then than if you were a professional.' The essence of winning is to do so effortlessly, without having to try. And, at the end of the day, winning is not everything; *style* is everything. Not, of course, that Bertie is a loser. He wins the Moselle Cup (on his beloved Forest King) and would have won the Prix de Dames at Baden were it not for his horse being nobbled by a 'welsher' and a villainous 'Hebrew' money lender (there is, the reader should be warned, an unpleasant vein of anti-Semitism in the early sections of *Under Two Flags*).

What is most perplexing for the modern reader in the portrait of Bertie is a kind of Wildean sexual ambiguity. Take, for example, our first view of him as he emerges from behind the clouds of his great meerschaum, like Venus from the waves. He has, we are told:

a face of as much delicacy as a woman's; handsome, thoroughbred, languid, non-chalant, with a certain latent recklessness under the impassive calm of habit, and a singular softness given to the large, dark hazel eyes by the unusual length of the lashes over them. His features were exceedingly fair—fair as the fairest girl's; his hair was of the softest, silkiest, brightest chestnut; his mouth very beautifully shaped; on the whole, with a certain gentle, mournful love-me look that his eyes had with them, it was no wonder that great ladies and gay *lionnes* alike gave him the palm as the handsomest man in all the Household Regiments. (Vol. 1, Chapter 1)

Bertie's nickname, earned at Eton, is 'Beauty'.

Bertie is 'one of the cracks of the Household', the son and heir of Viscount Royallieu (with whom he is at odds). He is adored by women, by

[3] For a social history of smoking at mid-century see R. D. Altick, ch. 8, 'The Favourite Vice of the Nineteenth Century', *The Presence of the Present* (Columbus: Ohio, 1991).

his fellow officers, and by his factotum, Mr Rake (who dies loyally on the desert sands, respectfully asking to be remembered to the Hon. Bertie's horse, should his master by some happy chance ever return to England). For all his apparent sexual ambiguity, Bertie has many mistresses, although whether they serve his bodily needs or are merely ornaments is not clear. His ménage includes: Zu-Zu, 'the last *coryphée* whom Bertie had translated from a sphere of garret bread and cheese to a sphere of villa champagne and chicken'; the Lady Regalia (to whom he has made love with a servant in the room—regarding that functionary as no more sentient than a piece of furniture); and the ominous Lady Guenevere.

In Africa Bertie, ex-officer in the Life Guards, is transmuted into 'Louis Victor', Private (and after twelve arduous years Corporal) in the Chasseurs. None the less, his extraordinary gallantry and his wholesale slaughter of Arabs (which Ouida keeps offstage) earn him the regimental title of '*Bel à faire peur*', and 'Beau Lion'. Although he seeks death and oblivion his past seeks him out. Once again, he faces the impossible demands of 'honour'—what should a gentleman do when his superior officer casts a 'vile' aspersion against a pure English lady? Swallow the insult, or revenge it, with the certainty that the mutinous act will mean death by firing squad? The question admits of only one answer in Ouida's fiction.

Like most three-volume novels, *Under Two Flags* is at least one volume too long. Characters are garrulous (Cigarette, for instance, talks interminably with the firing squad's point-blank hail of bullets lodged in her breast.) None the less, Ouida contrives to compensate in brio for the longueurs of her middle and later narrative. Not to spoil the reader's pleasure, it may be said that the extraordinary epilogue, in which the hero is clearly more interested in reunion with his old horse, Forest King, than with the adoring Princess Venetia, has its prophetic aspects. In 1871 Ouida moved on from London to Italy—a country she discovered she loved more than any other (even France). Here she was to live until her death. Her principal companion was her mother, until the old lady's death in 1893. Thereafter, Ouida devoted herself to her beloved dogs. Her popularity had long been eclipsed by faster, less innocent writers (including the abominated Marie Corelli). She ended her days destitute, dying at Viareggio in 1908. Bury St Edmunds, determined against her wish to remember her, erected a fountain for dogs and horses with an inscription composed by her friend Lord Curzon: 'Here may God's creatures whom she loved assuage her tender soul as they drink.'

SELECT BIBLIOGRAPHY

Under Two Flags was serialized in *The British Army and Navy Review* beginning in August 1865. In June 1866 the journal ceased publication with the climax of the novel unpublished. The present edition is set from the first book edition of 1867, published by Chapman and Hall, Piccadilly, London.

Biographies and biographical studies of Ouida include:

Yvonne ffrench, *Ouida: A Study in Ostentation* (New York, 1938).
Elizabeth Lee, *Ouida: A Memoir* (London, 1914).
Eileen Bigland, *Ouida: The Passionate Pilgrim* (London, 1950).
Monica Stirling, *The Fine and the Wicked: The Life and Times of Ouida* (London, 1958).

Stirling's book contains a useful bibliography of secondary, cultural-historical sources. The best compendium of (and directory to) criticism of Ouida's fiction is found in *Twentieth-Century Literary Criticism*, vol. 43, ed. Laurie DiMauro (Detroit and London, 1992), 336–77. Celia Phillips (whose article on *Under Two Flags* is cited in the Introduction here) has also written the definitive study on 'Ouida and her Publishers', *The Bulletin of Research in the Humanities*, 81: 2 (Summer, 1978), 210–15. An excellent introduction to Victorian society in Ouida's day will be found in R. D. Altick, *The Presence of the Present* (Columbus: Ohio, 1991). The 1936 film version of *Under Two Flags* is available in video stores specializing in Hollywood 'classics'. Material relating to other film versions can readily be found in the British Film Institute.

Under Two Flags

To Colonel Poulett Cameron,

CB, KCT & S, &C.,

WHOSE FAMILY

HAS GIVEN SO MANY BRILLIANT SOLDIERS TO THE ARMIES OF

FRANCE AND ENGLAND,

AND MADE THE BATTLE-FIELDS OF EUROPE RING WITH

'THE WAR-CRY OF LOCHIEL',

THIS

STORY OF A SOLDIER'S LIFE

IS DEDICATED IN

SINCERE FRIENDSHIP

AVIS AU LECTEUR

This story was originally written for a military periodical. It has been fortunate enough to receive much commendation from military men, and for them it is now specially issued in its present form. For the general public it may be as well to add, that where translations are appended to the French phrases, those translations follow the idiomatic and special meaning attached to those expressions in the *argot* of the Army of Algeria, and not the correct or literal one given to such words or sentences in ordinary grammatical parlance.

OUIDA

VOLUME I

CHAPTER I

'Beauty of the Brigades'

'I don't say but what he's difficult to please with his Tops,' said Mr Rake, factotum to the Hon. Bertie Cecil, of the First Life Guards, with that article of hunting toggery suspended in his right hand as he paused, before going upstairs, to deliver his opinions with characteristic weight and vivacity to the stud-groom, 'he *is* uncommon particular about 'em; and if his leathers ain't as white as snow he'll never touch 'em, tho' as soon as the pack come nigh him at Royallieu, the leathers might just as well never have been cleaned, them hounds jump about him so; old Champion's at his saddle before you can say Davy Jones. Tops are trials, I ain't denying that, specially when you've jacks, and moccasins, and moor boots, and Russia-leather crickets, and turf hacks, and Hythe boots, and waterproofs, and all manner of varnish things for dress, that none of the boys will do right unless you look after 'em yourself. But is it likely that *he* should know what a worry a Top's complexion is, and how hard it is to come right with all the Fast Brown polishing in the world? How should *he* guess what a piece of work it is to get 'em all of a colour, and how like they are to come mottled, and how a'most sure they'll ten to one go off dark just as they're growing yellow, and put you to shame, let you do what you will to make 'em cut a shine over the country? How should *he* know? *I* don't complain of that; bless you, he never thinks. It's "do this, Rake", "do that", and *he* never remember 't isn't done by magic. But he's a true gentleman, Mr Cecil; never grudge a guinea, or a fiver to you; never out of temper neither; always have a kind word for you if you want; thoro'-bred every inch of him; see him bring down a rocketer, or lift his horse over the Broad Water! He's a gentleman—not like your snobs that have nothing sound about 'em but their cash, and swept out their shops before they bought their fine feathers!—and I'll be d——d if I care what I do for him.'

With which peroration to his born-enemy the stud-groom, with whom he waged a perpetual and most lively feud, Rake flourished the tops that had been under discussion, and triumphant, as he invariably was, ran up the back stairs of his master's lodgings in Piccadilly, opposite the Green Park, and with a rap on the panels entered his master's bedroom.

A Guardsman at home is always, if anything, rather more luxuriously accommodated than a young Duchess, and Bertie Cecil was never behind his fellows in anything; besides, he was one of the 'cracks' of the Household, and women sent him pretty things enough to fill the Palais Royal. The dressing-table was littered with Bohemian glass and gold-stoppered bottles, and all the perfumes of Araby represented by Breidenbach and Rimmel.

The dressing-case was of silver, with the name studded on the lid in turquoises; the brushes, boot-jacks, boot-trees, whip-stands, were of ivory and tortoiseshell; a couple of tiger-skins were on the hearth, with a retriever and blue greyhound in possession; above the mantelpiece were crossed swords in all the varieties of gilt, gold, silver, ivory, aluminum, chiselled and embossed hilts; and on the walls were a few perfect French pictures, with the portraits of a greyhound drawn by Landseer, of a steeplechaser by Harry Hall, one or two of Herring's hunters, and two or three fair women in crayons.

The hangings of the room were silken and rose-coloured, and a delicious confusion prevailed through it pell-mell, box spurs, hunting stirrups, cartridge-cases, curb chains, muzzle-loaders, hunting-flasks, and white gauntlets, being mixed up with Paris novels, pink notes, point-lace ties, bracelets and bouquets to be dispatched to various destinations, and velvet and silk bags for banknotes, cigars, or vesuvians, embroidered by feminine fingers, and as useless as those pretty fingers themselves. On the softest of sofas, half dressed, and having half an hour before splashed like a water dog out of the bath, as big as a small pond, in the dressing-chamber beyond, was the Hon. Bertie himself, second son of Viscount Royallieu, known generally in the Brigades as 'Beauty'. The appellative, gained at Eton, was in no way undeserved. When the smoke cleared away that was circling round him out of a great meerschaum-bowl, it showed a face of as much delicacy and brilliancy as a woman's, handsome, thoroughbred, languid, nonchalant, with a certain latent recklessness under the impassive calm of habit, and a singular softness given to the large dark hazel eyes by the unusual length of the lashes over them. His features were exceedingly fair—fair as the fairest girl's; his hair was of the softest, silkiest, brightest chesnut; his mouth very beautifully shaped; on the whole, with a certain gentle, mournful love-me look that his eyes had with them, it was no wonder that great ladies and gay lionnes alike gave him the palm as the handsomest man in all the Household Regiments—not even excepting that splendid golden-haired Colossus, his oldest friend and closest comrade, known as 'the Seraph'.

He looked now at the tops that Rake swung in his hand and shook his head.

'Better, Rake, but not right yet. *Can't* you get that tawny colour in the tiger's skin there? You go so much to brown.'

Rake shook his head in turn, as he set down the incorrigible tops beside six pairs of their fellows and six times six of every other sort of boots that the covert-side, the heather, the flat, or the 'sweet shady side of Pall Mall' ever knew.

'Do my best, sir; but Polish don't come nigh Nature, Mr Cecil.'

'Goes beyond it, the ladies say; and to do them justice, they favour it much the most,' laughed Cecil to himself, floating fresh clouds of turkish about him. 'Willon up?'

'Yes, sir. Come in this minute for orders.'

'How'd Forest King stand the train?'

'Bright as a bird, sir; *he* never mind nothing. Mother o' Pearl she worreted a little, he says; she always do, along of the engine noise; but the King walked in and out just as if the stations were his own stable-yard.'

'He gave them gruel and chilled water after the shaking before he let them go to their corn?'

'He *says* he did, sir.'

Rake would by no means take upon himself to warrant the veracity of his sworn foe the stud-groom; unremitting feud was between them; Rake considered that he knew more about horses than any other man living, and the other functionary proportionately resented back his knowledge and his interference, as utterly out of place in a body-servant.

'Tell him I'll look in at the stable after duty and see the screws are all right; and that he's to be ready to go down with them by my train tomorrow—noon, you know. Send that note there, and the bracelets, to St John's Wood: and that white bouquet to Mrs Delamaine. Bid Willon get some Banbury bits—I prefer the revolving mouths—and some of Wood's double mouths and Nelson gags; we want new ones. Mind that lever-snap breech-loader comes home in time. Look in at the Commission stables, and, if you see a likely black charger as good as Black Douglas, tell me. Write about the stud fox-terrier, and buy the blue Dandy Dinmont; Lady Guenevere wants him. I'll take him down with me. But first put me into harness, Rake; it's getting late.'

Murmuring which multiplicity of directions, for Rake to catch as he could, in the softest and sleepiest of tones, Bertie Cecil drank a glass of curaçoa, put his tall lithe limbs indolently off his sofa, and surrendered himself to the martyrdom of cuirass and gorget, standing six feet one without his spurred jacks, but light-built and full of grace as a deer, or his weight would not have been what it was in gentleman-rider races from the Hunt steeplechase at La Marche to the Grand National in the Shires.

'As if Parliament couldn't meet without dragging us through the dust! The idiots write about "the swells in the Guards," as if we had all fun and no work, and knew nothing of the rough of the Service. I should like to learn what they call sitting motionless in your saddle through half a day, while a London mob goes mad round you, and lost dogs snap at your charger's nose, and dirty little beggars squeeze against your legs, and the sun broils you, or the fog soaks you, and you sit sentinel over a gingerbread coach till you're deaf with the noise, and blind with the dust, and sick with the crowd, and half dead for want of sodas and brandies, and, from going a whole morning without one cigarette!—not to mention the inevitable apple-woman who invariably entangles herself between your horse's legs, and the certainty of your riding down somebody and having a summons about it the next day! If all that isn't the rough of the Service, I should like to know what is? Why, the hottest day in the batteries, or the sharpest rush into Ghoorkahs or Bhoteahs, would be light work compared!' murmured Cecil, with the most plaintive pity for the hardships of life in the Household, while Rake, with the rapid proficiency of long habit, braced, and buckled, and buttoned, knotted the sash with the knack of professional genius, girt on the brightest of all glittering, polished, silver steel 'Cut-and-Thrusts', with its rich gilt mountings, and contemplated with flattering self-complacency leathers white as snow, jacks brilliant as black varnish could make them, and silver spurs of glittering radiance, until his master stood full harnessed, at length, as gallant a Life Guardsman as ever did duty at the Palace by making love to the handsomest lady-in-waiting.

'To sit wedged in with one's troop for five hours, and in a drizzle, too! Houses oughtn't to meet until the day's fine; I'm sure *they* are in no hurry,' said Cecil to himself, as he pocketed a dainty, filmy handkerchief, all perfume, point, and embroidery, with the interlaced B. C., and the crest on the corner, while he looked hopelessly out of the window. He was perfectly happy, drenched to the skin on the moors after a royal, or in a fast thing with the Melton men from Thorpe Trussels to Ranksborough; but three drops of rain when on duty were a totally different matter, to be resented with any amount of dandy's lamentations and epicurean diatribes.

'Ah, young one, how are you? Is the day *very* bad?' he asked, with languid wistfulness as the door opened.

But indifferent and weary—on account of the weather—as the tone was, his eyes rested with a kindly, cordial light on the newcomer, a young fellow of scarcely twenty, like himself in feature, though much smaller and slighter in build, a graceful boy enough, with no fault in his face, except a certain weakness in the mouth, just shadowed only, as yet, with down.

A celebrity, the Zu-Zu, the last coryphée whom Bertie had translated from a sphere of garret bread-and-cheese to a sphere of villa champagne and chicken (and who, of course, in proportion to the previous scarcity of her bread-and-cheese grew immediately intolerant of any wine less than 90*s.* the dozen), said that Cecil cared for nothing longer than a fortnight, unless it were his horse, Forest King. It was very ungrateful in the Zu-Zu, since he cared for her at the least a whole quarter, paying for his fidelity at the tone of a hundred a month; and also, it was not true, for besides Forest King, he loved his young brother Berkeley: which, however, she neither knew nor guessed.

'Beastly!' replied that young gentleman, in reference to the weather, which was indeed pretty tolerable for an English morning in February. 'I say, Bertie—are you in a hurry?'

'The very deuce of a hurry, little one: why?' Bertie never was in a hurry, however, and he said this as lazily as possible, shaking the white horsehair over his helmet, and drawing in deep draughts of Turkish previous to parting with his pipe for the whole of four or five hours.

'Because I am in a hole—no end of a hole—and I thought you'd help me,' murmured the boy, half penitently, half caressingly; he was very girlish in his face and his ways. On which confession, Rake retired into the bathroom; he could hear just as well there, and a sense of decorum made him withdraw, though his presence would have been wholly forgotten by them. In something the same spirit as the French Countess accounted for her employing her valet to bring her her chocolate in bed—'Est ce que vous appelez *cette chose-là* un homme?'—Bertie had, on occasion, so wholly regarded servants as necessary furniture, that he had gone through a love scene with that handsome coquette, Lady Regalia, totally oblivious of the presence of the groom of the chambers, and the possibility of that person's appearance in the witness-box of the Divorce Court. It was in no way his passion that blinded him—he did not put the steam on like that, and never went in for any disturbing emotion—it was simply habit and forgetfulness that those functionaries were not born mute, deaf, and sightless.

He tossed some essence over his hands, and drew on his gauntlets.
'What's up, Berk?'

The boy hung his head, and played a little uneasily with an ormolu
terrier-pot, upsetting half the tobacco in it; he was trained to his brother's
nonchalant impenetrable school, and used to his brother's set, a cool,
listless, reckless, thoroughbred, and impassive set, whose first canon was
that you must lose your last thousand in the world without giving a sign
that you winced, and must win half a million without showing that you
were gratified; but he had something of girlish weakness in his nature,
and a reserve in his temperament that was with difficulty conquered.

Bertie looked at him, and laid his hand gently on the young one's
shoulder.

'Come, my boy, out with it! It's nothing very bad, I'll be bound?'

'I want some more money; a couple of ponies,' said the boy, a little
huskily; he did not meet his brother's eyes, that were looking straight
down on him.

Cecil gave a long low whistle, and drew a meditative whiff from his
meerschaum.

'*Très cher*, you're always wanting money. So am I. So is everbody. The
normal state of man is to want money. Two ponies. What's it for—eh?'

'I lost it at chicken-hazard last night. Poulteney lent it me, and I told
him I would send it him in the morning. The ponies were gone before I
thought of it, Bertie, and I haven't a notion where to get them to pay him
again.'

'Heavy stakes, young one, for *you*,' murmured Cecil, while his hand
dropped from the boy's shoulder, and a shadow of gravity passed over his
face; money was very scarce with himself. Berkeley gave him a hurried
appealing glance. He was used to shift all his anxieties on to his elder
brother, and to be helped by him under any difficulty. Cecil never allotted
two seconds' thought to his own embarrassments, but he would multiply
them tenfold by taking other people's on him as well with an unremitting
and thoughtless good nature.

'I couldn't help it,' pleaded the lad, with coaxing and almost piteous
apology. 'I backed Grosvenor's play, and you know he's always the most
wonderful luck in the world. I couldn't tell he'd have such cards as he
had. How shall I get the money, Bertie? I daren't ask the governor; and
besides, I told Poulteney he should have it this morning. What do you
think if I sold the mare? But then I couldn't sell her in a minute——'

Cecil laughed a little, but his eyes, as they rested on the lad's young,
fair, womanish face, were very gentle under the long shade of their lashes.

'Sell the mare! Nonsense! How should anybody live without a hack? I can pull you through, I dare say. Ah! by George, there's the quarters chiming. I shall be too late, as I live.'

Not hurried still, however, even by that near prospect, he sauntered to his dressing-table, took up one of the pretty velvet and gold-filigreed absurdities, and shook out all the banknotes there were in it. There were fives and tens enough to count up £45. He reached over and caught up a five from a little heap lying loose on a novel of Du Terrail's, and tossed the whole across the room to the boy.

'There you are, young one! But don't borrow of any but your own people again, Berk. *We* don't do that. No, no!—no thanks. Shut up all that. If ever you get in a hole, I'll take you out if I can. Goodbye. Will you go to the Lords'? Better not—nothing to see, and still less to hear. All stale. That's the only comfort for us—we *are* outside!' he said, with something that almost approached hurry in the utterance, so great was his terror of anything approaching a scene, and so eager was he to escape his brother's gratitude. The boy had taken the notes with delighted thanks indeed, but with that tranquil and unprotesting readiness with which spoiled childishness, or unhesitating selfishness, accepts gifts and sacrifices from another's generosity, which have been so general that they have ceased to have magnitude. As his brother passed him, however, he caught his hand a second, and looked up with a mist before his eyes, and a flush, half of shame, half of gratitude, on his face.

'What a trump you are!—how good you are, Bertie!'

Cecil laughed and shrugged his shoulders.

'First time I ever heard it, my dear boy,' he answered, as he lounged down the staircase, his chains clashing and jingling, while pressing his helmet on to his forehead and pulling the chin-scale over his moustaches, he sauntered out into the street where his charger was waiting.

'The deuce!' he thought, as he settled himself in his stirrups, while the raw morning wind tossed his white plume hither and thither. 'I never remembered!—I don't believe I've left myself money enough to take Willon and Rake and the cattle down to the Shires tomorrow. If I shouldn't have kept enough to take my own ticket with!—that would be no end of a sell. On my word, I don't know how much there's left on the dressing-table. Well! I can't help it, Poulteney had to be paid; I can't have Berk's name show in anything that looks shady.'

The £50 had been the last remnant of a bill, done under great difficulties with a sagacious Jew, and Cecil had no more certainty of possessing any more money until next pay-day should come round than he had of possessing the moon; lack of ready money, moreover, is a serious inconvenience when you belong to clubs where 'pounds and fives' are the lowest points, and live with men who take the odds on most events in thousands; but the thing was done, he would not have undone it at the boy's loss if he could, and Cecil, who never was worried by the loss of the most stupendous 'crusher', and who made it a rule never to think of disagreeable inevitabilities two minutes together, shook his charger's bridle and cantered down Piccadilly towards the barracks, while Black Douglas reared, curvetted, made as if he would kick, and finally ended by 'passaging' down half the length of the road, to the prominent peril of all passers-by, and looking eminently glossy, handsome, stalwart, and foam-flecked, while he thus expressed his disapprobation of forming part of the escort from Palace to Parliament.

'Home Secretary should see about it; it's abominable! If we must come among them they ought to be made a little odoriferous first. A couple of fire-engines now, playing on them continuously with rose-water and bouquet d'Ess, for an hour before we come up, might do a little good. I'll get some men to speak about it in the House; call it "Bill for the Purifying of the Unwashed, and Prevention of their Suffocating Her Majesty's Brigades,"' murmured Cecil to the Earl of Broceliande, next him, as they sat down in their saddles with the rest of the 'First Life' in front of St Stephen's, with a hazy fog steaming round them, and a London mob crushing against their chargers' flanks, while Black Douglas stood like a rock, though a butcher's tray was pressed against his withers, a mongrel was snapping at his hocks, and the inevitable apple-woman, of Cecil's prophetic horror, was wildly plunging between his legs, as the hydra-headed rushed down in insane headlong haste to stare at, and crush on to, that superb body of Guards.

'I would give a kingdom for a soda and brandy. Bah! ye gods! what a smell of fish and fustian,' sighed Bertie, with a yawn of utter famine for want of something to drink and something to smoke, were it only a glass of brown sherry and a little papelito, while be glanced down at the snow-white and jet-black masterpieces of Rake's genius, all smirched, and splashed, and smeared.

He had given fifty pounds away, and scarcely knew whether he should have enough to take his ticket next day into the Shires, and he owed fifty hundred without having the slightest grounds for supposing he should ever be able to pay it, and he cared no more about either of these things than he cared about the Zu-Zu's throwing the half-guinea peaches into the river after a Richmond dinner, in the effort to hit dragon-flies with them; but to be half a day without a cigarette, and to have a disagreeable odour of apples and corduroys wafted up to him, was a calamity that made him insupportably depressed and unhappy.

Well, why not? It is the trifles of life that are its bores after all. Most men can meet ruin calmly, for instance, or laugh when they lie in a ditch with their own knee-joint and their hunter's spine broken over the double-post-and-rails; it is the mud that has choked up your horn just when you wanted to rally the pack, it's the county member who catches you by the button in the lobby, it's the whip who carries you off to a division just when you've sat down to your turbot, it's the ten seconds by which you miss the train, it's the dust that gets in your eyes as you go down to Epsom, it's the pretty little rose-note that went by accident to your house instead of your club, and raised a storm from Madame, it's the dog that always will run wild into the birds, it's the cook who always will season the white soup wrong—it is these that are the bores of life, and that try the temper of your philosophy.

An acquaintance of mine told me the other day of having lost heavy sums through a swindler, with as placid an indifference as if he had lost a toothpick; but he swore like a trooper because a thief had stolen the steel-mounted hoof of a dead pet hunter.

'Insufferable!' murmured Cecil, hiding another yawn behind his gauntlet; 'the Line's nothing half so bad as this; one day in a London mob beats a year's campaigning. What's charging a pah to charging an oyster-stall, or a parapet of fascines to a bristling row of umbrellas?'

Which questions as to the relative hardships of the two Arms was a question of military interest never answered, as Cecil scattered the umbrellas right and left, and dashed from the Houses of Parliament full trot with the rest of the escort on the return to the Palace, the afternoon sun breaking out with a brightened gleam from the clouds, and flashing off the drawn swords, the streaming plumes, the glittering breastplates, the gold embroideries, and the fretting chargers.

But a mere sun-gleam just when the thing was over, and the escort was pacing back to the barracks, could not console Cecil for fog, wind, mud, oyster-vendors, bad odours, and the uproar and riff-raff of the streets; specially when his throat was as dry as a limekiln, and his longing for the sight of a cheroot approaching desperation. Unlimited sodas, three pipes smoked silently over Delphine Demirep's last novel, a bath well dashed with eau-de-cologne, and some glasses of anisette after the fatigue-duty of unharnessing, restored him a little; but he was still weary and depressed into gentler languor than ever through all the courses at a dinner party at the Austrian Embassy, and did not recover his dejection at a reception of the Duchess of Lydiard-Tregoze, where the prettiest French Countess of her time asked him if anything was the matter?

'Yes!' said Bertie, with a sigh, and a profound melancholy, in what the woman called his handsome Spanish eyes, 'I have had a great misfortune; we have been on duty all day!'

He did not thoroughly recover tone, light and careless though his temper was, till the Zu-Zu, in her diamond-edition of a villa, prescribed Crême de Bouzy and Parfait Amour in succession, with a considerable amount of pineapple ice at three o'clock in the morning, which restorative prescription succeeded.

Indeed, it took something as tremendous as divorce from all forms of smoking for five hours, to make an impression on Bertie. He had the most serene insouciance that ever a man was blessed with; in worry he did not believe, he never let it come near him; and beyond a little difficulty sometimes in separating too many entangled rose-chains caught round him at the same time, and the annoyance of a miscalculation on the flat or the ridge-and-furrow, when a Maldon or Danebury favourite came 'nowhere', or his book was wrong for the Grand National, Cecil had no cares of any sort or description.

True, the Royallieu Peerage, one of the most ancient and almost one of the most impoverished in the kingdom, could ill afford to maintain its sons in the expensive career on which it had launched them, and the chief there was to spare usually went between the eldest, a Secretary of Legation in that costly and charming city of Vienna, and to the young one, Berkeley, through the old Viscount's partiality, so that had Bertie ever gone so far as to study his actual position, he would have probably confessed that it was, to say the least, awkward. But then he never did this; certainly never did it thoroughly. Sometimes he felt himself near the wind when settling-day came, or the Jews appeared utterly impracticable; but, as a rule, things had always trimmed *somehow*, and though his debts were considerable, and he was literally as penniless as a man can be to stay in the Guards at all, he had never in any shape realized the want of money. He might not be able to raise a guinea to go towards that long-standing account, his army tailor's bill, and post-obits had long ago forestalled the few hundreds a year that, under his mother's settlements, would come to him at the Viscount's death; but Cecil had never known in his life what it was not to have a first-rate stud, not to live as luxuriously as a Duke, not to order the costliest dinners at the clubs, and be amongst the first to lead all the splendid entertainments and extravagances of the Household; he had never been without his Highland shooting, his Baden gaming, his prize-winning schooner amongst the RVY Squadron, his September battues, his Pytchley hunting, his pretty expensive Zu-Zus and other toys, his drag for Epsom and his trap and hack for the Park, his crowd of engagements through the season, and his bevy of fair leaders of the fashion to smile on him, and shower their invitation-cards on him, like a rain of rose-leaves, as one of their 'best men'.

'Best', that is in the sense of fashion, flirting, waltzing, and general social distinction; in no other sense, for the newest of débutantes knew well that 'Beauty', though the most perfect of flirts, would never be 'serious', and had nothing to be serious with, on which understanding he was allowed by the sex to have the run of their boudoirs and drawing-rooms much as if he were a little lion-dog; they counted him quite 'safe', he made love to the married women to be sure, but he was quite certain not to run away with the marriageable daughters.

Hence, Bertie had never felt the want of all that is bought by and represents money, and imbibed a vague indistinct impression that all these things that made life pleasant came by Nature, and were the natural inheritance and concomitants of anybody born in a decent station, and endowed with a tolerable tact; such a matter-of-fact difficulty as not having gold enough to pay for his own and his stud's transit to the Shires had very rarely stared him in the face, and when it did, he trusted to chance to lift him safely over such a social 'yawner', and rarely trusted in vain.

According to all the canons of his Order he was never excited, never disappointed, never exhilarated, never disturbed, and also of course never by any chance embarrassed. '*Votre imperturbabilité*', as the Prince de Ligne used to designate La Grande Catherine, would have been an admirable designation for Cecil; he was imperturbable under everything; even when an heiress, with feet as colossal as her fortune, made him a proposal of marriage, and he had to retreat from all the offered honours and threatened horrors, courteously, but steadily declined them. Nor in more interesting adventures was he less happy in his coolness. When my Lord Regalia, who never knew when he was not wanted, came in inopportunely in a very tender scene of the young Guardsman's (then but a Cornet) with his handsome Countess, Cecil lifted his lashes lazily, turning to him a face of the most *plait-il*? and innocent demureness—or consummate impudence, whichever you like. 'We're playing Solitaire. Interesting game. Queer fix though, the ball's in, that's left all alone in the middle, don't you think?' Lord Regalia felt his own similarity to the 'ball in a fix' too keenly to appreciate the interesting character of the amusement, or the coolness of the chief performer in it; but 'Beauty's Solitaire' became a synonym thenceforth among the Household to typify any very tender passages '*sotto quartr' occhi*'.

This made his reputation on the town; the ladies called it very wicked, but were charmed by the Richelieu-like impudence all the same and petted the sinner; and from then till now he had held his own with them; dashing through life very fast as became the first riding man in the Brigades, but enjoying it very fully, smoothly, and softly, liking the world, and being liked by it.

To be sure, in the background there was always that ogre of money, and the beast had a knack of gnawing bigger and darker every year; but then, on the other hand, Cecil never looked at him, never thought about him, knew, too, that he stood just as much behind the chairs of men whom the world accredited as millionaires, and whenever the ogre gave him a cold grip that there was for the moment no escaping, washed away the touch of in a warm fresh draught of pleasure.

CHAPTER II

The Loose Box, and the Tabagie

'How long before the French can come up?' asked Wellington, hearing of the pursuit that was thundering close on his rear in the most critical hours of the short, sultry, Spanish night. 'Half an hour at least,' was the answer. 'Very well, then, I will turn in and get some sleep,' said the Commander-in-Chief, rolling himself in a cloak, and lying down in a ditch to rest as soundly for the single half-hour as any tired drummer-boy.

Serenely as Wellington, another hero slept profoundly, on the eve of a great event, of a great contest to be met when the day should break, of a critical victory, depending on him alone to save the Guards of England from defeat and shame; their honour and their hopes rested on his solitary head, by him they would be lost or saved; but, unharassed by the magnitude of the stake at issue, unhaunted by the past, unfretted by the future, he slumbered the slumber of the just.

Not Sir Tristram, Sir Calidore, Sir Launcelot, no, nor Arthur himself, was ever truer knight, was ever gentler, braver, bolder, more staunch of heart, more loyal of soul, than he to whom the glory of the Brigades was trusted now; never was there spirit more dauntless and fiery in the field, never temper kindlier and more generous with friends and foes. Miles of the ridge and furrow, stiff fences of terrible blackthorn, double posts and rails, yawners and croppers both, tough as Shire and Stewards could make them, awaited him on the morrow; on his beautiful lean head capfuls of money were piled by the Service and the Talent; and in his stride all the fame of the Household would be centred on the morrow; but he took his rest like the cracker he was—standing as though he were on guard, and steady as a rock, a hero every inch of him. For he was Forest King, the great steeplechaser, on whom the Guards had laid all their money for the Grand Military—the Soldiers' Blue Riband.

His quarters were a loose box, his camp-bed a litter of straw fresh shaken down, his clothing a very handsome rug, hood, and quarter-piece buckled on and marked B. C.; above the manger and the door was lettered his own name in gold, FOREST KING; and in the panels of the latter were miniatures of his sire and of his dam: Lord of the Isles, one of the greatest hunters that the grass countries ever saw sent across them; and Bayadere, a wild-pigeon-blue mare of Circassia. How further more he stretched up to his long line of ancestry by The Sovereign, out of Queen of Roses, by Belted Earl, out of Fallen Star, by Marmion, out of Court Coquette, and straight up to the White Cockade blood, etc., etc., etc., is it not written in the mighty and immortal chronicle, precious as the Koran, patrician as the Peerage, known and beloved to mortals as the 'Stud-Book'?

Not an immensely large or unusually powerful horse, but with *race* in every line of him; steel-grey in colour, darkening well at all points, shining and soft as satin, with the firm muscles quivering beneath at the first touch of excitement to the high mettle and finely strung organiz-ation; the head small, lean, racer-like, 'blood' all over, with the delicate taper ears, almost transparent in full light; well ribbed-up, fine shoulders, admirable girth and loins; legs clean, slender, firm, promising splendid knee action; sixteen hands high, and up to thirteen stone; clever enough for anything, trained to close and open country, a perfect brook jumper, a clipper at fencing, taking a great deal of riding, as any one could tell by the set-on of his neck, but docile as a child to a well-known hand; such was Forest King with his English and Eastern strains, winner at Chertsey, Croydon, the National, the Granby, the Belvoir Castle, the Curragh, and all the gentleman-rider steeplechases and military sweepstakes in the kingdom, and entered now, with tremendous bets on him, for the Gilt Vase.

It was a crisp cold night outside, starry and wintry, but open weather, and clear; the ground would be just right on the morrow, neither hard as the slate of a billiard-table, nor wet as the slush of a quagmire. Forest King slept steadily on in his warm and spacious box, dreaming doubtless of days of victory, cub-hunting in the reedy October woods and pastures, of the ringing notes of the horn, and the sweet music of the pack, and the glorious quick burst up-wind, breasting the icy cold water, and showing the way over fence and bullfinch. Dozing and dreaming pleasantly; but alert for all that; for he awoke suddenly, shook himself, had an hilarious roll in the straw, and stood 'at attention'.

Awake only, could you tell the generous and gallant promise of his perfect temper; for there are no eyes that speak more truly, none on earth that are so beautiful, as the eyes of a horse. Forest King's were dark as a gazelle's, soft as a woman's, brilliant as stars, a little dreamy and mournful, and as infinitely caressing when he looked at what he loved, as they could blaze full of light and fire when danger was near and rivalry against him. How loyally such eyes have looked at me over the paddock fence, as a wild happy gallop was suddenly broken for a gentle head to be softly pushed against my hand with the gentlest of welcomes! They sadly put to shame the million human eyes that so fast learn the lie of the world, and utter it as falsely as the lips.

The steeplechaser stood alert, every fibre of his body strung to pleasurable excitation; the door opened, a hand held him some sugar, and the voice he loved best said fondly, 'All right, old boy?'

Forest King devoured the beloved dainty with true equine unction, rubbed his forehead against his master's shoulder, and pushed his nose into the nearest pocket in search for more of his sweetmeat.

'You'd eat a sugar-loaf, you dear old rascal. Put the gas up, George,' said his owner, while he turned up the body clothing to feel the firm, cool skin, loosened one of the bandages, passed his hand from thigh to fetlock, and glanced round the box to be sure the horse had been well suppered and littered down.

'Think we shall win, Rake?'

Rake, with a stable-lantern in his hand and a forage-cap on one side of his head, standing a little in advance of a group of grooms and helpers, took a bit of straw out of his mouth, and smiled a smile of sublime scorn and security. '*Win, sir*? I should be glad to know as when was that ere King ever beat yet, or you either, sir, for that matter?'

Bertie Cecil laughed a little languidly.

'Well, we take a good deal of beating, I think, and there are not very many who can give it us; are there, old fellow?' he said to the horse, as he passed his palm over the withers; 'but there are some crushers in the lot tomorrow; you'll have to do all you know.'

Forest King caught the manger with his teeth, and kicked in a bit of play and ate some more sugar, with much licking of his lips to express the nonchalance with which *he* viewed his share in the contest, and his tranquil certainty of being first past the flags. His master looked at him once more and sauntered out of the box.

'He's in first-rate form, Rake, and right as a trivet.'

'In course he is, sir; nobody ever laid leg over such cattle as all that White Cockade blood, and he's the very best of the strain,' said Rake, as he held up his lantern across the stable-yard, that looked doubly dark in the February night after the bright gas glare of the box.

'So he need be,' thought Cecil, as a bull terrier, three or four Gordon setters, an Alpine mastiff, and two wiry Skyes dashed at their chains, giving tongue in frantic delight at the sound of his step, while the hounds echoed the welcome from their more distant kennels, and he went slowly across the great stone yard, with the end of a huge cheroot glimmering through the gloom. 'So he need be, to pull *me* through. The Ducal and the October let me in for it enough; I never was closer in my life. The deuce, if I don't do the distance tomorrow, I shan't have sovereigns enough to play pound-points at night! *I* don't know what a man's to do; if he's put into this life he must go the pace of it. Why did Royal send me into the Guards, if he meant to keep the screw on in this way; he'd better have drafted me into a marching regiment at once, if he wanted me to live upon nothing.'

Nothing meant anything under £6,000 a year with Cecil, as the minimum of monetary necessities in this world, and a look of genuine annoyance and trouble, most unusual there, was on his face, the picture of carelessness and gentle indifference habitually, though shadowed now as he crossed the courtyard after his after-midnight visit to his steeplechaser. He had backed Forest King heavily, and stood to win or lose a cracker on his own riding on the morrow; and though he had found sufficient to bring him into the Shires, he had barely enough lying on his dressing-table, up in the bachelor suite within, to pay his groom's book, or a notion where to get more, if the King should find his match over the ridge and furrow in the morning!

It was not pleasant: a cynical, savage world-disgusted Timon derives on the whole a good amount of satisfaction from his breakdown, in the fine philippics against his contemporaries that it is certain to afford, and the magnificent grievances with which it furnishes him; but when life is very pleasant to a man, and the world very fond of him; when existence is

perfectly smooth—bar that single pressure of money—and is an incessantly changing kaleidoscope of London seasons, Paris winters, ducal
houses in the hunting months, dinners at the Pall Mall Clubs, dinners at
the Star and Garter, dinners irreproachable everywhere, cottage for Ascot
week, yachting with the RVY Club, Derby handicaps at Hornsey, pretty
chorus-singers set up in Bijou villas, dashing rosières taken over to Baden,
warm corners in Belvoir, Savernake, and Longeat battues, and all the rest
of the general programme, with no drawback to it except the duties at the
Palace, the heat of a review, or the extravagance of a pampered lionne,
then to be pulled up in that easy swinging gallop for sheer want of a golden
shoe, as one may say, is abominably bitter, and requires far more philosophy to endure than Timon would ever manage to muster. It is a bore, an
unmitigated bore, a harsh, hateful, unrelieved martyrdom that the world
does not see, and that the world would not pity if it did.

'Never mind! Things will come right. Forest King never failed me yet;
he is as full of running as a Derby winner, and he'll go over the yawners
like a bird,' thought Cecil, who never confronted his troubles with more
than sixty seconds' thought, and who was of that light, impassable, half-
levity, half-languor of temperament that both throws off worry easily, and
shirks it persistently. 'Sufficient for the day,' etc., was the essence of his
creed; and if he had enough to lay a fiver at night on the rubber, he was
quite able to forget for the time that he wanted five hundred for settling-
day in the morning, and had not an idea how to get it. There was not a
trace of anxiety on him when he opened a low-arched door, passed down
a corridor, and entered the warm full light of that chamber of liberty, that
sanctuary of the persecuted, that temple of refuge, thrice blessed in all its
forms throughout the land, that consecrated Mecca of every true believer
in the divinity of the meerschaum, and the paradise of the narghilé—the
smoking-room.

A spacious easy chamber, too, lined with the laziest of divans, seen just
now through a fog of smoke, and tenanted by nearly a score of men
in every imaginable loose velvet costume, and with faces as well known
in the Park at six o'clock in May, and on the Heath in October, in Paris in
January, and on the Solent in August, in Pratts' of a summer's night, and
on the Moors in an autumn morning, as though they were features that
came round as regularly as the 'July' or the Waterloo Cup. Some were
puffing away in calm meditative comfort, in silence that they would not
have broken for any earthly consideration; others were talking hard and
fast, and through the air heavily weighted with the varieties of tobacco,
from tiny cigarettes to giant cheroots, from rough bowls full of cavendish
to sybaritic rose-water hookahs, a Babel of sentences rose together: 'Gave

him too much riding, the idiot.' 'Take the field, bar one.' 'Nothing so good for the mare as a little nitre and antimony in her mash.' 'Not at all! the Regent and Rake cross in the old strain, always was black-tan with a white frill.' 'The Earl's as good a fellow as Lady Flora; always give you a mount.' 'Nothing like a Kate Terry, though, on a bright day for salmon.' 'Faster thing I never knew; found at twenty minutes past eleven, and killed just beyond Longdown Water at ten to twelve.' All these various phrases were rushing in among each other, and tossed across the eddies of smoke in the conflicting of tongues loosened in the *tabagie* and made eloquent, though slightly inarticulate, by pipe-stems; while a tall, fair man, with the limbs of a Hercules, the chest of a prize-fighter, and the face of a Raphael Angel, known in the Household as Seraph, was in the full flood of a story of whist played under difficulties in the Doncaster express.

'I wanted a monkey; I wanted monkeys awfully,' he was stating as Forest King's owner came into the smoking-room.

'Did you, Seraph? The "Zoo" or the Clubs could supply you with apes fully developed to any amount,' said Bertie, as he threw himself down.

'You be hanged!' laughed the Seraph, known to the rest of the world as the Marquis of Rockingham, son of the Duke of Lyonnesse. 'I wished monkeys, but the others wished ponies and hundreds, so I gave in; Vandeleur and I won two rubbers, and we'd just begun the third, when the train stopped with a crash; none of us dropped the cards though, but the tricks and the scores all went down with the shaking. "Can't play in that row," said Charlie, for the women were shrieking like mad, and the engine was roaring like my mare Philippa—I'm afraid she'll never be cured, poor thing!—so I put my head out and asked what was up? We'd run into a cattle train. Anybody hurt? No, nobody hurt; but we were to get out. "I'll be shot if I get out," I told 'em, "till I've finished the rubber." "But you must get out," said the guard; "carriages must be moved". "Nobody says 'must' to him," said Van (he'd drank more Perles du Rhin than was good for him at Doncaster); "don't you know the Seraph?" Man stared. "Yes, sir, know the Seraph, sir; leastways, did sir, afore he died; see him once at Moulsey Mill, sir; his 'one, two' was amazin'. Waters soon threw up the sponge." We were all dying with laughter, and I tossed him a tenner. "There, my good fellow," said I, "shunt the carriage and let us finish the game. If another train comes up, give it Lord Rockingham's compliments and say he'll thank it to stop, because collisions shake his trumps together." Man thought us mad—took tenner though—shunted us to one side out of the noise, and we played two rubbers more before they'd repaired the damage and sent us on to town.'

And the Seraph took a long-drawn whiff from his silver meerschaum, and then a deep draught of soda and brandy to refresh himself after the narrative; biggest, best-tempered, and wildest of men in or out of the Service, despite the angelic character of his fair-haired head, and blue eyes that looked as clear and as innocent as those of a six-year-old child.

'Not the first time, by a good many, that you've "shunted off the straight", Seraph?' laughed Cecil, substituting an amber mouthpiece for his half-finished cheroot. 'I've been having a good-night look at the King. He'll stay.'

'Of course he will,' chorused half a dozen voices.

'With all our pots on him,' added the Seraph. 'He's too much of a gentleman to put us all up a tree; he knows he carries the honour of the Household.'

'There are some good mounts, there's no denying that,' said Chester-field of the Blues (who was called Tom for no other reason than that it was entirely unlike his real name of Adolphus), where he was curled up almost invisible, except for the movement of the jessamine stick of his chibouque. 'That brute, Day Star, is a splendid fencer, and for a brook jumper, it would be hard to beat Wild Geranium, though her shoulders are not quite what they ought to be. Montecute, too, can ride a good thing, and he's got one in Pas de Charge.'

'I'm not much afraid of Monti, he makes too wild a burst first; he never *saves* one atom,' yawned Cecil, with the coils of his hookah bubbling among the rose-water; 'the man I'm afraid of is that fellow from the Twelfth; he's as light as a feather and as hard as steel. I watched him yesterday going over the water, and the horse he'll ride for Trelawney is good enough to beat even the King if he's properly piloted.'

'You haven't kept yourself in condition, Beauty,' growled 'Tom', with the chibouque in his mouth, 'else nothing could give you the go-by. Its tempting Providence to go in for the Gilt Vase after such a December and January as you spent in Paris. Even the week you've been in the Shires you haven't trained a bit; you've been waltzing or playing baccarat till five in the morning, and taking no end of sodas after to bring you right for the meet at nine. If a man will drink champagnes and burgundies as you do, and spend his time after women, I should like to know how he's to be in hard riding condition, unless he expects a miracle.'

With which Chesterfield, who weighed fourteen stone himself, and was, therefore, out of all but welter-races, and wanted a weight carrier of tremendous power even for them, subsided under a heap of velvet and cashmere, and Cecil laughed: lying on a divan just under one of the gas branches, the light fell full on his handsome face, with its fair hue and its gentle languor on which there was not a single trace of the *outre cuidance* attributed to him. Both he and the Seraph could lead the wildest life of any men in Europe without looking one shadow more worn than the brightest beauty of the season, and could hold wassail in riotous rivalry till the sun rose, and then throw themselves into saddle as fresh as if they had been sound asleep all night, to keep up with the pack the whole day in a fast burst or on a cold scent, or in whatever sport Fortune and the coverts gave them, till their second horses wound their way homeward, through muddy leafless lanes, when the stars had risen.

'Beauty don't believe in training. No more do I. Never would train for anything,' said the Seraph, now, pulling the long tawny moustaches that were not altogether in character with his seraphic cognomen. 'If a man can ride—let him. If he's born to the pigskin he'll be in at the distance safe enough, whether he smoke or don't smoke, drink or don't drink. As for training on raw chops, giving up wine, living like the very deuce and all, as if you were in a monastery, and changing yourself into a mere bag of bones—it's utter bosh! You might as well be in purgatory; besides, it's no more credit to win then than if you were a professional.'

'But you must have trained at Christ Church, Rock, for the Eight?' asked another Guardsman, Sir Vere Bellingham, 'Severe', as he was christened, chiefly because he was the easiest-going giant in existence.

'Did I! Men came to me; wanted me to join the Eight; coxswain came, awful strict little fellow, docked his men of all their fun—took plenty himself, though! Coxswain said I must begin to train, do as all his crew did. I threw up my sleeve and showed him my arm'; and the Seraph stretched out an arm magnificent enough for a statue of Milo. 'I said, "There, sir, I'll help you thrash Cambridge if you like, but train I *won't*, for you or for all the University. I've been Captain of the Eton Eight, but I didn't keep my crew on tea and toast. I fattened 'em regularly three times a week on venison and champagne at Christopher's. Very happy to feed yours, too, if you like; game comes down to me every Friday from the Duke's moors; they look uncommonly as if they wanted it!" You should have seen his face!—Fatten the Eight! He didn't let me do that, of course, but he was very glad of my oar in his rowlocks, and I helped him beat Cambridge without training an hour myself except so far as rowing hard went.'

And Philip, Marquis of Rockingham, made thirsty by the recollection, dipped his fair moustaches into a foaming seltzer.

'Quite right, Seraph!' said Cecil. 'When a man comes up to the weights, looking like a homonunculus after he's been getting every atom of flesh off him like a jockey, he ought to be struck out for the stakes, to my mind. 'Tisn't a question of riding, then, nor yet of pluck, or of management; it's nothing but a question of pounds, and of who can stand the tamest life the longest.'

'Well, beneficial for one's morals, at any rate,' suggested Sir Vere.

'Morals be hanged!' said Bertie, very immorally. 'I'm glad *you* remind us of them, Vere, you're such a quintessence of decorum and respectability yourself! I say—anybody know anything of this fellow of the Twelfth that's to ride Trelawney's chesnut?'

'Jimmy Delmar? Oh yes; I know Jimmy,' answered Lord Cosmo Wentworth, of the Scots Fusiliers, from the far depths of an armchair. 'Knew him at Aldershot. Fine rider; give you a good bit of trouble, Beauty. Hasn't been in England for years; troop been such a while at Calcutta. The Fancy take to him rather; offering very freely on him this morning in the Village; and he's got a rare good thing in the chesnut.'

'Not a doubt of it. The White Lily blood, out of that Irish mare D'Orleans Diamonds, too.'

'Never mind! Twelfth won't beat *us*. The Household will win safe enough, unless Forest King goes and breaks his back over Brixworth— eh, Beauty?' said the Seraph, who believed devoutly in his comrade, with all the loving loyalty characteristic of the House of Lyonnesse, that to monarchs and to friends had often cost it very dear.

'You put your faith in the wrong quarter, Rock; I *may* fail you, *he* never will,' said Cecil, with ever so slight a dash of sadness in his words. The thought crossed him of how boldly, how straightly, how gallantly the horse always breasted and conquered his difficulties—did he himself deal half so well with his own?

'Well! you both of you carry all our money and all our credit; so for the fair fame of the Household do "all you know". I haven't hedged a shilling, not laid off a farthing, Bertie; I stand on you and the King, and nothing else. See what a sublime faith I have in you.'

'I don't think you're wise, then, Seraph; the field will be very strong,' said Cecil, languidly. The answer was indifferent, and certainly thankless; but under his drooped lids a glance, frank and warm, rested for the moment on the Seraph's leonine strength and Raphaelesque head; it was not his way to say it, or to show it, or even much to think it; but in his heart he loved his old friend wonderfully well.

And they talked on of little else than of the great steeplechase of the Service, for the next hour in the Tabâk-Parliament, while the great clouds of scented smoke circled heavily round, making a halo of turkish above the gold locks of the Titanic Seraph, steeping Chesterfield's velvets in strong odours of cavendish, and drifting a light rose-scented mist over

Bertie's long lithe limbs, light enough and skilled enough to disdain all 'training for the weights'.

'*That*'s not the way to be in condition,' growled 'Tom', getting up with a great shake as the clock clanged the strokes of five; they had only returned from a ball three miles off when Cecil had paid his visit to the loose box. Bertie laughed; his laugh was like himself, rather languid but very light-hearted, very silvery, very engaging.

'Sit and smoke till breakfast-time if you like, Tom; it won't make any difference to *me*.'

But the Smoke-Parliament wouldn't hear of the champion of the Household over the ridge and furrow risking the steadiness of his wrist and the keenness of his eye by any such additional tempting of Providence, and went off itself in various directions, with good-night iced drinks, yawning considerably like most other Parliaments after a sitting.

It was the old place in the Shires of the Royallieu Family in which he had congregated half the Guardsmen in the Service for the great event, and consequently the bachelor chambers in it were of the utmost comfort and spaciousness, and when Cecil sauntered into his old quarters, familiar from boyhood, he could not have been better off in his own luxurious haunts in Piccadilly. Moreover, the first thing that caught his eye was a dainty scarlet silk riding jacket broidered in gold and silver, with the motto of his house, 'Cœur Vaillant Se Fait Royaume', all circled with oak and laurel leaves on the collar.

It was the work of very fair hands, of very aristocratic hands, and he looked at it with a smile. 'Ah, my lady, my lady!' he thought half aloud, 'do you really love me? Do I really love you?'

There was a laugh in his eyes as he asked himself what might be termed an interesting question; then something more earnest came over his face, and he stood a second with the pretty costly embroideries in his hand, with a smile that was almost tender, though it was still much more amused.

'I suppose we do,' he concluded at last; 'at least, quite as much as is ever worth while. Passions don't do for the drawing-room, as somebody says in "Coningsby"; besides—I would not feel a strong emotion for the universe. Bad style always, and more detrimental to "condition", as Tom would say, than three bottles of brandy!'

He was so little near what he dreaded, at present at least, that the scarlet jacket was tossed down again, and gave him no dreams of its fair and titled embroideress. He looked out, the last thing, at some ominous clouds drifting heavily up before the dawn, and the state of the weather, and the chance of its being rainy, filled his thoughts, to the utter exclusion of the donor of that bright gold-laden dainty gift. 'I hope to goodness there won't be any drenching shower. Forest King can stand ground as hard as a slate, but if there's one thing he's weak in, it's slush!' was Bertie's last conscious thought as he stretched his limbs out and fell sound asleep.

CHAPTER III

The Soldiers' Blue Riband

'Take the Field bar one.' 'Two to one on Forest King.' 'Two to one on Bay Regent.' 'Fourteen to seven on Wild Geranium.' 'Seven to two against Brother to Fairy.' 'Three to five on Pas de Charge.' 'Nineteen to six on Day Star.' 'Take the Field bar one,' rose above the hoarse tumultuous roar of the Ring on the clear, crisp, sunny morning that was shining on the Shires on the day of the famous steeplechase.

The talent had come in great muster from London; the great bookmakers were there with their stentor lungs and their quiet quick entry of thousands; and the din and the turmoil, at the tiptop of their height, were more like a gathering on the Heath or before the Red House, than the local throngs that usually mark steeplechase meetings, even when they be the Grand Military or the Grand National.

There were keen excitement and heavy stakes on the present event; the betting had never stood still a second in Town or the Shires; and even the 'knowing ones', the worshippers of the 'flat' alone, the professionals who ran down gentlemen races, and the hypercritics who affirmed that there is not such a thing as a steeplechaser to be found on earth (since, to be a fencer, a water-jumper, *and* a racer, were to attain an equine perfection impossible on earth, whatever it may be in 'the happy hunting-ground' of immortality)—even these, one and all of them, came eager to see the running for the Gilt Vase.

For it was known very well that the Guards had backed their horse tremendously, and the county laid most of its money on him, and the bookmakers were shy of laying off much against one of the first cross-country riders of the Service, who had landed his mount at the Grand National Handicap, the Billesdon Coplow, the Ealing, the Curragh, the

Prix du Donjon, the Rastatt, and almost every other for which he had entered. Yet, despite this, the 'Fancy' took most to Bay Regent; they thought he would 'cut the work out'; his sire had won the Champagne Stakes at Doncaster, and the Drawing-room at 'glorious Goodwood', and that racing strain through the White Lily blood, coupled with a magnificent reputation which he brought from Leicestershire as a fencer, found him chief favour among the Fraternity.

His jockey, Jimmy Delmar, too, with his bronzed, muscular, sinewy frame, his low stature, his light weight, his sunburnt, acute face, and a way of carrying his hands as he rode that was precisely like Aldcroft's, looked a hundred times more professional than the brilliance of 'Beauty', and the reckless dash of his well-known way of 'sending the horse along with all he had in him', which was undeniably much more like a fast kill over the Melton country than like a weight-for-age race anywhere. 'You see the Service in his *stirrups*,' said an old nobbler who had watched many a trial spin, lying hidden in a ditch or a drain; and indisputably you did: Bertie's riding was superb, but it was still the riding of a cavalryman, not of a jockey. The mere turn of the foot in the stirrups told it, as the old man had the shrewdness to know.

So the King went down at one time two points in the morning betting.

'Know them flash cracks of the Household,' said Tim Varnet, as sharp a little Leg as ever 'got on' a dark thing, and 'went halves' with a jock who consented to rope a favourite at the Ducal. 'Them swells, ye see, they give any money for blood. They just go by Godolphin heads, and little feet, and winners' strains, and all the rest of it; and so long as they get pedigree never look at substance; and their bone comes no bigger than a deer's. Now, its *force* as well as pace that tells over a bit of plough; a critter that would win the Derby on the flat would knock up over the first spin over the clods; and that King's legs are too light for my fancy, 'andsome as 't is ondeniable he looks—for a little 'un, as one may say.'

And Tim Varnet exactly expressed the dominant mistrust of the talent; despite all his race and all his exploits, the King was not popular in the Ring, because he was like his backers—'a swell'. They thought him 'showy—very showy', 'a picture to frame', 'a lustre to look at'; but they disbelieved in him, almost to a man, as a *stayer*, and they trusted him scarcely at all with their money.

'It's plain that he's "meant", though,' thought little Tim, who was so used to the 'shady' in stable matters, that he could hardly persuade himself that even the Grand Military could be run fair, and would have thought a Guardsman or a Hussar only exercized his just privilege as a jockey in 'roping' after selling the race, if so it suited his book. 'He's "meant", that's clear, 'cause the swells have put all their pots on him—but if the pots don't bile over, strike me a loser!' a contingency he knew he might very well invoke, his investments being invariably so matchlessly arranged, that let what would be 'bowled over', Tim Varnet never could be.

Whatever the King might prove, however, the Guards, the Flower of the Service, must stand or fall by him; they had entered nothing else for the race, so complete was the trust that, like the Seraph, they put in 'Beauty' and his grey. But there was no doubt as to the tremendousness of the struggle lying before him. The running ground covered four miles and a half, and had forty-two jumps in it, exclusive of the famous Brixworth: half was grassland, and half ridge and furrow; a lane with a very awkward double, fences laced in and in with the memorable blackthorn, a laid hedge with thick growers in it, and many another 'teaser', coupled with the yawning water, made the course a severe one; while thirty-two starters of unusual excellence gave a good field and promised a close race. Every fine bit of steeplechase blood that was to be found in their studs had been brought together by the Service for the great event; and if the question could ever be solved, whether it is possible to find a strain that shall combine pace over the flat, with a heart to stay over an enclosed country, the speed to race, with the bottom to fence and the force to clear water, it seemed likely to be settled now. The Service and the Stable had done their uttermost to reach its solution.

The clock of the course pointed to half-past one; the saddling-bell would ring at a quarter to two, for the days were short and darkened early; the Stewards were all arrived, except the Marquis of Rockingham, and the Ring was in the full rush of excitement, some 'getting on' hurriedly to make up for lost time, some 'peppering' one or other of the favourites hotly, some laying off their moneys in a cold fit of caution, some putting capfuls on the King, or Bay Regent, or Pas de Charge, without hedging a shilling. The London talent, the agents from the great commission stables, the local betting men, the shrewd wiseacres from the Ridings, all the rest of the brotherhood of the Turf were crowding together with the deafening shouting common to them, which sounds so tumultuous, so insane, and so unintelligible to outsiders. Amidst them, half the titled heads of England, all the great names known on the flat, and men in the Guards, men in the Rifles, men in the Light Cavalry, men in the Heavies, men in the Scots Greys, men in the Horse Artillery, men in all the Arms and all the Regiments, were backing their horses with crackers, and jotting down figure after figure, with jewelled pencils, in dainty books, taking long odds with the fielders. Carriages were standing in long lines along the course, the stands were filled with almost as bright a bevy of fashionable loveliness as the Ducal brings together under the park trees of Goodwood; the horses were being led into the enclosure for saddling, a brilliant sun shone for the nonce on the freshest of February noons; beautiful women were fluttering out of their barouches in furs and velvets, wearing the colours of the jockey they favoured, and more predominant than any were Cecil's scarlet and white, only rivalled in prominence by the azure of the Heavy Cavalry champion, Sir Eyre Montacute.

A drag with four bays—with fine hunting points about them—had dashed up, late of course; the Seraph had swung himself from the roller-bolt into the saddle of his hack (one of those few rare hacks that are *perfect*, and combine every excellence of pace, bone, and action under their modest appellative), and had cantered off to join the Stewards, while Cecil had gone up to a group of ladies in the Grand Stand, as if he had no more to do with the morning's business than they. Right in front of that Stand was an artificial bullfinch which promised to treat most of the field to a 'purler', a deep ditch dug and filled with water, with two towering blackthorn fences on either side of it, as awkward a leap as the most cramped country ever showed; some were complaining of it; it was too severe, it was unfair, it would break the back of every horse sent at it. The other Stewards were not unwilling to have it tamed down a little, but the Seraph, generally the easiest of all sweet-tempered creatures, refused resolutely to let it be touched.

'Look here,' said he, confidentially, as he wheeled his hack round to the Stand and beckoned Cecil down—'look here, Beauty, they're want-ing to alter that teaser, make it less awkward, you know, but I wouldn't, because I thought it would look as if I lessened it for *you*, you know. Still it *is* a cracker and no mistake; Brixworth itself is nothing to it, and if you'd like it toned down I'll let them do it?——'

'My dear Seraph, not for worlds! You were quite right not to have a thorn taken out. Why *that*'s where I shall thrash Bay Regent,' said Bertie, serenely, as if the winning of the stakes had been forecast in his horoscope.

The Seraph whistled, stroking his moustaches. 'Between ourselves, Cecil, that fellow is going up no end. The Talent fancy him so——'

'Let them,' said Cecil, placidly, with a great cheroot in his mouth, lounging into the centre of the Ring to hear how the betting went on his own mount, perfectly regardless that he would keep them waiting at the weights while he dressed. Everybody there knew him by name and sight; and eager glances followed the tall form of the Guards' champion as he moved through the press, in a loose brown sealskin coat, with a little strip of scarlet ribbon round his throat, nodding to this Peer, taking evens with that, exchanging a whisper with a Duke, and squaring his book with a Jew. Murmurs followed about him as if he were the horse himself: 'Looks in racing form'—'Looks used up, to *me*'—'Too little hands sure*ly* to hold in long in a spin'—'Too much length in the limbs for a light weight, bone's always awfully heavy'—'Dark under the eye, been going too fast for trainin' '—'A swell all over, but rides no end'; with other innumerable contradictory phrases, according as the speaker was 'on' him or against him, buzzed about him from the riffraff of the Ring, in no way disturbing his serene equanimity.

One man, a big fellow, 'ossy' all over, with the genuine sporting cut-away coat, and a superabundance of showy necktie and bad jewellery, eyed him curiously, and slightly turned so that his back was towards Bertie, as the latter was entering a bet with another Guardsman well known on the Turf, and he himself was taking long odds with little Berk Cecil, the boy having betted on his brother's riding as though he had the Bank of England at his back. Indeed, save that the lad had the hereditary Royallieu instinct of extravagance, and, with a half thoughtless, half wilful improvidence, piled debts and difficulties on his rather brainless and boyish head, he had much more to depend on than his elder; for the old Lord Royallieu doted on him, spoilt him, and denied him nothing, though himself a stern, austere, passionate man, made irascible by ill-health, and, in his fits of anger, a very terrible personage indeed, no more to be conciliated by persuasion than iron is to be bent by the hand; so terrible, that even his pet dreaded him mortally, and came to Bertie to get his imprudences and peccadilloes covered from the Viscount's sight.

Glancing round at this moment as he stood in the Ring, Cecil saw the betting-man with whom Berkeley was taking long odds on the race; he raised his eyebrows and his face darkened for a second, though resuming his habitual listless serenity almost immediately.

'You remember that case of welshing after the Ebor St Leger, Con?' he said in a low tone to the Earl of Constantia, with whom he was talking. The Earl nodded assent, every one had heard of it, and a very flagrant case it was.

'There's the fellow,' said Cecil, laconically, and strode towards him with his long, lounging cavalry-swing. The man turned pallid under his florid skin, and tried to edge imperceptibly away; but the density of the throng prevented his moving quickly enough to evade Cecil, who stooped his head, and said a word in his ear. It was briefly:

'Leave the Ring.'

The rascal, half bully, half coward, rallied from the startled fear into which his first recognition by the Guardsman (who had been the chief witness against him in a very scandalous matter at York, and who had warned him that if he ever saw him again in the Ring he would have him turned out of it) had thrown him, and, relying on insolence and the numbers of his fraternity to back him out of it, stood his ground.

'I've as much right here as you swells,' he said, with a horse-laugh. 'Are you the whole Jockey Club that you come it to a honest gentleman like that?'

Cecil looked down on him slightly amused, immeasurably disgusted—of all earth's terrors there was not one so great for him as a scene, and the eager bloodshot eyes of the Ring were turning on them by the thousand, and the loud shouting of the bookmakers was thundering out, 'What's up?'

'My "honest gentleman,"' he said, wearily, 'leave this, I tell you; do you hear?'

'Make me!' retorted the 'Welsher', defiant in his stout-built square strength, and ready to brazen the matter out. 'Make me, my cock o' fine feathers! Put me out of the Ring if you can, Mr Dainty-Limbs! I've as much business here as you.'

The words were hardly out of his mouth, before, light as a deer and close as steel, Cecil's hand was on his collar, and without any seeming effort, without the slightest passion, he calmly lifted him off the ground as though he were a terrier, and thrust him through the throng; Ben Davis, as the Welsher was named, meantime being so utterly amazed at such unlooked-for might in the grasp of the gentlest, idlest, most gracefully made, and indolently tempered of his born foes and prey 'the swells', that he let himself be forced along backward in sheer passive paralysis of astonishment. Bertie, profoundly insensible to the tumult that began to rise and roar about him, from those who were not too absorbed in the business of the morning to note what took place, thrust him along in the single clasp of his right hand, pushed him outward to where the running ground swept past the Stand, and threw him, lightly, easily, just as one may throw a lapdog to take his bath, into the artificial ditch filled with water that the Seraph had pointed out as 'a teaser'. The man fell unhurt, unbruised, so gently was he dropped on his back among the muddy chilly water and the overhanging brambles; and as he rose from the ducking a shudder of ferocious and filthy oaths poured from his lips, increased tenfold by the uproarious laughter of the crowd, who knew him as 'a Welsher', and thought him only too well served.

Policemen rushed in at all points, rural and metropolitan, breathless, austere, and, of course, too late. Bertie turned to them with a slight wave of his hand to sign them away.

'Don't trouble yourselves! It's nothing *you* could interfere in. Take care that person does not come into the betting-ring again, that's all.'

The Seraph, Lord Constantia, Wentworth, and many others of his set, catching sight of the turmoil and of 'Beauty,' with the great square-set figure of Ben Davis pressed before him through the mob, forced their way up as quickly as they could; but before they reached the spot Cecil was sauntering back to meet them, cool and listless, and a little bored with so much exertion, his cheroot in his mouth, and his ear serenely deaf to the clamour about the ditch.

He looked apologetically at the Seraph and the others; he felt some apology was required for having so far wandered from all the canons of his Order as to have approached 'a row', and run the risk of a scene.

'Turf *must* be cleared of these scamps, you see,' he said, with a half sigh. 'Law can't do anything. Fellow was trying to "get on" with the young one too. Don't bet with those riffraff, Berke. The great bookmakers will make you dead money, and the little Legs will do worse to you.'

The boy hung his head, but looked sulky rather than thankful for his brother's interference with himself and the Welsher.

'You have done the Turf a service, Beauty, a very great service; there's no doubt about that,' said the Seraph. 'Law can't do anything, as you say; opinion must clear the Ring of such rascals; a Welsher ought not to dare to show his face here, but, at the same time, *you* oughtn't to have gone unsteadying your muscle, and risking the firmness of your hand, at such a minute as this, with pitching that fellow over. Why couldn't you wait till afterwards? Or have let me do it?'

'My dear Seraph,' murmured Bertie, languidly, 'I've gone in today for exertion; a little more or less is nothing. Besides, Welshers are slippery dogs, you know.'

He did not add that it was having seen Ben Davis taking odds with his young brother which had spurred him to such instantaneous action with that disreputable personage, who, beyond doubt, only received a tithe part of his deserts, and merited to be double-thonged off every course in the kingdom.

Rake at that instant darted panting like a hot retriever out of the throng. 'Mr Cecil, sir, will you please come to the weights—the saddling-bell's a-going to ring, and——'

'Tell them to wait for me; I shall only be twenty minutes dressing,' said Cecil, quietly, regardless that the time at which the horses should have been at the starting-post was then clanging from the clock within the Grand Stand. Did you ever go to a gentleman-rider race where the jocks were not at least an hour behind time, and considered themselves, on the whole, very tolerably punctual? At last, however, he consented to saunter into the dressing-shed, and was aided by Rake into tops that had at length achieved a spotless triumph, and the scarlet gold-broidered jacket of his fair friend's art with white hoops, and the 'Cœur Vaillant Se Fait Royaume' on the collar, and the white gleaming sash to be worn across it, fringed by the same fair hands with silver.

Meanwhile, the 'Welsher', driven off the course by a hooting and indignant crowd, shaking the water from his clothes, with bitter oaths, and livid with a deadly passion at his exile from the harvest-field of his lawless gleanings, went his way, with a savage vow of vengeance against the 'd——d dandy', the 'Guards' swell', who had shown him up before his world as the scoundrel he was.

The bell was clanging and clashing passionately, as Cecil at last went down to the weights, all his friends of the Household about him, and all standing 'crushers' on their champion, for their stringent *esprit du corps* was involved, and the Guards are never backward in putting their gold down, as all the world knows. In the enclosure, the cynosure of devouring eyes, stood the King, with the sang froid of a superb gentleman, amidst the clamour raging round him, one delicate ear laid back now and then but otherwise indifferent to the din, with his coat glistening like satin, the beautiful tracery of vein and muscle like the veins of vine-leaves standing out on the glossy clear-carved neck that had the arch of Circassia, and his dark antelope eyes gazing with a pensive earnestness on the shouting crowd.

His rivals, too, were beyond par in fitness and in condition, and there were magnificent animals among them. Bay Regent was a huge raking chesnut, upwards of sixteen hands, and enormously powerful, with very fine shoulders, and an all-over-like-going head; he belonged to a Colonel in the Hussars, but was to be ridden by Jimmy Delmar of the Twelfth Lancers, whose colours were violet with orange hoops. Montacute's horse, Pas de Charge, which carried most of the money of the English Heavy Cavalry, Montacute himself being in the Dragoon Guards, was of much the same order, a black hunter with racing-blood in him, loins and withers that assured any amount of force, and no fault but that of a rather coarse head, traceable to a slur on his 'scutcheon on the distaff side from a plebeian great-grandmother, who had been a cart mare, the only stain in his otherwise faultless pedigree. However, she had given him her massive shoulders, so that he was in some sense a gainer by her after all. Wild

Geranium was a beautiful creature enough, a bright bay Irish mare, with that rich red gloss that is like the glow of a horse-chesnut, very perfect in shape, though a trifle light perhaps, and with not quite strength enough in neck or barrel; she would jump the fences of her own paddock half a dozen times a day for sheer amusement, and was game to anything.[1] She was entered by Cartouche of the Royal Irish Dragoons, to be ridden by 'Baby Grafton', of the same corps, a featherweight, and quite a boy, but with plenty of science in him. These were the three favourites; Day Star ran them close, the property of Durham Vavassour, of the Inniskillings, and to be ridden by his owner, a handsome flea-bitten-grey sixteen-hander, with ragged hips, and action that looked a trifle string-halty, but noble shoulders, and great force in the loins and withers; the rest of the field, though unusually excellent, did not find so many 'sweet voices' for them, and were not so much to be feared: each starter was of course much backed by his party, but the betting was tolerably even on these four—all famous steeplechasers—the King at one time, and Bay Regent at another, slightly leading in the Ring.

Thirty-two starters were hoisted up on the telegraph board, and as the field got at last under weigh, uncommonly handsome they looked, while the silk jackets of all the colours of the rainbow glittered in the bright noon-sun. As Forest King closed in, perfectly tranquil still, but beginning to glow and quiver all over with excitement, knowing as well as his rider the work that was before him, and longing for it in every muscle and every limb, while his eyes flashed fire as he pulled at the curb and tossed his head aloft, there went up a general shout of 'Favourite!' His beauty told on the populace, and even somewhat on the professionals, though the Legs still kept a strong business prejudice against the working powers of 'the Guards' crack'. The ladies began to lay dozens in gloves on him; not altogether for his points, which, perhaps, they hardly appreciated, but for his owner and rider, who, in the scarlet and gold, with the white sash across his chest, and a look of serene indifference on his face, they considered the handsomest man of the field. The Household is usually safe to win the suffrages of the Sex.

In the throng on the course Rake instantly bonneted an audacious dealer who had ventured to consider that Forest King was 'light and curby in the 'ock'. 'You're a wise 'un, you are!' retorted the wrathful and ever eloquent Rake; 'there's more strength in his clean flat legs, bless him! than in all the round thick mill-posts of *your* half-breds, that have no

---

[1] The portrait of this lady is that of a very esteemed young Irish beauty of my acquaintance; she this season did seventy-six miles on a warm June day, and ate her corn and tares afterwards as if nothing had happened. She is six years old.

more tendon than a bit of wood, and are just as flabby as a sponge!' Which hit the dealer home just as his hat was hit over his eyes; Rake's arguments being unquestionably in their force.

The thoroughbreds pulled and fretted, and swerved in their impatience; one or two over-contumacious bolted incontinently, others put their heads between their knees in the endeavours to draw their riders over their withers; Wild Geranium reared straight upright, fidgeted all over with longing to be off, passaged with the prettiest wickedest grace in the world, and would have given the world to neigh if she had dared, but she knew it would be very bad style, so, like an aristocrat as she was, restrained herself; Bay Regent almost sawed Jimmy Delmar's arms off, looking like a Titan Bucephalus; while Forest King, with his nostrils dilated till the scarlet tinge on them glowed in the sun, his muscles quivering with excitement as intense as the little Irish mare's, and all his Eastern and English blood on fire for the fray, stood steady as a statue for all that, under the curb of a hand light as a woman's, but firm as iron to control, and used to guide him by the slightest touch.

All eyes were on that throng of the first mounts in the Service; brilliant glances by the hundred gleamed down behind hothouse bouquets of their chosen colour, eager ones by the thousand stared thirstily from the crowded course, the roar of the Ring subsided for a second, a breathless attention and suspense succeeded it; the Guardsmen sat on their drags, or lounged near the ladies with their race-glasses ready, and their habitual expression of gentle and resigned weariness in no wise altered, because the Household, all in all, had from sixty to seventy thousand on the event; and the Seraph murmured mournfully to his cheroot, 'That chesnut's no end *fit*,' strong as his faith was in the champion of the Brigades.

A moment's good start was caught—the flag dropped—off they went sweeping out for the first second like a line of Cavalry about to charge.

Another moment, and they were scattered over the first field, Forest King, Wild Geranium, and Bay Regent leading for two lengths, when Montacute, with his habitual 'fast burst', sent Pas de Charge past them like lightning. The Irish mare gave a rush and got alongside of him; the King would have done the same, but Cecil checked him and kept him in that cool swinging canter which covered the grassland so lightly; Bay Regent's vast thundering stride was Olympian, but Jimmy Delmar saw his worst foe in the 'Guards' crack', and waited on him warily, riding superbly himself.

The first fence disposed of half the field, they crossed the second in the same order, Wild Geranium racing neck to neck with Pas de Charge; the

King was all athirst to join the duello, but his owner kept him gently back, saving his pace and lifting him over the jumps as easily as a lapwing. The second fence proved a cropper to several, some awkward falls took place over it, and 'tailing' commenced; after the third field, which was heavy plough, all knocked off but eight, and the real struggle began in sharp earnest: a good dozen who had shown a splendid stride over the grass being done up by the terrible work on the clods. The five favourites had it now all to themselves; Day Star pounding onward at tremendous speed, Pas de Charge giving slight symptoms of distress owing to the madness of his first burst, the Irish mare literally flying ahead of him, Forest King and the chesnut waiting on one another.

In the Grand Stand the Seraph's eyes strained after the Scarlet and White, and he muttered in his moustaches, 'Ye Gods, what's up! The world's coming to an end!—Beauty's turned cautious!'

Cautious indeed—with that giant of Pytchley fame running neck to neck by him; cautious—with two-thirds of the course unrun, and all the yawners yet to come; cautious—with the blood of Forest King lashing to boiling heat, and the wondrous greyhound stride stretching out faster and faster beneath him, ready at a touch to break away and take the lead: but he would be reckless enough by-and-by; reckless, as his nature was, under the indolent serenity of habit.

Two more fences came, laced high and stiff with the Shire thorn, and with scarce twenty feet between them, the heavy ploughed land leading to them black and hard, with the fresh earthy scent steaming up as the hoofs struck the clods with a dull thunder. Pas de Charge rose to the first: distressed too early, his hind feet caught in the thorn, and he came down rolling clear of his rider; Montacute picked him up with true science, but the day was lost to the English Heavy Cavalry. Forest King went in and out over both like a bird and led for the first time; the chesnut was not to be beat at fencing and ran even with him; Wild Geranium flew still as fleet as a deer; true to her sex she would not bear rivalry; but little Grafton, though he rode like a professional, was but a young one, and went too wildly; her spirit wanted cooler curb.

And now only, Cecil loosened the King to his full will and his full speed. Now only, the beautiful Arab head was stretched like a racer's in the run-in for the Derby, and the grand stride swept out till the hoofs seemed never to touch the dark earth they skimmed over; neither whip nor spur was needed; Bertie had only to leave the gallant temper and the generous fire that were roused in their might to go their way and hold their own. His hands were low; his head a little back; his face very calm, the eyes only had a daring, eager, resolute will lighting in them; Brixworth lay before him. He knew well what Forest King could do; but he did not know how great the chesnut Regent's powers might be.

The water gleamed before them, brown and swollen, and deepened with the meltings of winter snows a month before; the brook that has brought so many to grief over its famous banks, since cavaliers leapt it with their falcon on their wrist, or the mellow note of the horn rang over the woods in the hunting days of Stuart reigns. They knew it well, that long dark line, shimmering there in the sunlight, the test that all must pass who go in for the Soldiers' Blue Riband. Forest King scented water, and went on with his ears pointed, and his greyhound stride lengthening, quickening, gathering up all its force and its impetus for the leap that was before—then like the rise and the swoop of a heron he spanned the stream, and, landing clear, launched forward with the lunge of a spear darted through air. Brixworth was passed—the Scarlet and White, a mere gleam of bright colour, a mere speck in the landscape, to the breathless crowds in the Stand, sped on over the brown and level grassland; two and a quarter miles done in four minutes and twenty seconds. Bay Regent was scarcely behind him; the chesnut abhorred the water, but a finer trained hunter was never sent over the Shires, and Jimmy Delmar rode like Grimshaw himself. The giant took the leap in magnificent style, and thundered on neck and neck with the 'Guards' crack'. The Irish mare followed, and with miraculous gameness landed safely; but her hind-legs slipped on the bank, a moment was lost, and 'Baby' Grafton scarce knew enough to recover it, though he scoured on nothing daunted.

Pas de Charge, much behind, refused the yawner; his strength was not more than his courage, but both had been strained too severely at first. Montacute struck the spurs into him with a savage blow over the head; the madness was its own punishment; the poor brute rose blindly to the jump, and missed the bank with a reel and a crash; Sir Eyre was hurled out into the brook, and the hope of the Heavies lay there with his breast and fore-legs resting on the ground, his hind-quarters in the water, and his back broken. Pas de Charge would never again see the starting-flag waved, or hear the music of the hounds, or feel the gallant life throb and glow through him at the rallying notes of the horn. His race was run.

Not knowing, or looking, or heeding what happened behind, the trio tore on over the meadow and the plough; the two favourites neck by neck, the game little mare hopelessly behind, through that one fatal moment over Brixworth. The turning-flags were passed; from the crowds on the course a great hoarse roar came louder and louder, and the shouts rang, changing every second: 'Forest King wins'—'Bay Regent wins'— 'Scarlet and White's ahead'—'Violet's up with him'—'Violet's past him'—'Scarlet recovers'—'Scarlet beats'—'A cracker on the King'—'Ten to one on the Regent'—'Guards are over the fence first'—'Guards are winning'—'Guards are losing'—'Guards are beat!!'

Were they!

As the shout rose, Cecil's left stirrup-leather snapped and gave way; at the pace they were going most men, ay, and good riders too, would have been hurled out of their saddle by the shock; he scarcely swerved; a moment to ease the King and to recover his equilibrium, then he took the pace up again as though nothing had chanced. And his comrades of the Household when they saw this through their race-glasses, broke through their serenity and burst into a cheer that echoed over the grasslands and the coppices like a clarion, the grand rich voice of the Seraph leading foremost and loudest—a cheer that rolled mellow and triumphant down the cold bright air like the blast of trumpets, and thrilled on Bertie's ear where he came down the course a mile away. It made his heart beat quicker with a victorious headlong delight, as his knees pressed closer into Forest King's flanks, and, half stirrupless like the Arabs, he thundered forward to the greatest riding feat of his life. His face was very calm still, but his blood was in tumult, the delirium of pace had got on him, a minute of life like this was worth a year, and he knew that he would win, or die for it, as the land seemed to fly like a black sheet under him, and, in that killing speed, fence and hedge and double and water all went by him like a dream, whirling underneath him as the grey stretched stomach to earth over the level, and rose to leap after leap.

For that instant's pause, when the stirrup broke, threatened to lose him the race.

He was more than a length behind the Regent, whose hoofs as they dashed the ground up sounded like thunder, and for whose Herculean strength the ploughed lands had no terrors; it was more than the lead to keep now, there was ground to cover, and the King was losing like Wild Geranium. Cecil felt drunk with that strong keen west wind that blew so strongly in his teeth, a passionate excitation was in him, every breath of winter air that rushed in its bracing currents round him seemed to lash him like a stripe: the Household to look on and see him beaten!

Certain wild blood that lay latent in him under the tranquil gentleness of temper and of custom woke and had the mastery; he set his teeth hard, and his hands clenched like steel on the bridle. 'Oh! my beauty, my beauty!' he cried, all unconsciously, half aloud, as they cleared the thirty-sixth fence. 'Kill me if you like, but don't *fail* me!'

As though Forest King heard the prayer and answered it with all his heart, the splendid form launched faster out, the stretching stride stretched farther yet with lightning spontaneity, every fibre strained, every nerve struggled, with a magnificent bound like an antelope the grey recovered the ground he had lost, and passed Bay Regent by a quarter-length. It was a neck to neck race once more, across the three meadows with the last and lower fences that were between them and the final leap of all; that ditch of artificial water with the towering double hedge of oak rails and of blackthorn which was reared black and grim and well-nigh hopeless just in front of the Grand Stand. A roar like the roar of the sea broke up from the thronged course as the crowd hung breathless on the even race; ten thousand shouts rang as thrice ten thousand eyes watched the closing contest, as superb a sight as the Shires ever saw while the two ran together, the gigantic chesnut, with every massive sinew swelled and strained to tension, side by side with the marvellous grace, the shining flanks, and the Arab-like head of the Guards' horse.

Louder and wilder the shrieked tumult rose: 'The chesnut beats!' 'The grey beats!' 'Scarlet's ahead!' 'Bay Regent's caught him!' 'Violet's winning, Violet's winning!' 'The King's neck by neck!' 'The King's beating!' 'The Guards will get it.' 'The Guards' crack has it!' 'Not yet, not yet!' 'Violet will thrash him at the jump!' 'Now for it!' 'The Guards, the Guards, the Guards!' 'Scarlet will win!' 'The King has the finish!' 'No, no, no, NO!'

Sent along at a pace that Epsom flat never eclipsed, sweeping by the Grand Stand like the flash of electric flame, they ran side to side one moment more, their foam flung on each other's withers, their breath hot in each other's nostrils, while the dark earth flew beneath their stride. The blackthorn was in front behind five bars of solid oak, the water yawning on its farther side, black and deep, and fenced, twelve feet wide if it were an inch, with the same thorn wall beyond it; a leap no horse should have been given, no Steward should have set.

Cecil pressed his knees closer and closer, and worked the gallant hero for the test; the surging roar of the throng, though so close, was dull on his ear; he heard nothing, knew nothing, saw nothing but that lean chesnut head beside him, the dull thud on the turf of the flying gallop, and the black wall that reared in his face. Forest King had done so much, could he have stay and strength for this?

Cecil's hands clenched unconsciously on the bridle, and his face was very pale—pale with excitation—as his foot where the stirrup was broken crushed closer and harder against the grey's flank. 'Oh, my darling, my beauty—*now!*'

One touch of the spur—the first—and Forest King rose at the leap, all the life and power there were in him gathered for one superhuman and crowning effort; a flash of time, not half a second in duration, and he was lifted in the air higher, and higher, and higher in the cold, fresh, wild winter wind; stakes and rails and thorn and water lay beneath him black and gaunt and shapeless, yawning like a grave; one bound, even in mid-air, one last convulsive impulse of the gathered limbs, and Forest King was over!

And as he galloped up the straight run-in he was alone.

Bay Regent had refused the leap.

As the grey swept to the Judge's chair, the air was rent with deafening cheers that seemed to reel like drunken shouts from the multitude. 'The Guards win, the Guards win'; and when his rider pulled up at the distance with the full sun shining on the scarlet and white, with the gold glisten of the embroidered 'Cœur Vaillant Se Fait Royaume', Forest King stood in all his glory, winner of the Soldiers' Blue Riband, by a feat without its parallel in all the annals of the Gold Vase.

But as the crowd surged about him, and the mad cheering crowned his victory, and the Household in the splendour of their triumph and the fulness of their gratitude rushed from the drags and the stands to cluster to his saddle, Bertie looked as serenely and listlessly nonchalant as of old, while he nodded to the Seraph with a gentle smile.

'Rather a close finish, eh? Have you any Moselle Cup going there? I'm a little thirsty.'

Outsiders would much sooner have thought him defeated than triumphant; no one, who had not known him, could possibly have imagined that he had been successful; an ordinary spectator would have concluded that, judging by the resigned weariness of his features, he had won the race greatly against his own will to his now infinite ennui. No one could have dreamt that he was thinking in his heart of hearts how passionately he loved the gallant beast that had been victor with him, and that, if he had followed out the momentary impulse in him, he could have put his arms round the noble bowed neck and kissed the horse like a woman!

The Moselle Cup was brought to refresh the tired champion, and before he drank it Bertie glanced at a certain place in the Grand Stand and bent his head as the cup touched his lips: it was a dedication of his victory to his Queen of Beauty. Then he threw himself lightly out of saddle, and, as Forest King was led away for the after ceremony of bottling, rubbing, and clothing, his rider, regardless of the roar and hubbub of the course, and of the tumultuous cheers that welcomed both him and his horse from the men who pressed round him, into whose pockets he had put thousands on thousands, and whose ringing hurrahs greeted the 'Guards' crack', passed straight up towards Jimmy Delmar and held out his hand.

'You gave me a close thing, Major Delmar. The Vase is as much yours as mine; if your chesnut had been as good a water jumper as he is a fencer we should have been neck to neck at the finish.'

The browned Indian-sunned face of the Lancer broke up into a cordial smile, and he shook the hand held out to him warmly; defeat and disappointment had cut him to the core, for Jimmy was the first riding man of the Light Cavalry, but he would not have been the frank campaigner that he was if he had not responded to the graceful and generous overture of his rival and conqueror.

'Oh! I can take a beating,' he said, good humouredly; 'at any rate, I am beat by the Guards, and it is very little humiliation to lose against such riding as yours and such a magnificent brute as your King. I congratulate you most heartily, most sincerely.'

And he meant it, too. Jimmy never canted, nor did he ever throw the blame, with paltry savage vindictiveness, on the horse he had ridden. Some men there are—their name is legion—who never allow that it is *their* fault when they are 'nowhere'; oh no! it is the 'cursed screw' always, according to them. But a very good rider will not tell you that.

Cecil, while he talked, was glancing up at the Grand Stand, and when the others dispersed to look over the horses, and he had put himself out of his shell into his sealskin in the dressing-shed, he went up thither without a moment's loss of time.

He knew them all; those dainty beauties with their delicate cheeks just brightened by the western winterly wind, and their rich furs and laces glowing among the colours of their respective heroes; he was the pet of them all; 'Beauty' had the suffrages of the sex without exception; he was received with bright smiles and graceful congratulations, even from those who had espoused Eyre Montacute's cause, and still fluttered their losing azure, though the poor hunter lay dead, with his back broken, and a pistol-ball mercifully sent through his brains—the martyr to a man's hot haste, as the dumb things have ever been since creation began.

Cecil passed them as rapidly as he could for one so well received by them, and made his way to the centre of the Stand, to the same spot at which he had glanced when he had drunk the Moselle.

A lady turned to him; she looked like a rose camellia in her floating scarlet and white, just toned down and made perfect by a shower of Spanish lace; a beautiful brunette, dashing yet delicate, a little fast yet intensely thoroughbred, a coquette who would smoke a cigarette, yet a peeress who would never lose her dignity.

'*Au cœur vaillant rien d'impossible!*' she said, with an *envoi* of her lorgnon, and a smile that should have intoxicated him—a smile that might have rewarded a Richepanse for a Hohenlinden. 'Superbly ridden! I absolutely trembled for you as you lifted the King to that last leap. It was terrible!'

It was terrible; and a woman, to say nothing of a woman who was in love with him, might well have felt a heart-sick fear at sight of that yawning water and those towering walls of blackthorn, where one touch of the hoofs on the topmost bough, one spring too short of the gathered limbs, must have been death to both horse and rider. But as she said it, she was smiling, radiant, full of easy calm and racing interest, as became her ladyship, who had had 'bets at even' before now on Goodwood, and could lead the first flight over the Belvoir and the Quorn countries. It was possible that her ladyship was too thoroughbred not to see a man killed over the oak-rails without deviating into unseemly emotion, or being capable of such bad style as to be agitated.

Bertie, however, in answer, threw the tenderest eloquence into his eyes; very learned in such eloquence.

'If I could not have been victorious while *you* looked on, I would at least not have lived to meet you here!'

She laughed a little, so did he; they were used to exchange these passages in an admirably artistic masquerade, but it was always a little droll to each of them to see the other wear the domino of sentiment, and neither had much credence in the other.

'What a preux chevalier!' cried his Queen of Beauty. 'You would have died in a ditch out of homage to me. Who shall say that chivalry is past? Tell me, Bertie, is it so very delightful that desperate effort to break your neck? It looks pleasant, to judge by its effects. It is the only thing in the world that amuses you!'

'Well—there is a great deal to be said for it,' replied Cecil, musingly. 'You see, until one *has* broken one's neck, the excitement of the thing isn't totally worn out; can't be, naturally, because the—what-do-you-call-it?—consummation isn't attained till then. The worst of it is, it's getting common-place, getting vulgar, such a number break *their* necks, doing Alps and that sort of thing, that we shall have nothing at all left to ourselves soon.'

'Not even the monopoly of sporting suicide! Very hard,' said her ladyship, with the lowest, most languid laugh in the world, very like 'Beauty's' own, save that it had a considerable inflection of studied affectation, of which he, however much of a dandy he was, was wholly guiltless. 'Well! you won magnificently; that little black man, who is he?—Lancers, somebody said—ran you so fearfully close. I really thought at one time that the Guards had lost.'

'Do you suppose that a man happy enough to wear Lady Guenevere's colours could lose? An embroidered scarf given by such hands has been a gage of victory ever since the days of tournaments!' murmured Cecil with the softest tenderness, but just enough laziness in the tone and laughter in the eye to make it highly doubtful whether he was not laughing both at her and at himself, and was not wondering why the deuce a fellow had to talk such nonsense.

Yet she was Lady Guenevere, with whom he had been in love ever since they had stayed together at Belvoir for the Croxton Park week last autumn; and who was beautiful enough to make their 'friendship' as enchanting as a page out of the 'Decamerone'. And while he bent over her, flirting in the fashion that made him the darling of the drawing-rooms, and looking down into her superb Velasquez eyes, he did not know, and, if he had known, would have been careless of it, that afar off, white with rage, and with his gaze straining on to the course through his race-glass, Ben Davis, 'the Welsher', who had watched the finish—watched the 'Guards' crack' landed at the distance—muttered, with a mastiff's savage growl:

'He wins, does he? Curse him! The d——d swell—he shan't win long.'

CHAPTER IV

Love à la Mode

Life was very pleasant at Royallieu.

It lay in the Melton country, and was almost equally well placed for Pytchley, Quorn, and Belvoir, besides possessing its own small but very perfect pack of 'little ladies', or the 'demoiselles', as they were severally nicknamed; the game was closely preserved, pheasants were fed on Indian corn till they were the finest birds in the country, and in the little winding paths of the elder and bilberry coverts thirty first-rate shots, with two loading-men to each, could find flock and feather to amuse them till dinner, with rocketers and warm corners enough to content the most insatiate of <u>knickerbockered</u> gunners. The stud was superb; the cook a French artist of consummate genius, who had a <u>brougham</u> to his own use, and wore diamonds of the first water; on the broad beech-studded grassy lands no lesser thing than doe and deer ever swept through the thick ferns in the sunlight and the shadow; a retinue of powdered servants filled the old halls, and guests of highest degree dined in its stately banqueting-room, with its scarlet and gold, its Vandykes and its Vernets, and yet—there was terribly little money at Royallieu with it all. Its present luxury was purchased at the cost of the future, and the parasite of extravagance was constantly sapping, unseen, the gallant old Norman-planted oak of the family-tree. But then who thought of that? Nobody. It was the way of the House never to take count of the morrow.

True, any one of them would have died a hundred deaths rather than have had one acre of the beautiful green diadem of woods felled by the axe of the timber contractor, or passed to the hands of a stranger; but no one among them ever thought that this was the inevitable end to which they surely drifted with blind and unthinking improvidence. The old Viscount, haughtiest of haughty nobles, would never abate one jot of his accustomed magnificence; and his sons had but imbibed the teaching of all that surrounded them; they did but do in manhood what they had been unconsciously moulded to do in boyhood, when they were sent to Eton at ten, with gold dressing-boxes to grace their Dame's tables, embryo-Dukes for their co-fags, and tastes that already knew to a nicety the worth of the champagnes at the Christopher. The old, old story—how it repeats itself! Boys grow up amidst profuse prodigality, and are launched into a world where they can no more arrest themselves, than the featherweight can pull in the lightning-stride of the two-year old, who defies all check, and takes the flat as he chooses. They are brought up like young Dauphins and tossed into the costly whirl to float as best they can—on nothing. Then, on the lives and deaths that follow; on the graves where a dishonoured alien lies forgotten by the dark Austrian lake-side, or under the monastic shadow of some crumbling Spanish crypt; where a red cross chills the lonely traveller in the virgin solitudes of Amazonian forest aisles, or the wild scarlet creepers of Australia trail over a nameless mound above the trackless stretch of sun-warmed waters—then at them the world 'shoots out its lips with scorn'. Not on *them* lies the blame.

A wintry, watery sun was shining on the terraces as Lord Royallieu paced up and down the morning after the Grand Military; his step and limbs excessively enfeebled, but the carriage of his head and the flash of his dark hawk's eyes as proud and untameable as in his earliest years. He never left his own apartments; and no one, save his favourite 'little

Berke', ever went to him without his desire; he was too sensitive a man to thrust his age and ailing health in amongst the young leaders of fashion, the wild men of pleasure, the good wits and the good shots of his son's set; he knew very well that his own day was past, that they would have listened to him out of the patience of courtesy, but that they would have wished him away as 'no end of a bore'. He was too shrewd not to know this; but he was too quickly galled ever to bear to have it recalled to him.

He looked up suddenly and sharply; coming towards him he saw the figure of the Guardsman. For 'Beauty' the Viscount had no love; indeed, well-nigh a hatred, for a reason never guessed by others, and never betrayed by him.

Bertie was not like the Royallieu race; he resembled his mother's family. She, a beautiful and fragile creature whom her second son had loved, for the first years of his life, as he would have thought it now impossible that he could love any one, had married the Viscount with no affection towards him, while he had adored her with a fierce and jealous passion that her indifference only inflamed. Throughout her married life, however, she had striven to render loyalty and tenderness towards a lord into whose arms she had been thrown, trembling and reluctant; of his wife's fidelity he could not entertain a doubt, though that he had never won her heart he could not choose but know. He knew more, too; for she had told it him with a noble candour before he wedded her; knew that the man she did love was a penniless cousin, a cavalry officer, who had made a famous name among the wild mountain tribes of Northern India. This cousin, Alan Bertie—a fearless and chivalrous soldier, fitter for the days of knighthood than for these—had seen Lady Royallieu at Nice, some three years after her marriage; accident had thrown them across each other's path; the old love, stronger, perhaps, now than it had ever been, had made him linger in her presence, had made her shrink from sending him to exile. Evil tongues at last had united their names together; Alan Bertie had left the woman he idolized lest slander should touch her through him, and fallen two years later under the dark dank forests on the desolate moorside of the hills of Hindostan, where long before he had rendered 'Bertie's Horse' the most famous of all the wild Irregulars of the East.

After her death, Lord Royallieu found Alan's miniature among her papers, and recalled those winter months by the Mediterranean till he cherished, with the fierce, eager, self-torture of a jealous nature, doubts and suspicions that, during her life, one glance from her eyes would have disarmed and abashed. Her second and favourite child bore her family name, her late lover's name; and, in resembling her race, resembled the dead soldier. Moreover, Bertie had been born in the spring following that Nice winter, and it sufficed to make the Viscount hate him with a cruel and savage detestation which he strove indeed to temper—for he was by nature a just man, and, in his better moments, knew that his doubts wronged both the living and the dead—but which coloured, too strongly to be dissembled, all his feelings and his actions towards his son, and might both have soured and wounded any temperament less nonchalantly gentle and supremely careless than Cecil's. As it was, Cecil was sometimes surprised at his father's dislike to him: but never thought much about it, and attributed it, when he did think of it, to the caprices of a tyrannous old man. To be envious of the favour shown to his boyish brother could never for a moment have come into his imagination. Lady Royallieu, with her last words, had left the little fellow, a child of three years old, to the affection and the care of Bertie—himself then a boy of twelve or fourteen—and little as he thought of such things now, the trust of his dying mother had never been wholly forgotten.

A heavy gloom came now over the Viscount's still handsome saturnine aquiline face as his second son approached up the terrace; Bertie was too like the cavalry soldier whose form he had last seen standing against the rose light of a Mediterranean sunset. The soldier had been dead eight-and-twenty years; but the jealous hate was not dead yet.

Cecil took off his hunting-cap with a certain courtesy that sat very well on his habitual languid nonchalance; he never called his father anything but 'Royal'; rarely saw, still less rarely consulted him, and cared not a straw for his censure or opinion, but he was too thoroughbred by nature to be able to follow the under-bred indecorum of the day which makes disrespect to old age the fashion. 'You sent for me?' he asked, taking the cigarette out of his mouth.

'No, sir,' answered the old Lord, curtly, 'I sent for your brother. The fools can't take even a message right now, it seems.'

'Shouldn't have named us so near alike; it's often a bore!'

'I didn't name you, sir, your mother named you,' answered his father, sharply; the subject irritated him.

'It's of no consequence which!' murmured Cecil, with an expostulatory wave of his cigar. 'We're not even asked whether we like to come into the world; we can't expect to be asked what we like to be called in it. Good day to you, sir.'

He turned to move away to the house; but his father stopped him; he knew that he had been discourteous, a far worse crime in Lord Royallieu's eyes than to be heartless.

'So you won the Vase yesterday?' he asked, pausing in his walk with his back bowed, but his stern, silver-haired head erect.

'*I* didn't—the King did.'

'That's absurd, sir,' said the Viscount, in his resonant and yet melodious voice. 'The finest horse in the world may have his back broke by bad riding, and a screw has won before now when it's been finely handled. The finish was tight, wasn't it?'

'Well—rather. I have ridden closer spins, though. The fallows were light.'

Lord Royallieu smiled grimly.

'I know what the Shire "plough" is like,' he said, with a flash of his falcon eyes over the landscape, where, in the days of his youth, he had led the first flight so often, George Rex, and Waterford, and the Berkeleys, and the rest following the rally of his hunting-horn. 'You won much in bets?'

'Very fair. Thanks.'

'And won't be a shilling richer for it this day next week!' retorted the Viscount, with a rasping, grating irony; he could not help darting savage thrusts at this man who looked at him with eyes so cruelly like Alan Bertie's. 'You play £5 points, and lay £500 on the odd trick, I've heard, at your whist in the Clubs—pretty prices for a younger son!'

'Never bet on the odd trick; spoils the game; makes you sacrifice play *to* the trick. We always bet on the game,' said Cecil, with gentle weariness; the sweetness of his temper was proof against his father's attacks upon his patience.

'No matter *what* you bet, sir; you live as if you were a Rothschild while you are a beggar!'

'Wish I were a beggar: fellows always have no end in stock, they say; and your tailor can't worry you very much when all you have to think about is an artistic arrangement of tatters!' murmured Bertie, whose impenetrable serenity was never to be ruffled by his father's bitterness.

'You will soon have your wish, then,' retorted the Viscount, with the unprovoked and reasonless passion which he vented on every one, but one none so much as the son he hated. 'You are on a royal road to it. I live out of the world, but I hear from it, sir. I hear that there is not a man in the Guards—not even Lord Rockingham—who lives at the rate of imprudence you do; that there is not a man who drives such costly horses, keeps such costly mistresses, games to such desperation, fools gold away with such idiocy as you do. You conduct yourself as if you were a millionaire, sir, and what are you? A pauper on my bounty, and on your brother

Montagu's after me—a pauper with a tinsel fashion, a gilded beggary, a
Queen's commission to cover a sold-out poverty, a dandy's reputation to
stave off a defaulter's future! A pauper, sir—and a Guardsman!'

The coarse and cruel irony flashed out with wicked scorching
malignity, lashing and upbraiding the man who was the victim of his own
unwisdom and extravagance.

A slight tinge of colour came on his son's face as he heard; but he gave
no sign that he was moved, no sign of impatience or anger. He lifted his
cap again, not in irony, but with a grave respect in his action that was
totally contrary to his whole temperament.

'This sort of talk is very exhausting, very bad style,' he said, with his
accustomed gentle murmur. 'I will bid you good morning, my Lord.'

And he went without another word. Crossing the length of the old-
fashioned Elizabethan terrace, little Berk passed him; he motioned the
lad towards the Viscount. 'Royal wants to see you, young one.'

The boy nodded and went onward; and as Bertie turned to enter the
low door that led out to the stables he saw his father meet the lad—meet
him with a smile that changed the whole character of his face, and
pleasant kindly words of affectionate welcome, drawing his arm about
Berkeley's shoulder, and looking with pride upon his bright and gracious
youth.

More than an old man's preference would be thus won by the young
one; a considerable portion of their mother's fortune, so left that it could
not be dissipated, yet could be willed to which son the Viscount chose,
would go to his brother by this passionate partiality; but there was not a
tinge of jealousy in Cecil; whatever else his faults he had no mean ones,
and the boy was dear to him, by a quite unconscious yet unvarying
obedience to his dead mother's wish.

'Royal hates me as game birds hate a red dog. Why the deuce, I
wonder?' he thought, with a certain slight touch of pain despite his
idle philosophies and devil-may-care indifference. 'Well—I *am* good for
nothing, I suppose. Certainly I am not good for much, unless it's riding
and making love.'

With which summary of his merits, 'Beauty', who felt himself to be a
master in those two arts, but thought himself a bad fellow out of them,
sauntered away to join the Seraph and the rest of his guests. His father's
words pursuing him a little despite his carelessness, for they had borne an
unwelcome measure of truth.

'Royal can hit hard,' his thoughts continued. 'A pauper and a Guardsman! By Jove! it's true enough; but he made me so. They brought me up

as if I had a million coming to me, and turned me out among the cracks to take my running with the best of them;—and they give me just about what pays my groom's book! Then they wonder that a fellow goes to the Jews. Where the deuce else can he go?'

And Bertie, whom his gains the day before had not much benefited, since his play-debts, his young brother's needs, and the Zu-Zu's insatiate little hands were all stretched ready to devour them without leaving a sovereign for more serious liabilities, went, for it was quite early morning, to act the MFH in his father's stead, at the meet on the great lawns before the house, for the Royallieu 'lady-pack' were very famous in the Shires, and hunted over the same country alternate days with the Quorn.

They moved off ere long to draw the Holt Wood, in as open a morning, and as strong a scenting wind, as ever favoured Melton Pink.

A whimper and 'gone away!' soon echoed from Beeby-side, and the pack, not letting the fox hang a second, dashed after him, making straight for Scraptoft. One of the fastest things up wind that hounds ever ran took them straight through the Spinnies, past Hamilton Farm, away beyond Burkby village, and down into the valley of the Wreake without a check, where he broke away, was headed, tried earths, and was pulled down scarce forty minutes from the find. The pack then drew Hungerton foxholes blank, drew Carver's spinnies without a whimper; and lastly, drawing the old familiar Billesden Coplow, had a short quick burst with a brace of cubs, and returning, settled themselves to a fine dog fox that was raced an hour and half, hunted slowly for fifty minutes, raced again another hour and quarter, sending all the field to their 'second horses'; and, after a clipping chase through the cream of the grass country, nearly saved his brush in the twilight when scent was lost in a rushing hailstorm, but had the 'little ladies' laid on again like wildfire, and was killed with the 'who-whoop!' ringing far and away over Glenn Gorse, after a glorious run—thirty miles in and out—with pace that tried the best of them.

A better day's sport even the Quorn had never had in all its brilliant annals, and faster things the Melton men themselves had never wanted: both those who love the 'quickest thing you ever knew; thirty minutes without a check; *such* a pace!' and care little whether the finale be 'killed' or 'broke away', and those of older fashion, who prefer 'long day, you know, steady as old time, the beauties stuck like wax through fourteen parishes as I live; six hours if it were a minute; horses dead beat; positively *walked*, you know, no end of a day!' but must have the fatal 'who-whoop' as conclusion—both of these, the 'new style and the old', could not but be content with the doings of the Demoiselles from start to finish.

Was it likely that Bertie remembered the caustic lash of his father's ironies while he was lifting Mother of Pearl over the posts and rails, and sweeping on, with the halloo ringing down the wintry wind as the grasslands flew beneath him? Was it likely that he recollected the difficulties that hung above him while he was dashing down the Gorse happy as a king, with the wild hail driving in his face, and a break of stormy sunshine just welcoming the gallant few who were landed at the death as twilight fell? Was it likely that he could unlearn all the lessons of his life, and realize in how near a neighbourhood he stood to ruin, when he was drinking Regency sherry out of his gold flask as he crossed the saddle of his second horse, or, smoking, rode slowly homeward, chatting with the Seraph through the leafless muddy lanes in the gloaming.

Scarcely—it is very easy to remember our difficulties when we are eating and drinking them, so to speak, in bad soups and worse wines in continental impecuniosity, sleeping on them as rough Australian shake-downs, or wearing them perpetually in Californian rags and tatters, it were impossible very well to escape from them then; but it is very hard to remember them when every touch and shape of life is pleasant to us— when everything about us is symbolical and redolent of wealth and ease—when the art of enjoyment is the only one we are called on to study, and the science of pleasure all we are asked to explore.

It is well-nigh impossible to believe yourself a beggar while you never want sovereigns for whist; and it would be beyond the powers of human nature to conceive your ruin irrevocable, while you still eat turbot and terrapin with a powdered giant behind your chair daily. Up in his garret a poor wretch knows very well what he is, and realises in stern fact the extremities of the last sou, the last shirt, and the last hope; but in these devil-may-care pleasures—in this pleasant, reckless, velvet-soft rush down-hill—in this club-palace, with every luxury that the heart of man can devise and desire, yours to command at your will—it is hard work, then, to grasp the truth that the crossing-sweeper yonder, in the dust of Pall Mall, is really not more utterly in the toils of poverty than you are!

'Beauty' was never, in the whole course of his days, virtually or physically, or even metaphorically, reminded that he was not a millionaire; much less still was he ever reminded so painfully. Life petted him, pampered him, caressed him, gifted him, though of half his gifts he never made use; lodged him like a prince, dined him like a king, and never recalled to him by a single privation or a single sensation that he was not as rich a man as his brother-in-arms, the Seraph, future Duke of Lyonnesse. How could he then bring himself to understand, as nothing less than truth, the grim and cruel insult his father had flung at him in that brutally bitter phrase—'A Pauper and a Guardsman'?

If he had ever been near a comprehension of it, which he never was, he must have ceased to realise it when—pressed to dine with Lord Guenevere, near whose house the last fox had been killed, while grooms dashed over to Royallieu for their changes of clothes—he caught a glimpse, as they passed through the hall, of the ladies taking their pre-prandial cups of tea in the library, an enchanting group of lace and silks, of delicate hue and scented hair, of blonde cheeks and brunette tresses, of dark velvets and gossamer tissue; and when he had changed the scarlet for dinner-dress, went down amongst them to be the darling of that charmed circle, to be smiled on and coquetted with by those soft, languid aristocrats, to be challenged by the lustrous eyes of his châtelaine, and to be spoiled as women will spoil the privileged pet of their drawing-rooms whom they have made 'free of the guild', and endowed with a flirting commission, and acquitted of anything 'serious'.

He was the recognized darling, and permitted property, of the young married beauties; the unwedded knew he was hopeless for *them*, and tacitly left him to the more attractive conquerors; who hardly prized the Seraph so much as they did Bertie, to sit in their barouches and opera boxes, ride and drive and yacht with them, conduct a Boccaccio intrigue through the height of the season, and make them really believe themselves actually in love while they were at the Moors or down the Nile, and would have given their diamonds to get a new distraction.

Lady Guenevere was the last of these, his titled and wedded captors; and perhaps the most resistless of all of them. Neither of them believed very much in their attachment, but both of them wore the masquerade dress to perfection. He had fallen in love with her as much as he ever fell in love, which was just sufficient to amuse him, and never enough to disturb him. He let himself be fascinated, not exerting himself either to resist or to advance the affair, till he was, perhaps, a little more entangled with her than it was according to his canons expedient to be; and they had the most enchanting—friendship.

Nobody was ever so indiscreet as to call it anything else; and my Lord was too deeply absorbed in the Alderney beauties that stood knee-deep in the yellow straw of his farmyard, and the triumphant conquests that he gained over his brother Peers' Short-horns and Suffolks, to trouble his head about Cecil's attendance on his beautiful Countess.

They corresponded in Spanish; they had a thousand charming cyphers; they made the columns of the *Times* and the *Post* play the unconscious rôle of medium to appointments; they eclipsed all the pages of Calderon's or Congreve's comedies in the ingenuities with which they met, wrote, got invitations together to the same country-houses, and arranged signals for mute communication: but there was not the slightest occasion for it all. It passed the time, however, and went far to persuade them that they really were in love, and had a mountain of difficulties and dangers to contend with; it added the 'spice to the sauce', and gave them the 'relish of being forbidden'. Besides, an open scandal would have been very shocking to her brilliant ladyship, and there was nothing on earth, perhaps, of which he would have had a more lively dread than a 'scene'; his present 'friendship', however, was delightful, and presented no such dangers, while his fair 'friend' was one of the greatest beauties and the greatest coquettes of her time. Her smile was honour; her fan was a sceptre; her face was perfect; and her heart never troubled herself or her lovers: if she had a fault, she was a trifle exacting, but that was not to be wondered at in one so omnipotent, and her chains, after all, were made of roses.

As she sat in the deep ruddy glow of the library fire, with the light flickering on her white brow and her violet velvets; as she floated to the head of her table, with opals shining amongst her priceless point laces, and some tropical flower with leaves of glistening gold crowning her bronze hair; as she glided down in a waltz along the polished floor, or bent her proud head over écarté in a musing grace that made her opponent utterly forget to mark the king or even play his cards at all; as she talked in the low music of her voice of European imbrogli, and consols and coupons, for she was a politician and a speculator; when she lapsed into a beautifully-tinted study of *la femme incomprise*, when time and scene suited, when the stars were very clear above the terraces without, and the conservatory very solitary, and a touch of Musset or Owen Meredith chimed in well with the light and shade of the oleanders and the brown lustre of her own eloquent glance; in all these various moments how superb she was! And if in truth her bosom only fell with the falling of Shares and rose with the rising of Bonds, if her soft shadows were only taken up like the purple tinting under her lashes to embellish her beauty; if in her heart of hearts she thought Musset a fool, and wondered why *Lucille* was not written in prose, in her soul far preferring *Le Follet*; why— it did not matter, that I can see; all great ladies gamble in stock now-a-days under the rose; and women are for the most part as cold, clear, hard, and practical as their adorers believe them the contrary; and a *femme incomprise* is so charming when she avows herself comprehended by *you*, that you would never risk spoiling the confidence by hinting a doubt of its truth.

If she and Bertie only played at love, if neither believed much in the other, if each trifled with a pretty gossamer soufflet of passion much as they trifled with their soufflets at dinner, if both tried it to trifle away ennui much as they tried staking a Friedrich d'Or at Baden, this light, surface, fashionable, philosophic form of a passion they both laughed at in its hot and serious follies, suited them admirably. Had it ever mingled a grain of bitterness in her ladyship's Souchong before dinner, or given an aroma of bitterness to her lover's Naples punch in the smoking-room, it would have been out of all keeping with themselves and their world.

Nothing on earth is so pleasant as being a little in love; nothing on earth so destructive as being too much so; and as Cecil, in the idle enjoyment of the former gentle luxury, flirted with his liege lady that night, lying back in the softest of lounging-chairs, with his dark dreamy handsome eyes looking all the eloquence in the world, and his head drooped till his moustaches were almost touching her laces, his Queen of Beauty listened with charmed interest, and to judge by his attitude he might have been praying after the poet—

> 'How is it under our control
> To love or not to love?'

In real truth he was gently murmuring.

'Such a pity that you missed today! Hounds found directly; three of the fastest things I ever knew, one after another; you should have seen the "little ladies" head him just above the Gorse! Three hares crossed us and a fresh fox; some of the pack broke away after the new scent, but old Bluebell, your pet, held on like death, and most of them kept after her—you had your doubts about Silver Trumpet's shoulders; they're not the thing, perhaps, but she ran beautifully all day, and didn't show a symptom of rioting.'

Cecil could, when needed, do the Musset and Meredith style of thing to perfection, but on the whole he preferred love à la mode; it is so much easier and less exhausting to tell your mistress of a ringing run, or a close finish, than to turn perpetual periods on the lustre of her eyes, and the eternity of your devotion.

Nor did it at all interfere with the sincerity of his worship, that the Zu-Zu was at the prettiest little box in the world, in the neighbourhood of Market Harborough, which he had taken for her, and had been at the

meet that day in her little toy trap (with its pair of snowy ponies and its
bright blue liveries, that drove so desperately through his finances), and
had ridden his hunter Maraschino with immense dash and spirit for a
young lady, who had never done anything but pirouette till the last six
months, and a total and headlong disregard of 'purlers', very reckless in a
white-skinned bright-eyed illiterate avaricious little beauty, whose face
was her fortune, and who most assuredly would have been adored no
single moment longer had she scarred her fair tinted cheek with the
blackthorn, or started as a heroine with a broken nose like Fielding's
cherished Amelia.

The Zu-Zu might rage, might sulk, might pout, might even swear
all sorts of naughty Mabille oaths, most villanously pronounced, at the
ascendancy of her haughty unapproachable patrician rival; she did do all
these things; but Bertie would not have been the consummate tactician,
the perfect flirt, the skilled and steeled campaigner in the boudoirs that
he was, if he had not been equal to the delicate task of managing both the
Peeress and the Ballet-dancer with inimitable ability, even when they
placed him in the seemingly difficult dilemma of meeting them both with
twenty yards between them on the neutral ground of the gathering to
see the Pytchley or the Tailby throw off—a task he had achieved with
victorious brilliance more than once already this season.

'You drive a team, Beauty—never drive a team,' the Seraph had said on
occasion over a confidential 'sherry-peg' in the mornings, meaning by the
metaphor of a team, Lady Guenevere, the Zu-Zu, and various other
contemporaries in Bertie's affections. 'Nothing on earth so dangerous:
your leader will bolt, or your off-wheeler will turn sulky, or your young
one will passage and make the very deuce of a row; they'll never go quiet
till the end, however clever your hand is on the ribbons. Now, I'll drive
six-in-hand as soon as any man, drove a ten-hander last year in the Bois,
when the team comes out of the stables; but I'm hanged if I'd risk my
neck with managing even a *pair* of women. Have one clean out of the
shafts before you trot out another!'

To which salutary advice Cecil only gave a laugh, going on his own
ways with the 'team' as before, to the despair of his fidus Achates; the
Seraph, being a quarry so incessantly pursued by dowager-beaters,
chaperone-keepers, and the whole hunt of the Matrimonial Pack, with
those clever hounds Belle and Fashion ever leading in full cry after him,
that he dreaded the sight of a ballroom meet; and, shunning the rich
preserves of the Salons, ran to earth persistently in the shady woods of
St John's, and got—at some little cost and some risk of trapping, it is
true, but still efficiently—preserved from all other hunters or poachers by

the lawless Robin Hoods *aux yeux noirs* of those welcome and familiar coverts.

CHAPTER V

Under the Keeper's Tree

'You're a lad o' wax, my beauty!' cried Mr Rake, enthusiastically, surveying the hero of the Grand Military with adoring eyes as that celebrity, without a hair turned or a muscle swollen from his exploit, was having a dressing-down after a gentle exercise. 'You've pulled it off, haven't you? You're cut the work out for 'em! You've shown 'em what a lustre is! Strike me a loser, but what a deal there is in blood. The littlest pippin that ever threw a leg across the pigskin knows that *in* the stables; then why the dickens do the world run against such a plain fact *out of* it?'

And Rake gazed with worship at the symmetrical limbs of the champion of the 'First Life', and plunged into speculation on the democratic tendencies of the age as clearly contradicted by all the evidences of the flat and furrow, while Forest King drank a dozen go-downs of water, and was rewarded for the patience with which he had subdued his inclination to kick, fret, spring, and break away throughout the dressing by a full feed thrown into his crib, which Rake watched him with adoring gaze eat to the very last grain.

'You precious one!' soliloquized that philosopher, who loved the horse with a sort of passion since his victory over the Shires. 'What a lot o' enemies you've been and gone and made!—that's where it is, my boy; nobody can't never forgive Success. All them fielders have lost such a sight of money by you; them bookmakers have had such a lot of pots upset by you; bless you! if you were on the flat you'd be doctored or roped in no time. You've won for the gentlemen, my lovely—for your own cracks, my boy—and that's just what they'll never pardon you.'

And Rake, rendered almost melancholy by his thoughts (he liked the 'gentlemen' himself), went out of the box to get into saddle and ride off on an errand of his master's to the Zu-Zu at her tiny hunting-lodge, where the snow-white ponies made her stud, and where she gave enchanting little hunting-dinners, at which she sang equally enchanting little hunting-songs, and arrayed herself in the Fontainebleau hunting-costume, gold-hilted knife and all, and spent Cecil's winnings for him with a rapidity that threatened to leave very few of them for the London season.

She was very pretty; sweetly pretty; with fair hair that wanted no gold powder, the clearest, sauciest eyes, and the handsomest mouth in the world; but of grammar she had not a notion, of her aspirates she had never a recollection, of conversation she had not an idea, of slang she had, to be sure, a repertoire, but to this was her command of language limited. She dressed perfectly, but she was a vulgar little soul; drank everything, from Bass's ale to rum-punch, and from cherry-brandy to absinthe; thought it the height of wit to stifle you with cayenne slid into your vanille ice, and the climax of repartee to cram your hat full of peach-stones and lobster-shells; was thoroughly avaricious, thoroughly insatiate, thoroughly heartless, pillaged with both hands and then never had enough; had a coarse good nature when it cost her nothing, and was 'as jolly as a grig', according to her phraseology, so long as she could stew her pigeons in champagne, drink wines and liqueurs that were beyond price, take the most dashing trap in the Park up to Flirtation Corner, and laugh and sing and eat Richmond dinners, and show herself at the Opera with Bertie or some other 'swell' attached to her, in the very box next to a Duchess.

The Zu-Zu was perfectly happy; and as for the pathetic pictures that novelists and moralists draw, of vice sighing amidst turtle and truffles for childish innocence in the cottage at home where honeysuckles blossomed and brown brooks made melody, and passionately grieving on the purple cushions of a barouche for the time of straw pallets and untroubled sleep, why—the Zu-Zu would have vaulted herself on the box-seat of a drag, and told you to 'stow all that trash'; her childish recollections were of a stifling lean-to with the odour of pigsty and straw-yard, pork for a feast once a week, starvation all the other six days, kicks, slaps, wrangling, and a general atmosphere of beer and wash-tubs: she hated her past, and loved her cigar on the drag. The Zu-Zu is fact; the moralists' pictures are moonshine.

The Zu-Zu is an openly acknowledged fact, moreover, daily becoming more prominent in the world, more brilliant, more frankly recognized, and more omnipotent. Whether this will ultimately prove for the better or the worse, it would be a bold man who should dare say; there is at least one thing left to desire in it—i.e. that the synonym of 'Aspasia', which serves so often to designate in journalistic literature these Free Lances of life, were more suitable in artistic and intellectual similarity, and that when the Zu-Zu and her co-brigands plunge their white arms elbow-deep into so many fortunes, and rule the world right and left as they do, they could also sound their H's properly, and know a little <u>orthography</u>, if they could not be changed into such queens of grace, of intellect, of sovereign mind and splendid wit as were their prototypes when she whose name they debase held her rule in the City of the Violet Crown, and gathered about her Phidias the divine, haughty and eloquent Antipho, the gay Crates, the subtle Protagorus, Cratinus so acrid and yet so jovial, Damon of the silver lyre, and the great poets who are poets for all time. Author and artist, noble and soldier, court the Zu-Zu order now as the Athenians courted their brilliant ἑταιραι; but it must be confessed that the Hellenic idols were of a more exalted type than are the Hyde Park goddesses!

However, the Zu-Zu was the rage, and spent Bertie's money when he got any just as her wilful sovereignty fancied, and Rake rode on now with his master's note, bearing no very good will to her; for Rake had very strong prejudices, and none stronger than against these pillagers who went about seeking whom they should devour, and laughing at the wholesale ruin that they wrought, while the sentimentalists babbled in 'Social Science' of 'pearls lost' and 'innocence betrayed'.

'A girl that used to eat tripe and red herring in a six-pair back, and dance for a shilling a night in gauze, coming it so grand that she'll only eat asparagus in March, and drink the best Brands with her truffles! Why, she ain't worth sixpence thrown away on her, unless it's worth while to hear how hard she can swear at you!' averred Rake, in his eloquence; and he was undoubtedly right for that matter, but then—the Zu-Zu was the rage, and if ever she should be sold up, great ladies would crowd to her sale, as they have done ere now to that of celebrities of her sisterhood, and buy with eager curiosity, at high prices, her most trumpery pots of pomatum, her most flimsy gewgaws of marqueterie!

Rake had seen a good deal of men and manners, and, in his own opinion at least, was 'up to every dodge on the cross' that this iniquitous world could unfold. A bright, lithe, animated, vigorous, yellow-haired, and sturdy fellow, seemingly with a dash of the Celt in him that made him vivacious and peppery, Mr Rake polished his wits quite as much as he polished the tops, and considered himself a philosopher. Of whose son he was he had not the remotest idea; his earliest recollections were of the tender mercies of the workhouse; but even that chill foster-mother, the parish, had not damped the liveliness of his temper or the independence of his opinions, and as soon as he was fifteen, Rake had run away and joined a circus, distinguishing himself there by his genius for standing on his head, and tying his limbs into a porter's knot.

From the circus he migrated successively into the shape of a comic singer, a tapster, a navvy, a bill-sticker, a guacho in Mexico (working his passage out), a fireman in New York, a ventriloquist in Maryland, a vaquero in Spanish California, a lemonade-seller in San Francisco, a revolutionist in the Argentine (without the most distant idea what he fought for), a boatman on the Bay of Mapiri, a blacksmith in Santarem, a trapper in the Wilderness, and finally, working his passage home again, took the Queen's shillng in Dublin, and was drafted into a light cavalry regiment. With the —th he served half a dozen years in India, a rough-rider, a splendid fellow in a charge or a pursuit, with an astonishing power over horses, and the clearest backhanded sweep of a sabre that ever cut down a knot of natives; *but*—insubordinate. Do his duty whenever fighting was in question, he did most zealously, but to kick over the traces at other times was a temptation that at last became too strong for that lawless lover of liberty.

From the moment that he joined the regiment, a certain Corporal Warne and he had conceived an antipathy to one another, which Rake had to control as he might, and which the Corporal was not above indulg-ing in every petty piece of tyranny that his rank allowed him to exercise. On active service Rake was, by instinct, too good a soldier not to manage to keep the curb on himself tolerably well, though he was always re-garded in his troop rather as a hound that *will* 'riot' is regarded in the pack; but when the —th came back to Brighton and to barracks, the evil spirit of rebellion began to get a little hotter in him under the Corporal's 'Idées Napoliennes' of justifiable persecution. Warne indisputably pro-voked his man in a cold, iron, strictly lawful sort of manner, moreover, all the more irritating to a temper like Rake's.

'Hanged if I care how the officers come it over me; they're gentlemen, and it don't try a fellow,' would Rake say in confidential moments over purl and a penn'orth of bird's-eye, his experience in the Argentine Republic having left him with strongly aristocratic prejudices; 'but when it comes to a duffer like that, that knows no better than me, what *ain't* a bit better than me, and what is as clumsy a duffer about a horse's plates as ever I knew, and would a'most let a young 'un buck him out of his saddle, why then I do cut up rough, I ain't denying it, and I don't see what there is in his Stripes to give him such a licence to be aggravating.'

With which Rake would blow the froth off his pewter with a puff of concentrated wrath, and an oath against his non-commissioned officers that might have let some light in upon the advocates for 'promotion from the ranks' had they been there to take the lesson. At last, in the leisure of Brighton, the storm broke. Rake had a Scotch hound that was the pride of his life, his beer-money often going instead to buy dainties for the dog, who became one of the channels through which Warne could annoy and thwart him. The dog did no harm, being a fine, well-bred deerhound; but it pleased the Corporal to consider that it did, simply because it belonged to Rake, whose popularity in the corps, owing to his good nature, his good spirits, and his innumerable tales of American experiences and amorous adventures, increased the jealous dislike which his knack with an unbroken colt and his abundant stable science had first raised in his superior.

One day in the chargers' stables the hound ran out of a loose box with a rush to get at Rake, and upset a pailful of warm mash. The Corporal, who was standing by in harness, hit him over the head with a heavy whip he had in his hand; infuriated by the pain, the dog flew at him, tearing his overalls with a fierce crunch of his teeth. 'Take the brute off, and string him up with a halter; I've put up with him too long!' cried Warne to a couple of privates working near in their stable dress. Before the words were out of his mouth, Rake threw himself on him with a bound like lightning, and wrenching the whip out of his hand, struck him a slashing, stinging blow across the face.

'Hang my hound, you cur! If you touch a hair of him I'll double-thong you within an inch of your life!'

And assuredly he would have kept his word had he not been made a prisoner, and marched off to the guard-room.

Rake learnt the stern necessity of the law, which, for the sake of morale, must make the soldiers, whose blood is wanted to be like fire on the field, patient, pulseless, and enduring of every provocation, cruelty, and insolence in the camp and the barrack, as though they were statues of stones—a needful law, a wise law, an indispensable law, doubtless, but a very hard law to be obeyed by a man full of life and all life's passions.

At the court-martial on his mutinous conduct which followed, many witnesses brought evidence, on being pressed, to the unpopularity of Warne in the regiment, and to his harshness and his tyranny to Rake. Many men spoke out what had been chained down in their thoughts for years; and, in consideration of the provocation received, the prisoner, who was much liked by the officers, was condemned to six months' imprisonment for his insubordination and blow to his non-commissioned officer, without being tied up to the triangles. At the court-martial, Cecil, who chanced to be in Brighton after Goodwood, was present one day with some other Guardsmen, and the look of Rake, with his cheerfulness under difficulties, his love for the hound, and his bright, sunburnt, shrewd, humorous countenance, took his fancy.

'Beauty' was the essence of good nature. Indolent himself, he hated to see anything or anybody worried; lazy, gentle, wayward, and spoilt by his own world, he was still never so selfish and philosophic as he pretended, but what he would do a kindness if one came in his way; it is not a very great virtue, perhaps, but it is a rare one.

'Poor devil! struck the other because he wouldn't have his dog hanged. Well, on my word I should have done the same in his place, if I could have got up the pace for so much exertion,' murmured Cecil to his cheroot, careless of the demoralizing tendency of his remarks for the Army in general. Had it occurred in the Guards, and he had 'sat' on the case, Rake would have had one very lenient judge.

As it was, Bertie actually went the lengths of thinking seriously about the matter; he liked Rake's devotion to his dumb friend, and he heard of his intense popularity in his troop; he wished to save, if he could, so fine a fellow from the risks of his turbulent passion, and from the stern fetters of a trying discipline; hence, when Rake found himself condemned to his cell, he had a message sent him by Bertie's groom that when his term of punishment should be over Mr. Cecil would buy his discharge from the Service and engage him as extra body-servant, having had a good account of his capabilities: he had taken the hound to his own kennels.

Now the fellow had been thoroughly devil-me-care throughout the whole course of the proceedings, had heard his sentence with sublime impudence, and had chaffed his sentinels with an utterly reckless nonchalance; but somehow or other, when that message reached him, a vivid sense that he was a condemned and disgraced man suddenly flooded in on him; a passionate gratitude seized him to the young aristocrat who had thought of him in his destitution and condemnation, who had even thought of his dog; and Rake, the philosophic and the undauntable, could have found it in his heart to kneel down in the dust and kiss the stirrup-leather when he held it for his new master, so strong was the loyalty he bore from that moment to Bertie.

Martinets were scandalized at a Life-Guardsman taking as his private valet a man who had been guilty of such conduct in the Light Cavalry; but Cecil never troubled his head about what people said; and so invaluable did Rake speedily become to him, that he had kept him about his person wherever he went from then until now, two years after.

Rake loved his master with a fidelity very rare in these days; he loved his horses, his dogs, everything that was his, down to his very rifle and boots, slaved for him cheerfully, and was as proud of the deer he stalked, of the brace he bagged, of his innings when the Household played the Zingari, or his victory when his yacht won the Cherbourg Cup, as though those successes had been Rake's own.

'My dear Seraph,' said Cecil himself once on this point to the Marquis, 'if you want generosity, fidelity, and all the rest of the cardinal what-d'ye-call-ems—sins, ain't it?—go to a noble-hearted Scamp; *he*'ll stick to you till he kills himself. If you want to be cheated, get a Respectable Immaculate; *he*'ll swindle you piously, and decamp with your Doncaster Vase.'

And Rake, who assuredly had been an out-and-out scamp, made good Bertie's creed; he 'stuck to him' devoutly, and no terrier was ever more alive to an otter than he was to the Guardsman's interests. It was that very vigilance which made him, as he rode back from the Zu-Zu's in the twilight, notice what would have escaped any save one who had been practised as a trapper in the red Canadian woods, namely, the head of a man almost hidden among the heavy though leafless brushwood and the yellow gorse of a spinney which lay on his left in Royallieu Park. Rake's eyes were telescopic and microscopic; moreover, they had been trained to know such little signs as a marsh from a hen harrier in full flight, by the length of wing and tail, and a widgeon or a coot from a mallard or a teal, by the depth each swam out of the water. Grey and foggy as it was, and high as was the gorse, Rake recognized his born-foe, Willon.

'What's he up to there?' thought Rake, surveying the place, which was wild, solitary, and an unlikely place enough for a head groom to be found in. 'If he ain't a rascal, I never see one; it's my belief he cheats the stable thick and thin, and gets on Mr Cecil's mounts to a good tune—ay, and would nobble 'em as soon as not, if it just suited his book; that blessed King hates the man: how he lashes his heels at him!'

It was certainly possible that Willon might be passing an idle hour in potting rabbits, or be otherwise innocently engaged enough; but the sight of him there among the gorse was a sight of suspicion to Rake. Instantaneous thoughts darted through his mind of tethering his horse, and making a reconnaissance safely and unseen with the science at stalking brute or man that he had learnt of his friends the Sioux. But second thoughts showed him this was impossible. The horse he was on was a mere colt just breaking in, who had barely had so much as a 'dumb jockey' on his back, and stand for a second the colt would not.

'At any rate, I'll unearth him,' mused Rake, with his latent animosity to the head groom, and his vigilant loyalty to Cecil overruling any scruple as to his right to overlook his foe's movements; and with a gallop that was muffled on the heather'd turf he dashed straight at the covert unperceived till he was within ten paces. Willon started and looked up hastily; he was talking to a square-built man very quietly dressed in shepherds' plaid, chiefly remarkable by a red-hued beard and whiskers.

The groom turned pale and laughed nervously as Rake pulled up with a jerk.

'You on that young 'un again? Take care you don't get bucked out o' saddle in the shape of a cocked-hat.'

'*I* ain't afraid of going to grass, if you are!' retorted Rake, scornfully; boldness was not his enemy's strong point. 'Who's your pal, old fellow?'

'A cousin o' mine, out o' Yorkshire,' vouchsafed Mr Willon, looking anything but easy, while the cousin aforesaid nodded sulkily on the introduction.

'Ah! looks like a Yorkshire tyke,' muttered Rake, with a volume of meaning condensed in these innocent words. 'A nice dry, cheerful sort of place to meet your cousin in, too; uncommon lively; hope it'll raise his spirits to see all *his* cousins a grinning there; his spirits don't seem much in sorts now,' continued the ruthless inquisitor, with a glance at the 'keeper's tree' by which they stood, in the middle of dank undergrowth, whose branches were adorned with dead cats, curs, owls, kestrels, stoats, weasels, and martens. To what issue the passage of arms might have come it is impossible to say, for at that moment the colt took matters into his own hands, and bolted with a rush, that even Rake could not pull in till he had had a mile-long 'pipe opener'.

'Something up there,' thought that sagacious rough-rider; 'if that red-haired chap ain't a rum lot, I'll eat him. I've seen his face, too, some-where: where the deuce was it? Cousin; yes, cousins in Queer-street, I dare say! Why should he go and meet his "cousin" out in the fog there, when if you took twenty cousins home to the servants' hall nobody'd ever say anything? If that Willon ain't as deep as Old Harry——'

And Rake rode into the stable-yard, thoughtful and intensely sus-picious of the rendezvous under the keeper's tree in the outlying coverts. He would have been more so had he guessed that Ben Davis's red beard and demure attire, with other as efficient disguises, had prevented even his own keen eyes from penetrating the identity of Willon's 'cousin' with the Welsher he had seen thrust off the course the day before by his master.

CHAPTER VI

The End of a Ringing Run

'Tally-ho! is the word, clap spurs and let's follow,
The world has no charm like a rattling view-halloa!'

Is hardly to be denied by anybody in this land of fast bursts and gallant MFHs, whether they 'ride to hunt', or 'hunt to ride', in the immortal distinction of Assheton Smith's old whip: the latter class, by-the-by, becoming far and away the larger, in these days of rattling gallops and desperate breathers. Who cares to patter after a sly old dog-fox, that, fat and wary, leads the pack a tedious interminable wind in and out through gorse and spinney, bricks himself up in a drain, and takes an hour to be dug out, dodges about till twilight, and makes the hounds pick the scent slowly and wretchedly, over marsh and through water? Who would not give fifty guineas a second for the glorious thirty minutes of *racing* that shows steam and steel over fence and fallow in a clipping rush without a check from find to finish? So be it ever! The riding that graces the Shires, that makes Tedworth and Pytchley, the Duke's, and the Fitzwilliam's, household words and 'names beloved,' that fills Melton and Market Harborough, and makes the best flirts of the ballroom gallop fifteen miles to covert, careless of hail or rain, mire or slush, mist or cold, so long as it is a fine scenting wind, is the same riding that sent the Six Hundred down into the blaze of the Muscovite guns, that in their fathers' days gave to Grant's Hussars their swoop, like eagles, on to the rearguard at Morales, and that in the grand old East and the rich trackless West makes exiled campaigners with high English names seek and win an *aristeia* of their own at the head of their wild Irregular Horse, who would charge hell itself at their bidding.

Now in all the Service there was not a man who loved hunting better than Bertie. Though he was incorrigibly lazy, and inconceivably effeminate in every one of his habits, though he suggested a portable lounging-chair as an improvement at battues so that you might shoot sitting, drove to every breakfast and garden party in the season in his brougham with the blinds down lest a grain of dust should touch him, thought a waltz too exhaustive, and a saunter down Pall Mall too tiring, and asked to have the end of a novel told him in the clubs because it was too much trouble to read on a warm day—though he was more indolent than any spoiled Creole, 'Beauty' never failed to head the first flight, and adored a hard day cross country, with an east wind in his eyes, and the sleet in his teeth. The only trouble was to make him get up in time for it.

'Mr Cecil, sir, if you please, the drag will be round in ten minutes,' said Rake, with a dash of desperation for the seventh time into his chamber, one fine scenting morning.

'I don't please,' answered Cecil, sleepily, finishing his cup of coffee, and reading a novel of La Demirep's.

'The other gentlemen are all down, sir, and you will be too late.'

'Not a bit. They must wait for me,' yawned Bertie.

Crash came the Seraph's thunder on the panels of the door, and a strong volume of Turkish through the keyhole: 'Beauty, Beauty, are you dead?'

'Now, what an inconsequent question!' expostulated Cecil, with appealing rebuke. 'If a fellow *were* dead, how the devil could he say he was? Do be logical, Seraph.'

'Get up!' cried the Seraph with a deafening rataplan, and a final dash of his colossal stature into the chamber. 'We've all done breakfast; the traps are coming round; you'll be an hour behind time at the meet.'

Bertie lifted his eyes with plaintive resignation from the Demirep's yellow-papered romance.

'I'm really in an interesting chapter: Aglae has just had a marquis kill his son, and two brothers kill each other in the Bois, about her, and is on the point of discovering a man she's in love with to be her own grandfather; the complication is absolutely thrilling,' murmured Beauty, whom nothing could ever 'thrill', not even plunging down the Matterhorn, losing 'long odds in thou'' over the Oaks, or being sunned in the eyes of the fairest women of Europe.

The Seraph laughed, and tossed the volume straight to the other end of the chamber.

'Confound you, Beauty, get up!'

'Never swear, Seraph, not ever so mildly,' yawned Cecil; 'it's gone out, you know; only the cads and the clergy can damn one nowadays; it's such bad style to be so impulsive. Look! you have broken the back of my Demirep!'

'You deserve to break the King's back over the first cropper,' laughed the Seraph. 'Do get up!'

'Bother!' sighed the victim, raising himself with reluctance, while the Seraph disappeared in a cloud of Turkish.

Neither Bertie's indolence nor his insouciance were assumed; utter carelessness was his nature, utter impassability was his habit, and he was truly for the moment loth to leave his bed, his coffee, and his novel; he must have his leg over the saddle, and feel the strain on his arms of that 'pulling' pace with which the King always went when once he settled into his stride, before he would really think about winning.

The hunting breakfasts of our forefathers and of our present squires found no favour with Bertie; a slice of game and a glass of curaçoa were all he kept the drag waiting to swallow, and the four bays going at a pelting pace, he and the rest of the Household who were gathered at Royallieu were by good luck in time for the throw-off of the Quorn, where the hero of the Blue Riband was dancing impatiently under Willon's hand, scenting the fresh, keen, sunny air, and knowing as well what all those bits of scarlet straying in through field and lane, gate and gap, meant, as well as though the merry notes of the master's horn were winding over the gorse. The meet was brilliant and very large; showing such a gathering as only the Melton country can; and foremost among the crowd of carriages, hacks, and hunters, were the beautiful roan mare Vivandière of the Lady Guenevere, mounted by that exquisite Peeress in her violet habit, and her tiny velvet hat; and the pony equipage of the Zu-Zu, all glittering with azure and silver, leopard rugs, and snowy reins: the breadth of half an acre of grassland was between them, but the groups of men about them were tolerably equal for number and for rank.

'Take Zu-Zu off my hands for this morning, Seraph, there's a good fellow,' murmured Cecil, as he swung himself into saddle. The Seraph gave a leonine growl, sighed, and acquiesced. He detested women in the hunting-field, but that sweetest tempered giant of the Brigades never refused anything to anybody—much less to 'Beauty'.

To an uninitiated mind it would have seemed marvellous and beautiful in its combination of simplicity and intricacy, to have noted the delicate tactics with which Bertie conducted himself between his two claimants; bending to his Countess with a reverent devotion that assuaged whatever of incensed perception of her unacknowledged rival might be silently lurking in her proud heart; wheeling up to the pony-trap under cover of speaking to the men from Egerton Lodge, and restoring the Zu-Zu from sulkiness, by a propitiatory offer of a little gold sherry-flask, studded with turquoises, just ordered for her from Regent-street, which, however, she ungraciously contemned, because she thought it had only cost twenty guineas; anchoring the victimized Seraph beside her by an adroit 'Ah! by the way, Rock, give Zu-Zu one of your rose-scented *papelitos*; she's been wild to smoke them'; and leaving the Zu-Zu content at securing a future Duke, was free to canter back and flirt on the off-side of Vivandière, till the 'signal', the 'cast', made with consummate craft, the waving of the white sterns among the brushwood, the tightening of girths, the throwing away of cigars, the challenge, the whimper, and the 'stole away!' sent the field headlong down the course after as fine a long-legged greyhound fox as ever carried a brush.

Away he went in a rattling spin, breaking straight at once for the open, the hounds on the scent like mad: with a tally-ho that thundered through the cloudless, crisp, cold, glittering noon, the field dashed off pell-mell, the violet habit of her ladyship, and the azure skirts of the Zu-Zu foremost of all in the rush through the spinneys; while Cecil on the King, and the Seraph on a magnificent white weight-carrier, as thoroughbred and colossal as himself, led the way with them. The scent was hot as death in the spinneys, and the pack raced till nothing but a good one could live with them; few but good ones, however, were to be found with the Quorn, and the field held together superbly over the first fence, and on across the grassland, the game old fox giving no sign of going to covert, but running straight as a crow flies, while the pace grew terrific.

'Beats cock-fighting!' cried the Zu-Zu, while her blue skirts fluttered in the wind, as she lifted Cecil's brown mare, very cleverly, over a bilberry hedge, and set her little white teeth with a will on the Seraph's otto-of-rose cigarette. Lady Guenevere heard the words as Vivandière rose in the air with the light bound of a roe, and a slight superb dash of scorn came into her haughty eyes for the moment; she never seemed to know that 'that person' in the azure habit even existed, but the contempt awoke in her, and shone in her glance, while she rode on as that fair leader of the Belvoir and Pytchley alone could ride over the fallows.

The steam was on at full pressure, the hounds held close to his brush, heads up, sterns down, running still straight as an arrow over the open, past coppice and covert, through gorse and spinney, without a sign of the fox making for shelter. Fence, and double, hedge and brook, soon scattered the field; straying off far and wide, and coming to grief with lots of 'downers', it grew select, and few but the crack men could keep the hounds in view. 'Catch 'em who can,' was the one *mot d'ordre*, for they were literally racing, the line-hunters never losing the scent a second, as the fox, taking to dodging, made all the trouble he could for them through the rides of the woods. Their working was magnificent, and, heading him, they ran him round and round in a ring, viewed him for a second, and drove him out of covert once more into the pastures, while they laid on at a hotter scent and flew after him like staghounds.

Only half a dozen were up with them now; the pace was tremendous, though all over grass; here a flight of posts and rails tried the muscle of the boldest; there a bullfinch yawned behind the blackthorn; here a big fence towered; there a brook rushed angrily among its rushes; while the keen, easterly wind blew over the meadows, and the pack streamed along like the white trail of a plume. Cecil 'showed the way' with the self-same stride and the self-same fencing as had won him the Vase. Lady Guenevere and the Seraph were running almost even with him; three of the Household farther down; the Zu-Zu and some Melton men two meadows off; the rest of the field, nowhere. Fifty-two minutes had gone by in that splendid running, without a single check, while the fox raced as gamely and as fast as at the find; the speed was like lightning past the brown woods, the dark-green pine plantations, the hedges, bright with scarlet berries; through the green low-lying grasslands, and the winding drives of coverts, and the boles of ash-hued beech trunks, whose roots the violets were just purpling with their blossom; while far away stretched the blue haze of the distance, and above-head a flight of rooks cawed merrily in the bright air, soon left far off as the pack swept onward in the most brilliant thing of the hunting year.

'Water! take care!' cried Cecil, with a warning wave of his hand as the hounds with a splash like a torrent dashed up to their necks in a broad brawling brook that Reynard had swam in first-rate style, and struggled as best they could after him. It was an awkward bit, with bad taking-off and a villanous mud-bank for landing; and the water, thickened and swollen with recent rains, had made all the land that sloped to it miry and soft as sponge. It was the risk of life and limb to try it; but all who still viewed the hounds catching Bertie's shout of warning worked their horses up for it, and charged towards it as hotly as troops charge a square. Forest King was over like a bird; the winner of the Grand Military was not to be daunted by all the puny streams of the Shires; the artistic riding of the Countess landed Vivandière, with a beautiful clear spring, after him by a couple of lengths: the Seraph's handsome white hunter, brought up at a headlong gallop with characteristic careless dash and fine science mingled, cleared it; but, falling with a mighty crash, gave him a purler on the opposite side, and was within an ace of striking him dead with his hoof in frantic struggles to recover. The Seraph, however, was on his legs with a rapidity marvellous in a six-foot-three son of Anak, picked up the horse, threw himself into saddle, and dashed off again quick as lightning, with his scarlet stained all over, and his long fair moustaches floating in the wind. The Zu-Zu turned Mother of Pearl back with a fiery French oath; she hated to be 'cut down', but she liked still less to risk her neck; and two of the Household were already treated to 'crackers' that disabled them for the day, while one Melton man was pitched head-foremost into the brook, and another was sitting dolorously on the bank with his horse's head in his lap, and the poor brute's spine broken. There were only three of the first riders in England now alone with the hounds, who, with a cold scent as the fox led them through the angular corner of a thick pheasant covert, stuck like wax to the line, and working him out, viewed him once more, for one wild, breathless, tantalizing second, and, on a scent breast-high, raced him with the rush of an express through the straggling street of a little hamlet, and got him out again on the level pastures and across a fine line of hunting country, with the leafless woods and the low gates of a park far away to their westward.

'A guinea to a shilling that we kill him!' cried the flute-voice of her brilliant ladyship, as she ran a moment side by side with Forest King, and flashed her rich eyes on his rider; she had scorned the Zu-Zu, but on occasion she would use betting slang and racing slang with the daintiest grace in the world herself without their polluting her lips. As though the old fox heard the wager, he swept in a bend round towards the woods on the right, making with all the craft and the speed there were in him for the deep shelter of the boxwood and laurel. 'After him, my beauties, my beauties—if he run there he'll go to ground and save his brush!' thundered the Seraph, as though he were hunting his own hounds at Lyonnesse, who knew every tone of his rich clarion notes as well as they knew every wind of his horn. But the young ones of the pack saw Reynard's move and his meaning as quickly as he did; having run fast before, they flew now: the pace was terrific. Two fences were crossed as though they were paper; the meadows raced with lightning speed, a ha-ha leaped, a gate cleared with a crashing jump, and in all the furious excitement of 'view' they tore down the mile-long length of an avenue, dashed into a flower-garden, and smashing through a gay trellis-work of scarlet creeper, plunged into the home-paddock and killed with as loud a shout ringing over the country in the bright sunny day as ever was echoed by the ringing cheers of the Shire; Cecil, the Seraph, and her victorious ladyship alone coming in for the glories of 'finish'.

'Never had a faster seventy minutes up-wind,' said Lady Guenevere, looking at the tiny jewelled watch, the size of a sixpence, that was set in the handle of her whip, as the brush, with all the compliments customary, was handed to her. She had won twenty before.

The park, so unceremoniously entered, belonged to a baronet who, though he hunted little himself, honoured the sport and scorned a vulpicide; he came out naturally and begged them to lunch. Lady Guenevere refused to dismount, but consented to take a biscuit and a little Lafitte, while clarets, liqueurs, and ales, with anything they wanted to eat, were brought to her companions. The stragglers strayed in; the MFH came up just too late; the men getting down gathered about the

Countess or lounged on the grey stone steps of the Elizabethan house. The sun shone brightly on the oriel casements, the antique gables, the twisted chimneys, all covered with crimson parasites and trailing ivy; the horses, the scarlet, the pack in the paddock adjacent, the shrubberies of laurel and auraucaria, the sun-tinted terraces, made a bright and picturesque grouping. Bertie, with his hand on Vivandière's pommel, after taking a deep draught of sparkling Rhenish, looked on at it all with a pleasant sigh of amusement.

'By Jove!' he murmured softly, with a contented smile about his lips; 'that *was* a ringing run!'

At the very moment, as the words were spoken, a groom approached him hastily; his young brother, whom he had scarcely seen since the find, had been thrown and taken home on a hurdle; the injuries were rumoured to be serious.

Bertie's smile faded; he looked very grave: world-spoiled as he was, reckless in everything, and egotist though he had long been by profession, he loved the lad.

When he entered the darkened room, with its faint chloroform odour, the boy lay like one dead, his bright hair scattered on the pillow, his chest bare, and his right arm broken and splintered. The death-like coma was but the result of the chloroform; but Cecil never stayed to ask or remember that, he was by the couch in a single stride, and dropped down by it, his head bent on his arms.

'It is my fault. I should have looked to him.'

The words were very low; he hated that any should see he could still be such a fool as to *feel*. A minute, and he conquered himself; he rose, and with his hand on the boy's fair tumbled curls, turned calmly to the medical men who, attached to the household, had been on the spot at once.

'What is the matter?'

'Fractured arm, contusion, nothing serious, nothing at all, at his age,' replied the surgeon; 'when he wakes out of the lethargy he will tell you so himself, Mr Cecil.'

'You are certain?'—do what he would his voice shook a little; his hand had not shaken, two days before, when nothing less than ruin or ransom had hung on his losing or winning the race.

'Perfectly certain,' answered the surgeon, cheerfully. 'He is not over strong, to be sure, but the contusions are slight; he will be out of that bed in a fortnight.'

'How did he fall?'

But while they told him he scarcely heard; he was looking at the handsome Antinous-like form of the lad stretched helpless and stricken before him; and he was remembering the death-bed of their mother, when the only voice he had ever reverenced had whispered, as she pointed to the little child of three summers: 'When you are a man, take care of him, Bertie.' How had he fulfilled the injunction? Into how much brilliantly-tinted evil had he not led him—by example at least?

The surgeon touched his arm apologetically, after a lengthened silence:

'Your brother will be best unexcited when he comes to himself, sir; look—his eyes are unclosing now. Could you do me the favour to go to his Lordship? His grief made him perfectly wild—so dangerous to his life at his age. We could only persuade him to retire, a few minutes ago, on the plea of Mr Berkeley's safety. If you could see him——'

Cecil went, mechanically almost, and with a grave, weary depression on him; he was so unaccustomed to think at all, so utterly unaccustomed to think painfully, that he scarcely knew what ailed him. Had he had his old tact about him, he would have known how worse than useless it would be for *him* to seek his father in such a moment.

Lord Royallieu was lying back exhausted as Cecil opened the door of his private apartments, heavily darkened and heavily perfumed; at the turn of the lock he started up eagerly.

'What news of him?'

'Good news, I hope,' said Cecil, gently, as he came forward. 'The injuries are not grave, they tell me. I am so sorry that I never watched his fencing, but——'

The old man had not recognized him till he heard his voice, and he waved him off with a fierce contemptuous gesture; the grief for his favourite's danger, the wild terrors that his fears had conjured up, his almost frantic agony at the sight of the accident, had lashed him into passion well-nigh delirious.

'Out of my sight, sir!' he said, fiercely, his mellow tones quivering with rage. 'I wish to God you had been dead in a ditch before a hair of my boy's had been touched. You live, and he lies dying there!'

Cecil bowed in silence; the brutality of the words wounded, but they did not offend him, for he knew his father was in that moment scarce better than a maniac, and he was touched with the haggard misery upon the old Peer's face.

'Out of my sight, sir!' re-echoed Lord Royallieu, as he strode forward, passion lending vigour to his emaciated frame, while the dignity of his grand carriage blent with the furious force of his infuriated blindness. 'If you had had the heart of a man you would have saved such a child as that from his peril; warned him, watched him, succoured him at least when he fell. Instead of that, you ride on and leave him to die, if death come to him! *You* are safe; you are always safe. You try to kill yourself with every vice under heaven, and only get more strength, more grace, more pleasure from it—you are always safe because I hate you. Yes! I hate you, sir!'

No words can give the force, the malignity, the concentrated meaning with which the words were hurled out, as the majestic form of the old Lord towered in the shadow, with his hands outstretched as if in imprecation.

Cecil heard him in silence, doubting if he could hear aright, while the bitter phrases scathed and cut like scourges, but he bowed once more with the manner that was as inseparable from him as his nature.

'Hate is so very exhausting; I regret I give you the trouble of it. May I ask why you favour me with it?'

'You may!' thundered his father, while his hawk's eyes flashed their glittering fire. 'You are like the man I cursed living and curse dead. You look at me with Alan Bertie's eyes, you speak to me with Alan Bertie's voice; I loved your mother, I worshipped her; but—you are his son, not mine!'

The secret doubt, treasured so long, was told at last. The blood flushed Bertie's face a deep and burning scarlet; he started with an irrepressible tremor, like a man struck with a shot; he felt like one suddenly stabbed in the dark by a sure and a cruel hand. The insult and the amazement of the words seemed to paralyse him for the moment, the next he recovered himself, and lifted his head with as haughty a gesture as his father's, his features were perfectly composed again, and sterner than in all his careless, easy life they ever yet had looked.

'You lie, and you know that you lie. My mother was pure as the angels. Henceforth you can be only to me a slanderer who has dared to taint the one name holy in my sight.'

And without another word he turned and went out of the chamber. Yet, as the door closed, old habit was so strong on him, that, even in his hot and bitter pain, and his bewildered sense of sudden outrage, he almost smiled at himself. 'It is a mania; he does not know what he says,' he thought. 'How could *I* be so melodramatic? We were like two men at the Porte St Martin. Inflated language *is* such a bad form!'

But the cruel stroke had not struck the less closely home, and gentle though his nature was, beyond all forgiveness from him was the dishonour of his mother's memory.

CHAPTER VII

After a Richmond Dinner

It was the height of the season, and the duties of the Household were proportionately and insupportably heavy. The Brigades were fairly worked to death, and the Indian service, in the heat of the Afghan war, was never more onerous than the campaigns that claimed the Guards from Derby to Ducal.

Escorts to Levees, guards of honour to Drawing-rooms, or field-days in the Park and the Scrubs, were but the least portion of it. Far more severe, and still less to be shirked, were the morning exercise in the Ride; the daily parade in the Lady's Mile; the reconnaissances from club windows, the videttes at Flirtation Corner; the long campaigns at mess-breakfasts, with the study of dice and baccarat tactics, and the fortifications of Strasburg pâté against the invasions of Chartreuse and Chambertin; the breathless, steady charges up Belgravian staircases when a fashionable drum beat the rataplan; the skirmishes with sharpshooters of the bright-armed Irregular Lances; the foraging-duty when fair commanders wanted ices or strawberries at garden parties; the ball-practice at Hornsey Handicaps; the terrible risk of crossing into the enemy's lines, and being made to surrender as prisoners of war at the jails of St George's, or of St Paul's Knightsbridge; the constant inspections of the Flying Battalions of the Ballet, and the pickets afterwards in the Wood of St John; the anxieties of the Club commissariats, and the close vigilance over the mess wines; the fatigue duty of ballrooms, and the continual unharnessing consequent on the clause in the Regulations never to wear the same gloves twice; all these, without counting the close battles of the Corner and the unremitting requirements of the Turf, worked the First Life and the rest of the Brigades, Horse and Foot, so hard and incessantly, that some almost thought of changing into the dreary depot of St Stephen's; and one mutinous Coldstreamer was even rash enough and false enough to his colours to meditate deserting to the enemy's camp, and giving himself up at St George's—'because a fellow once hanged is let alone, you know!'

The Household were very hard pressed through the season—a crowded and brilliant one; and Cecil was in request most of all. Bertie, somehow or other, was the fashion—marvellous and indefinable word, that gives a more powerful crown than thrones, blood, beauty, or intellect can ever bestow. And no list was 'the thing' without his name, no reception, no garden party, no opera-box, or private concert, or rose-shadowed boudoir, fashionably *affiché* without being visited by him. How he, in especial, had got his reputation it would have been hard to say, unless it were that he dressed a shade more perfectly than any one, and with such inimitable carelessness in the perfection, too, and had an almost unattainable matchlessness in the sang-froid of his soft languid insolence, and incredible though ever gentle effrontery. However gained, he had it: and his beautiful hack Sahara, his mail-phaeton with two blood greys dancing in impatience over the stones, or his little dark-green brougham for night-work, were, one or another of them, always seen from two in the day till four or five in the dawn about the Park or the town.

And yet this season, while he made a prima donna by a bravissima, introduced a new tie by an evening's wear, gave a cook the cordon with his praise, and rendered a fresh-invented liqueur the rage by his recommendation, Bertie knew very well that he was ruined.

The breach between his father and himself was irrevocable. He had left Royallieu as soon as his guests had quitted it, and young Berkeley was out of all danger. He had long known he could look for no help from the old Lord, or from his elder brother, the heir; and now every chance of it was hopelessly closed; nothing but the whim or the will of those who held his floating paper, and the tradesmen who had his name on their books at compound interest of the heaviest, stood between him and the fatal hour when he must 'send in his papers to sell' and be 'nowhere' in the great race of life.

He knew that a season, a month, a day, might be the only respite left him, the only pause for him betwixt his glittering luxurious world and the fiat of outlawry and exile. He knew that the Jews might be down on him any night that he sat at the Guards' mess, flirted with foreign Princesses, or laughed at the gossamer gossip of the town over iced drinks in the clubs. His liabilities were tremendous, his resources totally exhausted; but such was the latent recklessness of the careless Royallieu blood, and such the languid devil-may-care of his training and his temper, that the knowledge scarcely ever seriously disturbed his enjoyment of the moment. Somehow, he never realised it.

If any weatherwise had told the Lisbon people of the coming of the great earthquake, do you think they could have brought themselves to realise that midnight darkness, that yawning desolation which were nigh, while the sun was still so bright and the sea so tranquil, and the bloom so sweet on purple pomegranate and amber grape, and the scarlet of odorous flowers, and the blush of a girl's kiss-warmed cheek?

A sentimental metaphor with which to compare the difficulties of a dandy of the Household, because his 'stiff' was floating about in too many directions at too many high figures, and he had hardly enough till next pay-day came round to purchase the bouquets he sent, and meet the club-fees that were due! But, after all, may it not well be doubted if a sharp shock and a second's blindness, and a sudden sweep down under the walls of the Cathedral or the waters of the Tagus, were not, on the whole, a quicker and pleasanter mode of extinction than that social earthquake—'gone to the bad with a crash?'

And the Lisbonites did not more disbelieve in, and dream less of their coming ruin, than Cecil did his, while he was doing the season, with engagements enough in a night to spread over a month, the best horses in the town, a dozen rose-notes sent to his clubs or his lodgings in a day, and the newest thing in soups, colts, beauties, neckties, perfumes, tobaccos, or square dances, waiting his dictum to become the fashion.

'How you *do* go on with those women, Beauty,' growled the Seraph, one day after a morning of fearful hard work consequent on having played the Foot Guards at Lord's, and, in an unwary moment, having allowed himself to be decoyed afterwards to a private concert, and very nearly proposed to in consequence, during a Symphony in A; an impending terror from which he could hardly restore himself by puffing turkish like a steam-engine, to assure himself of his jeopardised safety. 'You're horribly imprudent!'

'Not a bit of it,' rejoined Beauty, serenely. 'That is the superior wisdom and beautiful simplicity of making love to your neighbour's wife—she can't marry you!'

'But she may get you into the DC,' mused the Seraph, who had gloomy personal recollection of having been twice through that phase of law and life, and of having been enormously mulcted in damages because he was a Duke *in futuro*, and because, as he piteously observed on the occasion, 'You couldn't make that fellow Cresswell see that it was *they* ran away with *me* each time!'

'Oh! everybody goes through the DC somehow or other,' answered Cecil, with philosophy. 'It's like the Church, the Commons, and the Gallows, you know—one of the popular Institutions.'

'And it's the only Law Court where the robber cuts a better figure than the robbed,' laughed the Seraph, consoling himself that he had escaped the future chance of showing in the latter class of marital defrauded, by shying that proposal during the Symphony in A, on which his thoughts ran as the thoughts of one who has just escaped from an Alpine crevasse run on the past abyss in which he has been so nearly lost for ever. 'I say, Beauty, were you ever near doing anything serious—asking anybody to marry you, eh? I suppose you have been—they do make such awful hard running on one!' and the poor hunted Seraph stretched his magnificent limbs with the sigh of a martyred innocent.

'I was once—only once!'

'Ah, by Jove! and what saved you?'

The Seraph lifted himself a little, with a sort of pitying sympathising curiosity towards a fellow-sufferer.

'Well, I'll tell you,' said Bertie, with a sigh as of a man who hated long sentences, and who was about to plunge into a painful past. 'It's ages ago; day I was at a Drawing-room—year Blue Ruin won the Clearwell for Royal, I think. Wedged up there, in that poking place, I saw *such* a face— the deuce, it almost makes me feel enthusiastic now. She was just out— an angel with a train! She had delicious eyes—like a spaniel's, you know—a cheek like this peach, and lips like that strawberry, there, on the top of your ice. She looked at me, and I was in love! I knew who she was—Irish Lord's daughter—girl I could have had for the asking; and I vow that I thought I *would* ask her—I actually was as far gone as that. I actually said to myself, I'd hang about her a week or two, and then propose. You'll hardly believe it, but I did! Watched her presented; such grace, such a smile, such a divine lift of the lashes. I was really in love, and with a girl who would marry me! I was never so near a fatal thing in my life——'

'Well?' asked the Seraph, pausing to listen till he let the ice in his sherry-cobbler melt away; when you have been so near breaking your neck down the Matrimonial Matterhorn, it is painfully interesting to hear how your friend escaped the same risks of descent.

'Well,' resumed Bertie, 'I was *very* near it. I did nothing but watch her; she saw me, and I felt she was as flattered and as touched as she ought to be. She blushed most enchantingly; just enough, you know; she was conscious I followed her; I contrived to get close to her as she passed out; so close, that I could see those exquisite eyes lighten and gleam, those exquisite lips part with a sigh, that beautiful face beam with the sunshine of a radiant smile. It was the dawn of love I had taught her! I pressed nearer and nearer, and I caught her soft whisper as she leaned to her mother: "*Mamma, I'm so hungry! I could eat a whole chicken!*" The sigh, the smile, the blush, the light, were for her dinner—not for me! The spell was broken for ever. A girl whom *I* had looked at could think of wings and merry-thoughts and white sauce! I have never been near a proposal again.'

The Seraph, with the clarion roll of his gay laughter, flung a hautboy at him.

'Hang you, Beauty! If I didn't think you were going to tell one how you really got out of a serious thing; it is so awfully difficult to keep clear of them now-a-days. Those before-dinner teas are only just so many new traps! What became of her—eh?'

'She married a Scotch laird and became socially extinct, somewhere among the Hebrides. Serve her right,' murmured Cecil, sententiously. 'Only think what she lost just through hungering for a chicken; if I hadn't have proposed for her, for one hardly keeps the screw up to such self-sacrifice as *that* when one is cool the next morning, I would have made her the fashion!'

With which masterly description in one phrase of all he could have done for the ill-starred débutante who had been hungry in the wrong place, Cecil lounged out of the club to drive with half a dozen of his set to a water-party—a Bacchanalian water-party, with the Zu-Zu and her sisters for the Naiads, and the Household for their Tritons.

A water-party whose water element apparently consisted in driving down to Richmond, dining at nine, being three hours over the courses, contributing seven guineas apiece for the repast, listening to the songs of the Café Alcazar, reproduced with matchless *élan* by a pretty French actress, being pelted with brandy cherries by the Zu-Zu, seeing their best cigars thrown away half-smoked by pretty pillagers, and driving back again to town in the soft starry night, with the gay rhythms ringing out from the box-seat as the leaders dashed along in a stretching gallop down the Kew road. It certainly had no other more aquatic feature in it save a little drifting about for twenty minutes before dining, in toy boats and punts, as the sun was setting, while Laura Lelas, the brunette actress, sang a barcarolle that would have been worthy of medieval

'Venice, and her people, only born to bloom and drop.'

It did not set Cecil thinking, however, after Browning's fashion,

'Where be all those
Dear dead women, with such hair too, what's become of all the gold
Used to hang and brush their bosoms? I feel chilly and grown old.'

because, in the first place, it was a canon with him never to think at all; in the second, if put to it he would have averred that he knew nothing of Venice, except that it was a musty old bore of a place, where they worried you about visas and luggage and all that, chloride of lime'd you if you came from the East, and couldn't give you a mount if it were ever so; and in the third, instead of longing for the dear dead women, he was entirely contented with the lovely living ones who were at that moment puffing the smoke of his scented cigarettes into his eyes, making him eat lobster drowned in Chablis, or pelting him with bonbons.

As they left the Star and Garter, Laura Lelas, mounted on Cecil's box-seat, remembered she had dropped her cashmere in the dining-room. A cashmere is a Parisian's soul, idol, and fetish; servants could not find it; Cecil, who, to do him this justice, was always as courteous to a Comédienne as to a Countess, went himself. Passing the open windows of another room, he recognized the face of his little brother among a set of young Civil Service fellows, attachés, and cornets. They had no women with them; but they had brought what was perhaps worse—dice for hazard—and were turning the unconscious Star and Garter into an impromptu Crockford's over their wine.

Little Berke's pretty face was very flushed; his lips were set tight, his eyes were glittering; the boy had the gambler's passion of the Royallieu blood in its hottest intensity. He was playing with a terrible eagerness that went to Bertie's heart with the same sort of pang of remorse with which he had looked on him when he had been thrown like dead on his bed at home.

Cecil stopped and leaned over the open window.

'Ah, young one, I did not know you were here. We are going home; will you come?' he asked, with a careless nod to the rest of the young fellows.

Berkeley looked up with a wayward, irritated annoyance.

'No, I can't,' he said, irritably; 'don't you see we are playing, Bertie.'

'I see,' answered Cecil, with a dash of gravity, almost of sadness in him, as he leaned further over the window-sill with his cigar in his teeth.

'Come away,' he whispered, kindly, as he almost touched the boy, who chanced to be close to the casement. 'Hazard is the very deuce for anybody; and you know Royal hates it. Come with us, Berke: there's a capital set here, and I'm going to half a dozen good houses tonight, when we get back. I'll take you with me. Come! you like waltzing, and all that sort of thing, you know.'

The lad shook himself peevishly; a sullen cloud over his fair, picturesque, boyish face.

'Let me alone before the fellows,' he muttered, impatiently. 'I won't come, I tell you.'

'*Soit!*'

Cecil shrugged his shoulders, left the window, found the Lelas' cashmere, and sauntered back to the drags without any more expostulation. The sweetness of his temper could never be annoyed, but also he never troubled himself to utter useless words. Moreover, he had never been in his life much in earnest about anything; it was not worth while.

'A pretty fellow I am to turn preacher, when I have sins enough on my own shoulders for twenty,' he thought, as he shook the ribbons and started the leaders off to the gay music of Laura Lelas' champagne-tuned laughter.

The thoughts that had crossed his mind when he had looked on his brother's inanimate form, had not been wholly forgotten since; he felt something like self-accusation whenever he saw, in some grey summer dawn, as he had seen now, the boy's bright face, haggard and pale with the premature miseries of the gamester, or heard his half-piteous, half-querulous lamentations over his losses; and he would essay, with all the consummate tact the world had taught him, to persuade him from his recklessness, and warn him of its consequences. But little Berke, though he loved his elder after a fashion, was wayward, selfish, and unstable as water. He would be very sorry sometimes, very repentant, and would promise anything under the sun; but five minutes afterwards he would go his own way just the same, and be as irritably resentful of interference as a proud, spoiled, still-childish temper can be. And Cecil—the last man in the world to turn mentor—would light a cheroot, as he did tonight, and forget all about it. The boy would be right enough when he had had his swing, he thought. Bertie's philosophy was the essence of *laissez-faire*.

He would have defied a Manfred, or an Aylmer of Aylmer's Field, to be long pursued by remorse or care if he drank the right *cru*, and lived in the right set. 'If it be *very* severe,' he would say, 'it may give him a pang once a twelvemonth—say the morning after a whitebait dinner. Repentance is generally the fruit of indigestion, and contrition may generally be traced to too many truffles or olives.'

Cecil had no time or space for thought; he never thought; would not have thought seriously for a kingdom. A novel, idly skimmed over in bed, was the extent of his literature; he never bored himself by reading the papers, he heard the news earlier than they told it; and as he lived, he was too constantly supplied from the world about him with amusement and variety to have to do anything beyond letting himself be amused: quietly fanned, as it were, with the lulling punkah of social pleasure, without even the trouble of pulling the strings. He had naturally considerable talents, and an almost dangerous facility in them; but he might have been as brainless as a mollusc for any exertion he gave his brain.

'If I were a professional diner-out, you know, I'd use such wits as I have: but why should I now?' he said on one occasion, when a fair lady reproached him with this inertia. 'The best style is only just to say yes and no—and be bored even in saying that—and a very comfortable style it is, too. You get amused without the trouble of opening your lips.'

'But if everybody were equally monosyllabic, how then? you would not get amused,' suggested his interrogator, a brilliant Parisienne.

'Well—everybody *is* pretty nearly,' said Bertie; 'but there are always a lot of fellows who give their wits to get their dinners—social rockets, you know—who will always fire themselves off to sparkle instead of you if you give them a white ball at the clubs, or get them a card for good houses. It saves you so much trouble; it is such a bore to have to talk.'

He went that night, as he had said, to half a dozen good houses, midnight receptions, and after-midnight waltzes, making his bow in a Cabinet Minister's vestibule, and taking up the thread of the same flirtation at three different balls, showing himself for a moment at a Premier's At-home, and looking eminently graceful and pre-eminently weary in an Ambassadress' drawing-room, and winding up the series by a dainty little supper in the grey of the morning, with a sparkling party of French actresses, as bright as the bubbles of their own Clicquot.

When he went upstairs to his own bedroom, in Piccadilly, about five o'clock, therefore, he was both sleepy and tired, and lamented to that cherished and ever-discreet confidant, a cheroot, the brutal demands of the Service, which would drag him off, in five hours' time, without the slightest regard to his feelings, to take share in the hot, heavy, dusty, scorching work of a field-day up at the Scrubs.

'Here—get me to perch as quick as you can, Rake,' he murmured, dropping into an armchair: astonished that Rake did not answer, he saw standing by him instead the boy Berkeley. Surprise was a weakness of raw inexperience that Cecil never felt; his gazette as Commander-in-Chief, or the presence of the Wandering Jew in his lodgings, would never have excited it in him. In the first place, he would have merely lifted his eyebrows and said, 'Be a fearful bore!' in the second he would have done the same, and murmured, 'Queer old cad!'

Surprised, therefore, he was not, at the boy's untimely apparition; but his eyes dwelt on him with a mild wonder, while his lips dropped but one word:

'Amber-Amulet?'

Amber-Amulet was a colt of the most marvellous promise at the Royallieu establishment, looked on to win the next Clearwell, Guineas, and Derby as a certainty. An accident to the young chesnut was the only thing that suggested itself as of possibly sufficient importance to make his brother wait for him at five o'clock on a June morning.

Berkeley looked up confusedly, impatiently.

'You are never thinking but of horses or women,' he said, peevishly; 'there may be other things in the world, surely.'

'Indisputably there are other things in the world, dear boy, but none so much to my taste,' said Cecil, composedly, stretching himself with a yawn. 'With every regard to hospitality, and the charms of your society, might I hint that five o'clock in the morning is not precisely the most suitable hour for social visits and ethical questions?'

'For God's sake be serious, Bertie! I am the most miserable wretch in creation.'

Cecil opened his closed eyes, with the sleepy indifference vanished from them, and a look of genuine and affectionate concern on the serene insouciance of his face.

'Ah, you *would* stay and play that chicken hazard,' he thought, but he was not one who would have reminded the boy of his own advice and its rejection; he looked at him in silence a moment, then raised himself with a sigh.

'Dear boy, why didn't you sleep upon it? I never think of disagreeable things till they wake me with my coffee; then I take them up with the cup and put them down with it. You don't know how well it answers; it disposes of them wonderfully.'

The boy lifted his head with a quick, reproachful anger, and in the gaslight his cheeks were flushed, his eyes full of tears.

'How brutal you are, Bertie! I tell you I am ruined, and you care no more than if you were a stone. You only think of yourself; you only live for yourself!'

He had forgotten the money that had been tossed to him off that very table the day before the Grand Military; he had forgotten the debts that had been paid for him out of the winnings of that very race. There is a childish, wayward, wailing temper, which never counts benefits received save as title-deeds by which to demand others. Cecil looked at him with just a shadow of regret, not reproachful enough to be rebuke, in his glance, but did not defend himself in any way against the boyishly passionate accusation, nor recall his own past gifts into remembrance.

'"Brutal!" What a word, little one. Nobody's brutal now; you never see that form nowadays. Come, what is the worst this time?'

Berkeley looked sullenly down on the table where his elbows leaned, scattering the rose-notes, the French novels, the cigarettes, and the gold essence-bottles with which it was strewn; there was something dogged yet agitated, half-insolent, yet half-timidly irresolute upon him that was new there.

'The worst is soon told,' he said, huskily, and his teeth chattered together slightly as though with cold as he spoke: 'I lost two hundred tonight; I must pay it, or be disgraced for ever; I have not a farthing; I cannot get the money for my life; no Jews will lend to me, I am under age; and—and——' his voice sank lower and grew more defiant, for he knew that the sole thing forbidden him peremptorily by both his father and his brothers was the thing he had now to tell, 'and—I borrowed three ponies of Granville Lee yesterday, as he came from the Corner with a lot of banknotes after settling-day. I told him I would pay them tomorrow; I made sure I should have won tonight.'

The piteous unreason of the born gamester, who clings so madly to the belief that luck must come to him, and acts on that belief as though a bank were his to lose his gold from, was never more utterly spoken in all its folly, in all its pitiable optimism, than now in the boy's confession.

Bertie started from his chair, his sleepy languor dissipated, on his face the look that had come there when Lord Royallieu had dishonoured his mother's name. In his code there was one shameless piece of utter and unmentionable degradation—it was to borrow of a friend.

'You will bring some disgrace on us before you die, Berkeley,' he said, with a keener infliction of pain and contempt than had ever been in his voice. 'Have you no common knowledge of honour?'

The lad flushed under the lash of the words, but it was a flush of anger rather than of shame; he did not lift his eyes, but gazed sullenly down on the yellow paper of a Paris romance he was irritably dog-earing.

'You are severe enough,' he said, gloomily, and yet insolently. 'Are you such a mirror of honour yourself? I suppose my debts at the worst are about one-fifth of yours.'

For a moment even the sweetness of Cecil's temper almost gave way. Be his debts what they would, there was not one among them to his friends, or one for which the law could not seize him. He was silent; he did not wish to have a scene of dissension with one who was but a child to him; moreover, it was his nature to abhor scenes of any sort, and to avert even a dispute at any cost.

He came back and sat down without any change of expression, putting his cheroot in his mouth.

'*Très cher*, you are not courteous,' he said, wearily, 'but it may be that you are right. I am not a good one for you to copy from in anything, except the fit of my coats; I don't think I ever told you I was. I am not altogether so satisfied with myself as to suggest myself as a model for anything, unless it were to stand in a tailor's window in Bond Street to show the muffs how to dress. That isn't the point, though; you say you want near

£300 by tomorrow—today rather. I can suggest nothing except to take the morning mail to the Shires, and ask Royal straight out; he never refuses you.'

Berkeley looked at him with a bewildered terror that banished at a stroke his sullen defiance; he was irresolute as a girl, and keenly moved by fear.

'I would rather cut my throat,' he said, with a wild exaggeration that was but the literal reflexion of the trepidation on him; 'as I live I would! I have had so much from him lately—you don't know how much—and now of all times when they threaten to foreclose the mortgage on Royallieu——'

'What? Foreclose what?'

'The mortgage!' answered Berkeley, impatiently; to his childish egotism it seemed cruel and intolerable that any extremities should be considered save his own. 'You know the lands are mortgaged as deeply as Monti and the entail would allow them. They threatened to foreclose— I think that's the word—and Royal has had God knows what work to stave them off; I no more dare face him, and ask him for a sovereign now, than I dare ask him to give me the gold plate off the sideboard.'

Cecil listened gravely; it cut him more keenly than he showed to learn the evils and the ruin that so closely menaced his house; and to find how entirely his father's morbid mania against him severed him from all the interests and all the confidence of his family, and left him ignorant of matters even so nearly touching him as these.

'Your intelligence is not cheerful, little one,' he said, with a languid stretch of his limbs; it was his nature to glide off painful subjects. 'And— I really am sleepy! You think there is no hope Royal would help you?'

'I tell you I will shoot myself through the brain rather than ask him.'

Bertie moved restlessly in the soft depths of his lounging-chair; he shunned worry, loathed it, escaped it at every portal, and here it came to him just when he wanted to go to sleep. He could not divest himself of the feeling that, had his own career been different, less extravagant, less dissipated, less indolently spendthrift, he might have exercised a better influence, and his brother's young life might have been more prudently launched upon the world. He felt, too, with a sharper pang than he had ever felt it for himself, the brilliant beggary in which he lived, the utter inability he had to raise even the sum that the boy now needed—a sum, so trifling in his set and with his habits, that he had betted it over and over again in a club-room on a single game of whist. It cut him with a bitter impatient pain; he was as generous as the winds, and there is no trial keener to such a temper than the poverty that paralyses its power to give.

'It is no use to give you false hopes, young one,' he said, gently. 'I can do nothing! You ought to know me by this time; and, if you do, you know, too, that if the money were mine it should be yours at a word;—if you don't, no matter! Frankly, Berke, I am all downhill. My bills may be called in any moment; when they are, I must send in my papers to sell, and cut the country, if my duns don't catch me before—which they probably will—in which event I shall be to all intents and purposes—dead. This is not lively conversation, but you will do me the justice to say that it was not I who introduced it. Only—one word for all, my boy, understand this: if I could help you I would, cost what it might; but, as matters stand—I *cannot*.'

And with that Cecil puffed a great cloud of smoke to envelop him; the subject was painful; the denial wounded him by whom it had to be given full as much as it could wound him whom it refused. Berkeley heard it in silence, his head still hung down, his colour changing, his hands nervously playing with the bouquet-bottles, shutting and opening their gold tops.

'No—yes—I know,' he said, hurriedly; 'I have no right to expect it, and have been behaving like a cur, and—and—all that I know. But——there is one way you *could* save me, Bertie, if it isn't too much for a fellow to ask.'

'I can't say I see the way, little one,' said Cecil, with a sigh. 'What is it?'

'Why—look here. You see, I'm not of age; my signature is of no use; they won't take it; else I could get money in no time on what must come to me when Royal dies; though 'tisn't enough to make the Jews "melt" at a risk. Now—now—look here. I can't see that there could be any harm in it. You are such chums with Lord Rockingham, and he's as rich as all the Jews put together. What could there be in it if you just asked him to lend you a monkey for me? He'd do it in a minute, because he'd give his head away to you—they all say so—and he'd never miss it. Now, Bertie——will you?'

In his boyish incoherence and its disjointed inelegance the appeal was panted out rather than spoken, and while his head drooped and the hot colour burned in his face, he darted a swift look at his brother, so full of dread and misery that it pierced Cecil to the quick as he rose from his chair and paced the room, flinging his cheroot aside; the look disarmed the reply that was on his lips, but his face grew dark.

'What you ask is impossible,' he said, briefly. 'If I did such a thing as that, I should deserve to be hounded out of the Guards tomorrow.'

The boy's face grew more sullen, more haggard, more evil, as he still bent his arms on the table, his glance not meeting his brother's.

'You speak as if it would be a crime,' he muttered, savagely, with a plaintive moan of pain in the tone; he thought himself cruelly dealt with, and unjustly punished.

'It would be the trick of a swindler, and it would be the shame of a gentleman,' said Cecil, as briefly still. 'That is answer enough.'

'Then you will not do it?'

'I have replied already.'

There was that in the tone, and in the look with which he paused before the table, that Berkeley had never heard or seen in him before; something that made the supple, childish, petulant, cowardly nature of the boy shrink and be silenced; something for a single instant of the haughty and untameable temper of the Royallieu blood that awoke in the too feminine softness and sweetness of Cecil's disposition.

'You said that you would aid me at any cost, and now that I ask you so wretched a trifle, you treat me as if I were a scoundrel,' he moaned, passionately. 'The Seraph would give you the money at a word. It is your pride—nothing but pride. Much pride is worth to us who are penniless beggars!'

'If we are penniless beggars, by what right should we borrow of other men?'

'You are wonderfully scrupulous all of a sudden!'

Cecil shrugged his shoulders slightly and began to smoke again. He did not attempt to push the argument. His character was too indolent to defend itself against aspersion, and horror of a quarrelsome scene far greater than his heed of misconstruction.

'You are a brute to me!' went on the lad with his querulous and bitter passion rising almost to tears like a woman's. 'You pretend you can refuse me nothing; and the moment I ask you the smallest thing, you turn round on me, and speak as if I were the greatest blackguard on earth. You'll let me go to the bad to-morrow, rather than bend your pride to save me; you live like a Duke, and don't care if I should die in a debtors' prison! You only brag about "honour" when you want to get out of helping a fellow, and if I were to cut my throat tonight you would only shrug your shoulders, and sneer at my death in the club-room, with a jest picked out of your cursed French novels!'

'Melodramatic, and scarcely correct,' murmured Bertie.

The ingratitude to himself touched him indeed but little, he was not given to making much of anything that was due to himself, partly through carelessness, partly through generosity; but the absence in his brother of that delicate, intangible, indescribable sensitive-nerve which men call Honour, an absence that had never struck on him so vividly as it did tonight, troubled him, surprised him, oppressed him.

There is no science that can supply this defect to the temperament created without it; it may be taught a counterfeit, but it will never own a reality.

'Little one, you are heated, and don't know what you say,' he began, very gently, a few moments later, as he leaned forward and looked straight in the boy's eyes; 'don't be down about this; you will pull through, never fear. Listen to me; go down to Royal, and tell him all frankly. I know him better than you; he will be savage for a second, but he would sell every stick and stone on the land for your sake; he will see you safe through this. Only bear one thing in mind—tell him *all*. No half measures, no half confidences; tell him the worst, and ask his help. You will not come back without it.'

Berkeley listened, his eyes shunning his brother's, the red colour darker on his face.

'Do as I say,' said Cecil, very gently still. 'Tell him, if you like, that it is through following my follies that you have come to grief; he will be sure to pity you then.'

There was a smile, a little sad, on his lips as he said the last words, but it passed at once as he added:

'Do you hear me? Will you go?'

'If you want me—yes.'

'On your word, now?'

'On my word.'

There was an impatience in the answer, a feverish eagerness in the way he assented, that might have made the consent rather a means to evade the pressure than a genuine pledge to follow the advice; that darker, more evil, more defiant look, was still upon his face, sweeping its youth away and leaving in its stead a wavering shadow. He rose with a sudden movement; his tumbled hair, his disordered attire, his bloodshot eyes, his haggard look of sleeplessness and excitement, in strange contrast with the easy perfection of Cecil's dress, and the calm languor of his attitude. The boy was very young, and was not seasoned to his life and acclimatized to his ruin like his elder brother. He looked at him with a certain petulant envy; the envy of every young fellow for a man of the world. 'I beg your pardon for keeping you up, Bertie,' he said, huskily. 'Good night.'

Cecil gave a little yawn.

'Dear boy, it *would* have been better if you could have come in with the coffee. Never be impulsive; don't do a bit of good, and *is* such a bad form!'

He spoke lightly, serenely, both because such was as much his nature as it was to breathe, and because his heart was heavy that he had to send away the young one without help, though he knew that the course he had made him adopt would serve him more permanently in the end. But he leant his hand a second on Berke's shoulder, while for one single moment in his life he grew serious.

'You must know I could not do what you asked; I could not meet any man in the Guards face to face if I sunk myself and sunk them so low. Can't you see that, little one?'

There was a wishfulness in the last words; he would gladly have believed that his brother had at length some perception of his meaning.

'You say so, and that is enough,' said the boy, pettishly. 'I cannot understand that I asked anything so dreadful, but I suppose you have too many needs of your own to have any resources left for mine.'

Cecil shrugged his shoulders slightly again, and let him go. But he could not altogether banish a pang of pain at his heart, less even for his brother's ingratitude, than at his callousness to all those finer better instincts of which honour is the concrete name. For the moment, thought, grave, weary, and darkened, fell on him; he had passed through what he would have suffered any amount of misconstruction to escape—a disagreeable scene; he had been as unable as though he were a Commissionaire in the streets, to advance a step to succour the necessities for which his help had been asked; and he was forced, despite all his will, to look for the first time blankly in the face the ruin that awaited him. There was no other name for it: it would be ruin complete and wholly inevitable. His signature would have been accepted no more by any bill-discounter in London; he had forestalled all to the uttermost farthing; his debts pressed heavier every day; he could have no power to avert the crash that must in a few weeks, or at most a few months, fall upon him. And to him an utter blankness and darkness lay beyond.

Barred out from the only life he knew, the only life that seemed to him endurable or worth the living; severed from all the pleasures, pursuits, habits, and luxuries of long custom; deprived of all that had become to him as second nature from childhood; sold up, penniless, driven out from all that he had known as the very necessities of existence, his very name forgotten in the world of which he was now the darling, a man without a career, without a hope, without a refuge—he could not realise that this was what awaited him then; this was the fate that must within so short space be his. Life had gone so smoothly with him, and his world was a world from whose surface every distasteful thought was so habitually excluded, that he could no more understand this desolation lying in wait for him, than one in the fulness and elasticity of health can believe the doom that tells him he will be a dead man before the sun has set.

As he sat there, with the gas of the mirror branches glancing on the gold and silver hilts of the crossed swords above the fireplace, and the smoke of his cheroot curling amongst the pile of invitation-cards to all the best houses in the town, Cecil could not bring himself to believe that things were really come to this pass with him; it is so hard for a man who has the magnificence of the fashionable clubs open to him day and night to beat into his brain the truth that in six months hence he may be lying in the debtors' prison at Baden; it is so difficult for a man who has had no greater care on his mind than to plan the courtesies of a Guards' Ball or of a yacht's summer-day banquet, to absolutely conceive the fact that in a year's time he will thank God if he have a few francs left to pay for a wretched dinner in a miserable estaminet in a foreign bathing-place.

'It mayn't come to that,' he thought; 'something may happen. If I could get my troop now, that would stave off the Jews; or if I should win some heavy pots on the Prix de Dames, things would swim on again. I *must* win; the King will be as fit as in the Shires, and there will only be the French horses between us and an absolute "walk over". Things mayn't come to the worst after all.'

And so careless and quickly oblivious, happily or unhappily, was his temperament, that he read himself to sleep with Terrail's *Club des Valets de Cœur*, and slept in ten minutes' time as composedly as though he had heard he had inherited fifty thousand a year.

That evening, in the loose-box down at Royallieu, Forest King stood without any body-clothing, for the night was close and sultry; a lock of the sweetest hay unnoticed in his rack, and his favourite wheaten-gruel standing uncared-for under his very nose: the King was in the height of excitation, alarm, and haughty wrath. His ears were laid flat to his head, his nostrils were distended, his eyes were glancing uneasily with a nervous angry fire rarely in him, and ever and anon he lashed out his heels with a tremendous thundering thud against the opposite wall with a force that reverberated through the stables, and made his companions start and edge away. It was precisely these companions that the aristocratic hero of the Soldiers' Blue Riband scornfully abhorred.

They had just been looking him over—to their own imminent peril—and the patrician winner of the Vase, the brilliant six-year-old of Paris, and Shire and Spa steeplechase fame, the knightly descendant of the White Cockade blood, and of the coursers of Circassia, had resented the familiarity proportionately to his own renown and dignity. The King was a very sweet-tempered horse, a perfect temper, indeed, and ductile to a touch from those he loved; but he liked very few, and would suffer liberties from none. And of a truth his prejudices were very just; and if his clever heels had caught—as it was not his fault that they did not—the heads of his two companions, instead of coming with that ponderous crash into the panels of his box, society would certainly have been no loser, and his owner would have gained more than had ever before hung in the careless balance of his life.

But the iron heels with their shining plates only caught the oak of his box-door; and the tête-à-tête in the sultry oppressive night went on as the speakers moved to a prudent distance, one of them thoughtfully chewing a bit of straw, after the immemorial habit of grooms, who ever seem as if they had been born into this world with a corn-stalk ready in their mouths.

'It's a'most a pity—he's in such perfect condition. Tiptop. Cool as a cucumber after the longest pipe-opener; licks his oats up to the last grain; leads the whole string such a rattling spin as never *was* spun but by a Derby cracker before him. It's a'most a pity,' said Willon, meditatively, eyeing his charge, the King, with remorseful glances.

'Prut—tush—tish!' said his companion, with a whistle in his teeth that ended with a 'damnation!' 'It'll only knock him over for the race; he'll be right as a trivet after it. What's your little game, coming it soft like that all of a sudden? You hate that ere young swell like pison.'

'Ay,' assented the head groom, with a tigerish energy, viciously consuming his bit of straw. 'What for am I—head groom come nigh twenty years; and to Markisses and Wiscounts afore him—put aside in that ere way for a fellow as he's took into his service out of the dregs of a regiment, what was tied up at the triangles and branded D, as I know on, and sore suspected of even worse games than that, and now is that set up with pride and sich-like, that nobody's woice ain't heard here except his; I say what am I called on to bear it for?' and the head groom's tones grew hoarse and vehement, roaring louder under his injuries. 'A man what's attended on Dukes' 'osses ever since he was a shaver, to be put aside for that workhus blackguard! A 'oss has a cold—it's Rake's mash what's to be given . A 'oss is off his feed—it's Rake what's to weigh out the nitre and steel. A 'oss is a buck-jumper—it's Rake what's to cure him. A 'oss is entered for a race—it's Rake what's to order his mornin' gallops, and his go-downs o' water. It's past bearin' to have a rascally chap what's been and gone and turned walet, set up over one's head in one's own establishment, and let to ride the high 'oss over one roughshod like that!'

And Mr Willon, in his disgust at the equestrian contumely thus heaped on him, bit the straw savagely in two, and made an end of it, with a vindictive '*Will* yer be quiet there: blow yer,' to the King, who was protesting with his heels against the conversation.

'Come, then, no gammon,' growled his companion—the 'cousin out o' Yorkshire' of the keeper's tree.

'What's yer figure you say?' relented Willon, meditatively.

'Two thousand to nothin'—come!—can't no handsomer,' retorted the Yorkshire cousin, with the air of a man conscious of behaving very nobly.

'For the race in Germany?' pursued Mr Willon, still meditatively.

'Two thousand to nothin'—come!' reiterated the other, with his arms folded to intimate that this and nothing else was the figure to which he would bind himself.

Willon chewed another bit of straw, glanced at the horse as though he were a human thing to hear, to witness, and to judge; grew a little pale; and stooped forward.

'Hush! Somebody'll spy on us. It's a bargain.'

'Done. And you'll paint him, eh?'

'Yes—I'll—paint him.'

The assent was very husky, and dragged slowly out, while his eyes glanced with a furtive, frightened glance over the loose-box. Then—still with that cringing, terrified look backward to the horse as an assassin may steal a glance before his deed at his unconscious victim—the head groom and his comrade went out and closed the door of the loose-box and passed into the hot lowering summer night.

Forest King, left in solitude, shook himself with a neigh; took a refreshing roll in the straw, and turned with an appetite to his neglected gruel. Unhappily for himself, his fine instincts could not teach him the conspiracy that lay in wait for him and his; and the gallant beast, content to be alone, soon slept the sleep of the righteous.

CHAPTER VIII

A Stag Hunt au Clair de la Lune

'Seraph—I've been thinking'—said Cecil, musingly, as they paced homeward together from the Scrubs, with the long line of the First Life stretching before and behind their chargers, and the bands of the Household Cavalry playing mellowly in their rear.

'You don't mean it. Never let it ooze out, Beauty; you'll ruin your reputation!'

Cecil laughed a little, very languidly; to have been in the sun for four hours, in full harness, had almost taken out of him any power to be amused at anything.

'I've been thinking,' he went on, undisturbed, pulling down his chin-scale. 'What's a fellow to do when he's smashed?'

'Eh?' The Seraph couldn't offer a suggestion; he had a vague idea that men who were smashed never *did* do anything except accept the smashing; unless, indeed, they turned up afterwards as touts, of which he had an equally vague suspicion.

'What *do* they do?' pursued Bertie.

'Go to the bad,' finally suggested the Seraph, lighting a great cigar, without heeding the presence of the Duke, a Field-Marshal, and a Serene Highness far on in front.

Cecil shook his head.

'Can't go where they are already. I've been thinking what a fellow might do that was up a tree; and on my honour there are lots of things one might turn to——'

'Well, I suppose there are,' assented the Seraph, with a shake of his superb limbs in his saddle, till his cuirass, and chains, and scabbard rang again. '*I* should try the PR, only they will have you train.'

'One might do better than the PR. Getting yourself into prime condition, only to be pounded *out* of condition and into a jelly, seems hardly logical or satisfactory—specially to your looking-glass; though, of course, it's a matter of taste. But now, if I had a cropper, and got sold up——'

'You, Beauty?' The Seraph puffed a giant puff of amazement from his havannah, opening his blue eyes to their widest.

'Possible!' returned Bertie, serenely, with a nonchalant twist to his moustaches. 'Anything's possible. If I do now, it strikes me there are vast fields open.'

'Gold fields?' suggested the Seraph, wholly bewildered——

'Gold fields? No! I mean a field for—what d'ye call it—genius. Now, look here; nine-tenths of creatures in this world don't know how to put on a glove. It's an art, and an art that requires long study. If a few of us were to turn glove-fitters when we are fairly crashed, we might civilize the whole world, and prevent the deformity of an ill-fitting glove ever blotting creation and prostituting Houbigant. What do you say?'

'Don't be such a donkey, Beauty,' laughed the Seraph, while his charger threatened to passage into an oyster-cart.

'You don't appreciate the majesty of great plans,' rejoined Beauty, reprovingly. 'There's an immense deal in what I'm saying. Think what e might do for society—think how we might extinguish snobbery, if we just dedicated our smash to Mankind. We might open a College, where the traders might go through a course of polite training before they blossomed out as millionaires; the world would be spared an agony of dropped H's and bad bows. We might have a Bureau where we registered all our social experiences, and gave the Plutocracy a map of Belgravia, with all the pitfalls marked, all the inaccessible heights coloured red, and all the hard-up great people dotted with gold to show the amount they'd be bought for, with directions to the ignoramuses whom to know, court, and avoid. We might form a Courier Company, and take Brummagem abroad under our guidance, so that the Continent shouldn't think Englishwomen always wear blue veils and grey shawls, and hear every Englishman shout for porter and beefsteak in Tortoni's. We might teach them to take their hats off to women, and not to prod pictures with sticks, and to look at statues without poking them with an umbrella, and to be persuaded that all foreigners don't want to be bawled at, and won't understand bad French any the better for its being shouted. Or we might have a Joint-Stock Toilette Association, for the purposes of national art, and receive Brummagem to show it how to dress; we might even succeed in making the feminine British Public drape itself properly, and the BP masculine wear boots that won't creak, and coats that don't wrinkle, and take off its hat without a jerk as though it were a wooden puppet hung on very stiff strings. Or one might——'

'Talk the greatest nonsense under the sun!' laughed the Seraph. 'For mercy's sake, are you mad, Bertie?'

'Inevitable question addressed to Genius!' yawned Cecil. 'I'm showing you plans that might teach a whole nation good style if we just threw ourselves into it a little. I don't mean *you*, because you'll never smash, and one don't turn bear-leader, even to the BP, without the primary impulse of being hard-up. And I don't talk for myself, because, when I go to the dogs, I have my own project.'

'And what's that?'

'To be groom of the chambers at Meurice's or Claridge's,' responded Bertie, solemnly. 'Those sublime creatures with their silver chains round their necks, and their ineffable supremacy over every other mortal!—one would feel in a superior region still. And when a snob came to poison the air, how exquisitely one could annihilate him with showing him his ignorance of claret; and when an epicure dined, how delightfully, as one carried in a turbot, one could test him with the *éprouvette positive*, or crush him by the *éprouvette negative*. We have been Equerries at the Palace, both of us, but I don't think we know what true dignity is till we shall have risen to head waiters at a Grand Hotel.'

With which Bertie let his charger pace onward, while he reflected thoughtfully on his future state. The Seraph laughed till he almost swayed out of saddle, but he shook himself into his balance again with another clash of his brilliant harness, while his eyes lightened and glanced with a fiery gleam down the line of the Household Cavalry.

'Well, if I went to the dogs, I wouldn't go to Grand Hotels, but I'll tell you where I would go, Beauty.'

'Where's that?'

'Into hot service, somewhere. By Jove, I'd see some good fighting under another flag—out in Algeria, there, or with the Poles, or after Garibaldi. I would, in a day—I'm not sure I won't now, and I bet you ten to one the life would be better than this.'

Which was ungrateful in the Seraph, for his happy temper made him the sunniest and most contented of men, with no cross in his life save the dread that somebody would manage to marry him some day. But Rock had the true dash and true steel of the soldier in him, and his blue eyes flashed over his Guards as he spoke, with a longing wish that he were leading them on to a charge instead of pacing with them towards Hyde Park.

Cecil turned in his saddle and looked at him with a certain wonder and pleasure in his glance, and did not answer aloud. 'The deuce—that's not a bad idea,' he thought to himself; and the idea took root and grew with him.

Far down, very far down, so far that nobody had ever seen it, nor himself ever expected it, there was a lurking instinct in 'Beauty'—the instinct that had prompted him, when he sent the King at the Grand Military cracker, with that prayer, 'Kill me if you like, but don't *fail* me!'—which, out of the languor and pleasure-loving temper of his un-ruffled life, had a vague, restless impulse towards the fiery perils and nervous excitement of a sterner and more stirring career.

It was only vague, for he was naturally very indolent, very gentle, very addicted to taking all things passively, and very strongly of persuasion that to rouse yourself for anything was a *niaiserie* of the strongest possible folly; but it was there. It always is there with men of Bertie's order, and only comes to light when the match of danger is applied to the touchhole. Then, though 'the Tenth don't dance', perhaps, with graceful indolent dandy insolence, they can fight as no others fight when Boot and Saddle rings through the morning air, and the slashing charge sweeps down with lightning speed and falcon swoop.

'In the case of a Countess, sir, the imagination is more excited,' says Dr Johnson, who had, I suppose, little opportunity of putting that doctrine for amatory intrigues to the test in actual practice. Bertie, who had many opportunities, differed with him. He found love-making in his own polished tranquil circles apt to become a little dull, and was more amused by Laura Lelas. However, he was sworn to the service of the Guenevere, and he drove his mail-phaeton down that day to another sort of Richmond dinner, of which the Lady was the object instead of the Zu-Zu.

She enjoyed thinking herself the wife of a jealous and inexorable lord, and arranged her flirtations to evade him with a degree of skill so great, that it was lamentable it should be thrown away on an agricultural husband, who never dreamt that the 'Fidelio—III—TstnegeR,' which met his eyes in the innocent face of his *Times*, referred to an appointment at a Regent Street modiste's, or that the advertisement—'White wins—Twelve', meant that if she wore white camellias in her hair at the opera, she would give 'Beauty' a meeting after it.

Lady Guenevere was very scrupulous never to violate conventionalities. And yet she was a little fast—very fast, indeed, and was a queen of one of the fastest sets; but then—O sacred shield of a wife's virtue—she could not have borne to lose her very admirable position, her very magnificent jointure, and, above all, the superb Guenevere diamonds!

I don't know anything that will secure a husband from an infidelity so well as very fine family jewels, when such an infidelity would deprive his wife of them for ever. Many women will leave their homes, their lords, their children, and their good name, if the fancy take them; but there is not one in a million who will so far forget herself as to sever from pure rose-diamonds.

So, for sake of the diamonds, she and Bertie had their rendezvous under the rose.

This day she went down to see a Dowager Baroness aunt, out at Hampton Court—really went: she was never so imprudent as to falsify her word, and with the Dowager, who was very deaf and purblind, dined at Richmond, while the world thought her dining at Hampton Court. It was nothing to any one, since none knew it to gossip about, that Cecil joined her there: that over the Star and Garter repast they arranged their meeting at Baden next month: that while the Baroness dozed over the grapes and peaches—she had been a beauty herself, in her own day, and still had her sympathies—they went on the river, in the little toy that he kept there for his fair friends' use, floating slowly along in the coolness of evening, while the stars loomed out in the golden trail of the sunset, and doing a graceful scene *à la* Musset and Meredith, with a certain languid amusement in the assumption of those poetic guises, for they were of the world worldly; and neither believed very much in the other.

When you have just dined well, and there has been no fault in the clarets, and the scene is pretty, if it be not the Nile in the after-glow, the Arno in the moonlight, or the Loire in vintage-time, but only the Thames above Richmond, it is the easiest thing in the world to feel a touch of sentiment when you have a beautiful woman beside you who expects you to feel it. The evening was very hot and soft. There was a low south wind, the water made a pleasant murmur, wending among its sedges. She was very lovely, moreover, lying back there among her laces and Indian shawls, with the sunset in the brown depths of her eyes and on her delicate cheek. And Bertie, as he looked on his liege lady, really had a glow of the old, real, foolish, forgotten feeling stir at his heart, as he gazed on her in the half-light, and thought, almost wistfully, 'If the Jews were down on me tomorrow, would she really care, I wonder?'

Really care? Bertie knew his world and its women too well to deceive himself in his heart about the answer. Nevertheless, he asked the question. 'Would you care much, *chère belle*?'

'Care what?'

'If I came to grief—went to the bad, you know; dropped out of the world altogether?'

She raised her splendid eyes in amaze, with a delicate shudder through all her lace. 'Bertie! you would break my heart! What can you dream of?'

'Oh, lots of us end so. How *is* a man to end?' answered Bertie, philosophically, while his thoughts still ran off in a speculative scepticism. 'Is there a heart to break?'

Her ladyship looked at him, and laughed.

'A Werter in the Guards! I don't think the role will suit either you or your corps, Bertie; but if you do it, pray do it artistically. I remember, last year, driving through Asnières, when they had found a young man in the Seine; he was very handsome, beautifully dressed, and he held fast in his clenched hand a gold lock of hair. Now, there was a man who knew how to die gracefully, and make his death an idyl!'

'Died for a woman?—ah!' murmured Bertie, with the Brummel nonchalance of his order. 'I don't think I should do that, even for *you*, not, at least, while I had a cigar left.'

And then the boat drifted backward, while the stars grew brighter and the last reflection of the sun died out; and they planned to meet tomorrow, and talked of Baden, and sketched projects for the winter in Paris, and went in and sat by the window, taking their coffee, and feeling, in a half vague pleasure, the heliotrope-scented air blowing softly in from the garden below, and the quiet of the starlit river in the summer evening, with a white sail gleaming here and there, or the gentle splash of an oar following on the swift trail of a steamer: the quiet, so still and so strange after the crowded rush of the London season.

'Would she really care?' thought Cecil, once more. In that moment he could have wished to think she would.

But heliotrope, stars, and a river, even though it had been tawny and classical Tiber instead of ill-used and inodorous Thames, were not things sufficiently in the way of either of them to detain them long. They had both seen the Babylonian sun set over the ruins of the Birs Nimrud, and had talked of Paris fashions while they did so; they had both leaned over the terraces of Bellosguardo, while the moon was full on Giotto's tower, and had discussed their dresses for the Veglione masquerade. It was not their style to care for these matters; they were pretty, to be sure, but they had seen so many of them.

The Dowager went home in her brougham; the Countess drove in his mail-phaeton, objectionable, as she might be seen, but less objectionable than letting her servants know he had met her at Richmond. Besides, she obviated danger by bidding him set her down at a little villa across the park, where dwelt a confidential protégée of hers, whom she patronized; a former French governess, married tolerably well, who had the Countess's confidences, and kept them religiously for sake of so aristocratic a patron, and of innumerable reversions of Spanish point and shawls that had never been worn, and rings, of which her lavish ladyship had got tired.

From here, she would take her ex-governess's little brougham, and get quietly back to her own house in Eaton Square in due time for all the drums and crushes at which she must make her appearance. This was the sort of little devices which really made them think themselves in love, and gave the salt to the whole affair. Moreover, there was this ground for it, that had her lord once roused from the straw-yards of his prize cattle, there was a certain stubborn, irrational, old-world prejudice of pride and temper in him that would have made him throw expediency to the winds, then and there, with a blind and brutal disregard to slander and to the fact that none would ever adorn his diamonds as she did. So that Cecil had not only her fair fame, but her still more valuable jewels, in his keeping when he started from the Star and Garter in the warmth of the bright summer's evening.

It was a lovely night; a night for lonely highland tarns, and southern shores by Baiæ; without a cloud to veil the brightness of the stars; a heavy dew pressed the odours from the grasses, and the deep glades of the avenues were pierced here and there with a broad beam of silvery moonlight, slanting through the massive boles of the trees, and falling white and serene across the turf. Through the park, with the gleam of the water ever and again shining through the branches of the foliage, Cecil started his horses; his groom he had sent away on reaching Richmond, for the same reason as the Countess had dismissed her barouche, and he wanted no servant, since, as soon as he had set down his liege-lady at her protégée's, he would drive straight back to Piccadilly. But he had not noticed what he noticed now, that instead of one of his carriage-greys, who had fallen slightly lame, they had put into harness the young one, Maraschino, who matched admirably for size and colour, but who, being really a hunter, though he had been broken to shafts as well, was not the horse with which to risk driving a lady.

However, Beauty was a perfect whip, and had the pair perfectly in hand, so that he thought no more of the change, as the greys dashed at a liberal half-speed through the park, with their harness flashing in the moonlight, and their scarlet rosettes fluttering in the pleasant air. The eyes beside him, the Titian-like mouth, the rich, delicate cheek, these were, to be sure, rather against the coolness and science that such a five-year-old as Maraschino required; they were distracting even to Cecil, and he had not prudence enough to deny his sovereign lady when she put her hands on the ribbons.

'The beauties! give them to me, Bertie. Dangerous? How absurd you are; as if I could not drive *anything*! Do you remember my four roans at Longchamps?'

She could, indeed, with justice, pique herself on her skill; she drove matchlessly, but as he resigned them to her, Maraschino and his companion quickened their trot, and tossed their pretty thoroughbred heads, conscious of a less powerful hand on the reins.

'I shall let their pace out, there is nobody to run over here,' said her ladyship '*Va-t' en donc mon beau monsieur.*'

Maraschino, as though hearing the flattering conjuration, swung off into a light, quick canter, and tossed his head again; he knew that, good whip though she was, he could jerk his mouth free in a second if he wanted. Cecil laughed, prudence was at no time his virtue, and leant back contentedly, to be driven at a slashing pace through the balmy summer's night, while the ring of the hoofs rang merrily on the turf, and the boughs were tossed aside with a dewy fragrance. As they went, the moonlight was shed about their path in the full of the young night, and at the end of a long aisle of boughs, on a grassy knoll, were some phantom forms, the same graceful shapes that stand out against the purple heather and the tawny gorse of Scottish moorlands, while the lean rifle-tube creeps up by stealth. In the clear starlight there stood the deer, a dozen of them, a clan of stags alone, with their antlers clashing like the clash of swords, and waving like swaying banners as they tossed their heads and listened.[1]

In an instant the hunter pricked his ears, snuffed the air, and twitched with passionate impatience at his bit; another instant and he had got his head, and launching into a sweeping gallop, rushed down the glade.

Cecil sprang forward from his lazy rest, and seized the ribbons that in one instant had cut his companion's gloves to stripes.

'Sit still,' he said, calmly, but under his breath. 'He has been always ridden with the Buck-hounds; he will race the deer as sure as we live!'

Race the deer he did.

Startled, and fresh for their favourite nightly wandering, the stags were off like the wind at the noise of alarm, and the horses tore after them; no skill, no strength, no science, could avail to pull them in; they had taken their bits between their teeth, and the devil that was in Maraschino lent the contagion of sympathy to the young carriage-mare, who had never gone at such a pace since she had been first put in her break.

Neither Cecil's hands nor any other force could stop them now; on they went, hunting as straight in line as though staghounds streamed in front of them, and no phaeton rocked and swayed in a dead and dragging weight behind them. In a moment he gauged the closeness and the vastness of the peril; there was nothing for it but to trust to chance, to keep his grasp on the reins to the last, and to watch for the first sign of exhaustion. Long ere that should be given death might have come to them both; but there was a gay excitation in that headlong rush through the summer night, there was a champagne-draught of mirth and mischief in that dash through the starlit woodland, there was a reckless, breathless pleasure in that neck-or-nothing moonlight chase!

Yet danger was so near with every oscillation; the deer were trooping in fast flight, now clear in the moonlight, now lost in the shadow, bounding with their lightning grace over sward and hillock, over briar and brushwood, at that speed which kills most living things that dare to race the 'Monarch of the Glens'. And the greys were in full pursuit; the hunting fire was in the fresh young horse; he saw the shadowy branches of the antlers toss before him, and he knew no better than to hunt down in their scenting line as hotly as though the field of the Queen's or the Baron's was after them. What cared he for the phaeton that rocked and reeled on his

---

[1] Let me here take leave to beg pardon of the gallant Highland stags for comparing them one instant with the shabby, miserable-looking wretches that travesty them in Richmond Park. After seeing these latter scrubby, meagre apologies for deer, one wonders why something better cannot be turned loose there. A hunting-mare I know well, nevertheless flattered them thus by racing them through the park; when in harness herself to her own great disgust.

traces, he felt its weight no more than if it were a wickerwork toy, and extended like a greyhound he swerved from the road, swept through the trees, and tore down across the grassland in the track of the herd.

Through the great boles of the trunks, bronze and black in the shadows, across the hilly rises of the turf, through the brushwood pell-mell, and crash across the level stretches of the sward, they raced as though the hounds were streaming in front; swerved here, tossed there, carried in a whirlwind over the mounds, wheeled through the gloom of the woven branches, splashed with a hiss through the shallow rain-pools, shot swift as an arrow across the silver radiance of the broad moonlight, borne against the sweet south wind, and down the odours of the trampled grass, the carriage was hurled across the park in the wild starlight chase. It rocked, it swayed, it shook, at every yard, while it was carried on like a paper toy; as yet the marvellous chances of accident had borne it clear of the destruction that threatened it at every step as the greys, in the height of their pace now, and powerless even to have arrested themselves, flew through the woodland, neither knowing what they did, nor heeding where they went; but racing down on the scent, not feeling the strain of the traces, and only maddened the more by the noise of the whirling wheels behind them.

As Cecil leaned back, his hands clenched on the reins, his sinews stretched almost to bursting in their vain struggle to recover power over the loosened beasts, the hunting zest woke in him too, even while his eyes glanced on his companion in fear and anxiety for her.

'Tally-ho! hark forward! As I live it is glorious!' he cried, half unconsciously. 'For God's sake sit still, Beatrice! I will save you.'

Inconsistent as the words were, they were true to what he felt: alone, he would have flung himself delightedly into the madness of the chase, for her he dreaded with horror the imminence of their peril.

On fled the deer, on swept the horses; faster in the gleam of the moonlight the antlered troop darted on through the gloaming; faster tore the greys in the ecstasy of their freedom; headlong and heedless they dashed through the thickness of leaves and the weaving of branches; neck to neck straining to distance each other, and held together by the gall of the harness. The broken boughs snapped, the earth flew up beneath their hoofs, their feet struck scarlet sparks of fire from the stones, the carriage was whirled, rocking and tottering, through the maze of tree-trunks, towering like pillars of black stone up against the steel-blue clearness of the sky. The strain was intense; the danger deadly: suddenly, straight ahead, beyond the darkness of the foliage, gleamed a line of light; shimmering, liquid, and glassy, here brown as gloom where the shadows fell on it, here light as life where the stars mirrored on it. That trembling line stretched right in their path. For the first time from the blanched lips beside him a cry of terror rang.

'The river!—oh, Heaven!—the river!'

There it lay in the distance, the deep and yellow water, cold in the moon's rays, with its further bank but a dull grey line in the mists that rose from it, and its swamp a yawning grave, as the horses, blind in their delirium, and racing against each other, bore down through all obstacles towards its brink. Death was rarely ever closer; one score yards more, one plunge, one crash down the declivity and against the rails, one swell of the noisome tide above their heads, and life would be closed and passed for both of them. For one breathless moment his eyes met hers—in that moment he loved her, in that moment their hearts beat with a truer, fonder impulse to each other than they had ever done. Before the presence of a threatening death life grows real, love grows precious, to the coldest and most careless.

No aid could come; not a living soul was nigh; the solitude was as complete as though a western prairie stretched around them; there were only the still and shadowy night, the chilly silence, on which the beat of the plunging hoofs shattered like thunder, and the glisten of the flowing water growing nearer and nearer every yard. The tranquillity around only jarred more horribly on ear and brain; the vanishing forms of the antlered deer only gave a weirder grace to the moonlight-chase whose goal was the grave. It was like the midnight hunt after Herne the Hunter; but here, behind them, hunted Death.

The animals neither saw nor knew what waited them, as they rushed down on to the broad, grey stream, veiled from them by the slope and the screen of flickering leaves; to save them there was but one chance, and that so desperate that it looked like madness. It was but a second's thought; he gave it but a second's resolve.

The next instant he stood on his feet, as the carriage swayed to and fro over the turf, balanced himself marvellously as it staggered in that furious gallop from side to side, clenched the reins hard in the grip of his teeth, measured the distance with an unerring eye, and crouching his body for the spring with all the science of the old playing-fields of his Eton days, cleared the dash-board and lighted astride on the back of the hunting five-year-old; how, he could never have remembered, or have told.

The tremendous pace at which they went swayed him with a lurch and a reel over the off-side; a woman's cry rang again, clear, and shrill, and agonized on the night; a moment more, and he would have fallen head downwards beneath the horses' feet. But he had ridden stirrup-less and saddle-less ere now; he recovered himself with the suppleness of an Arab, and firm-seated behind the collar, with one leg crushed between the pole and Maraschino's flanks, gathering in the ribbons till they were tight-drawn as a bridle, he strained with all the might and sinew that were in him to get the greys in hand before they could plunge down into the water. His wrists were wrenched like pulleys, the resistance against him was hard as iron, but as he had risked life and limb in the leap which had seated him across the harnessed loins of the now terrified beast, so he risked them afresh to get the mastery now; to slacken them, turn them ever so slightly, and save the woman he loved—loved, at least in this hour, as he had not loved her before. One moment more while the half-maddened beasts rushed through the shadows; one moment more, till the river stretched full before them in all its length and breadth, without a living thing upon its surface to break the still and awful calm; one moment—and the force of cool command conquered and broke their wills despite themselves. The hunter knew his master's voice, his touch, his pressure, and slackened speed by an irresistible, almost unconscious habit of obedience; the carriage-mare, checked and galled in the full height of her speed, stood erect, pawing the air with her forelegs, and flinging the white froth over her withers, while she plunged blindly in her nervous terror; then, with a crash, her feet came down upon the ground, the broken harness shivered together with a sharp, metallic clash; snorting, panting, quivering, trembling, the pair stood passive and vanquished.

The carriage was overthrown; but the high and fearless courage of the Peeress bore her unharmed, even as she was flung out on to the yielding fern-grown turf; fair as she was in every hour, she had never looked fairer than as he swung himself from the now powerless horses and threw himself beside her.

'My love—my love, you are saved!'

The beautiful eyes looked up half unconscious; the danger told on her now that it was passed, as it does most commonly with women.

'Saved!—lost! All the world must know, now, that you are with me this evening,' she murmured with a shudder. She lived for the world, and her first thought was of self.

He soothed her tenderly.

'Hush—be at rest. There is no injury but what I can repair, nor is there a creature in sight to have witnessed the accident. Trust in me, no one shall ever know of this. You shall reach town safely and alone.'

And while he promised, he forgot that he thus pledged his honour to leave four hours of his life so buried that, however much he needed, he neither should nor could account for them.

CHAPTER IX

The Painted Bit

Baden was at its brightest. The Victoria, the Badischer Hof, the Stephanie Bauer, were crowded. The Kurliste had a dazzling string of names. Imperial grandeur sauntered in slippers, chiefs used to be saluted with 'Ave Cæsar Imperator', smoked a papelito in peace over *Galignani*. Emperors gave a good-day to ministers who made their thrones beds of thorns, and little kings elbowed great capitalists who could have bought them all up in a morning's work in the money market. Statecraft was in its slippers, and diplomacy in its dressing-gown. Statesmen who had just been outwitting each other at the hazard of European politics, laughed good humouredly as they laid their gold down on the colour. Rivals, who had lately been quarrelling over the knotty points of national frontiers, now only vied for a twenty-franc rosebud from the bouquetière. Knights of the Garter and Knights of the Golden Fleece, who had hated each other to deadliest rancour with the length of the Continent between them, got friends over a mutually good book on the Rastadt or Forêt Noir. Brains that were the powder depot of one half of the universe, let themselves be lulled with the monotone of '*Faites votre jeu!*' or fanned to tranquil amusement by a fair idiot's coquetry. And lips that, with a whisper, could loosen the coursing slips of the wild hell-dogs of war, murmured love to a Princess, led the laugh at a supper at five in the morning, or smiled over their own caricatures done by Tenniel or Cham.

Baden was full. The supreme empires of demi-monde sent their sovereigns diamond-crowned and resistless to outshine all other principalities and powers, while in breadth of marvellous skirts, in costliness of cobweb laces, in unapproachability of Indian shawls, and gold embroideries, and mad fantasies, and Cleopatra extravagances, and jewels fit for a Maharajah, the Zu-Zu was distanced by none.

Among the kings and heroes and celebrities who gathered under the pleasant shadow of the pine-crowned hills, there was not one in his way greater than the steeplechaser, Forest King—certes, there was not one half so honest.

The Guards' crack was entered for the Prix de Dames, the sole representative of England. There were two or three good things out of French stables, specially a killing little bay, L'Etoile, and there was an Irish sorrel, the property of an Austrian of rank, of which fair things were whispered; but it was scarcely possible that anything could stand against the King, and that wonderful stride of his which spread-eagled his field like magic, and his countrymen were well content to leave their honour and their old renown to 'Beauty' and his six-year-old.

Beauty himself, with a characteristic philosophy, had a sort of conviction that the German race would set everything square. He stood either to make a very good thing on it, or to be very heavily hit. There could be no medium. He never hedged in his life; and as it was almost a practical impossibility that anything the foreign stables could get together would even be able to land within half a dozen lengths of the King, Cecil, always willing to console himself, and invariably too careless to take the chance of adverse accidents into account, had come to Baden, and was amusing himself there dropping a Friedrich d'Or on the rouge, flirting in the shady alleys of the Lichtenthal, waltzing Lady Guenevere down the ballroom, playing écarté with some Serene Highness, supping with the Zu-Zu and her set, and occupying rooms that a Russian Prince had had before him, with all the serenity of a millionaire as far as memory of money went. With much more than the serenity in other matters of most millionaires, who, finding themselves uncommonly ill at ease in the pot-pourri of monarchs and ministers, of beau-monde and demi-monde, would have given half their newly-turned thousands to get rid of the odour of Capel Court and the Bourse, and to attain the calm, negligent assurance, the easy tranquil insolence, the nonchalance with Princes, and the supremacy amongst the Free Lances, which they saw and coveted in the indolent Guardsman.

Bertie amused himself. He might be within a day of his ruin, but that was no reason why he should not sip his iced sherbet and laugh with a pretty French actress tonight. His epicurean formulary was the same as old Herrick's, and he would have paraphrased this poet's famous quatrain into

> Drink a pure claret while you may,
>   Your 'stiff' is still a flying;
> And he who dines so well today
>   To-morrow may be lying,
> Pounced down upon by Jews *tout net*,
>   Or outlawed in a French *guinguette*!

Bertie was a great believer—if the words are not too sonorous and too earnest to be applied to his very inconsequent views upon any and everything—in the philosophy of happy accident. Far as it was in him to have a conviction at all, which was a thorough-going serious sort of thing not by any means his 'form', he had a conviction that the doctrine of 'Eat, drink, and enjoy, for tomorrow we die,' was an universal panacea. He was reckless to the uttermost stretch of recklessness, all serene and quiet though his pococurantism and his daily manner were; and while subdued to the undeviating monotone and languor of his peculiar set in all his temper and habits, the natural daredevil in him took out its inborn instincts in a wildly careless and gamester-like imprudence with that most touchy-tempered and inconsistent of all coquettes—Fortune.

Things, he thought, could not well be worse with him than they were now. So he piled all on one *coup*, and stood to be sunk or saved by the Prix de Dames. Meanwhile, all the same, he murmured Mussetism to the Guenevere under the ruins of the Alte Schloss, lost or won a rouleau at the roulette wheel, gave a bank-note to the famous Isabel for a tea-rose, drove the Zu-Zu four in hand to see the Flat races, took his guinea tickets for the Concerts, dined with Princes, lounged arm-in-arm with Grand Dukes, gave an Emperor a hint as to the best cigars, and charmed a Monarch by unfolding the secret of the aroma of a Guards' Punch, sacred to the Household.

'*Si on ne meurt pas de désespoir ou finit par manger des huitres,*' said the witty Frenchwoman; Bertie, who believed in bivalves but not in heroics, thought it best to take the oysters first, and eschew the despair entirely.

He had one unchangeable quality—insouciance; and he had, more-over, one unchangeable faith—the King. Lady Guenevere had reached home unnoticed after the accident of their moonlight stag-hunt. His brother meeting him a day or two after their interview had nodded affirmatively, though sulkily, in answer to his inquiries, and had murmured that it was 'all square now'. The Jews and the tradesmen had let him leave for Baden without more serious measures than a menace, more or less insolently worded. In the same fashion he trusted that the King's running at the Bad, with the moneys he had on it, would set all things right for a little while, when, if his family interest, which was great, would get him his step in the First Life, he thought, desperate as things were, they might come round again smoothly, without a notorious crash.

'You are sure the King will "stay", Bertie?' asked Lady Guenevere, who had some hundreds in gloves (and even under the rose 'sported a pony' or so more seriously) on the event.

'Certain! But if he don't, I promise you as pretty a tableau as your Asnières one; for your sake, I'll make the finish as picturesque as possible; wouldn't it be well to give me a lock of hair in readiness?'

Her ladyship laughed, and shook her head; if a man killed himself she did not desire that her gracious name should be entangled with the folly.

'No, I don't do those things,' she said, with captivating waywardness. 'Besides, though the Oos looks cool and pleasant, I greatly doubt that under any pressure you would trouble it; suicides are too pronounced for your style, Bertie.'

'At all events, a little morphia in one's own rooms would be quieter, and better taste,' said Cecil, while he caught himself listlessly wondering, as he had wondered at Richmond, if this badinage were to turn into serious fact—how much would she care?

'May your sins be forgiven you!' cried Chesterfield, the apostle of training, as he and the Seraph came up to the table where Cecil and Cos Wentworth were breakfasting in the garden of the Stephanien on the race day itself. 'Liqueurs, truffles, and every devilment under the sun?—cold beef, and nothing to drink, Beauty, if you've any conscience left!'

'Never had a grain, dear boy, since I can remember,' murmured Bertie, apologetically. '*You* took all the rawness off me at Eton.'

'And you've been taking coffee in bed, I'll swear?' pursued the cross-examiner.

'What if he have? Beauty's condition can't be upset by a little mocha, nor mine either,' said his universal defender; and the Seraph shook his splendid limbs with a very pardonable vanity.

'Ruteroth trains; Ruteroth trains awfully,' put in Cos Wentworth, looking up out of a great silver flagon of Badminton, with which he was ending his breakfast; and, referring to the Austrian who was to ride the Paris favourite, 'Remember him at La Marche last year, and the racing at Vincennes—didn't take a thing that could make flesh—muscles like iron, you know—never touched a soda even——'

'I've trained too,' said Bertie, submissively; 'look how I've been waltzing! There isn't harder work than that for any fellow. A deuxtemps with the Duchess takes it out of you like any spin over the flat.'

His censurers laughed, but did not give in their point.

'You've run shocking risks, Beauty,' said Chesterfield; 'the King's in fine running-form, don't say he isn't; but you've said scores of times what a deal of riding he takes. Now, can you tell us yourself that you're in as hard condition as you were when you won the Military, eh?'

Cecil shook his head with a sigh:

'I don't think I am; I've had things to try me, you see. There was that Verschoyle's proposal. I did absolutely think at one time she'd marry me before I could protest against it! Then there was that shock to one's whole nervous system, when that indigo man, who took Lady Laura's house, asked *us* to dinner, and actually thought we should go!—and there was a scene, you know, of all earthly horrors, when Mrs Gervase was so near eloping with me, and Gervase cut up rough, instead of pitying me;

and then the field-days were so many, and so late into the season; and I exhausted myself so at the Belvoir theatricals at Easter; and I toiled so atrociously playing *Almaviva* at your place, Seraph—a private opera's galley-slave's work!—and, altogether, I've had a good many things to pull me down since the winter,' concluded Bertie, with a plaintive self-condolence over his truffles.

The rest of his condemning judges laughed, and passed the plea out of sympathy; the Coldstreamer alone remained censorious and untouched.

'Pull you *down!* You'll never pull *off* the race if you sit drinking liquors all the morning,' growled that censor. 'Look at that!'

Bertie glanced at the London telegram tossed across to him, sent from a private and confidential agent.

'Betting here—2 to 1 on L'Etoile; Irish Roan offered and taken freely. Slight decline in closing prices for the King; getting on French bay rather heavily at midnight. Fancy there's a commission out against the King. Looks suspicious.' Cecil shrugged his shoulders and raised his eyebrows a little.

'All the better for us. Take all they'll lay against me. It's as good as *our* having a "Commission out"; and if any cads get one against us it can't mean mischief, as it would with professional jocks.'

'*Are* you so sure of yourself, Beauty?'

Beauty shook his head repudiatingly.

'Never am sure of anything, much less of myself; I'm a chameleon, a perfect chameleon!'

'Are you so sure of the King, then?'

'My dear fellow, no! I ask you in reason, how can I be sure of what isn't proved? I like that country fellow the old story tells of, he believed in fifteen shillings because he'd once had it in his hand; others, he'd heard, believed in a pound; but, for his part, he didn't, because he'd never seen it. Now that was a man who'd never commit himself; he might have had the Exchequer! I'm the same; I believe, the King can win at a good many things because I've seen him do 'em; but I can't possibly tell whether he can get this, because I've never ridden him for it. I shall be able to tell you at three o'clock—but that you don't care for——'

And Bertie, exhausted with making such a lengthened exposition—the speeches he preferred were monosyllabic—completed his sins against training with a long draught of claret-cup.

'Then, what the devil do you mean by telling us to pile our pots on you?' asked the outraged Coldstreamer with natural wrath.

'Faith is a beautiful sight!' said Bertie, with solemnity. 'If *I'm* bowled over, *you'll* be none the less sublime instances of heroic devotion——'

'Offered on the altar of the Jews!' laughed the Seraph, as he turned him away from the breakfast-table by the shoulders. 'Thanks, Beauty; I've "four figures" on you, and you'll be good enough to win them for me. Let's have a look at the King. They are just going to walk him over.'

Cecil complied; while he lounged away with the others to the stables, with a face of the most calm, gentle, weary indifference in the world, the thought crossed him for a second of how *very* near he was to the wind. The figures in his betting-book were to the tune of several thousands, one way or another. If he won this morning it would be all right, of course: if he lost—even Beauty, odd mixture of devil-may-care and languor though he was, felt his lips grow, for the moment, hot and cold by turns as he thought of that possible contingency.

The King looked in splendid condition; he knew well enough what was up again, knew what was meant by that extra sedulous dressing-down, that setting muzzle that had been buckled on him some nights previous, the limitation put to his drink, the careful trial spins in the grey of the mornings, the conclusive examination of his plates by a skilful hand; he knew what was required of him, and a horse in nobler condition never stepped out in body-clothing, as he was ridden slowly down on to the plains of Iffesheim. The Austrian Dragoon, a Count and a Chamberlain likewise, who was to ride his only possible rival, the French horse L'Etoile, pulled his tawny silken moustaches, as he saw the great English hero come up the course, and muttered to himself, '*L'affaire est finie.*' L'Etoile was a brilliant enough bay in his fashion, but Count Ruteroth knew the measure of his pace and powers too thoroughly to expect him to live against the strides of the Guards' grey.

'My beauty, won't you cut those German fellows down!' muttered Rake, the enthusiast, in the saddling enclosure. 'As for those fools what go agin you, you'll put *them* in a hole, and no mistake. French horse, indeed! Why, you'll spread-eagle all them Mossoos' and Meinherrs' cattle in a brace of seconds——'

Rake's foe, the head groom, caught him up savagely,

'Won't you never learn decent breedin'? When *we* wins we wins on the quiet, and when *we* loses we loses as if we liked it; all that brayin', and flauntin', and boastin', is only fit for cads. The 'oss is in tip-top condition; let him show what he can do over furren ground.'

'Lucky for him, then, that he hasn't got *you* across the pigskin; you'd rope him, I believe, as soon as look at him, if it was made worth your while,' retorted Rake, in caustic wrath; his science of repartee chiefly lay in a successful 'plant', and he was here uncomfortably conscious that his opponent was in the right of the argument, as he started through the throng to put his master into the 'shell' of the Shire-famous scarlet and white.

'Tip-top condition, my boy—tip-top, and no mistake,' murmured Mr Willon, for the edification of those around them as the saddle-girths were buckled on, and the Guards' crack stood the cynosure of every eye at Iffesheim.

Then, in his capacity as head attendant on the hero, he directed the exercise-bridle to be taken off, and with his own hands adjusted a new and handsome one, slung across his arm.

' 'Tis a'most a pity. 'Tis a'most a pity,' thought the worthy, as he put the curb on the King; 'but I shouldn't have been haggravated with that hinsolent soldiering chap. There, my boy, if you'll win with a painted quid, I'm a Dutchman.'

Forest King champed his bit between his teeth a little; it tasted bitter; he tossed his head and licked it with his tongue impatiently; the taste had got down his throat and he did not like its flavour: he turned his deep, lustrous eyes with a gentle patience on the crowd about him as though asking them what was the matter with him. No one moved his bit; the only person who could have had such authority was busily giving the last polish to his coat with a fine handkerchief— that glossy neck which had been so dusted many a time with the cobweb coronet-broidered handkerchiefs of great ladies—and his instincts, glorious as they were, were not wise enough to tell him to kick his head groom down, then and there, with one mortal blow, as his poisoner and betrayer.

The King chafed under the taste of that 'painted quid'; he felt a nausea as he swallowed, and he turned his handsome head with a strange, pathetic astonishment in his glance: at that moment a familiar hand stroked his mane, a familiar foot was put into his stirrup, Bertie threw himself into saddle, the lightest weight that ever gentleman-rider rode, despite his six-foot length of limb. The King, at the well-known touch, the well-loved voice, pricked his delicate ears, quivered in all his frame with eager excitation, snuffed the air restlessly through his distended nostrils, and felt every vein under his satin skin thrill and swell with pleasure; he was all impatience, all power, all longing, vivid, intensity of life. If only that nausea would go! He felt a restless sickliness stealing on him that his young and gallant strength had never known since he was foaled. But it was not in the King to yield to a little; he flung his head up, champing angrily at the bit, then walked down to the starting-post with his old calm, collected grace; and Cecil, looking at the glossy bow of the neck, and feeling the width of the magnificent ribs beneath him, stooped from his saddle a second as he rode out of the enclosure and bent to the Seraph.

'Look at him, Rock! the thing's as good as won.'

The day was very warm and brilliant; all Baden had come down to the race-course, continuous strings of carriages, with their four or six horses and postilions, held the line far down over the plains; mob, there was none, save of women in matchless toilettes, and men with the highest names in the 'Almanac de Gotha': the sun shone cloudlessly on the broad, green plateau of Iffesheim, on the white amphitheatre of chalk hills, and on the glittering, silken folds of the flags of England, France, Prussia, and of the Grand Duchy itself, that floated from the summits of the Grand Stand, Pavilion, and Jockey Club.

The ladies, descending from the carriages, swept up and down on the green course that was so free from 'cads' and 'legs', their magnificent skirts trailing along without the risk of a grain of dust, their costly laces side by side with the Austrian uniforms of the military men from Rastadt. The betting was but slight; the Paris formulas, 'Combien contre l'Etoile?' 'Six cents francs sur le cheval Anglais?' echoing everywhere in odd contrast with the hubbub and striking clamour of English betting rings; the only approach to anything like 'real business', being transacted between the members of the Household and those of the Jockey Clubs. Iffesheim was pure pleasure, like every other item of Baden existence, and all aristocratic, sparkling, rich, amusement-seeking Europe seemed gathered there under the sunny skies, and on everyone's lips in the titled throng was but one name—Forest King's. Even the coquettish bouquet-sellers, who remembered the dresses of his own colours which Cecil had given them last year when he had won the Rastadt, would sell nothing except little twin scarlet and white moss rosebuds, of which thousands were gathered and died that morning in honour of the English Guards' champion.

A slender event usually, the presence of the renowned crack of the Household Cavalry made the Prix de Dames the most eagerly watched for entry on the card, and the rest of the field were scarcely noticed as the well-known gold-broidered jacket came up at the starting-post.

The King saw that blaze of light and colour over course and stands that he knew so well by the time; he felt the pressure round him of his foreign rivals, as they reared, and pulled, and fretted, and passaged; the old longing quivered in all his eager limbs, the old fire wakened in all his dauntless blood; like the charger at sound of the trumpet-call, he lived in his past victories, and was athirst for more. But yet—between him and the sunny morning there seemed a dim, hazy screen; on his delicate ear the familiar clangour smote with something dulled and strange; there seemed a numbness stealing down his frame, he shook his head in an unusual and irritated impatience, he did not know what ailed him. The hand he loved so loyally told him the work that was wanted of him, but he felt its guidance dully too, and the dry, hard, hot earth, as he struck it with his hoof, seemed to sway and heave beneath him; the opiate had stolen into his veins, and was creeping stealthily and surely to the sagacious brain, and over the clear, bright senses.

The signal for the start was given; the first mad headlong rush broke away with the force of a pent-up torrent suddenly loosened; every instinct of race and custom, and of that obedience which rendered him flexible as silk to his rider's will, sent him forward with that stride which made the Guards' crack a household word in all the Shires. For a moment he shook himself clear of all his horses, and led off in the old grand sweeping canter before the French bay three lengths in the one single effort.

Then into his eyes a terrible look of anguish came; the numb and sickly nausea was upon him, his legs trembled, before his sight was a blurred whirling mist; all the strength and force and mighty life within him felt ebbing out, yet he struggled bravely. He strained, he panted, he heard the thundering thud of the first flight gaining nearer and nearer upon him, he felt his rivals closing hotter and harder in on him, he felt the steam of his opponent's smoking foam-dashed withers burn on his own flanks and shoulders, he felt the maddening pressure of a neck to neck struggle, he felt what in all his victorious life he had never known—the paralysis of *defeat*.

The glittering throngs spreading over the plains gazed at him in the sheer stupor of amazement; they saw that the famous English hero was dead beat as any used-up knacker.

One second more he strove to wrench himself through the throng of his horses, through the headlong crushing press, through—worst foe of all!— the misty darkness curtaining his sight; one second more he tried to wrestle back the old life into his limbs, the unworn power and freshness into nerve and sinew. Then the darkness fell utterly; the mighty heart failed; he could do no more;—and his rider's hand slackened and turned him gently backward, his rider's voice sounded very low and quiet to those who, seeing that every effort was hopeless, surged and clustered round his saddle.

'Something ails the King,' said Cecil, calmly; 'he is fairly knocked off his legs. Some Vet must look to him; ridden a yard further he will fall.'

Words so gently spoken!—yet in the single minute that alone had passed since they had left the Starter's Chair, a lifetime seemed to have been centred alike to Forest King and to his owner.

The field swept on with a rush without the favourite; and the Prix de Dames was won by the French bay, L'Etoile.

CHAPTER X

'Petite Reine'

When a young Prussian had shot himself the night before for Roulette losses, the event had not thrilled, startled, and impressed the gay Baden gathering one tithe so gravely and so enduringly as did now the unaccountable failure of the great Guards' crack.

Men could make nothing of it save the fact that there was 'something dark' somewhere. The 'painted quid' had done its work more thoroughly than Willon and the Welsher had intended; they had meant that the opiate should be just sufficient to make the favourite off his speed, but not to take effects so palpable as these. It was, however, so deftly prepared, that under examination no trace could be found of it, and the results of veterinary investigation, while it left unremoved the conviction that the horse had been doctored, could not explain when or how, or by what medicines. Forest King had simply 'broken down'; favourites do this on the flat and over the furrow from an over-strain, from a railway journey, from a touch of cold, from a sudden decay of power, from spasm, or from vertigo; those who lose by them may think what they will of 'roping', or 'painting', or 'nobbling', but what can they prove?

Even in the great scandals that come before the autocrats of the Jockey Club, where the tampering is clearly known, can the matter ever be really proved and sifted? Very rarely: the trainer affects stolid unconsciousness, or unimpeachable respectability; the hapless stable-boy is cross-examined to protest innocence and ignorance, and most likely protest them rightly; he is accused, dismissed, and ruined, or some young jock has a 'caution' out everywhere against him, and never again can get a mount even for the commonest handicap; but as a rule the real criminals are never unearthed, and by consequence are never reached and punished.

The Household, present and absent, were heavily hit; they cared little for the 'crushers' they incurred, but their champion's failure when he was in the face of Europe cut them down more terribly. The fame of the English riding-men had been trusted to Forest King and his owner, and they who had never before betrayed the trust placed in them had broken down like any screw out of a livery-stable, like any jockey bribed to 'pull' at a suburban selling-race. It was fearfully bitter work, and unanimous to a voice the indignant murmur of 'doctored' ran through the titled fashionable crowds on the Baden course in deep and ominous anger.

The Seraph's grand wrath poured out fulminations against the wicked-doer whosoever he was, or wheresoever he lurked; and threatened with a vengeance that would be no empty words, the direst chastisement of the 'Club', of which both his father and himself were stewards, upon the unknown criminal. The Austrian and French nobles, while winners by the event, were scarce in less angered excitement; it seemed to cast the foulest slur upon their honour, that upon foreign ground the renowned English steeplechaser should have been tampered with thus; and the fair ladies of either world added the influence of their silver tongues, and were eloquent in the vivacity of their sympathy and resentment with an unanimity women rarely show in savouring defeat, but usually reserve for the fairer opportunity of swaying the censer before success.

Cecil alone, amidst it all, was very quiet; he said scarcely a word, nor could the sharpest watcher have detected an alteration in his counten-ance. Only once, when they talked around him of the investigations of the Club, and of the institution of inquiries to discover the guilty traitor, he looked up with a sudden, dangerous lighting of his soft, dark, hazel eyes, under the womanish length of their lashes: 'When you find him, leave him to me.'

The light was gone again in an instant, but to those who knew the wild strain that ran in the Royallieu blood, knew by it that, despite his gentle temper, a terrible reckoning for the evil done his horse might come some day from the Quietist.

He said little or nothing else, and to the sympathy and indignation expressed for him on all sides he answered with his old, listless calm. But in truth he barely knew what was saying or doing about him; he felt like a man stunned and crushed with the violence of some tremendous fall; the excitation, the agitation, the angry amazement around him (growing as near clamour and tumult as was possible in those fashionable betting-circles, so free from roughs and almost free from bookmakers), the con-flicting opinions clashing here and there, even, indeed, the graceful condolence of the brilliant women, were insupportable to him. He longed to be out of this world which had so well amused him; he longed passionately for the first time in his life to be alone.

For he knew that with the failure of Forest King had gone the last plank that saved him from ruin; perhaps the last chance that stood between him and dishonour. He had never looked on it as within the possibilities of hazard that the horse could be defeated; now, little as those about him knew it, an absolute and irremediable disgrace fronted him. For, secure in the issue of the Prix de Dames, and compelled to weight his chances in it very heavily that his winnings might be wide enough to relieve some of the debt-pressure upon him, his losses now were great, and he knew no more how to raise the moneys to meet them than he would have known how to raise the dead.

The blow fell with crashing force; the fiercer, because his indolence had persisted in ignoring his danger, and because his whole character was so naturally careless, and so habituated to ease and to enjoyment.

A bitter, heartsick misery fell on him; the tone of honour was high with him; he might be reckless of everything else, but he could never be reckless in what infringed, or went nigh to infringe, a very stringent code. Bertie never reasoned in that way; he simply followed the instincts of his breeding without analysing them; but these led him safely and surely right in all his dealings with his fellow-men, however open to censure his life might be in other matters. Careless as he was, and indifferent, to levity, in many things, his ideas of honour were really very pure and elevated; he suffered proportionately now, that through the follies of his own imprudence, and the baseness of some treachery he could neither sift nor avenge, he saw himself driven down into as close a jeopardy of disgrace as ever befell a man who did not wilfully, and out of guilty coveting of its fruits, seek it.

For the first time in his life the society of his troops of acquaintance became intolerably oppressive; for the first time in his life he sought refuge from thought in the stimulus of drink, and dashed down neat cognac as though it were iced Badminton, as he drove with his set off the disastrous plains of Iffesheim. He shook himself free of them as soon as he could; he felt the chatter round him insupportable; the men were thoroughly good-hearted, and though they were sharply hit by the day's issue, never even by implication hinted at owing the disaster to their faith in him, but the very cordiality and sympathy they showed cut him the keenest—the very knowledge of their forbearance made his own thoughts darkest.

Far worse to Cecil than the personal destruction the day's calamity brought him was the knowledge of the entire faith these men had placed in him, and the losses to which his own mistaken security had caused them. Granted he could neither guess nor avert the trickery which had brought about his failure; but none the less did he feel that he had failed them; none the less did the very generosity and magnanimity they showed him sting him like a scourge.

He got away from them at last, and wandered out alone into the gardens of the Stephanien, till the green trees of an alley shut him in in solitude, and the only echo of the gay world of Baden was the strain of a band, the light mirth of a laugh, or the roll of a carriage sounding down the summer air.

It was eight o'clock; the sun was slanting to the west in cloudless splendour, bathing the bright scene in a rich golden glow, and tinging to bronze the dark masses of the Black Forest. In another hour he was the expected guest of a Russian Prince at a dinner-party, where all that was highest, fairest, greatest, most powerful, and most bewitching of every nationality represented there would meet; and in the midst of this radiant whirlpool of extravagance and pleasure, where every man worth owning as such was his friend, and every woman whose smile he cared for welcomed him, he knew himself as utterly alone, as utterly doomed, as the lifeless Prussian lying in the dead-house. No aid could serve him, for it would have been but to sink lower yet to ask or to take it; no power could save him from the ruin which in a few days later at the furthest would mark him out for ever an exiled, beggared, perhaps dishonoured, man—a debtor and an alien.

Where he had thrown himself on a bench beneath a mountain-ash, trying vainly to realize this thing which had come upon him, and to meet which not training, nor habit, nor a moment's grave reflection had ever done the slightest to prepare him, gazing blankly and unconsciously at the dense pine woods and rugged glens of the Forest that sloped upward and around above the green and leafy nest of Baden, he watched mechanically the toiling passage of a charcoal-burner going up the hillside in the distance through the firs.

'Those poor devils envy *us*!' he thought. 'Better be one of them ten thousand times than be trained for the Great Race, and started with the cracks, dead weighted with the penalty-shot of Poverty!'

A soft touch came on his arm as he sat there; he looked up, surprised: before him stood a dainty, delicate, little form, all gay with white lace, and broideries, and rose ribands, and floating hair fastened backward with a golden fillet; it was that of the little Lady Venetia, the only daughter of the House of Lyonnesse, by a late marriage of his Grace, the eight-year-old sister of the colossal Seraph; the plaything of a young and lovely mother, who had flirted in Belgravia with her future stepson before she fell sincerely and veritably in love with the gallant and still handsome Duke.

Cecil roused himself and smiled at her; he had been by months together at Lyonnesse most years of the child's life, and had been gentle to her as he was to every living thing, though he had noticed her seldom.

'Well, Petite Reine,' he said, kindly bitter as his thoughts were, calling her by the name she generally bore, 'all alone; where are your playmates?'

'Petite Reine', who, to justify her sobriquet, was a grand, imperial, little lady, bent her delicate head—a very delicate head, indeed, carrying itself royally, young though it was.

'Ah! you know I never care for children!'

It was said so disdainfully, yet so sincerely, without a touch of affectation, and so genuinely, as the expression of a matured and contemptuous opinion, that even in that moment it amused him. She did not wait an answer, but bent nearer with an infinite pity and anxiety in her pretty eyes.

'I want to know——; you are so vexed, are you not? They say you have lost all your money!'

'Do they? They are not far wrong, then. Who are "they", Petite Reine?'

'Oh, Prince Alexis, and the Duc de Lorance, and mamma, and everybody. Is it true?'

'Very true, my little lady.'

'Ah!' She gave a long sigh, looking pathetically at him, with her head no one side, and her lips parted. 'I heard the Russian gentlemen saying that you were ruined. Is *that* true, too?'

'Yes, dear,' he answered, wearily, thinking little of the child in the desperate excess to which his life had come.

Petite Reine stood by him, silent; her proud, imperial young ladyship had a very tender heart, and she was very sorry; she had understood what had been said before her of him vaguely indeed, and with no sense of its true meaning, yet still, with the quick perception of a brilliant and petted child. Looking at her, he saw with astonishment that her eyes were filled with tears; he put out his hand and drew her to him.

'Why, little one, what do you know of these things? How did you find me out here?'

She bent nearer to him, swaying her slender figure, with its bright gossamer muslins, like a dainty harebell, and lifting her face to his, earnest, beseeching, and very eager.

'I came—I came—*please* don't be angry—because I heard them say you had no money, and I want you to take mine. Do take it! Look, it is all bright gold, and it is my own, my *very* own. Papa gives it to me to do just what I like with. Do take it; pray do!'

Colouring deeply, for the Petite Reine had that true instinct of generous natures, a most sensitive delicacy for others, but growing ardent in her eloquence and imploring in her entreaty, she shook on to Cecil's knee, out of a little enamel sweetmeat-box, twenty bright Napoléons, that fell in a glittering shower on the grass.

He started, and looked at her in a silence that she mistook for offence. She leaned nearer, pale now with her excitement, and with her large eyes gleaming and melting with passionate entreaty.

'Don't be angry; pray take it; it is all my own, and you know I have bonbons, and books, and playthings, and ponies, and dogs, till I am tired of them. I never want the money, indeed I don't. Take it, please take it; and if you will only let me ask papa or Rock, they will give you thousands and thousands of pounds if that isn't enough. Do let me?'

Cecil, in silence still, stooped and drew her to him; when he spoke, his voice shook ever so slightly, and he felt his eyes dim with an emotion that he had not known in all his careless life; the child's words and action touched him deeply, the caressing generous innocence of the offered gift beside the enormous extravagance and hopeless bankruptcy of his career, smote him with a keen pang, yet moved him with a strange pleasure.

'Petite Reine,' he murmured gently, striving vainly for his old lightness—'Petite Reine, how some man will love you one day! Thank you from my heart, my little innocent friend.'

Her face flushed with gladness; she smiled with all a child's unshadowed joy.

'Ah! then you *will* take it? And if you want more, only let me ask them for it; papa and Philip never refuse me anything!'

His hand wandered gently over the shower of her hair, as he put back the Napoléons that he had gathered up into her azure bonbonnière.

'Petite Reine, you are a little angel; but I cannot take your money, my child, and you must ask for none for my sake from your father or from Rock. Do not look so grieved, little one; I love you none the less because I refuse it.'

Petite Reine's face was very pale and grave; a delicate face, in its miniature feminine childhood almost absurdly like the Seraph's; her eyes were full of plaintive wonder and of pathetic reproach.

'Ah,' she said, drooping her head with a sigh, 'it is no good to you because it is such a little. Do let me ask for more?'

He smiled, but the smile was very weary.

'No, dear, you must not ask for more; I have been very foolish, my little friend, and I must take the fruits of my folly; all men must. I can accept no one's money, not even yours; when you are older and remember this, you will know why; but I do not thank you the less from my heart.'

She looked at him, pained and wistful.

'You will not take *anything*, Mr Cecil?' she asked with a sigh, glancing at her rejected Napoléons.

He drew the enamel bonbonnière away.

'I will take that, if you will give it me, Petite Reine, and keep it in memory of you.'

As he spoke, he stooped and kissed her very gently; the act had moved him more deeply than he thought he had it in him to be moved by anything, and the child's face turned upwards to him was of a very perfect and aristocratic loveliness far beyond her years. She coloured as his lips touched hers, and swayed slightly from him. She was an extremely proud young sovereign, and never allowed caresses; yet she lingered by him troubled, grave, with something intensely tender and pitiful in the musing look of her eyes. She had a perception that this calamity which smote him was one far beyond the ministering of her knowledge.

He took the pretty Palais Royal gold-rimmed sweetmeat-box, and slipped it into his waistcoat-pocket. It was only a child's gift, a tiny Paris toy, but it had been brought to him in a tender compassion, and he did keep it—kept it through dark days and wild nights, through the scorch of the desert and the shadows of death, till the young eyes that questioned him now with such innocent wonder had gained the grander lustre of their womanhood, and had brought him a grief wider than he knew now.

At that moment, as the child stood beside him under the drooping acacia boughs, with the green sloping lower valley seen at glimpses through the wall of leaves, one of the men of the Stephanien approached him with an English letter, which, as it was marked 'Instant', they had laid apart from the rest of the visitors' pile of correspondence; Cecil took it wearily—nothing but fresh embarrassments could come to him from England—and looked at the little Lady Venetia.

'You will allow me?'

She bowed her graceful head: with all the naïf unconsciousness of a child she had all the manner of the *vielle cour*; together they made her enchanting.

He broke the envelope and read; a blurred, scrawled, miserable letter, the words erased with passionate strokes, and blotted with hot tears, and scored out in impulsive misery. It was long, yet at a glance he scanned its message and its meaning; at the first few words he knew its whole as well as though he had studied every line.

A strong tremor shook him from head to foot, a tremor at once of passionate rage and of as passionate pain; his face blanched to a deadly whiteness; his teeth clenched as though he were restraining some bodily suffering, and he tore the letter in two and stamped it down into the turf under his heel, with a gesture as unlike his common serenity of manner as the fiery passion that darkened in his eyes was unlike the habitual softness of his too pliant and too unresentful temper. He crushed the senseless paper again and again down into the grass beneath his heel, his lips shook under the silky abundance of his beard; the natural habit of long usage kept him from all utterance, and even in the violence of its shock he remembered the young Venetia's presence; but, in that one fierce unrestrained gesture, the shame and suffering upon him broke out despite himself.

The child watched him, startled and awed. She touched his hand softly.

'What is it? Is it anything worse?'

He turned his eyes on her with a dry, hot, weary anguish in them; he was scarcely conscious what he said or what he answered.

'Worse—worse?' he repeated, mechanically, while his heel still ground down in loathing the shattered paper into the grass. 'There can be nothing worse!—it is the vilest, blackest *shame.*'

He spoke to his thoughts, not to her. The words died in his throat; a bitter agony was on him; all the golden summer evening, all the fair green world about him, were indistinct and unreal to his senses; he felt as if the whole earth were of a sudden changed. He could not realize that this thing could come to him and his—that this foul dishonour could creep up and stain them—that this infamy could ever be of them and upon them. All the ruin that before had fallen on him today was dwarfed and banished; it looked nothing beside the unendurable horror that reached him now.

The gay laughter of children sounded down the air at that moment; they were the children of a French Princess seeking their playmate, Venetia, who had escaped from them and from their games to find her way to Cecil. He motioned her to them; he could not bear even the clear and pitying eyes of the Petite Reine to be upon him now.

She lingered wistfully; she did not like to leave him.

'Let me stay with you,' she pleaded, caressingly. 'You are vexed at something; I cannot help you, but Rock will—the Duke will. Do let me ask them?'

He laid his hand on her shoulder; his voice, as he answered, was hoarse and unsteady.

'No; go, dear. You will please me best by leaving me. Ask none—tell none; I can trust you to be silent, Petite Reine?'

She gave him a long earnest look.

'Yes,' she answered, simply and gravely, as one who accepts, and not lightly, a trust.

Then she went slowly and lingeringly, with the sun on the gold fillet binding her hair, but the tears heavy on the shadow of her silken lashes. When next they met, the lustre of a warmer sun that once burned on the white walls of the palaces of Phœnicia and the leaping flames of the Temple of the God of Healing, shone upon them, and through the veil of those sweeping lashes there gazed the resistless sovereignty of a proud and patrician womanhood.

Alone, his head sank down upon his hands, he gave reins to the fiery scorn, the acute suffering, which turn by turn seized him with every moment that seared the words of the letter deeper and deeper down into his brain. Until this, he had never known what it was to suffer; until this, his languid creeds had held that no wise man feels strongly, and that to glide through life untroubled and unmoved is as possible as it is politic. Now he suffered—he suffered dumbly as a dog, passionately as a barbarian; now he was met by that which, in the moment of its dealing, pierced his panoplies of indifference, and escaped his light philosophies.

'Oh God!' he thought, 'if it were anything—anything—except Disgrace!'

In a miserable den, an hour or so before—there are miserable dens even in Baden, that gold-decked rendezvous of Princes, where crowned heads are numberless as couriers, and great ministers must sometimes be content with a shakedown—two men sat in consultation. Though the chamber was poor and dark, their table was loaded with various expensive wines and liqueurs; of a truth they were flush of money, and selected this poor place from motives of concealment rather than of necessity. One of them was the 'Welsher', Ben Davis; the other, a smaller, quieter man, with a keen vivacious Hebrew eye and an olive-tinted skin, a Jew, Ezra Baroni. The Jew was cool, sharp, and generally silent; the 'Welsher', heated, eager, flushed with triumph and glowing with a gloating malignity. Excitement and the fire of very strong wines of whose vintage brandy formed a large part, had made him voluble in exultation; the monosyllabic sententiousness that had characterised him in the loose-box at Royallieu had been dissipated under the ardour of success; and Ben Davis, with his legs on the table, a pipe between his teeth, and his bloated face purple with a brutal contentment, might have furnished to a Teniers the personification of culminated cunning and of delighted tyranny.

'That precious Guards' swell!' he muttered, gloatingly, for the hundredth time. 'I've paid him out at last! *He* won't take a "Walk Over" again in a hurry. Cuss them swells! they allays die so game; it ain't half a go after all, giving 'em a facer; they just come up to time so cool under it all, and never show they're down, even when their backers throw up the sponge. You can't make 'em *give in* not even when they're mortal hit; that's the crusher of it.'

'Vell—vhat matter that ven you *have* hit 'em?' expostulated the more philosophic Jew.

'Why—it is a fleecin' of one,' retorted the Welsher, savagely even amidst his successes. 'A clear fleecin' of one. If one gets the better of a dandy chap like that, and brings him down neat and clean, one ought to have the spice of it. One ought to see him wince and——cuss 'em all!—that's just what they'll never do. No! not if it was ever so. You may pitch into 'em like Old Harry, and those d——d fine gentlemen 'll just look as if they liked it. You might strike 'em dead at your feet, and it's my belief, while they was cold as stones, they'd manage to look *not beaten yet*. It's a fleecin' of one—a fleecin' of one!' he growled afresh, draining down a great draught of brandy-heated Roussillon to drown the impatient conviction which possessed him that let him triumph as he would there would ever remain, in that fine intangible sense which his coarse nature could feel, though he could not have further defined it, a superiority in his adversary he could not conquer, a difference between him and his prey he could not bridge over.

The Jew laughed a little.

'Vot a shild you are, you Big Ben! Vot matter how he look so long as you have de success and pocket de monish?'

Big Ben gave a long growl like a mastiff tearing to reach a bone just held above him.

'Hang the blunt! The yellows ain't a quarter worth to me what it 'ud be to see him just *look* as if he knew he was knocked over. Besides, laying agin' him by that ere commission's piled up hatsful of the ready to be sure, I don't say it hain't, but there's two thou' knocked off for Willon, and the fool don't deserve a tizzy of it; he went and put the paint on so thick, that if the Club don't have a flare-up about the whole thing——'

'Let dem!' said the Jew, serenely. 'Dey can do vot dey like; dey von't get to de bottom of de vell. Dat Villon is sharp, he vill know how to keep his tongue still; dey can prove nothin'; dey may give de sack to a stable-boy, or dey may tink demselves mighty bright in seeing a mare's nest, but dey vill never come to *us*.'

The Welsher gave a loud horse-guffaw of relish and enjoyment.

'No! We know the ins and outs of Turf Law a trifle too well to be caught napping. A neater thing weren't ever done, if it hadn't been that the paint was put a trifle too thick. The 'oss should have just *run ill*, and not knocked over clean out o' time like that. However, there ain't no odds a cryin' over spilt milk. If the Club *do* come a inquiry, we'll show 'em a few tricks that'll puzzle 'em. But it's my belief they'll let it off on the quiet; there ain't a bit of evidence to show the 'oss was doctored, and the way he went stood quite as well for having been knocked off his feed and off his legs by the woyage and sich-like. And now, you go and put that swell to the grindstone for Act 2 of the comedy, will yer?'

Ezra Baroni smiled where he leant against the table looking over some papers.

'Dis is a delicate matter; don't you come putting your big paw in it; you'll spoil it all.'

Ben Davis growled afresh:

'No, I ain't a-goin'. You know as well as me I can't show in the thing. Hanged if I wouldn't a'most lief risk a lifer out at Botany Bay for the sake o' wringing my fine feathered bird myself, but I daresn't; if he was to see me in it all 'ud be up. You must do it. Get along; you look uncommon respectable. If your coat-tails was a little bit longer, you might right and away be took for a parson.'

The Jew laughed softly, the Welsher grimly, at the compliment they paid the Church. Baroni put up his papers into a neat Russia letter-book. Excellently dressed, without a touch of flashiness, he did look eminently respectable—and he lingered a moment——

'I say, dear shild—vat if de Marquis vant to buy off and hush up? Ten to von he vill, he care no more for monish than for dem macaroons, and he love his friend, dey say.'

Ben Davis took his legs off the table with a crash, and stood up, flushed, thirstily eager, almost aggressive in his peremptory excitement.

'Without wringing my dainty bird's neck? Not for a million paid out o'hand! Without crushing my fine gentleman down into powder? Not for all the blunt of every one o' the Rothschilds! Curse his woman's face! I've got to keep dark now, but when he's crushed, and smashed, and ruined, and pilloried, and druv' out of his fine world, and warned off of all his aristocratic racecourses, *then* I'll come in and take a look at him; then I'll see my brilliant gentleman a worn-out, broken-down swindler, a dyin' in a bagnio!'

The intense malignity, the brutal hungry lust for vengeance that inspired the words, lent their coarse vulgarity something that was for the moment almost tragical in its strength, almost horrible in its passion. Ezra Baroni looked at him quietly, then without another word went out—to a congenial task.

'Dat big shild is a fool,' mused the subtler and gentler Jew. 'Vengeance is but de breath of de vind, it blow for you one day, it blow against you de next: de only real good is monish.'

The Seraph had ridden back from Iffesheim to the Bad in company with some Austrian officers, and one or two of his own comrades. He had left the course late, staying to exhaust every possible means of enquiry as to the failure of Forest King, and to discuss with other members of the Newmarket and foreign jockey clubs the best methods—if method there were—of discovering what foul play had been on foot with the horse. That there was some, and very foul too, the testimony of men and angels would not have dissuaded the Seraph, and the event had left him most unusually grave and regretful.

The amount he had lost himself, in consequence, was of not the slightest moment to him, although he was extravagant enough to run almost to the end even of his own princely tether in money matters; but that 'Beauty' should be cut down was more vexatious to him than any evil accident that could have befallen himself, and he guessed pretty nearly the fatal influence the dead failure would have on his friend's position.

True, he had never heard Cecil breathe a syllable that hinted at embarrassment; but these things get known with tolerable accuracy about the town, and those who were acquainted, as most people in their set were, with the impoverished condition of the Royallieu Exchequer, however hidden it might be under an unabated magnificence of living, were well aware also that none of the old Viscount's sons could have any safe resources to guarantee them from as rapid a ruin as they liked to consummate. Indeed, it had of late been whispered that it was probable, despite the provisions of the entail, that all the green wealth and Norman beauty of Royallieu itself would come into the market. Hence the Seraph, the best-hearted and most generous-natured of men, was worried by an anxiety and a despondency which he would never have indulged most assuredly on his own account, as he rode away from Iffesheim after the defeat of his Corps' champion.

He was expected to dinner with one of the most lovely of foreign Ambassadresses, and was to go with her afterwards to the Vaudeville, at the pretty golden theatre, where a troupe from the Bouffes were playing;

but he felt anything but in the mood for even her bewitching and—in a marriageable sense—safe society, as he stopped his horse at his own hotel, the Badischer Hof.

As he swung himself out of saddle, a well-dressed, quiet, rather handsome little man drew near respectfully, lifting his hat—it was M. Baroni. The Seraph had never seen the man in his life that he knew of, but he was himself naturally frank, affable, courteous, and never given to hedging himself behind the pale of his high rank; provided you did not bore him, you might always get access to him easily enough—the Duke used to tell him, too easily.

Therefore, when Ezra Baroni deferentially approached 'the Most Noble the Marquis of Rockingham, I think?' the Seraph, instead of leaving the stranger there discomfited, nodded and paused with his inconsequent good nature, thinking how much less bosh it would be if everybody could call him like his family and his comrades, 'Rock'.

'That is my name,' he answered. 'I do not know you; do you want anything of me?'

The Seraph had a vivid terror of people who 'wanted him', in the subscription, not the police, sense of the word; and had been the victim of frauds innumerable.

'I wished,' returned Baroni, respectfully, but with sufficient independence to conciliate his auditor, whom he saw at a glance cringing subservience would disgust, 'to have the opportunity of asking your Lordship a very simple question.'

The Seraph looked a little bored, a little amused.

'Well, ask it, my good fellow; you have your opportunity!' he said, impatiently, yet good humoured still.

'Then would you, my Lord,' continued the Jew, with his strong Hebrew-German accent, 'be so good as to favour me by saying whether this signature be your own?'

The Jew held before him a folded paper, so folded that one line only was visible, across which was dashed in bold characters, *Rockingham*.

The Seraph put up his eye-glass, stooped, and took a steadfast look, then shook his head.

'No; that is not mine; at least, I think not. Never made my R half a quarter so well in my life.'

'Many thanks, my Lord,' said Baroni, quietly. 'One question more and we can substantiate the fact. Did your Lordship endorse any bill on the 15th of last month?'

The Seraph looked surprised, and reflected a moment. 'No, I didn't,' he said, after a pause; 'I have done it for men, but not on that day. I was shooting at Hornsey Wood most of it, if I remember right. Why do you ask?'

'I will tell you, my Lord, if you grant me a private interview.'

The Seraph moved away. 'Never do that,' he said, briefly. 'Private interviews,' thought he, acting on past experience, 'with women always mean proposals, and with men always mean extortion.'

Baroni made a quick movement towards him.

'An instant, my Lord! This intimately concerns yourself. The steps of an hotel is surely not the place in which to speak of it?'

'I wish to hear nothing about it,' replied Rock, putting him aside; while he thought to himself regretfully, 'That *is* "stiff", that bit of paper; perhaps some poor wretch is in a scrape. I wish I hadn't so wholly denied my signature. If the mischief's done, there's no good in bothering the fellow.'

The Seraph's good nature was apt to overlook such trifles as the Law.

Baroni kept pace with him as he approached the hotel door, and spoke very low.

'My Lord, if you do *not* listen, worse may befall the reputation both of your regiment and your friends.'

The Seraph swung round, his careless handsome face set stern in an instant, his blue eyes grave, and gathering an ominous fire.

'Step yonder,' he said, curtly, signing the Hebrew towards the grand staircase. 'Show that person to my rooms, Alexis.'

But for the publicity of the entrance of the Badischer Hof the mighty right arm of the Guardsman might have terminated the interview then and there in different fashion. Baroni had gained his point, and was ushered into the fine chambers set apart for the future Duke of Lyonnesse; the Seraph strode after him, and as the attendant closed the door and left them alone in the first of the great lofty suite, all glittering with gilding, and ormolu, and malachite, and rose velvet, and Parisian taste, stood like a tower above the Jew's small, slight form; while his words came curtly, and only by a fierce effort, through his lips.

'Substantiate what you dare to say, or my grooms shall throw you out of that window!—Now?'

Baroni looked up unmoved. The calm, steady, undisturbed glance sent a chill over the Seraph; he thought if this man came but for purposes of extortion, and were not fully sure that he could make good what he had said, this was not the look he would give.

'I desire nothing better, my Lord,' said Baroni, quietly, 'though I greatly regret to be the messenger of such an errand. This bill, which in a moment I will have the honour of showing you, was transacted by my house (I am one of the partners of a London discounting firm), endorsed thus by your celebrated name. Moneys were lent on it, the bill was made payable at two months' date, it was understood that you accepted it, there could be no risk with such a signature as yours. The bill was negotiated, I was in Leyden, Lubeck, and other places at the period, I heard nothing of the matter; when I returned to London, a little less than a week ago, I saw the signature for the first time. I was at once aware that it was not yours, for I had some paid bills, signed by you, at hand, with which I compared it. Of course, my only remedy was to seek you out, although I was nearly certain before your present denial that the bill was a forgery.'

He spoke quite tranquilly still, with a perfectly respectful regret, but with the air of a man who has his title to be heard, and is acting simply in his own clear right. The Seraph listened, restless, impatient, sorely tried to keep the passion in which had been awakened by the hint that this wretched matter could concern or attaint the honour of his corps.

'Well! speak out!' he said, impatiently. 'Details are nothing. Who drew it? Who forged my name, if it *be* forged? Quick! give me the paper.'

'With every trust and every deference, my Lord, I cannot let the bill pass out of my own hands until this unfortunate matter be cleared up— if cleared up it can be. Your Lordship shall see the bill, however, of course, spread here upon the table, but first let me warn you, my Lord Marquis, that the sight will be intensely painful to you.'

'Painful to me!'

'Very painful, my Lord,' added Baroni, impressively. 'Prepare yourself for——'

Rock dashed his hand down on the marble table with a force that made the lustres and statuettes on it ring and tremble.

'No more words! Lay the bill there.'

Baroni bowed and smoothed out upon the console the crumpled document, holding it with one hand, yet leaving visible with the counterfeited signature one other—the name of the forger in whose favour the bill was drawn. That other signature was——*Bertie Cecil*.

'I deeply regret to deal you such a blow from such a friend, my Lord,' said the Jew, softly. The Seraph stooped and gazed—one instant of horrified amazement kept him dumb there, staring at the written paper as at some ghastly thing; then all the hot blood rushed over his fair, bold face, he flung himself on the Hebrew, and ere the other could have breath or warning tossed him upward to the painted ceiling and hurled him down again upon the velvet carpet, as lightly as a retriever will catch up and let fall a wild duck or a grouse, and stood over Baroni where he lay.

'You hound!'

Baroni, lying passive and breathless with the violence of the shock and the surprise, yet keeping, even amidst the hurricane of wrath that had tossed him upward and downward as the winds toss leaves, his hold upon the document, and his clear, cool, ready self-possession.

'My Lord,' he said, faintly, 'I do not wonder at your excitement, aggressive as it renders you; but I cannot admit that false which I know to be a for——'

'Silence! Say that word once more and I shall forget myself, and hurl you out into the street like the dog of a Jew you are!'

'Have patience an instant, my Lord. Will it profit your friend and brother-in-arms if it be afterwards said that when this charge was brought against him, you, my Lord Rockingham, had so little faith in his power to refute it that you bore down with all your mighty strength in a personal assault upon one so weakly as myself, and sought to put an end to the evidence against him by bodily threats against my safety, and by—what will look legally, my Lord, like—an attempt to coerce me into silence, and to obtain the paper from my hands by violence?'

Faint and hoarse the words were, but they were spoken with quiet confidence, with admirable acumen; they were the very words to lash the passions of his listener into unendurable fire, yet to chain them powerless down; the Guardsman stood above him, his features flushed and dark with rage, his eyes literally blazing with fury, his lips working under his tawny leonine beard. At every syllable he could have thrown himself afresh upon the Jew and flung him out of his presence as so much carrion; yet the impotence that truth so often feels caught and meshed in the coils of subtlety, the desperate disadvantage at which Right is so often placed when met by the cunning science and sophistry of Wrong, held the Seraph in their net now. He saw his own rashness, he saw how his actions could be construed till they cast a slur even on the man he defended, he saw how legally he was in error, how legally the gallant vengeance of an indignant friendship might be construed into consciousness of guilt in the accused for whose sake the vengeance fell.

He stood silent, overwhelmed with the intensity of his own passion, baffled by the ingenuity of a serpent-wisdom he could not refute.

Ezra Baroni saw his advantage: he ventured to raise himself slightly.

'My Lord, since your faith in your friend is so perfect, send for him. If he be innocent, and I a liar, with a look I shall be confounded.'

The tone was perfectly impassive, but the words expressed a world. For a moment the Seraph's eyes flashed on him with a look that made him feel nearer his death than he had been near to it in all his days; but Rockingham restrained himself from force.

'I *will* send for him,' he said, briefly; in that answer there was more of menace and of meaning than in any physical action.

He moved, and let Baroni rise, shaken and bruised, but otherwise little seriously hurt, and still holding, in a tenacious grasp, the crumpled paper. He rang; his own servant answered the summons.

'Go to the Stephanien and enquire for Mr Cecil. Be quick; and request him, wherever he be, to be so good as to come to me instantly—here.'

The servant bowed and withdrew; a perfect silence followed between these two so strangely assorted companions; the Seraph stood with his back against the mantelpiece, with every sense on the watch to catch every movement of the Jew's, and to hear the first sound of Cecil's approach. The minutes dragged on, the Seraph was in an agony of probation and impatience. Once the attendants entered to light the chandeliers and candelabra; the full light fell on the dark, slight form of the Hebrew, and on the superb attitude, and the fair, frank, proud face of the standing Guardsman; neither moved. Once more they were left alone.

The moments ticked slowly away one by one, audible in the silence. Now and then the quarters chimed from the clock; it was the only sound in the chamber.

CHAPTER XI

For a Woman's Sake

The door opened. Cecil entered.

The Seraph crossed the room, with his hand held out; not for his life in that moment would he have omitted that gesture of friendship. Involuntarily he started and stood still one instant in amaze; the next he flung thought away, and dashed into swift, inconsequent words.

'Cecil, my dear fellow!—I'm ashamed to send for you on such a blackguard errand. Never heard of such a swindler's trick in all my life; couldn't pitch the fellow into the street because of the look of the thing, and can't take any other measures without *you*, you know. I only sent for you to expose the whole abominable business, never because I believe—— Hang it! Beauty, I can't bring myself to say it even! If a sound thrashing would have settled the matter, I wouldn't have bothered you about it, nor told you a syllable. Only you *are* sure, Bertie, aren't you, that I never listened to this miserable outrage on us both with a second's thought there could be truth in it? You know me? You trust me too well not to be certain of that?'

The incoherent address poured out from his lips in a breathless torrent; he had never been so excited in his life; and he pleaded with as imploring an earnestness as though he had been the suspected criminal, not to be accused with having one shadow of shameful doubt against his friend. His words would have told nothing except bewilderment to one who should have been a stranger to the subject on which he spoke; yet Cecil never asked even what he meant. There was no surprise upon his face, no flush of anger, no expression of amaze or indignation, only the look which had paralysed Rock on his entrance; he stood still and mute.

The Seraph looked at him, a great dread seizing him lest he should have seemed himself to cast this foul thing on his brother-in-arms; and in that dread all the fierce fire of his freshly loosened passion broke its bounds.

'Damnation! Cecil, can't you hear me? A hound has brought against you the vilest charge that ever swindlers framed: an infamy that he deserves to be shot for, as if he were a dog. He makes me stand before you as if *I* were your accuser; as if *I* doubted you; as if *I* lent an ear one second to his loathsome lie. I sent for you to confront him, and to give him up to the law. Stand out, you scoundrel, and let us see how you dare look at us now!'

He swung round at the last words, and signed to Baroni to rise from the couch where he sat. The Jew advanced slowly, softly.

'If your Lordship will pardon me, you have scarcely made it apparent what the matter is for which this gentleman is wanted. You have scarcely explained to him that it is on a charge of forgery.'

The Seraph's eyes flashed on him with a light like a lion's, and his right hand clenched hard.

'By my life! if you say that word again you shall be flung in the street, like the cur you are, let me pay what I will for it. Cecil, why don't you speak?'

Bertie had not moved; not a breath escaped his lips. He stood like a statue, deadly pale in the gaslight; when the figure of Baroni rose up and came before him, a great darkness stole on his face—it was a terrible bitterness, a great horror, a loathing disgust; but it was scarcely criminality, and it was not fear. Still he stood perfectly silent—a guilty man, any other than his loyal friend would have said; guilty, and confronted with a just accuser. The Seraph saw that look, and a deadly chill

passed over him, as it had done at the Jew's first charge—not doubt; such
heresy to his creeds, such shame to his comrade and his corps could not
be in him, but a vague dread hushed his impetuous vehemence. The
dignity of the old Lyonnesse blood asserted its ascendancy.

'Monsieur Baroni, make your statement. Later on, Mr Cecil can avenge
it.'

Cecil never moved; once his eyes went to Rockingham with a look of
yearning, grateful, unendurable pain, but it was repressed instantly; a
perfect passiveness was on him. The Jew smiled.

'My statement is easily made, and will not be so new to this gentleman
as it was to your Lordship. I simply charge the Honourable Bertie Cecil
with having negotiated a bill with my firm for £750, on the 15th of last
month, drawn in his own favour, and accepted at two months' date by
your Lordship, Your signature you, my Lord Marquis, admit to be a
forgery—with that forgery I charge your friend.'

'The 15th!'

The echo of those words alone escaped the dry white lips of Cecil; he
showed no amaze, no indignation; once only, as the charge was made, he
gave a sudden gesture, with a sudden gleam, so dark, so dangerous, in his
eyes, that his comrade thought and hoped that with one moment more
the Jew would be dashed down at his feet with the lie branded on his
mouth by the fiery blow of a slandered and outraged honour. The action
was repressed; the extraordinary quiescence, more hopeless, because
more resigned than any sign of pain or of passion, returned either by force
of self-control or by the stupor of despair.

The Seraph gazed at him with a fixed, astounded horror; he could not
believe his senses; he could not realise what he saw. His dearest friend
stood mute beneath the charge of lowest villany—stood powerless before
the falsehoods of a Jew extortioner!

'Bertie! Great Heaven!' he cried, well-nigh beside himself, 'how can
you stand silent there? Do you hear—do you hear aright? Do you know
the accursed thing this conspiracy has tried to charge you with? Say
something, for the love of God! *I* will have vengeance on your slanderer, if
you take none.'

He had looked for the rise of the same passion that rang in his own
imperious words, for the fearless wrath of an insulted gentleman, the
instantaneous outburst of a contemptuous denial, the fire of scorn, the
lightning flash of fury—all that he gave himself, all that must be so
naturally given by a slandered man under a libel that brands him with
disgrace. He had looked for these as surely as he looked for the setting of
one sun and the rise of another; he would have staked his life on the

course of his friend's conduct as he would upon his own, and a ghastly terror sent a pang to his heart.

Still—Cecil stood silent; there was a strange, set, repressed anguish on his face that made it chill as stone; there was an unnatural calm upon him; yet he lifted his head with a gesture haughty for the moment as any action that his defender could have wished.

'I am not guilty,' he said, simply.

The Seraph's hands were on his own in a close, eager grasp almost ere the words were spoken.

'Beauty, Beauty! never say that to *me*. Do you think *I* can ever doubt you?'

For a moment Cecil's head sank, the dignity with which he had spoken remained on him, but the scorn of his defiance and his denial faded.

'No, *you* cannot; *you* never will.'

The words were spoken almost mechanically, like a man in a dream. Ezra Baroni, standing calmly there with the tranquillity that an assured power alone confers, smiled slightly once more.

'You are not guilty, Mr Cecil? I shall be charmed if we can find it so. Your proofs?'

'Proof? I give you *my word*.'

Baroni bowed, with a sneer at once insolent but subdued.

'We men of business, sir, are—perhaps inconveniently for gentlemen—given to a preference in favour of something more substantial. Your word, doubtless, is your bond among your acquaintance; it is a pity for you that your friend's name should have been added to the bond you placed with us. Business men's pertinacity is a little wearisome, no doubt, to officers and members of the aristocracy like yourself; but all the same I must persist—how can you disprove this charge?'

The Seraph turned on him with the fierceness of a bloodhound.

'You dog! If you use that tone again in my presence, I will double-thong you till you cannot breathe!'

Baroni laughed a little; he felt secure now, and could not resist the pleasure of braving and of torturing the 'aristocrats'.

'I don't doubt your will or your strength, my Lord; but neither do I doubt the force of the law to make you account for any brutality of the prize-ring your Lordship may please to exert on me.'

The Seraph ground his heel into the carpet.

'We waste words on that wretch,' he said abruptly to Cecil. 'Prove his insolence the lie it is, and we will deal with him later on.'

'Precisely what I said, my Lord,' murmured Baroni. 'Let Mr Cecil prove his innocence.'

Into Bertie's eyes came a hunted, driven desperation. He turned them on Rockingham with a look that cut him to the heart; yet the abhorrent thought crossed him—was it thus that men guiltless looked?

'Mr Cecil was with my partner at 7.50 on the evening of the 15th. It was long over business hours, but my partner to oblige him stretched a point,' pursued the soft, bland, malicious voice of the German Jew. 'If he were not at our office—where was he? That is simple enough.'

'Answered in a moment!' said the Seraph, with impetuous certainty. 'Cecil!—to prove this man what he is, not for an instant to satisfy me— where were you at that time on the 15th?'

'The 15th!'

'Where *were* you?' pursued his friend. 'Were you at mess? at the clubs? dressing for dinner?—where? where? There must be thousands of ways of remembering—thousands of people who'll prove it for you?'

Cecil stood mute still; his teeth clenched on his under lip; he could not speak; a woman's reputation lay in his silence.

'*Can't* you remember?' implored the Seraph. 'You will think—you must think!'

There was a feverish entreaty in his voice. That hunted helplessness with which a question so slight yet so momentous was received, was forcing in on him a thought that he flung away like an asp.

Cecil looked both of them full in the eyes—both his accuser and his friend. He was held as speechless as though his tongue were paralysed; he was bound by his word of honour; he was weighted with a woman's secret.

'Don't look at me so, Bertie, for mercy's sake! Speak! *where* were you?'

'I cannot tell you; but I was not there.'

The words were calm; there was a great resolve in them, moreover; but his voice was hoarse and his lips shook. He paid a bitter price for the butterfly pleasure of a summer-day love.

'Cannot tell me?—*cannot*? You mean you have forgotten!'

'I cannot tell you; it is enough.'

There was an almost fierce and sullen desperation in the answer; its firmness was not shaken, but the ordeal was severe. A woman's reputation, a thing so lightly thrown away with an idler's word, a Lovelace's smile! that was all he had to sacrifice to clear himself from the toils gathering around him. That was all! And his word of honour.

Baroni bent his head with an ironic mockery of sympathy.

'I feared so, my Lord. Mr Cecil "cannot tell". As it happens, my partner *can* tell. Mr Cecil was with him at the hour and on the day I specify; and Mr Cecil transacted with him the bill that I have had the honour of showing you——'

'Let *me* see it.'

The request was peremptory to imperiousness, yet Cecil would have faced his death far sooner than he would have looked upon that piece of paper.

Baroni smiled.

'It is not often that we treat gentlemen under misfortune in the manner we treat you, sir; they are usually dealt with more summarily, less mercifully. You must excuse altogether my showing you the document; both you and his Lordship are officers skilled, I believe, in the patrician science of fist-attack.'

He could not deny himself the pleasure and the rarity of insolence to the men before him, so far above him in social rank, yet at that juncture so utterly at his mercy.

'You mean that we should fall foul of you and seize it?' thundered Rockingham in the magnificence of his wrath. 'Do you judge the world by your own wretched villanies? Let him see the paper; lay it there, or, as there is truth on earth, I will kill you where you stand.'

The Jew quailed under the fierce flashing of those leonine eyes. He bowed with that tact which never forsook him.

'I confide it to *your* honour, my Lord Marquis,' he said, as he spread out the bill on the console. He was an able diplomatist.

Cecil leaned forward and looked at the signatures dashed across the paper; both who saw him saw also the shiver, like a shiver of intense cold, that ran through him as he did so, and saw his teeth clench tight, in the extremity of rage, in the excess of pain, or—to hold in all utterance that might be on his lips.

'Well?' asked the Seraph, in a breathless anxiety. He knew not what to believe, what to do, whom to accuse, or how to unravel this mystery of villany and darkness; but he felt, with a sickening reluctance which drove him wild, that his friend did not act in this thing as he should have acted; not as men of assured innocence and secure honour act beneath such a charge. Cecil was unlike himself, unlike every deed and word of his life, unlike every thought of the Seraph's fearless expectance, when he had looked for the coming of the accused as the signal for the sure and instant unmasking, condemnation, and chastisement of the false accuser.

'Do you still persist in denying your criminality in the face of that bill, Mr Cecil?' asked the bland, sneering, courteous voice of Ezra Baroni.

'I do. I never wrote either of these signatures; I never saw that document until tonight.'

The answer was firmly given, the old blaze of scorn came again in his weary eyes, and his regard met calmly and unflinchingly the looks fastened on him; but the nerves of his lips twitched, his face was haggard as by a night's deep gambling; there was a heavy dew on his forehead; it was not the face of a wholly guiltless, of a wholly unconscious man; often even as innocence may be unwittingly betrayed into what wears the semblance of self-condemnation.

'And yet you equally persist in refusing to account for your occupation of the early evening hours of the 15th? Unfortunate!'

'I do; but in your account of them you lie.'

There was a sternness inflexible as steel in the brief sentence. Under it for an instant, though not visibly, Baroni flinched; and a fear of the man he accused smote him, more deep, more keen than that with which the sweeping might of the Seraph's fury had moved him. He knew now why Ben Davis had hated with so deadly a hatred the latent strength that slept under the Quietist languor and nonchalance of 'the d——d Guards' swell'.

What he felt, however, did not escape him by the slightest sign.

'As a matter of course, you deny it!' he said, with a polite wave of his hands. 'Quite right; you are not required to criminate yourself. I wish sincerely we were not compelled to criminate you.'

The Seraph's grand rolling voice broke in; he had stood chafing, chained, panting, in agonies of passion and of misery.

'M. Baroni!' he said, hotly, the furious vehemence of his anger and his bewilderment obscuring in him all memory of either law or fact, 'you have heard his signature and your statements alike denied once for all by Mr Cecil. Your document is a libel and a conspiracy, like your charge ; it is false, and you are swindling; it is an outrage, and you are a scoundrel; you have schemed this infamy for the sake of extortion; not a sovereign will you obtain through it. Were the accusation you dare to make true, I am the only one whom it can concern, since it is my name which is involved. Were it true, could it possibly be true, I should forbid any steps to be taken in it; I should desire it ended once and for ever. It shall be so now, by God!'

He scarcely knew what he was saying, yet what he did say, utterly as it defied all checks of law or circumstance, had so gallant a ring, had so kingly a wrath, that it awed and impressed even Baroni in the instant of its utterance.

'They say that those fine gentlemen fight like a thousand lions when they are once roused,' he thought. 'I can believe it.'

'My Lord,' he said, softly, 'you have called me by many epithets, and menaced me with many threats, since I have entered this chamber; it is not a wise thing to do with a man who knows the law. However, I can allow for the heat of your excitement. As regards the rest of your speech, you will permit me to say that its wildness of language is only equalled by the utter irrationality of your deductions, and your absolute ignorance of all legalities. Were you alone concerned and alone the discoverer of this fraud, you could prosecute or not, as you please; but we are the subjects of its imposition, ours is the money that he has obtained by that forgery, and we shall in consequence open the prosecution.'

'Prosecution?' The echo rang in an absolute agony from his hearer; he had thought of it as, at its worst, only a question between himself and Cecil.

The accused gave no sign, the rigidity and composure he had sustained throughout did not change; but at the Seraph's accent the hunted and pathetic misery which had once before gleamed in his eyes came there again; he held his comrade in a loyal and exceeding love. He would have let all the world stone him, but he could not have borne that his friend should cast even a look of contempt.

'Prosecution!' replied Baroni, quietly, 'It is a matter of course, my Lord, that Mr Cecil denies the accusation; it is very wise; the law specially cautions the accused to say nothing to criminate themselves. But we waste time in words; and, pardon me, if you have your friend's interest at heart, you will withdraw this very stormy championship, this utterly useless opposition to an inevitable line of action. I *must* arrest Mr Cecil; but I am willing—for I know to high families these misfortunes are terribly distressing—to conduct everything with the strictest privacy and delicacy. In a word, if you and he consult his interests he will accompany me unresistingly; otherwise I must summon legal force. Any opposition will only compel a very unseemly encounter of physical force, and with it the publicity I am desirous for the sake of his relatives and position to spare him.'

A dead silence followed his words, the silence that follows on an insult that cannot be averted or avenged, on a thing too hideously shameful for the thoughts to grasp it as reality.

In the first moment of Baroni's words, Cecil's eyes had gleamed again with that dark and desperate flash of a passion that would have been worse to face even than his comrade's wrath; it died, however ,well-nigh instantly, repressed by a marvellous strength of control, whatever its motive. He was simply, as he had been throughout, passive; so passive that even Ezra Baroni, who knew what the Seraph never dreamt, looked at him in wonder, and felt a faint sickly fear of that singular unbroken calm. It perplexed him; the first thing which had ever done so in his own peculiar paths of finesse and of intrigue.

The one placed in ignorance between them, at once as it were the judge and champion of his brother-at-arms, felt wild and blind under this unutterable shame, which seemed to net them both in such close and hopeless meshes. He, heir to one of the greatest coronets in the world, must see his friend branded as a common felon, and could do no more to aid or to avenge him than if he were a charcoal-burner toiling yonder in the pine woods! His words were hoarse and broken as he spoke:

'Cecil, tell me, what is to be done? This infamous outrage cannot pass! cannot go on! I will send for the Duke, for——'

'Send for no one.'

Bertie's voice was slightly weaker, like that of a man exhausted by a long struggle, but it was firm and very quiet. Its composure fell on Rockingham's tempestuous grief and rage with a sickly, silencing awe, with a sense of some evil here beyond his knowledge and ministering, and of an impotence alike to act and to serve, to defend and to avenge— the deadliest thing his fearless life had ever known.

'Pardon me, my Lord,' interposed Baroni, 'I can waste time no more. You must be now convinced yourself of your friend's implication in this very distressing affair.'

'*I!*' The Seraph's majesty of haughtiest amaze and scorn blazed from his azure eyes on the man who dared say this thing to him. '*I!* If you dare hint such a damnable shame to my face again, I will wring your neck with as little remorse as I would a kite's. *I* believe in his guilt? Forgive me, Cecil, that I can even repeat the word! *I* believe in it? I would as soon believe in my own disgrace—in my father's dishonour!'

'How will your Lordship account, then, for Mr Cecil's total inability to tell us how he spent the hours between six and nine on the 15th?'

'Unable? He is not unable; he declines! Bertie, tell *me* what you did that one cursed evening? Whatever it was—wherever it was—say it for my sake and shame this devil.'

Cecil would more willingly have stood a line of levelled rifle-tubes aimed at his heart than that passionate entreaty from the man he loved best on earth. He staggered slightly, as if he were about to fall, and a faint white foam came on his lips; but he recovered himself almost instantly. It was so natural to him to repress every emotion, that it was simply old habit to do so now.

'I have answered,' he said very low, each word a pang—'I cannot.'

Baroni waved his hand again, with the same polite significant gesture.

'In that case, then, there is but one alternative. Will you follow me quietly, sir, or must force be employed?'

'I will go with you.'

The reply was very tranquil, but in the look that met his own as it was given, Baroni saw that some other motive than that of any fear was its spring; that some cause beyond the mere abhorrence of 'a scene' was at the root of the quiescence.

'It must be so,' said Cecil huskily to his friend. 'This man is right, so far as he knows. He is only acting on his own convictions. We cannot blame him. The whole is—a mystery, an error. But as it stands there is no resistance.'

'Resistance! By God! I would resist if I shot him dead, or shot myself. Stay—wait—one moment! If it be an error in the sense you mean, it must be a forgery of your name as of mine. You think that?'

'I did not say so.'

The Seraph gave him a rapid, shuddering glance; for once the suspicion crept in on him—*was* this guilt? Yet even now the doubt would not be harboured by him.

'Say so—you must mean so! You deny them as yours; what can they be but forgeries? There is no other explanation. *I* think the whole matter a conspiracy to extort money; but I may be wrong—let that pass. If it be, on the contrary, an imitation of both our signatures that has been palmed off upon these usurers, it is open to other treatment. Compensated for their pecuniary loss, they can have no need to press the matter further, unless they find out the delinquent. See here!'—he went to a writing-cabinet at the end of the room, flung the lid back, swept out a herd of papers, and wrenching a blank cheque from its book, threw it down before Baroni; 'here! fill it up as you like, and I will sign it, in exchange for the forged sheet.'

Baroni paused a moment. Money he loved with an adoration that excluded every other passion; that blank cheque, that limitless carte blanche, that vast exchequer from which to draw!—it was a sore temptation. He thought wistfully of the Welsher's peremptory forbiddance of all compromise—of the Welsher's inexorable command to 'wring the fine-feathered bird', lose whatever might be lost by it.

Cecil, ere the Hebrew could speak, leant forward, took the cheque, and tore it in two.

'God bless you, Rock,' he said, so low that it only reached the Seraph's ear, 'but *you* must not do that.'

'Beauty, are you mad?' cried the Marquis, passionately. 'If this villanous thing be a forgery, you are its victim as much as I—tenfold more than I. If this Jew choose to sell the paper to me, naming his own compensation, whose affair is it except his and mine? They have been losers, we indemnify them. It rests with us to find out the criminal. M. Baroni, there are a hundred more cheques in that book, name your price, and you shall have it; or, if you prefer my father's, I will send to him for it. His Grace will sign one without a question of its errand, if I ask him. Come!—your price?'

Baroni had recovered the momentary temptation, and was strong in the austerity of virtue, in the unassailability of social duty.

'You behave most nobly, most generously by your friend, my Lord,' he said, politely. 'I am glad such friendship exists on earth. But you really ask me what is not in my power. In the first place, I am but one of a firm, and have no authority to act alone; in the second, I most certainly, *were* I alone, should decline totally any pecuniary compromise. A great criminal action is not to be hushed up by any monetary arrangement. You, my Lord Marquis, may be ignorant, in the Guards, of a very coarse term used in law called "compounding with felony". That is to what you tempt me now.'

The Seraph, with one of those oaths that made the Hebrew's blood run cold, though he was no coward, opened his lips to speak; Cecil arrested him with that singular impassiveness, that apathy of resignation which had characterised his whole conduct throughout, save at a few brief moments.

'Make no opposition. The man is acting but in his own justification. I will wait for mine. To resist would be to degrade us with a bully's brawl; they have the law with them. Let it take its course.'

The Seraph dashed his hand across his eyes; he felt blind—the room seemed to reel with him.

'Oh, God! that you——'

He could not finish the words. That his comrade, his friend, one of his own corps, of his own world, should be arrested like the blackest thief in Whitechapel, or in the Rue du Temple!

Cecil glanced at him, and his eyes grew infinitely yearning—infinitely gentle; a shudder shook him all through his limbs. He hesitated a moment, then he stretched out his hand.

'Will you take it—still?'

Almost before the words were spoken, his hand was held in both of the Seraph's.

'Take it?—Before all the world—always—come what will.'

His eyes were dim as he spoke, and his rich voice rang clear as the ring of silver, though there was the tremor of emotion in it. He had forgotten the Hebrew's presence; he had forgotten all save his friend and his friend's extremity. Cecil did not answer; if he had done so, all the courage, all the calm, all the control that pride and breeding alike sustained in him, would have been shattered down to weakness; his hand closed fast in his companion's, his eyes met his once in a look of gratitude that pierced the heart of the other like a knife; then he turned to the Jew with a haughty serenity.

'M. Baroni, I am ready.'

'Wait!' cried Rockingham. 'Where you go, I come.'

The Hebrew interposed demurely.

'Forgive me, my Lord—not now. You can take what steps you will as regards your friend later on; and you may rest assured he will be treated with all delicacy compatible with the case, but you cannot accompany him now. I rely on his word to go with me quietly, but I now regard him, and you must remember this, as not the son of Viscount Royallieu—not the Honourable Bertie Cecil, of the Life Guards—not the friend of one so distinguished as yourself, but as simply an arrested forger.'

Baroni could not deny himself that last sting of his vengeance, yet, as he saw the faces of the men on whom he flung the insult, he felt for the moment that he might pay for his temerity with his life. He put his hand above his eyes with a quick, involuntary movement, like a man who wards off a blow.

'Gentlemen,' and his teeth chattered as he spoke, 'one sign of violence, and I shall summon legal force.'

Cecil caught the Seraph's lifted arm, and stayed it in its vengeance. His own teeth were clenched tight as a vice, and over the haggard whiteness of his face a deep red flush had come.

'We degrade ourselves by resistance. Let me go—they must do what they will. My reckoning must wait, and my justification. One word only: take the King, and keep him for my sake.'

Another moment, and the door had closed; he was gone out to his fate, and the Seraph, with no eyes on him, bowed down his head upon his arms where he leaned against the marble table, and, for the first time in all his life, felt the hot tears roll down his face like rain, as the weakness of woman mastered and unmanned him; he would sooner a thousand times have laid his friend down in his grave than have seen him live for this.

Cecil went slowly out beside his accuser. The keen bright eyes of the Jew kept vigilant watch and ward on him; a single sign of any effort to evade him would have been arrested by him in an instant with preconcerted skill. He looked, and saw that no thought of escape was in his prisoner's mind. Cecil had surrendered himself, and he went to his doom; he laid no blame on Baroni, and he scarce gave him a remembrance. The Hebrew did not stand to him in the colours he wore to Rockingham, who beheld this thing but on its surface: Baroni was to him only the agent of an inevitable shame, of a helpless fate that closed him in, netting him tight with the web of his own past actions; no more than the irresponsible executioner of what was in the Jew's sight and in knowledge a just sentence. He condemned his accuser in nothing; no more than the conscience of a guilty man can condemn the discoverers and the instruments of his chastisement.

Was he guilty?

Any judge might have said that he knew himself to be so as he passed down the staircase and outward to the entrance with that dead resignation on his face, that brooding, rigid look set on his features, and gazing almost in stupefaction out from the dark hazel depths of eyes that women had loved for their lustre, their languor, and the softness of their smile.

They walked out into the evening air unnoticed: he had given his consent to follow the bill-discounter without resistance, and he had no thought to break his word; he had submitted himself to the inevitable course of this fate that had fallen on him, and the whole tone of his temper and his breeding lent him the quiescence, though he had none of the doctrine, of a supreme fatalist. There were carriages standing before the hotel, waiting for those who were going to the ballroom, to the theatre, to an Archduke's dinner, to a Princess's entertainment; he looked at them with a vague, strange sense of unreality—these things of the life from which he was now barred out for ever. The sparkling tide of existence in Baden was flowing on its way, and he went out an accused felon, branded, and outlawed, and dishonoured from all place in the world that he had led, and been caressed by, and beguiled with for so long.

Tonight, at this hour, he should have been amongst all that was highest and gayest and fairest in Europe at the banquet of a Prince—and he went by his captor's side a convicted criminal.

Once out in the air, the Hebrew laid his hand on his arm: he started— it was the first sign that his *liberty was gone*! He restrained himself from all resistance still, and passed onward, down where Baroni motioned him out of the noise of the carriages, out of the glare of the light, into the narrow, darkened turning of a side-street. He went passively; for this man trusted to his honour.

In the gloom stood three figures, looming indistinctly in the shadow of the houses; one was a Huissier of the Staats-Procurator, beside whom stood the Commissary of Police of the district; the third was an English detective. Ere he saw them, their hands were on his shoulders, and the cold chill of steel touched his wrists. The Hebrew had betrayed him, and arrested him in the open street. In an instant, as the ring of the rifle rouses the slumbering tiger, all the life and the soul that were in him rose in revolt as the icy glide of the handcuffs sought their hold on his arms. In an instant, all the wild blood of his race, all the pride of his breeding, all the honour of his service, flashed into fire and leapt into action. Trusted, he would have been true to his accuser; deceived, the chains of his promise were loosened, and all he thought, all he felt, all he knew, were the lion impulses, the knightly instincts, the resolute choice to lose life rather than to lose freedom, of a soldier and a gentleman. All he remembered was that he would fight to the death rather than be taken alive; that they should kill him where he stood, in the starlight, rather than lead him in the sight of men as a felon.

With the strength that laid beneath all the gentle languor of his habits and with the science of the Eton Playing Fields of his boyhood, he wrenched his wrists free ere the steel had closed, and with the single straightening of his left arm felled the detective to earth like a bullock, with a crashing blow that sounded through the stillness like some heavy timber stove in; flinging himself like lightning on the Huissier, he twisted out of his grasp the metal weight of the handcuffs, and wrestling with him was woven for a second in that close-knit struggle which is only seen when the wrestlers wrestle for life and death. The German was a powerful and firmly built man, but Cecil's science was the finer and the more masterly. His long, slender, delicate limbs seemed to twine and writhe around the massive form of his antagonist like the coils of a cobra: they rocked and swayed to and fro on the stones, while the shrill, shrieking voice of Baroni filled the night with its clamour. The vice-like pressure of the stalwart arms of his opponent crushed him in till his ribs seemed to bend and break under the breathless oppression, the iron force; but desperation nerved him, the Royallieu blood, that never took defeat, was roused now, for the first time in his careless life; his skill and his nerve were unrivalled, and with a last effort he dashed the Huissier off him, and lifting him up—he never knew how—as he would have lifted a log of wood, hurled him down in the white streak of moonlight that alone slanted through the peaked roofs of the crooked by-street.

The cries of Baroni had already been heard; a crowd drawn by their shrieking appeals were bearing towards the place in tumult. The Jew had the quick wit to give them, as call-word, that it was a croupier who had been found cheating and fled; it sufficed to inflame the whole mob

Ouida

against the fugitive. Cecil looked round him once—such a glance as a Royal gives when the gaze-hounds are panting about him, and the fangs are in his throat; then with the swiftness of the deer itself he dashed downward into the gloom of the winding passage at the speed which had carried him, in many a foot-race, victor in the old green Eton meadows. There was scarce a man in the Queen's Service who could rival him for lightness of limb, for power of endurance in every sport of field and fell, of the moor and the gymnasium; and the athletic pleasures of many a happy hour stood him in good stead now, in the emergence of his terrible extremity.

Flight!—for the instant the word thrilled through him with a loathing sense. Flight!—the craven's refuge, the criminal's resource. He wished in the moment's agony that they would send a bullet through his brain as he ran, rather than drive him out to this. Flight!—he felt a coward and a felon as he fled; fled from every fairer thing, from every peaceful hour, from the friendship and good will of men, from the fame of his ancient race, from the smile of the women that loved him, from all that makes life rich and fair, from all that men call honour; fled, to leave his name disgraced in the service he adored; fled, to leave the world to think him a guilty dastard who dared not face his trial; fled, to bid his closest friend believe him low sunk in the depths of foulest felony, branded for ever with a criminal's shame—by his own act, by his own hand. Flight!—it has bitter pangs that make brave men feel cowards when they fly from tyranny and danger and death to a land of peace and promise; but in *his* flight he left behind him all that made life worth the living, and went out to meet eternal misery, renouncing every hope, yielding up all his future.

'It is for her sake—and his,' he thought: and without a moment's pause, without a backward look, he ran, as the stag runs with the bay of the pack behind it, down into the shadows of the night.

The hue and cry was after him; the tumult of a crowd's excitement raised it knows not why or wherefore, was on his steps, joined with the steadier and keener pursuit of men organized for the hunters' work, and trained to follow the faintest track, the slightest clue. The moon was out, and they saw him clearly, though the marvellous fleetness of his stride had borne him far ahead in the few moments' start he had gained. He heard the beat of their many feet on the stones, the dull thud of their running, the loud clamour of the mob, the shrill cries of the Hebrew offering gold with frantic lavishness to whoever should stop his prey. All the breathless excitation, all the keen and desperate straining, all the

tension of the neck-and-neck struggle that he had known so often over
the brown autumn country of the Shires at home, he knew now, intensi-
fied to horror, made deadly with despair, changed into a race for life and
death.

Yet, with it the wild blood in him woke; the recklessness of peril, the
daring and defiant courage that lay beneath his levity and languor heated
his veins and spurred his strength; he was ready to die if they chose to
slaughter him; but for his freedom he strove as men will strive for life; to
distance them, to escape them, he would have breathed his last at the
goal; they might fire him down if they would, but he swore in his teeth to
die *free*.

Some Germans in his path hearing the shouts that thundered after him
in the night, drew their mule-cart across the pent-up passage-way down
which he turned, and blocked the narrow road. He saw it in time: a
second later, and it would have been instant death to him at the pace he
went; he saw it, and gathered all the force and nervous impetus in his
frame to the trial as he came rushing downward along the slope of the
lane, with his elbows back, and his body straight, as prize-runners run.
The waggon, side-ways, stretched across—a solid barrier, heaped up with
fir-boughs brought for firing from the forests, the mules stood abreast,
yoked together. The mob following saw too, and gave a hoot and yell of
brutal triumph; their prey was in their clutches; the cart barred his
progress, and he must double like a fox faced with a stone wall.

Scarcely!—they did not know the man with whom they had to deal—
the daring and the coolness that the languid surface of indolent fashion
had covered. Even in the imminence of supreme peril, of breathless
jeopardy, he measured with unerring eye the distance and the need, rose
as lightly in the air as Forest King had risen with him over fence and
hedge, and with a single running leap cleared the width of the mules'
backs, and landing safely on the further side, dashed on, scarcely pausing
for breath. The yell that hissed in his wake, as the throng saw him escape,
by what to their slow Teutonic instincts seemed a devil's miracle, was on
his ear like the bay of the slot-hounds to the deer. They might kill him if
they could, but they should never take him captive.

And the moon was so brightly, so pitilessly clear, shining down in the
summer light, as though in love with the beauty of earth! He looked up
once; the stars seemed reeling round him in disordered riot; the chill face
of the moon looked unpitying as death. All this loveliness was round him;
this glory of sailing cloud and shadowy forest and tranquil planet, and
there was no help for him.

A gay burst of music broke on the stillness from the distance; he had left the brilliance of the town behind him, and was now in its by-streets and outskirts. The sound seemed to thrill him to the bone; it was like the echo of the lost life he was leaving for ever.

He saw, he felt, he heard, he thought; feeling and sense were quickened in him as they had never been before, yet he never slackened his pace save once or twice, when he paused for breath; he ran as swiftly, he ran as keenly, as ever stag or fox had run before him, doubling with their skill, taking the shadow as they took the covert, noting with their rapid eye the safest track, outracing with their rapid speed the pursuit that thundered in his wake.

The by-lanes he took were deserted, and he was now well-nigh out of the town, with the open country and forest lying before him. The people whom he met rushed out of his path; happily for him they were few, and were terrified, because they thought him a madman broken loose from his keepers. He never looked back; but he could tell that the pursuit was falling further and further behind him; that the speed at which he went was breaking the powers of his hunters; fresh throngs added indeed to the first pursuers as they tore down through the starlit night, but none had the science with which he went, the trained matchless skill of the University foot-race. He left them more and more behind him each second of the breathless chase, that endless as it seemed had lasted bare three minutes. If the night were but dark!—he felt that pitiless luminance glistening bright about him, everywhere, shining over all the summer world, and leaving scarce a shadow to fall athwart his way. The silver glory of the radiance was shed on every rood of ground; one hour of a winter night, one hour of the sweeping ink-black rain of an autumn storm, and he could have made for shelter as the stag makes for it across the broad brown highland water.

Before him stretched indeed the gloom of the masses of pine, the upward slopes of tree-stocked hills, the vastness of the Black Forest—but they were like the mirage to a man who dies in a desert; he knew at the pace he went he could not live to reach them. The blood was beating in his brain, and pumping from his heart; a tightness like an iron band seemed girt about his loins, his lips began to draw his breath in with loud gasping spasms; he knew that in a little space his speed must slacken;— he knew it by the roar like the noise of waters that was rushing on his ear, and the oppression like a hand's hard grip that seemed above his heart.

But he would go till he died; go till they fired on him; go though the skies felt swirling round like a sea of fire, and the hard hot earth beneath his feet jarred his whole frame as his feet struck it flying.

The angle of an old wood-house, with towering roof and high-peaked gables, threw a depth of shadow at last across his road; a shadow black and rayless, darker for the white glisten of the moon around. Built more in the Swiss than the German style, a massive balcony of wood ran round it, upon and beneath which in its heavy shade was an impenetrable gloom, while the twisted wooden pillars ran upward to the gallery, loggia-like. With rapid perception and intuition he divined rather than saw these things, and swinging himself up with noiseless lightness, he threw himself full length down on the rough flooring of the balcony. If they passed, he was safe for a brief time more at least; if they found him—his teeth clenched like a mastiff's where he lay—he had the strength in him still to sell his life dearly.

The pursuers came closer and closer, and by the clamours that floated up in indistinct and broken fragments, he knew that they had tracked him. He heard the tramp of their feet as they came under the loggia; he heard the click of the pistols—they were close upon him at last in the blackness of night.

CHAPTER XII

The King's last Service

'Is he up there?' asked a voice in the darkness.

'Not likely. A cat couldn't scramble up that woodwork,' answered a second.

'Send a shot, and try,' suggested a third.

There he lay, stretched motionless on the flat roof of the verandah. He heard the words as the thronging mob surged, and trampled, and swore, and quarrelled beneath him, in the blackness of the gloom, balked of their prey, and savage for some amends. There was a moment's pause, a hurried, eager consultation, then he heard the well-known sound of a charge being rammed down, and the sharp drawing out of a ramrod; there was a flash, a report, a line of light flamed a second in his sight, a ball hissed past him with a loud singing rush, and bedded itself in the timber a few inches above his uncovered hair. A dead silence followed; then the muttering of many voices broke out afresh.

'He's not there, at any rate,' said one, who seemed the chief; 'he couldn't have kept as still as that with a shot so near him. He's made for the open country and the forest, I'll take my oath.'

Then the treading of many feet trampled their way out from beneath the loggia; their voices and their rapid steps grew fainter and fainter as they hurried away through the night. For a while, at least, he was safe.

For some moments he lay prostrated there; the rushing of the blood on his brain, the beating of his heart, the panting of his breath, the quivering of his limbs after the intense muscular effort he had gone through, mastered him, and flung him down there beaten and powerless. He felt the foam on his lips, and he thought with every instant that the surcharged veins would burst; hands of steel seemed to crush in upon his chest, knotted cords to tighten in excruciating pain about his loins; he breathed in short convulsive gasps; his eyes were blind, and his head swam. A dreaming fancy that this was death vaguely came on him, and he was glad it should be so.

His eyelids closed unconsciously, weighed down as by the weight of lead; he saw the starry skies above him no more, and the distant noise of the pursuit waxed duller and duller on his ear; then he lost all sense and memory—he ceased even to feel the night air on his face. How long he lay there he never knew; when consciousness returned to him all was still; the moon was shining down clear as the day, the west wind was blowing softly amongst his hair. He staggered to his feet and leaned against the timber of the upper wall; the shelving, impenetrable darkness sloped below; above were the glories of a summer sky at midnight, around him the hills and woods were bathed in the silver light; he looked, and he remembered all.

He had escaped his captors; but for how long? While yet there were some hours of the night left, he must find some surer refuge, or fall into their hands again. Yet it was strange that in this moment his own misery and his own peril were less upon him than a longing to see once more— and for the last time—the woman for whose sake he suffered this. Their love had had the lightness and the languor of their world, and had had but little depth in it; yet in that hour of his supreme sacrifice to her he loved her as he had not loved in his life.

Recklessness had always been latent in him, with all his serenity and impassiveness; a reckless resolve entered him now—reckless to madness. Lightly and cautiously, though his sinews still ached, and his nerves still throbbed with the past strain, he let himself fall, hand over hand, as men go down a rope, along the woodwork to the ground. Once touching earth, off he glided, swiftly and noiselessly, keeping in the shadow of the walls all the length of the streets he took, and shunning every place where any sort of tumult could suggest the neighbourhood of those who were out and hunting him down. As it chanced, they had taken to the open country; he passed on unquestioned, and wound his way to the Kursaal. He remembered that tonight there was a masked ball, at which all the princely and titled world of Baden were present; to which he would himself have gone after the Russian dinner. By the look of the stars he saw that it must be midnight or past, the ball would be now at its height.

The daredevil wildness and the cool quietude that were so intimately and intricately mingled in his nature could alone have prompted and projected such a thought and such an action as suggested themselves to him now; in the moment of his direst extremity, of his utter hopelessness, of his most imminent peril, he went—to take a last look at his mistress! Baden, for aught he knew, might be but one vast network to mesh in and to capture him; yet he ran the risk with the dauntless temerity that had ever laid underneath the indifferentism and the indolence of his habits.

Keeping always in the shadow, and moving slowly, so as to attract no notice from those he passed, he made his way, deliberately, straight towards the blaze of light where all the gaiety of the town was centred; he reckoned, and rightly, as it proved, that the rumour of his story, the noise of his pursuit, would not have penetrated here as yet; his own world would be still in ignorance. A moment, that was all he wanted, just to look upon a woman's beauty; he went forward daringly and tranquilly to its venture. If any had told him that a vein of romance was in him, he would have stared, and thought them madmen; yet something almost as wild was in his instinct now. He had lost so much to keep her honour from attainder; he wished to meet the gaze of her fair eyes once more before he went out to his exile.

In one of the string of waiting carriages he saw a loose domino lying on the seat; he knew the liveries and the footmen, and he signed them to open the door. 'Tell Count Carl I have borrowed these,' he said to the servant, as he sprang into the vehicle, slipped the scarlet and black domino on, took the mask, and left the carriage. The man touched his hat, and said nothing; he knew Cecil well, as an intimate friend of his young Austrian master. In that masquerade guise he was safe, for the few minutes, at least, which were all he dared take.

He went on, mingled among the glittering throng, and pierced his way to the ballroom, the Venetian mask covering his features. Many spoke to him; by the scarlet and black colours they took him for the Austrian; he answered none, and threaded his way among the blaze of hues, the joyous echoes of the music, the flutter of the silk and satin dominoes, the mischievous challenge of whispers. His eyes sought only one; he soon saw her, in the white and silver mask-dress, with the spray of carmine-hued eastern flowers, by which he had been told days ago to recognise her. A crowd of dominoes were about her, some masked, some not. Her eyes glanced through the envious disguise, and her lips were laughing. He approached her with all his old tact in the art *d'arborer le cotillon*; not hurriedly, so as to attract notice, but carefully, so as to glide into a place near her.

'You promised me this waltz,' he said very gently in her ear. 'I have come in time for it.'

She recognized him by his voice, and turned from a French Prince to rebuke him for his truancy, with gay raillery and mock anger.

'Forgive me, and let me have this one waltz—please do!' She glanced at him a moment, and let him lead her out.

'No one has my step as you have it, Bertie,' she murmured, as they glided into the measure of the dance.

She thought his glance fell sadly on her as he smiled.

'No?—but others will soon learn it.'

Yet he had never threaded more deftly the maze of the waltzers, never trodden more softly, more swiftly, or with more science, the polished floor. The waltz was perfect; she did not know it was also a farewell. The delicate perfume of her floating dress, the gleam of the scarlet flower-spray, the flash of the diamonds studding her domino, the fragrance of her lips as they breathed so near his own; they haunted him many a long year afterwards.

His voice was very calm, his smile was very gentle, his step, as he swung easily through the intricacies of the circle, was none the less smooth and sure for the race that had so late strained his sinews to bursting; the woman he loved saw no change in him; but as the waltz drew to its end, she felt his heart beat louder and quicker on her own, she felt his hand hold her own more closely, she felt his head drooped over her till his lips almost touched her brow; it was his last embrace; no other could be given here, in the multitude of these courtly crowds. Then, with a few low-murmured words that thrilled her in their utterance, and echoed in her memory for years to come, he resigned her to the Austrian Grand Duke, who was her next claimant, and left her silently—for ever.

Less heroism has often proclaimed itself, with blatant trumpet to the world—a martyrdom.

He looked back once as he passed from the ballroom—back to the sea of colours, to the glitter of light, to the moving hues, amid which the sound of the laughing intoxicating music seemed to float; to the glisten of the jewels, and the gold, and the silver, to the scene, in a word, of the life that would be his no more. He looked back in a long, lingering look, such as a man may give the gladness of the earth before the gates of a prison close on him; then he went out once more into the night, threw the domino and the mask back again into the carriage, and took his way—alone.

He passed along till he had gained the shadow of a by-street, by a sheer unconscious instinct; then he paused, and looked round him. What could he do? He wondered vaguely if he were not dreaming; the air seemed to reel about him, and the earth to rock; the very force of control he had sustained made the reaction stronger; he began to feel blind and stupified. How could he escape? The railway station would be guarded by those on the watch for him; he had but a few pounds in his pocket, hastily slipped in as he had won them, 'money-down', at écarté that day; all avenues of escape were closed to him, and he knew that his limbs would refuse to carry him with any kind of speed further. He had only the short precious hours remaining of the night in which to make good his flight—and flight he must take to save those for whom he had elected to sacrifice his life. Yet how? And where?

A hurried, noiseless footfall came after him; Rake's voice came breathless on his ear, while the man's hand went up in the unforgotten soldier's salute:

'Sir!—no words. Follow me, and I'll save you.'

The one well-known voice was to him like water in a desert land; he would have trusted the speaker's fidelity with his life. He asked nothing, said nothing, but followed rapidly and in silence, turning and doubling down a score of crooked passages, and burrowing at the last like a mole in a still, deserted place on the outskirts of the town, where some close-set trees grew at the back of stables and out-buildings.

In a streak of the white moonlight stood two hunters, saddled; one was Forest King. With a cry, Cecil threw his arms round the animal's neck; he had no thought then except that he and the horse must part.

'Into saddle, sir! quick as your life!' whispered Rake. 'We'll be far away from this d—d den by morning.'

Cecil looked at him like a man in stupor—his arm still over the grey's neck.

'He can have no stay in him? He was dead-beat on the course.'

'I know he was, sir; but he ain't now; he was pisoned; but I've a trick with a 'oss that'll set that sort o' thing—if it ain't gone too far, that is to say—right in a brace of shakes. I doctored him; he's hisself agen; he'll take you till he drops.'

The King thrust his noble head closer in his master's bosom, and made a little murmuring noise, as though he said, 'Try me!'

'God bless you, Rake!' Cecil said, huskily. 'But I cannot take him; he will starve with me. And—how did you know of this?'

'Beggin' your pardon, your honour, he'll eat chopped furze with you better than he'll eat oats and hay along of a new master,' returned Rake, rapidly, tightening the girths. 'I don't know nothin', sir, save that I heard you was in a strait; I don't *want* to know nothin'; but I sees them cursed cads a runnin' of you to earth, and thinks I to myself, "Come what will, the King will be the ticket for him." So I ran to your room unbeknown, packed a little valise, and got out the passports, then back again to the stables, and saddled him like lightnin', and got 'em off, nobody knowing but Bill there. I seed you go by into the Kursaal, and laid in wait for you, sir. I made bold to bring Mother o' Pearl for myself.'

And Rake stopped, breathless and hoarse with passion and grief that he would not utter. He had heard more than he said.

'For yourself?' echoed Cecil. 'What do you mean? My good fellow, I am ruined. I shall be beggared from tonight—utterly. I cannot even help you, or keep you; but Lord Rockingham will do both, for my sake.'

The *ci-devant* soldier struck his heel into the earth with a fiery oath.

'Sir, there ain't time for no words. Where you goes I go. I'll follow you while there's a drop o' blood in me. You was good to me when I was a poor devil that every one scouted; you shall have me with you to the last, if I die for it. There!'

Cecil's voice shook as he answered. The fidelity touched him as adversity could not do.

'Rake, you are a noble fellow. I would take you, were it possible; but—in an hour I may be in a felon's prison. If I escape that, I shall lead a life of such wretchedness as——'

'That's not nothing to me, sir.'

'But it is much to me,' answered Cecil. 'As things have turned—life is over with me, Rake. What my own fate may be I have not the faintest notion—but let it be what it will, it must be a bitter one. I will not drag another into it.'

'If you send me away, I'll shoot myself through the head, sir, that's all.'

'You will do nothing of the kind. Go to Lord Rockingham, and ask him from me to take you into his service. You cannot have a kinder master.'

'I don't say nothing agen the Marquis, sir,' said Rake, doggedly; 'he's a right-on generous gentleman, but he aren't *you*. Let me go with you, if it's just to rub the King down. Lord, sir! you don't know what straits I've lived in—what a lot of things I can turn my hand to—what a one I am to fit myself into any rat-hole, and make it spicy. Why, sir, I'm that born scamp I am—I'm a deal happier on the cross and getting my bread just anyhow, than I am when I'm in clover like you've kep' me.'

Rake's eyes looked up wistfully and eager as a dog's when he prays to be let out of kennel to follow the gun; his voice was husky and agitated with a strong excitement. Cecil stood a moment, irresolute, touched and pained at the man's spaniel-like affection—yet not yielding to it.

'I thank you from my heart, Rake,' he said at length, 'but it must not be. I tell you my future life will be beggary——'

'You'll want me anyways, sir,' retorted Rake, ashamed of the choking in his throat. 'I ask your pardon for interruptin', but every second's that precious like. Besides, sir, I've got to cut and run for my own sake. I've laid Willon's head open down there in the loose box; and when he's come to himself a pretty hue and cry he'll raise after me. He painted the King, that's what he did; and I told him so, and I gev' it to him—one—two—amazin'! Get into saddle, sir, for the Lord's sake! and here, Bill—you run back, shut the door, and don't let nobody know the 'osses are out till the mornin'. Then look like a muff as you are, and say nothin'!'

The stable-boy stared, nodded assent, and sloped off. Rake threw himself across the brown mare.

'Now, sir! a steeplechase for our lives! We'll be leagues away by the day-dawn, and I've got their feed in the saddle-bags, so that they'll bait in the forests. Off, sir, for God's sake, or the blackguards will be down on you again!'

As he spoke, the clamour and tread of men of the town racing to the chase, were wafted to them on the night wind, drawing nearer and nearer; Rake drew the reins tight in his hand in fury.

'There they come—the d—d beaks! For the love of mercy, sir, don't check *now*. Ten seconds more and they'll be on you; off, off!—or by the Lord Harry, sir, you'll make a murderer of me, and I'll kill the first man that lays his hand on you!'

The blaze of bitter blood was in the ex-Dragoon's fiery face as the moon shone on it, and he drew out one of his holster-pistols, and swung round in his saddle facing the narrow entrance of the lane, ready to shoot down the first of the pursuit whose shadow should darken the broad stream of white light that fell through the archway.

Cecil looked at him, and paused no more; but vaulted into the old familiar seat, and Forest King bore him away through the starry night, with the brown mare racing her best by his side. Away—through the sleeping shadows, through the broad beams of the moon, through the odorous scent of the crowded pines, through the soft breaking grey of the dawn; away—to mountain solitudes and forest silence, and the shelter of lonely untracked ravines, and the woodland lairs they must share with wolf and boar; away—to flee with the flight of the hunted fox, to race with the wakeful dread of the deer; away—to what fate, who could tell?

Far and fast they rode through the night, never drawing rein. The horses laid well to their work, their youth and their mettle were roused, and they needed no touch of spur, but neck and neck dashed down through the sullen grey of the dawn and the breaking flush of the first sunrise. On the hard parched earth, on the dew-laden moss, on the stretches of wayside sward, on the dry white dust of the dunes the challenge of no pursuit stayed them, and they obeyed the call that was made on their strength with good and gallant willingness. Far and fast they rode, happily knowing the country well; now through the darkness of night, now through the glimmering daybreak. Tall walls of fir-crowned rocks passed by them like a dream; beetling cliffs and summer foliage swept past their eyes all fused and dim; grey piles of monastic buildings with the dull chimes tolling the hour, flashed on their sight to be lost in a moment; corn-lands yellowing for the sickle, fields with the sheaves set-up, orchards ruddy with fruit, and black barn-roofs lost in leafy nests, villages lying amongst their hills like German toys caught in the hollow of a guarding hand, masses of forest stretching wide, sombre and silent and dark as a tomb, the shine of water's silvery line where it flowed in a rocky channel—they passed them all in the soft grey of the waning night, in the white veil of the fragrant mists, in the stillness of sleep and of peace. Passed them, racing for more than life, flying with the speed of the wind.

'I failed him today through my foes and his,' Forest King thought, as he laid his length out in his mighty stride. 'But I love him well; I will save him tonight.' And save him the brave brute did. The grass was so sweet and so short, he longed to stop for a mouthful; the brooks looked so clear and so brown, he longed to pause for a drink; renewed force and reviving youth filled his loyal veins with their fire; he could have thrown himself down on that mossy turf, and had a roll in its thyme and its lichens for sheer joy that his strength had come back. But, he would yield to none of these longings, he held on for his master's sake, and tried to think as he ran that this was only a piece of play; only a steeplechase, for a silver vase and a lady's smile, such as he and his rider had so often run for, and so often won, in those glad hours of the crisp winter noons of English shires far away. He turned his eyes on the brown mare's, and she turned hers on his; they were good friends in the stables at home, and they understood one another now. 'If I were what I was yesterday, she wouldn't run even with *me*,' thought the King; but they were doing good work together, and he was too true a knight and too true a gentleman to be jealous of Mother o' Pearl. So they raced neck and neck through the dawn; with the noisy clatter of water-mill wheels, or the distant sound of a woodman's axe, or the tolling bell of a convent clock, the only sound on the air save the beat of the flying hoofs.

Away they went, mile on mile, league on league, till the stars faded out in the blaze of the sun, and the tall pines rose out of the gloom. Either his pursuers were baffled and distanced, or no hue and cry was yet after him; nothing arrested them as they swept on, and the silent land lay in the stillness of morning ere toil and activity awakened. It was strangely still, strangely lonely, and the echo of the gallop seemed to beat on the stirless breathless solitude. As the light broke and grew clearer and clearer, Cecil's face in it was white as death as he galloped through the mists, a hunted man, on whose head a price was set; but it was quite calm still, and very resolute; there was no '*harking back*' in it.

They had raced nigh twenty English miles by the time the chimes of a village were striking six o'clock; it was the only group of dwellings they had ventured near in their flight; the leaded lattices were thrust open with a hasty clang, and women's heads looked out as the iron tramp of the hunters' feet struck fire from the stones. A few cries were raised; one burgher called them to know their errand; they answered nothing, but traversed the street with lightning speed, gone from sight almost ere they were seen. A league further on was a wooded bottom, all dark and silent, with a brook murmuring through it under leafy shade of lilies and the tangle of water-plants; there Cecil checked the King and threw himself out of saddle.

'He is not quite himself yet,' he murmured, as he loosened the girths and held back the delicate head from the perilous cold of the water to which the horse stretched so eagerly; he thought more of Forest King than he thought, even in that hour, of himself. He did all that was needed with his own hands; fed him with the corn from the saddle-bags, cooled him gently, led him to drink a cautious draught from the bubbling little stream, then let him graze and rest under the shade of the aromatic pines and the deep bronze leaves of the copper beeches; it was almost dark, so heavy and thickly laced were the branches, and exquisitely tranquil in the heart of the hilly country, in the peace of the early day, with the rushing of the forest brook the sole sound that was heard, and the everlasting sighing of the pine-boughs overhead.

Cecil leaned awhile silently against one of the great gnarled trunks, and Rake affected to busy himself with the mare: in his heart was a tumult of rage, a volcano of curiosity, a pent-up storm of anxious amaze, but he would have let Mother o' Pearl brain him with a kick of her iron plates rather than press a single look that should seem like *doubt*, or seem like insult in adversity to his fallen master.

Cecil's eyes, drooped and brooding, gazed a long half-hour down in silence into the brook bubbling at his feet; then he lifted his head and spoke—with a certain formality and command in his voice, as though he gave an order on parade.

'Rake, listen, and do precisely what I bid you, neither more nor less. The horses cannot accompany me, nor you either; I must go henceforth where they would starve, and you would do worse. I do not take the King into suffering, nor you into temptation.'

Rake, who at the tone had fallen unconsciously into the attitude of 'attention', giving the salute with his old military instinct, opened his lips to speak in eager protestation. Cecil put up his hand.

'I have decided; nothing you can say will alter me. We are near a by-station now; if I find none there to prevent me, I shall get away by the first train; to hide in these woods is out of the question. You will return by easy stages to Baden, and take the horses at once to Lord Rockingham. They are his now. Tell him my last wish was that he should take you into his service; and he will be a better master to you than I have ever been; as for the King——' his lips quivered, and his voice shook a little despite himself, 'he will be safe with him. I shall go into some foreign service— Austrian, Russian, Mexican, whichever be open to me. I would not risk such a horse as mine to be sold, ill-treated, tossed from owner to owner, sent in his old age to a knacker's yard, or killed in a skirmish by a cannon-shot. Take both him and the mare back, and go back yourself. Believe me, I thank you from my heart for your noble offer of fidelity, but accept it I never shall.'

A dead pause came after his words; Rake stood mute: a curious look, half-dogged, half-wounded, but very resolute, had come on his face. Cecil thought him pained, and spoke with an infinite gentleness:

'My good fellow, do not regret it, or fancy I have no gratitude to you. I feel your loyalty deeply, and I know all you would willingly suffer for me; but it must not be. The mere offer of what you would do, has been quite testimony enough of your truth and your worth. It is impossible for me to tell you what has so suddenly changed my fortunes; it is sufficient that for the future I shall be, if I live, what you were—a private soldier in any army that needs a sword. But let my fate be what it will, I go to it *alone*. Spare me more speech, and simply obey my last command.'

Quiet as the words were, there was a resolve in them not to be disputed, an authority not to be rebelled against. Rake stared, and looked at him blankly; in this man who spoke to him with so subdued but so irresistible a power of command, he could scarcely recognize the gay, indolent, indulgent, *poco curante* Guardsman, whose most serious anxiety had been the set of a lace tie, the fashion of his hunting-dress, or the choice of the gold arabesque for his smoking-slippers.

Rake was silent a moment, then his hand touched his cap again.

'Very well, sir'; and without opposition or entreaty he turned to re-saddle the mare.

Our natures are oddly inconsistent. Cecil would not have taken the man into exile, and danger, and temptation, and away from comfort and an honest life, for any consideration; yet it gave him something of a pang that Rake was so soon dissuaded from following him, and so easily convinced of the folly of his fidelity. But he had dealt himself a far deadlier one when he had resolved to part for ever from the King. He loved the horse better than he loved anything, fed from his hand in foalhood, reared, broken, and trained under his own eye and his own care, he had had a truer welcome from those loving, lustrous eyes, than all his mistresses ever gave him. He had had so many victories, so many hunting-runs, so many pleasant days of winter and of autumn, with Forest King for his comrade and companion! He could better bear to sever from all other things than from the stable-monarch, whose brave heart never failed him, and whose honest love was always his.

He stretched his hand out with his accustomed signal, the King lifted his head where he grazed, and came to him with the murmuring noise of pleasure he always gave at his master's caress, and pressed his forehead against Cecil's breast, and took such tender heed, such earnest solicitude, not to harm him with a touch of the mighty fore-hoofs, as those only who care for and know horses well will understand in its relation.

Cecil threw his arm over his neck, and leant his own head down on it, so that his face was hidden. He stood motionless so many moments, and the King never stirred, but only pressed closer and closer against his bosom, as though he knew that this was his eternal farewell to his master. But little light came there, the boughs grew so thickly; and it was still and solitary as a desert in the gloom of the meeting trees.

There have been many idols, idols of gold, idols of clay, less pure, less true than the brave and loyal-hearted beast from whom he parted now.

He stood motionless awhile longer, and where his face was hidden, the grey silken mane of the horse was wet with great slow tears that forced themselves through his closed eyes; then he laid his lips on the King's forehead, as he might have touched the brow of a woman he loved; and with a backward gesture of his hand to his servant, plunged down into the deep slope of netted boughs and scarce penetrable leafage, that swung back into their places, and shrouded him from sight with their thick unbroken screen.

'He's forgot me right and away in the King,' murmured Rake, as he led Forest King away slowly and sorrowfully, while the hunter pulled and fretted to force his way to his master. 'Well, it's only natural like. I've cause to care for him, and plenty on it; but he ain't no sort of reason to think about me.'

That was the way the philosopher took his wound.

Alone, Cecil flung himself full length down on the turf beneath the beech woods, his arms thrown forward, his face buried in the grass, all gay with late summer forest blossoms; for the first time the whole might of the ruin that had fallen on him was understood by him; for the first time it beat him down beneath it as the overstrained tension of nerve and of self-restraint had their inevitable reaction. He knew what this thing was which he had done; he had given up his whole future.

Though he had spoken lightly to his servant of his intention to enter a foreign army, he knew himself how few the chances were that he could ever do so. It was possible that Rockingham might so exert his influence that he would be left unpursued, but unless this chanced so (and Baroni had seemed resolute to forego no part of his demands), the search for him would be in the hands of the law, and the wiles of secret police and of detectives' resources spread too far and finely over the world for him to have scarcely a hope of ultimate escape.

If he sought France, the Extradition Treaty would deliver him up; Russia—Austria—Prussia were of equal danger; he would be identified, and given up to trial. Into the Italian service he knew many a scoundrel was received unquestioned; and he might try the Western world; though he had no means to pay the passage, he might work it; he was a good sailor; yachts had been twice sunk under him, by steamers, in the Solent and the Spezzia, and his own schooner had once been fired at by mistake for a blockade runner, when he had brought to, and given them a broadside from his two shotted guns before he would signal them their error.

As these things swept disordered and aimless through his mind, he wondered if a nightmare were upon him; *he*, the darling of Belgravia, the Guards' champion, the lover of Lady Guenevere, to be here outlawed and friendless, wearily racking his brains to solve whether he had seamanship enough to be taken before the mast, or could stand before the tambourmajor of a French regiment, with a chance to serve the same Flag!

For a while he lay there like a drunken man, heavy and motionless, his brow resting on his arm, his face buried in the grass, he had parted more easily with the woman he loved than he had parted with Forest King. The chimes of some far-off monastery or castle-campanile swung lazily in the morning stillness; the sound revived him, and recalled to him how little time there was if he would seek the flight that had begun on impulse and was continued in a firm unshrinking resolve: he must go on, and on, and on; he must burrow like a fox, hide like a beaten cur; he must put leagues between him and all who had ever known him; he must sink his very name, and identity, and existence, under some impenetrable obscurity, or the burden he had taken up for others' sake would be uselessly borne. There must be action of some sort or other, instant and unerring.

'It don't matter,' he thought, with the old idle indifference, oddly becoming in that extreme moment the very height of stoic philosophy, without any thought or effort to be such; 'I was going to the bad of my own accord; I must have cut and run for the debts, if not for this; it would have been the same thing, anyway, so it's just as well to do it for *them*. Life's over, and I'm a fool that I don't shoot myself.'

But there was too imperious a spirit in the Royallieu blood to let him give in to disaster, and do this. He rose slowly, staggering a little, and feeling blinded and dazzled with the blaze of the morning sun as he went out of the beech wood. There were the marks of the hoofs on the damp, dewy turf; his lips trembled a little as he saw them; he would never ride the horse again!

Some two miles, more or less, lay between him and the railway. He was not certain of his way, and he felt a sickening exhaustion on him; he had been without food since his breakfast before the race. A gamekeeper's hut stood near the entrance of the wood; he had much recklessness in him, and no caution. He entered through the half-open door, and asked the keeper, who was eating his sausage and drinking his Läger, for a meal.

'I'll give you one if you'll bring me down that hen-harrier,' growled the man in south German, pointing to the bird that was sailing far off, a mere speck in the sunny sky.

Cecil took the rifle held out to him, and without seeming even to pause to take aim, fired. The bird dropped like a stone through the air into the distant woods. There was no tremor in his wrist, no uncertainty in his measure. The keeper stared; the shot was one he had thought beyond any man's range, and he set food and drink before his guest with a crestfallen surprise, oddly mingled with veneration.

'You might have let me buy my breakfast without making me do murder,' said Bertie, quietly, as he tried to eat. The meal was coarse—he could scarcely touch it, but he drank the beer down thirstily, and took a crust of bread. He slipped his ring, a great sapphire graven with his crest, off his finger, and held it out to the man.

'That is worth fifty double-Fredericks; will you take it in exchange for your rifle and some powder and ball?'

The German stared again, open-mouthed, and clenched the bargain eagerly. He did not know anything about gems, but the splendour of this dazzled his eye, while he had guns more than enough, and could get many others at his lord's cost. Cecil fastened a shot-belt round him, took a powder-flask and cartridge-case, and, with a few words of thanks, went on his way.

Now that he held the rifle in his hand, he felt ready for the work that was before him; if hunted to bay, at any rate he could now have a struggle for his liberty. The keeper stood bewildered, gazing blankly after him down the vista of pines.

'Hein! hein!' he growled, as he looked at the sapphire sparkling in his broad brown palm; 'I never saw such a with-lavishness-wasteful-and-with-courteous-speech-laconic gentleman! I wish I had not let him have the gun; he will take his own life, belike. Ach, Gott! he will take his own life!'

But Cecil had not bought it for that end—though he had called himself a fool for not sending a bullet through his brain, to quench in eternal darkness this ruined and wretched life that alone remained to him. He walked on through the still summer dawn, with the width of the country stretching sun-steeped around him. The sleeplessness, the excitement, the misery, the wild running of the past night had left him strengthless and racked with pain, but he knew that he must press onward or be caught, sooner or later, like netted game in the poacher's silken mesh. Where to go, what to do, he knew no more than if he were a child; everything had always been ready to his hand, the only thought required of him had been how to amuse himself and avoid being bored; now thrown alone on a mighty calamity, and brought face to face with the severity and emergency of exertion, he was like a pleasure-boat beaten under high billows, and driven far out to sea by the madness of a raging nor'-wester. He had no conception what to do; he had but one resolve— to keep his secret; if to do it he killed himself with the rifle his sapphire ring had bought.

Carelessly daring always, he sauntered now into the station for which he had made, without a sign on him that could attract observation. He wore still the violet velvet Spanish-like dress, the hessians, and the broad-leafed felt hat with an eagle's feather fastened in it, that he had worn at the races, and with the gun in his hand there was nothing to distinguish him from any tourist 'Milor', except that in one hand he carried his own valise. He cast a rapid glance around; no warrant for his apprehension, no announcement of his personal appearance had preceded him here; he was safe—safe in that; safer still in the fact that the train rushed in so immediately on his arrival there, that the few people about had no time to notice or speculate upon him. The coupé was empty, by a happy chance; he took it, throwing his money down with no heed that when the little he had left was once expended he would be penniless, and the train whirled on with him, plunging into the heart of forest and mountain, and the black gloom of tunnels, and the golden seas of corn-harvest. He was alone; and he leant his head on his hands, and thought, and thought, and thought, till the rocking, and the rushing, and the whirl, and the noise of the steam on his ear and the giddy gyrations of his brain in the exhaustion of over-strung exertion, conquered thought. With the beating of the engine seeming to throb like the great swinging of a pendulum through his mind, and the whirling of the country passing by him like a confused phantasmagoria, his eyes closed, his aching limbs stretched themselves out to rest, a heavy dreamless sleep fell on him, the sleep of intense bodily fatigue, and he knew no more.

Gendarmes awoke him to see his visa. He showed it them by sheer mechanical instinct, and slept again in that dead weight of slumber the moment he was alone. When he had taken his ticket and they had asked him to where it should be, he had answered to their amaze, 'To the furthest place it goes,' and he was borne on now unwitting where it went; through the rich champaign and the barren plains, through the reddening vintage, and over the dreary plateaux; through antique cities, and across broad flowing rivers; through the cave of riven rocks, and above nestling, leafy valleys; on and on, on and on, while he knew nothing, as the opium-like sleep of intense weariness held him in its stupor.

He awoke at last with a start. It was evening. The stilly twilight was settling over all the land, and the train was still rushing onward, fleet as the wind. His eyes, as they opened dreamily and blindly, fell on a face half obscured in the gloaming; he leaned forward, bewildered and doubting his senses.

'Rake!'

Rake gave the salute hurriedly and in embarrassment.

'It's I, sir!—yes, sir.'

Cecil thought himself dreaming still.

'You! You had my orders?'

'Yes, sir, I had your orders,' murmured the ex-soldier more confused than he had ever been in the whole course of his audacious life, 'and they was the first I ever disobeyed—they was. You see, sir, they was just what I *couldn't* swallow nohow—that's the real right down fact! Send me to the devil, Mr Cecil, for you, and I'll go at the first biddin', but leave you just when things are on the cross for you, damn *me* if I will!—beggin' your pardon, sir!'

And Rake, growing fiery and eloquent, dashed his cap down on the floor of the coupé with an emphatic declaration of resistance. Cecil looked at him in silence; he was not certain still whether this were not a fantastic folly he was dreaming.

'Damn *me* if I will, Mr Cecil! You won't keep me—very well; but you can't prevent me follerin' of you, and foller you I will; and so there's no more to be said about it, sir, but just to let me have my own lark, as one may say. You said you'd go to the station, I went there; you took your ticket, I took my ticket. I've been travelling behind you till about two hours ago, then I looked at you, you was asleep, sir. "I don't think my master's quite well," says I to Guard, "I'd like to get in there along of him." "Get in with you, then," says he (only we was jabbering that willainous tongue o' theirs), for he sees the name on my traps is the same as that on your traps—and in I get. Now, Mr Cecil, let me say one word for all, and don't think I'm a insolent ne'er-do-weel for having been and gone and disobeyed you; but you was good to me when I was sore in want of it; you was even good to my dog—rest his soul, the poor beast! there never were a braver!—and stick to you I *will*, till you kick me away like a cur. The truth is, it's only being near of you, sir, that keeps me straight; if I was to leave you, I should become a bad 'un again, right and away. Don't send me from you, sir, as you took mercy on me once!'

Rake's voice shook a little towards the close of his harangue, and in the shadows of evening light, as the train plunged through the gathering gloom, his ruddy bright bronzed face looked very pale and wistful.

Cecil stretched out his hand to him in silence that spoke better than words.

Rake hung his head.

'No, sir; you're a gentleman, and I've been an awful scamp! It's enough honour for me that you *would* do it. When I'm more worth it, p'rhaps—but that won't never be.'

'You are worth it now, my gallant fellow.' His voice was very low; the man's loyalty touched him keenly. 'It was only for yourself, Rake, that I ever wished you to leave me.'

'God bless you, sir,' said Rake, passionately, 'them words are better nor ten tosses of brandy! You see, sir, I'm so spry and happy in a wild life, I am, and if so be as you go to them American parts as you spoke on, why I know 'em just as well as I know Newmarket Heath, every bit! They're terrible rips in them parts, kill you as soon as look at you; it makes things uncommon larky out there, uncommon spicy. You aren't never sure but what there's a bowie-knife a-waiting for you.'

With which view of the delights of Western life, Rake, 'feeling like a fool', as he thought to himself, for which reason he had diverged into Argentine memories, applied himself to the touching and examining of the rifle with that tenderness which only gunnery love and lore produce.

Cecil sat silent awhile, his head drooped down on his hands, while the evening deepened to night. At last he looked up.

'The King? Where is he?'

Rake flushed shamefacedly under his tanned skin.

'Beggin' your pardon, sir, behind you.'

'Behind me?'

'Yes, sir; him and the brown mare. I couldn't do not nothin' else with 'em, you see, sir, so I shipped him along with us; they don't care for the train a bit, bless their hearts, and I've got a sharp boy a minding of 'em. You can easily send 'em on to England from Paris if you're determined to part with 'em, but you know the King always was fond of drums and trumpets and that like. You remember, sir, when he was a colt we broke him into it and taught him a bit of manoeuvring, 'cause till you found what pace he had in him, you'd thought of makin' a charger of him. He loves the noise of soldiering—he do; and if he thought you was goin' away without him, he'd break his heart, Mr Cecil, sir. It was all I could do to keep him from follerin' of you this morning, he sawed my arms off a'most.'

With which, Rake, conscious that he had been guilty of unpardonable disobedience and outrageous interference, hung his head over the gun, a little anxious and a good deal ashamed.

Cecil smiled a little despite himself.

'Rake, you will do for no service, I am afraid; you are terribly insubordinate!'

He had not the heart to say more; the man's fidelity was too true to be returned with rebuke; and stronger than all surprise and annoyance was a strange mingling of pain and pleasure in him to think that the horse he loved so well was still so near him, the comrade of his adversity as he had been the companion of his happiest hours.

'These things will keep him a few days,' he thought, as he looked at his hunting-watch and the single priceless pearl in each of his wristband-studs. He would have pawned every atom he had about him to have had the King with him a week longer.

The night fell, the stars came out, the storm-rack of a coming tempest drifted over the sky, the train rushed onward through the thickening darkness, through the spectral country—it was like his life, rushing head-long down into impenetrable gloom. The best, the uttermost, that he could look for was a soldier's grave, far away under some foreign soil.

A few evenings later the Countess Gueneverc stood alone in her own boudoir in her Baden suite; she was going to dine with a Grand-Duchess of Russia, and the splendid jewels of her House glittered through the black shower of her laces, and crowned her beautiful glossy hair, her delicate imperial head. In her hands was a letter; oddly written in pencil on a leaf torn out of a betting-book, but without a tremor or a change in the writing itself. And as she stood a shiver shook her frame in the solitude of her lighted and luxurious chamber, her cheek grew pale, her eyes grew dim.

'To refute the charge,' ran the last words of what was at best but a fragment, 'I must have broken my promise to you, and have compro-mised your name. Keeping silence myself, but letting the trial take place, law-inquiries, so execrable and so minute, would soon have traced through others that I was with you that evening. To clear myself, I must have attainted your name with public slander, and drawn this horrible ordeal on you before the world. Let me be *thought guilty*. It matters little. Henceforth I shall be dead to all who know me, and my ruin would have exiled me without this. Do not let an hour of grief for me mar your peace, my dearest; think of me with no pain, Beatrice, only with some memory of our past love. I have not strength yet to say—forget me; and yet, if it be for your happiness, blot out from your remembrance all thought of what we have been to one another; all thought of me and of my life, save to remember now and then that I was dear to you.'

The words grew indistinct before her sight, they touched the heart of the world-worn coquette, of the victorious sovereign, to the core; she trembled greatly as she read them. For, in her hands was his fate. Though no hint of this was breathed in his farewell letter, she knew that with a word she could clear him, free him, and call him back from exile and shame, give him once more honour and guiltlessness in the sight of the world. With a word she could do this: his life was in the balance that she held as utterly as though it were now hers to sign, or to destroy his death-warrant. It rested with her to speak, and to say he had no guilt!

But to do this she must sacrifice herself. She stood mute, irresolute, a shudder running through her till her diamonds shook in the light; the heavy tears stole slowly down one by one and fell upon the blurred and blackened paper, her heart ached with an exceeding bitterness. Then shudderingly still, and as though there were a coward crime in the action, her hand unclosed, and let the letter fall into the spirit flame of a silver lamp burning by; the words that were upon it merited a better fate, a fonder cherishing, but—they would have compromised her. She let them fall, and burn and wither. With them she gave up his life to its burden of shame, to its fate of exile.

She would hear his crime condemned, and her lips would not open; she would hear his name aspersed, and her voice would not be raised; she would know that he dwelt in misery, or died under foreign suns unhonoured and unmourned, while tongues around her would babble of his disgrace, and she would keep her peace.

She loved him—yes; but she loved better the dignity in which the world held her, and the diamonds from which the law would divorce her if their love were known.

She sacrificed him for her reputation and her jewels; the choice was thoroughly a woman's.

CHAPTER XIII

In the Café of the Chasseurs

The red hot light of the after-glow still burned on the waters of the bay, and shed its Egyptian-like lustre on the city that lies in the circle of the Sahel, with the Mediterranean so softly lashing with its violet waves the feet of the white sloping town. The sun had sunk down in fire—the sun that once looked over those waters on the legions of Scipio, and the iron brood of Hamilcar, and that now gave its lustre on the folds of the French flags as they floated above the shipping of the harbour, and on the glitter of the French arms, as a squadron of the army of Algeria swept back over the hills to their barracks. Pell-mell in its fantastic confusion, its incongruous blending, its forced mixture of two races, that will touch but never mingle, that will be chained together but will never assimilate, the Gallic-Moorish life of the city poured out; all the colouring of Haroun al Raschid scattered broadcast among Parisian fashion and French routine. Away, yonder on the spurs and tops of the hills, the green sea-pines seemed to pierce the transparent air; in the Casbah, old dreamy Arabian legends poetic as Hafiz seem still to linger, here and there, under the foliage of hanging gardens or the picturesque curves of broken terraces; in the distance the brown rugged Kabyl mountains lay like a couched camel, and far off, against the golden haze, a single palm rose, at a few rare intervals, with its drooped curled leaves, as though to recall amidst the shame of foreign domination, that this was once the home of Hannibal, the Africa that had made Rome tremble.

In the straight white boulevards, as in the winding ancient streets, under the huge barn-like walls of barracks, as beneath the marvellous mosaics of mosques, the strange bizarre conflict of European and Oriental life spread its panorama. Staff-officers, all a-glitter with crosses, galloped past; mules, laden with green maize and driven by lean brown Bedouins, swept past the plate-glass windows of bonbon shops; grave white-bearded Sheiks drank *petits verres* in the *guinguettes*; Sapeurs, Chasseurs, Zouaves, cantinières, all the varieties of French military life, mingled with jet-black Soudans, desert kings wrathful and silent, eastern women shrouded in haick and serroual, eagle-eyed Arabs flinging back snow-white burnous, and handling ominously the jewelled hilts of their cangiars. Alcazar chansons rang out from the cafés, while in their midst stood the mosque, that had used to resound with the Muezzin; Bijou-blondine and Belbée La-la and all the sister-heroines of demi-monde dragged their voluminous Paris-made dresses side by side with Moorish beauties, who only dared show the gleam of their bright black eyes through the yasmak; the *reverbères* were lit in the Place du Gouvernement, and a group fit for the days of Solyman the Magnificent sat under the white marble beauty of the Mahometan church. '*Rien n'est sacré pour un sapeur!*' was being sung to a circle of *sous-officiers*, close in the ear of a patriarch serenely majestic as Abraham; gas-lights were flashing, cigar-shops were filling, newspapers were being read, the Rigolboche was being danced, *commis-voyageurs* were chattering with grisettes, drums were beating, trumpets were sounding, bands were playing, and, amidst it all, grave men were dropping on their square of carpet to pray, brass trays of sweetmeats were passing, ostrich eggs were dangling, henna-tipped fingers were drawing the envious veil close, and noble Oriental shadows were gliding to and fro through the open doors of the mosques, like a picture of the 'Arabian Nights', like a poem of dead Islamism—in a word, it was Algiers at evening.

In one of the cafés there, a mingling of all the nations under the sun were drinking *demi-tasses*, absinthe, vermout, or old wines, in the comparative silence that had succeeded to a song, sung by a certain favourite of the Spahis, known as Loo-Loo-j'n-m'en soucie-guère from Mlle Loo-Loo's well-known habits of independence and bravado, which last had gone once so far as shooting a man through the chest in the Rue Bab-al-Oued, and setting all the gendarmes and sergents de ville at defiance afterwards. Half a dozen of that famous regiment the Chasseurs d'Afrique were gathered together, some with their feet resting on the little marble-topped tables, some reading the French papers, all smoking their inseparable companions—the *brûle-gueules*;—fine stalwart, sunburnt fellows, with faces and figures that the glowing colours of their uniform set off to the best advantage.

'Loo-Loo was in fine voice tonight,' said one.

'Yes, she took plenty of cognac before she sang; that always clears her voice,' said a second.

'And I think that did her spirits good, shooting that Kabyl,' said a third. 'By the way, did he die?'

'N'sais pas,' said the third, with a shrug of her shoulders; 'Loo-Loo's a good aim.'

'Sac à papier, yes! Rire-pour-tout taught her.'

'Ah! There never was a shot like Rire-pour-tout. When he went out, he always asked his adversary, "Where will you like it? your lungs, your heart, your brain? It is quite a matter of choice"; and whichever they chose, he shot there. *Le pauvre* Rire-pour-tout! he was always good-natured.'

'And did he never meet his match?' asked a sous-officier of the line.

The speaker looked down on the *piou-piou* with superb contempt, and twisted his moustaches. 'Monsieur! how could he? He was a Chasseur.'

'But if he never met his match, how did he die?' pursued the irreverent piou-piou—a little wiry man, black as a berry, agile as a monkey, tough and short as a pipe-stopper.

The magnificent Chasseur laughed in his splendid disdain. 'A piou-piou never killed him, that I promise you. He spitted half a dozen of you before breakfast, to give him a relish. How did Rire-pour-tout die? I will tell you.'

He dipped his long moustaches into a beaker of still champagne: Claude, Vicomte de Chanrellon, though in the ranks, could afford those luxuries.

'He died this way, did Rire-pour-tout! Dieu de Dieu! a very good way
to. Send us all the like when our time comes! We were out yonder' (and
he nodded his handsome head outward to where the brown seared
plateaux and the Kabyl mountains lay). 'We were hunting Arabs, of
course, pot-shooting rather, as we never got nigh enough to their main
body to have a clear charge at them. Rire-pour-tout grew sick of it. "This
won't do," he said; "here's two weeks gone by, and I haven't shot anything but kites and jackals. I shall get my hand out." For Rire-pour-tout,
as the army knows, somehow or other, generally potted his man every
day, and he missed it terribly. Well, what did he do? he rode off one
morning and found out the Arab camp, and he waved a white flag for a
parley. He didn't dismount, but he just faced the Arabs and spoke to their
Sheik. "Things are slow," he said to them. "I have come for a little
amusement. Set aside six of your best warriors, and I'll fight them one
after another for the honour of France, and a drink of brandy to the
conqueror." They demurred; they thought it unfair to him to have six to
one. "Ah!" he laughs, "you have heard of Rire-pour-tout, and you are
afraid!" That put their blood up: they said they would fight him before all
his Chasseurs. "Come, and welcome," said Rire-pour-tout; "and not a
hair of your beards shall be touched except by me." So the bargain was
made for an hour before sunset that night. Mort de Dieu! that was a grand
duel!'

He dipped his long moustaches again into another beaker of still.
Talking was thirsty work; the story was well known in all the African
army, but the *piou-piou*, having served in China, was new to the soil.

'The General was ill-pleased when he heard it, and half for arresting
Rire-pour-tout; but—sacré!—the thing was done; our honour was involved; he had engaged to fight these men, and engaged for us to let them
go in peace afterwards; there was no more to be said, unless we had
looked like cowards, or traitors, or both. There was a wide, level plateau
in front of our camp, and the hills were at our backs—a fine field for the
duello—and, true to time, the Arabs filed on to the plain, and fronted us
in a long line, with their standards, and their crescents, and their cymbals,
and reed-pipes, and kettle-drums, all glittering and sounding. *Sac á
papier!* there was a show, and we could not fight one of them! We were
drawn up in line—Rire-pour-tout all alone, some way in advance,
mounted, of course. The General and the Sheik had a conference; then
the play began. There were six Arabs picked out—the flower of the
army—all white and scarlet, and in their handsomest bravery, as if they
came to an *aouda*. They were fine men—*diable!*—they were fine men.
Now the duel was to be with swords; these had been selected; and each

Arab was to come against Rire-pour-tout singly, in succession. Our trumpets sounded the *pas de charge*, and their cymbals clashed; they shouted "*Fantasia!*" and the first Arab rode at him. Rire-pour-tout sat like a rock, and lunge went his steel through the Bedouin's lung, before you could cry holà!—a death-stroke, of course; Rire-pour-tout always killed: that was his perfect science. Another, and another, and another came, just as fast as the blood flowed. You know what the Arabs are—*vous autres?* how they wheel, and swerve, and fight flying, and pick up their sabre from the ground, while their horse is galloping *ventre à terre*, and pierce you here, and pierce you there, and circle round you like so many hawks? You know how they fought Rire-pour-tout then, one after another, more like devils than men. Mort de Dieu! it was a magnificent sight! He was gashed here, and gashed there; but they could never unseat him, try how they would; and one after another he caught them sooner or later, and sent them reeling out of their saddles, till there was a great red lake of blood all round him, and five of them lay dead or dying down in the sand. He had mounted afresh twice, three horses had been killed underneath him, and his jacket all hung in strips where the steel had slashed it. It was grand to see, and did one's heart good; but—*ventre-bleu!*—how one longed to go in too.

'There was only one left now; a young Arab, the Sheik's son, and down he came like the wind. He thought with the shock to unhorse Rire-pour-tout, and finish him then at his leisure. You could hear the crash as they met like two huge cymbals smashing together. Their chargers bit and tore at each other's manes, they were twined in together there as if they were but one man and one beast; they shook and they swayed, and they rocked; the sabres played about their heads so quick that it was like lightning as they flashed and twirled in the sun; the hoofs trampled up the sand till a yellow cloud hid their struggle, and out of it, all you could see was the head of a horse tossing up and spouting with foam, or a sword-blade lifted to strike. Then the tawny cloud settled down a little, the sand mist cleared away; the Arab's saddle was empty, but Rire-pour-tout sat like a rock. The old Chief bowed his head. "It is over! Allah is great!" And he knew his son lay there dead. Then we broke from the ranks, and we rushed to the place where the chargers and men were piled like so many slaughtered sheep. Rire-pour-tout laughed such a gay ringing laugh as the desert never had heard. "Vive la France!" he cried. "And now bring me my toss of brandy." Then down headlong out of his stirrups he reeled and fell under his horse; and when we lifted him up there were two broken sword-blades buried in him, and the blood was pouring fast as water out of thirty wounds and more. That was how Rire-pour-tout died, *piou-piou*, laughing to the last. Sacre-bleu! it was a splendid end; I wish I were sure of the like.'

And Claude de Chanrellon drank down his third beaker, for overmuch speech made him thirsty.

The men around him emptied their glasses in honour of the dead hero.

'Rire-pour-tout was a *croc-mitaine*,' they said solemnly, with almost a sigh, so tendering by their words the highest funeral oration.

'You have much of such sharp service here, I suppose?' asked a voice in very pure French. The speaker was leaning against the open door of the café; a tall, lightly built man, dressed in a velvet shooting tunic, much the worse for wind and weather, a loose shirt, and jackboots splashed and worn out.

'When we are at it, monsieur,' returned the Chasseur. 'I only wish we had more.'

'Of course. Are you in need of recruits?'

'They all want to come to us and to the Zouaves,' smiled Chanrellon, surveying the figure of the one who addressed him with a keen sense of its symmetry and its sinew. 'Still, a good sword brings its welcome. Do you ask seriously, monsieur?'

The bearded Arabs smoking their long pipes, the little *piou-piou* drowning his mortification in some curaçoa, the idlers reading the *Akbah* or the *Pressc*, the Chasseurs lounging over their drink, the écarté players lost in their game, all looked up at the new comer. They thought he looked a likely wearer of the dead honours of Rire-pour-tout.

He did not answer the questions literally, but came over from the doorway, and seated himself at the little marble table opposite Claude, leaning his elbows on it.

'I have a doubt,' he said. 'I am more inclined to your foes.'

'Dieu de Dieu!' ejaculated Chanrellon, pulling at his tawny moustaches. 'A bold thing to say before five Chasseurs.'

He smiled a little contemptuously, a little amusedly.

'I am not a croc-mitaine, perhaps; but I say what I think, with little heed of my auditors usually.'

Chanrellon bent his bright brown eyes curiously on him. 'He *is* a croc-mitaine,' he thought. 'He is not to be lost.'

'I prefer your foes,' went on the other, quite quietly, quite listlessly, as though the glittering, gaslit café were not full of French soldiers. 'In the first place, they are on the losing side; in the second, they are the lords of the soil; in the third, they live as free as air; and in the fourth, they have undoubtedly the right of the quarrel!'

'Monsieur!' cried the Chasseurs, laying their hands on their swords, fiery as lions. He looked indolently and wearily up from under the long lashes of his lids, and went on, as though they had not spoken.

'I will fight you all, if you like, as that worthy of yours, Rire-pour-tout, did, but I don't think it's worth while,' he said, carelessly, where he leaned over the marble table. 'Brawling's bad style; *we* don't do it. I was saying, I like your foes best; mere matter of taste; no need to quarrel over it—that *I* see. I shall go into their service, or into yours, monsieur—will you play a game of dice to decide?'

'Decide?—but how?'

'Why—this way,' said the other, with the weary listlessness of one who cares not two straws how things turn. 'If I win, I go to the Arabs—if you win, I come to your ranks.'

'Mort de Dieu! it is a droll gambling,' murmured Chanrellon. 'But—if you do win, do you think we shall let you go off to our enemies. *Pas si bête, monsieur!*'

'Yes, you will,' said the other, quietly. 'Men who knew what honour meant enough to redeem Rire-pour-tout's pledge of safety to the Bedouins, will not take advantage of an openly confessed and unarmed adversary.'

A murmur of ratification ran through his listeners.

Chanrellon swore a mighty oath.

'Pardieu! No. You are right. If you want to go, you shall go. Holà there! bring the dice. Champagne, monsieur? Vermout? Cognac?'

'Nothing, I thank you.'

He leant back with an apathetic indolence and indifference, oddly at contrast with the injudicious daring of his war-provoking words, and the rough campaigning that he sought. The assembled Chasseurs eyed him curiously; they liked his manner, and they resented his first speeches; they noted every particular about him, his delicate white hands, his weather-worn and travel-stained dress, his fair aristocratic features, his sweeping abundant beard, his careless, cool, tired, reckless way; and they were uncertain what to make of him.

The dice were brought.

'What stakes, monsieur?' asked Chanrellon.

'Ten Napoléons a side—and—the Arabs.'

He set ten Napoléons down on the table; they were the only coins he had in the world; it was very characteristic that he risked them.

They threw the main—two sixes.

'You see,' he murmured, with a half smile, 'the dice know it is a drawn duel between you and the Arabs.'

'*C'est un drôle, c'est un brave!*' muttered Chanrellon; and they threw again.

The Chasseur cast a five; his was a five again.

'The dice cannot make up their minds,' said the other, listlessly; 'they know you are Might and the Arabs are Right.'

The Frenchmen laughed; they could take a jest good-humouredly, and alone amidst so many of them he was made sacred at once by the very length of odds against him.

They rattled the boxes and threw again—Chanrellon's was three; his two.

'Ah!' he murmured. 'Right kicks the beam and loses; it always does, poor devil!'

The Chasseur leaned across the table, with his brown, fearless, sunny eyes full of pleasure.

'Monsieur! never lament such good fortune for France. You belong to us now; let me claim you!'

He bowed more gravely than he had borne himself hitherto.

'You do me much honour; fortune has willed it so. One word only in stipulation.'

Chanrellon assented courteously.

'As many as you choose.'

'I have a companion who must be brigaded with me, and I must go on active service at once.'

'With infinite pleasure. That doubtless can be arranged. You shall present yourself tomorrow morning; and for tonight, this is not the season here yet, and we are *triste à faire frémir*, still I can show you a little fun, though it is not Paris?'

But he rose, and bowed again.

'I thank you, not tonight. You shall see me at your barracks with the morning.'

'Ah, ah! monsieur!' cried the Chasseur, eagerly, and a little annoyed. 'What warrant have we that you will not dispute the decree of the dice, and go off to your favourites, the Arabs?'

He turned back and looked full in Chanrellon's face, his own eyes a little surprised, and infinitely weary.

'What warrant? My promise.'

Then, without another syllable, he lounged slowly out through the soldiers and the idlers, and disappeared in the confused din and chiar'oscuro of the gas-lit street without, through the press of troopers, grisettes, merchants, beggars, sweetmeat-sellers, lemonade-sellers, curaçoa-sellers, gaunt Bedouins, negro boys, shrieking muleteers, laughing lorettes, and glittering staff-officers.

'That is done!' he murmured to his own thoughts. 'Now for life under another Flag!'

Claude de Chanrellon sat mute and amazed awhile, gazing at the open door; then he drank a fourth beaker of champagne, and flung the emptied glass down with a mighty crash.

'Ventre-bleu! whoever he is, that man will eat fire, *bons garçons!*'

CHAPTER XIV

'De Profundis' before 'Plunging'

Three months later, it was guest-night in the mess-room of a certain famous light cavalry regiment, who bear the reputation of being the fastest corps in the English service. Of a truth, they do 'plunge' a little too wildly; and stories are told of bets over écarté in their ante-room that have been prompt extinction for ever and aye to the losers, for they rarely play money down, their stakes are too high, and moderate fortunes may go in a night with the other convenient but fatal system. But, this one indiscretion apart, they are a model corps for blood, for dash, for perfect social accord, for the finest horseflesh in the kingdom, and the best president at a mess-table that ever drilled the cook to matchlessness, and made the iced dry, and the old burgundies the admired of all new comers.

Just now they had pleasant quarters enough in York, had a couple of hundred hunters, all in all, in their stalls, were showing the Ridings that they could 'go like birds', and were using up their second horses with every day out, in the first of the season. A cracker over the best of the ground with the York and Ainsty, that had given two first-rate things quick as lightning, and both closed with a kill, had filled the day; and they were dining with a fair quantity of county guests, and all the splendour of plate, and ceremony, and magnificent hospitalities which characterize those *beaux sabreurs* wheresoever they go. At one part of the table a discussion was going on as the claret passed around; wines were perfection at the mess, but they drank singularly little; it was not their 'form' ever to indulge in that way; and the Chief, though lenient to looseness in all other matters, and very young for his command, would have been down like steel on 'the boys', had any of them taken to the pastime of overmuch drinking in any shape.

'I can't get the rights of the story,' said one of the guests, a hunting Baronet, and MFH. 'It's something very dark, isn't it?'

'Very dark,' assented a tall handsome man, with an habitual air of the most utterly exhausted apathy ever attained by the human features, but who, nevertheless, had been christened by the fiercest of the warrior nations of the Punjaub[1] as the Shumsheer-i-Shai-tan, or Sword of the Evil One, so terrible, when he was but a boy, had the circling sweep of one back stroke of his become to them.

'Guards cut up fearfully rough,' murmured one near him, known as 'The Dauphin'; 'such a low sort of thing, you know, that's the worst of it. Seraph's name, too.'

'Poor old Seraph! he's fairly bowled over about it,' added a third. 'Feels it awfully—by Jove he does! It's my belief he paid those Jew fellows the whole sum to get the pursuit slackened.'

'So Thelusson says. Thelusson says Jews have made a cracker by it. *I* dare say! Jews always do,' muttered a fourth. 'First Life would have given Beauty a million sooner than have him do it. Horrible thing for the Household.'

'But is he dead?' pursued their guest.

'Beauty?—Yes; smashed in that express, you know.'

'But there was no evidence?' suggested the Baronet.

'I don't know what you call evidence,' murmured the Dauphin. 'Horses are sent to England from Paris; clearly shows he went to Paris. Marseilles train smashes, twenty people ground into indistinguishable amalgamation; two of the amalgamated, jammed head-foremost in a carriage alone; only traps in carriage with them, Beauty's traps, with name clear on the brass outside, and crest clear on silver things inside; two men ground to atoms, but traps safe; two men of course, Beauty and servant; man was a plucky fellow, sure to stay with him.'

And having given the desired evidence in lazy little intervals of speech, he took some Rhenish.

'Well—yes; nothing could be more conclusive, certainly,' assented the Baronet, resignedly convinced. 'It was the best thing that could happen

---

[1] Although in Indian hill-countries no European troops (not even Zouaves) could be employed, the warfare being carried on by Irregular Levies and detachments of Native Infantry, it will be remembered that in the Afghan Campaigns and Sikh War the English Cavalry, specially the 16th Lancers and the 3rd and 4th Hussars, played a most brilliant part.

under the unfortunate circumstances, so Lord Royallieu thinks, I suppose. He allowed no one to wear mourning, and had his unhappy son's portrait taken down and burnt.'

'How melodramatic!' reflected Leo Charteris. 'Now what the deuce can it hurt a dead man to have his portrait made into a bonfire? Old Lord always did hate Beauty, though. Rock does all the mourning; he's cut up no end; never saw a fellow so knocked out of time. Vowed at first he'd sell out, and go into the Austrian service; swore he couldn't stay in the Household, but would get a command of some Heavies, and be changed to India.'

'Duke didn't like that—didn't want him shot; nobody else, you see, for the title. By George! I wish you'd seen Rock the other day on the Heath; little Pulteney came up to him.'

'What Pulteney?—Jimmy, or the Earl?'

'Oh, the Earl. Jimmy would have known better. Those new men never know anything. "You purchased that famous steeplechaser of his from Mr Cecil's creditors, didn't you?" asks Pulteney. Rock just looks him over. Such a look, by George! "I received Forest King as my dead friend's last gift." Pulteney never takes the hint, not he! On he blunders: "Because, if you were inclined to part with him, I want a good new hunting strain, with plenty of fencing power, and I'd take him for the stud at any figure you liked." I thought the Seraph would have knocked him down—I did, upon my honour! He was red as this wine in a second with rage, and then as white as a woman. "You are quite right," he says, quietly, and I swear each word cut like a bullet, "you *do* want a new strain with something like breeding in it, but—I hardly think you'll get it for the three next generations. You must learn to know what it means first." Then away he lounges, leaving Pulteney *planté-là*. By Jove! I don't think the Cotton-Earl will forget this Cambridgeshire in a hurry, or try horse-dealing on the Seraph again.'

Laughter loud and long greeted the story.

'Poor Beauty!' said the Dauphin, 'he'd have enjoyed that. He always put down Pulteney himself. I remember his telling me he was on duty at Windsor once when Pulteney was staying there. Pulteney's always horribly funked at Court; frightened out of his life when he dines with any royalties; makes an awful figure, too, in a public ceremony; can't walk backward for any money, and at his first levee tumbled down right in the Queen's face. Now at the Castle one night he just happened to come down a corridor as Beauty was smoking. Beauty made believe to take him for a servant, took out a sovereign, and tossed it to him. "Here, keep a still tongue about my cigar, my good fellow!" Pulteney turned hot and cold, and stammered out God knows what, about his mighty dignity being mistaken for a valet. Bertie just laughed a little, ever so softly. "Beg your pardon, thought you were one of the people; wouldn't have done it for worlds; I know you're never at ease with a sovereign!" Now Pulteney wasn't likely to forget that. If he wanted the King, I'll lay any money it was to give him to some wretched mount who'd break his back over a fence in a selling race.'

'Well, he won't have him; Seraph don't intend to have the horse ever ridden or hunted at all.'

'Nonsense!'

'By Jove, he means it! nobody's to cross the King's back; he wants weight carriers himself, you know, and precious strong ones, too. The King's put in the stud at Lyonnesse. Poor Bertie! nobody ever managed a close finish as he did at the Grand National—last but two—don't you remember?'

'Yes; waited so beautifully on Fly-by-Night, and shot by him like lightning just before the run-in. Pity he went to the bad!'

'Ah! what a hand he played at écarté; the very best of the French science.'

'But reckless at whist; a wild game there—uncommonly wild. Drove Cis Delareux half mad one night at Royallieu with the way he threw his trumps out. Old Cis dashed his cards down at last, and looked him full in the face. "Beauty, do you know, or do you *not* know, that a whist-table is not to be taken as you take timber in a hunting-field, on the principle of clear it, or smash it?"—"Faith!" said Bertie, "clear it, or smash it, is a very good rule for anything, but a trifle too energetic for me."'

'The deuce, he's had enough of "smashing" at last! I wish he hadn't come to grief in that style; it's shocking bore for the Guards—such an ugly story.'

'It was uncommonly like him to get killed just when he did—best possible taste.'

'Only thing he could do.'

'Better taste would have been to do it earlier. I always wondered he stopped for the row.'

'Oh, never thought it would turn up; trusted to a fluke.'

He whom the Punjaub knew as the Sword of the Evil One, but who held in polite society the title of Lord Kergenven, drank some hock slowly, and murmured as his sole quota to the conversation very lazily and languidly:

'Bet you he isn't dead at all.'

'The deuce you do? And why?' chorused the table; 'when a fellow's body's found with all his traps round him!'

'I don't believe he's dead,' murmured Kergenven, with closed slumberous eyes.

'But why? Have you heard anything?'

'Not a word.'

'Why do you say he's alive, then?'

My Lord lifted his brows ever so little.

'I think so, that's all.'

'But you must have a reason, Ker?'

Badgered into speech, Kergenven drank a little more hock, and dropped out slowly, in the mellowest voice in the world, the following:

'It don't follow one has reasons for anything; pray don't get logical. Two years ago I was out in a *chasse au sanglier*, central France; perhaps you don't know their work? It's uncommonly queer. Break up the Alps into little bits, scatter 'em pell-mell over a great forest, and then set a killing pack to hunt through and through it. Delightful chance for coming to grief; even odds that if you don't pitch down a ravine, you'll get blinded for life by a branch; that if you don't get flattened under a boulder, you'll be shot by a twig catching your rifle-trigger. Uncommonly good sport.'

Exhausted with so lengthened an exposition of the charms of the *vénérie* and the *hallali*, he stopped, and dropped a walnut into some Regency sherry.

'Hang it, Ker!' cried the Dauphin. 'What's that to do with Beauty?'

My Lord let fall a sleepy glance of surprise and of rebuke from under his black lashes, that said mutely, 'Do I, who hate talking, ever talk wide of any point?'

'Why this?' he murmured. 'He was with us, down at Veilleroc, Louis d'Auvrai's place, you know; and we were out after an old boar—not too old to *race*, but still tough enough to be likely to turn and trust to his tusks if the pace got very hot, and he was hard pressed at the finish. We hadn't found till rather late; the *limeurs* were rather new to the work, and the November day was short, of course; the pack got on the slot of a roebuck too, and were off the boar's scent a little while, running wild. Altogether we got scattered, and in the forest it grew almost as dark as pitch: you followed just as you could, and could only guide yourself by your ear when the hounds gave cry, or the horns sounded. On you blundered, hit or miss, headlong down the rocks and through the branches; horses warmed wonderfully to the business, scrambled like cats, slid down like otters, kept their footing where nobody'd have thought anything but a goat could stand. *Our* hunting bloods knock up over a cramped country like Monmouthshire; they wouldn't live an hour in a French forest: you see, we just look for pace and strength in the shoulders, we don't much want anything else—except good jumping power. What a lot of fellows—even in the crack packs—will always funk water! Horses will *fly*, but they can't swim. Now to my fancy, a clever beast ought to take even a swelling bit of water like a duck. How poor Standard breasted rivers till that fool staked him!'

He dropped more walnuts into his wine, wistfully recalling a mighty hero of Leicestershire fame, that had given him many a magnificent day out, and had been the idol of his stables, till in his twelfth year the noble old sorrel had been killed by a groom's recklessness; recklessness that met with such chastisement, as told how and why the Hill-tribes' sobriquet had been given to the hand that would lie so long in indolent rest, to strike with such fearful force when once raised.

'Well,' he went on once more. 'We were all of us scattered; scarcely two kept together anywhere; where the pack was, where the boar was, where the huntsmen were, nobody knew. Now and then I heard the hounds giving tongue at the distance, and I rode after that to the best of my science, and uncommonly bad was the best. That forest work perplexes one after the grass-country. You can't view the beauties two minutes together; and as for sinning by overriding 'em, you're very safe not to do that! At last I heard a crashing sound loud and furious; I thought they had got him to bay at last. There was a great oak thicket as hard as iron, and as close as a net, between me and the place; the boughs were all twisted together, God knows how, and grew so low down, that the naked branches had to be broken through at every step by the horse's fore-hoofs, before he could force a step. We did force it somehow at last, and came into a green open space, where there were fewer trees, and the moon was shining in; there, without a hound near, true enough was the boar rolling on the ground, and somebody rolling under him; they were locked in so close they looked just like one huge beast, pitching here and there, as you've seen the rhinos wallow in Indian jheels. Of course I levelled my rifle, but I waited to get a clear aim; for which was man and which was boar, the deuce a bit could I tell. Just as I had pointed, Beauty's voice called out to me: "Keep your fire, Ker! I want to have him myself." It was he that was under the brute. Just as he spoke they rolled towards me, the boar foaming and spouting blood, and plunging his tusks into Cecil; he got his right arm out from under the beast, and crushed under there as he was, drew it free with the knife well gripped; then down he dashed it three times into the veteran's hide, just beneath the ribs; it was the *coup de grâce*, the boar lay dead, and Beauty lay half dead too, the blood rushing out of him where the tusks had dived. Two minutes, though, and a draught of my brandy brought him all round; and the first words he spoke were: "Thanks, Ker; you did as you would be done by; a shot would have spoilt it all." The brute had crossed his path far away from the pack, and he had flung himself out of saddle and had a neck and neck struggle. And that night we played baccarat by his bedside to amuse him; and he played just as well as ever. Now this is why I don't think he's dead; a fellow who served a wild boar like that, won't have let a train knock him over. And I don't believe he forged that stiff, though all the evidence says so. Beauty hadn't a touch of the blackguard in him.'

With which declaration of his views Kergenven lapsed into immutable silence and slumberous apathy, from whose shelter nothing could tempt him afresh; and the Colonel, with all the rest, lounged into the ante-room, where the tables were set, and began 'plunging' in earnest at sums that might sound fabulous were they written here. The players staked heavily; but it was the *galérie* who watched around, making their bets, and backing their favourites, that lost on the whole the most.

'Horse Guards have heard of the plunging; think we're going too fast,' murmured the Chief to Kergenven, his Major, who lifted his brows, and murmured back with the demureness of a maiden:

'Tell 'em it's our only vice; we're models of propriety!'

Which possibly would not have been received with the belief desirable by the sceptics in authority.

So the De Profundis was said over Bertie Cecil; and 'Beauty of the Brigades' ceased to be named in the Service, and soon ceased to be even remembered. In the steeplechase of life there is no time to look back at the failures, who have broken down over a 'double and drop', and fallen out of the pace.

VOLUME II

CHAPTER I

'L'Amie du Drapeau'

'Did I not say he would eat fire?'

'Pardieu! *c'est un brave.*'

'Rides like an Arab.'

'Smokes like a Zouave.'

'Cuts off a head with that back circular sweep, ah—h—h! magnificent!'

'And dances like an Aristocrat; not like a tipsy Spahis!'

The last crown to the chorus of applause, and insult to the circle of applauders, was launched with all the piquance of inimitable canteen-slang and camp-assurance, from a speaker who had perched astride on a broken fragment of wall, with her barrel of wine set up on end on the stones in front of her, and her six soldiers, her *gros bébées* as she was given maternally to calling them, lounging at their ease on the arid dusty turf below. She was very pretty, audaciously pretty, though her skin was burnt to a bright sunny brown, and her hair was cut as short as a boy's, and her face had not one regular feature in it. But then—regularity! who wanted it, who would have thought the most pure classic type a change for the better, with those dark, dancing, challenging eyes, with that arch, brilliant, kitten-like face, so sunny, so *mignon*, and those scarlet lips like a bud of camellia that were never so handsome as when a cigarette was between them, or, sooth to say, not seldom a *brûle gueule* itself?

She was pretty, she was insolent, she was intolerably coquettish, she was mischievous as a marmoset, she would swear if need be like a Zouave, she could fire galloping, she could toss off her brandy or her vermout like a trooper, she would on occasion clench her little brown hand and deal a blow that the recipient would not covet twice, she was an *enfant de Paris*, and had all its wickedness at her fingers, she would sing you *guinguette* songs till you were suffocated with laughter, and she would dance the cancan at the Salle de Mars with the biggest giant of a Cuirassier there. And yet, with all that, she was not wholly unsexed, with all that she had the delicious fragrance of youth, and had not left a certain feminine grace behind her, though she wore a vivandière's uniform, and had been born in a barrack, and meant to die in a battle; it was the blending of the two that made her piquante, made her a notoriety in her own way; known at pleasure, and equally, in the Army of Africa as 'Cigarette', and 'L'Amie du Drapeau.'

'Not like a tipsy Spahis!' it was a cruel cut to her *gros bébées*, mostly Spahis, lying there at her feet, or rather at the foot of the wall, singing their praises—with magnanimity beyond praise—of a certain Chasseur d'Afrique.

'Ho, Cigarette!' growled a little Zouave, known as Tata Leroux. 'That is the way thou forsakest thy friends for the first fresh face.'

'Well, it is not a face like a tobacco-stopper, as thine is, Tata!' responded Cigarette, with a puff of her namesake; the repartee of the camp is apt to be rough. 'He is *Bel-à-faire-peur*, as you nickname him.'

'A woman's face!' growled the injured Tata; whose own countenance was of the colour and wellnigh of the flatness of one of the red bricks of the wall.

'Ouf!' said the Friend of the Flag with more expression in that single ejaculation than could be put in a volume. 'He does woman's deeds, does he? He has woman's hands, but they can fight, I fancy? Six Arabs to his own sword the other day in that skirmish! Superb!'

'Sapristi! And what did he say, this droll, when he looked at them lying there? Just shrugged his shoulders and rode away. "I'd better have killed myself, less mischief on the whole!" Now, who is to make anything of such a man as that?'

'Ah! he did not stop to cut their gold buttons off, and steal their cangiars as thou wouldst have done, Tata? Well! he has not learnt *la guerre*,' laughed Cigarette. 'It was a waste; he should have brought me their sashes at least. By the way—when did he join?'

'Ten—twelve—years ago, or thereabouts.'

'He should have learnt to strip Arabs by this time, then,' said the Amie du Drapeau, turning the tap of her barrel to replenish the wine-cup; 'and to steal from them too, living or dead. *Thou* must take him in hand, Tata!'

Tata laughed, considering that he had received a compliment.

'Diable! I did a neat thing yesterday. Out on the hills, there, was a shepherd; he'd got two live geese swinging by their feet. They were screeching—screeching—screeching!—and they looked so nice and so plump, that I could smell them, as if they were stewing in a casserole, till I began to get as hungry as a *gamin*. A lunge would just have cut the question at once; but the orders have got so strict potting about the natives, I thought I wouldn't have any violence, if the thing would go nice and smoothly. So I just walked behind him, and tripped him up before he knew where he was; it was a picture! He was down with his face in the sand before you could sing Tra-la-la! Then I just sat upon him; but gently—very gently: and what with the sand, and the heat, and the surprise, and, in truth, perhaps, a little too my own weight, he was half suffocated. He had never seen me; he did not know what it was that was sitting on him; and I sent my voice out with a roar—"I am a demon, and the fiend hath bidden me take him thy soul tonight!" Ah! how he began to tremble, and to kick, and to quiver. He thought it was the devil a-top of him; and he began to moan, as well as the sand would let him, that he was a poor man, and an innocent, and the geese were the only things he ever stole in all his life. Then I went through a little pantomime with him, and I was very terrible in my threats, and he was choking and choking with the sand, though he never let go of the geese. At last, I relented a little, and told him I would spare him that once, if he gave up the stolen goods, and never lifted his head for an hour. Sapristi! how glad he was of the terms! I dare say my weight was unpleasant; so the geese made us a divine stew that night, and the last thing I saw of my man was his lying flat as I left him, with his face still down in the sand-hole.'

Cigarette nodded and laughed.

'Pretty fair, Tata; but I have heard better. Bah! a grand thing certainly, to fright a peasant, and scamper off with a goose!'

'Sacre-bleu!' grumbled Tata, who was himself of opinion that his exploit had been worthy of the feats of Harlequin; 'thy heart is all gone to the Englishman.'

Cigarette laughed saucily and heartily, tickled at the joke. Sentiment has an exquisitely ludicrous side when one is a *vivandière aux yeux noirs*, perched astride on a wall, and dispensing brandy-dashed wine to half a dozen sun-baked Spahis.

> Vivandière du régiment,
>   C'est Catin qu'on me nomme;
> Je vends, je donne, je bois gaiment,
>   Mon vin et mon rogomme:
> J'ai le pied leste et l'œil mutin,
> Tintin, tintin, tintin, r'lin tintin,
> Soldats, voilà Catin!

she sang with the richest, freshest, mellowest voice that ever chanted the deathless refrains of the French Lucilius.

'My heart is a *réveil-matin*, Tata; it wakes fresh every day. An Englishman, perdie! Why dost thou think him that?'

'Because he is a giant,' said Tata.

Cigarette snapped her fingers:

'I have danced with Grenadiers and Cuirassiers quite as tall, and twice as heavy. *Après?*'

'Because he bathes—splash! like any water-dog.'

'Because he is silent.'

'Because he rises in his stirrups.'

'Because he likes the sea.'

'Because he knows *le boxe*.'

'Because he is so quiet, and blazes like the devil underneath.'

Under which mass of overwhelming proofs of nationality the Amie du Drapeau gave in.

'Yes, like enough. Besides, the other one is English. Lour-i-loo, of the *Chasses-marais*,[1] tells me that the other one waits on him like a slave when he can—cleans his harness, litters his horse, saves him all the hard work, when he can do it without being found out. Where did they come from?'

'They will never tell.'

Cigarette tossed her nonchalant head, with a pout of her cherry lips, and a slang oath, light as a bird, wicked as a rigolbochade.

'Paf!—they will tell it to me!'

'Chut! Thou mayst make a lion tame, a vulture leave blood, a drum beat its own rataplan, a dead man fire a *clarinette*[2] *à six pieds*; but thou wilt never make an Englishman speak when he is bent to be silent.'

Cigarette launched a choice missile of barrack slang at an array of metaphors which their propounder thought stupendous in their brilliancy.

'*Bécasse!* When you stole your geese, you did but take your brethren home! Englishmen are but men. Put the wine in their head, make them whirl in a waltz, promise them a kiss, and one turns such brains as they have inside out, as a *piou-piou* turns a dead soldier's wallet. When a woman is handsome, she is never denied. He shall tell me where he comes from. I doubt that it is from England; see here—why not?' and she checked the Noes off on her lithe brown fingers: 'first, he never says God-damn; second, he don't eat his meat raw; third, he speaks very soft; fourth, he waltzes so light, so light! fifth, he never grumbles in his throat like an angry bear; sixth, there is no fog in him. How can he be English with all that?'

'There are English, and English,' said the philosophic Tata, who piqued himself on being serenely cosmopolitan.

Cigarette blew a contemptuous puff of smoke.

---

[1] Chasseurs d'Afrique.   [2] A musket.

'There was never one yet that did not growl! *Pauvres diables!* if they don't use their tusks, they sit and sulk!—an Englishman is always boxing or grumbling;—the two make up his life.'

Which view of *Anglo-rabies* she had derived from a profound study of various vaudevilles, in which the traditional God-damn was pre-eminent in his usual hues; and having delivered it, she sprang down from her wall, strapped on her little *barillet*, nodded to her gros bébées, where they lounged full length in the shadow of the stone wall, and left them to resume their game at Boc, while she started on her way, as swift and as light as a chamois, singing, with gay ringing emphasis, that echoed all down the hot and silent air, the second verse of Béranger:

> Je fus chère à tous nos héros;
>   Hélas! combien j'en pleure,
> Ainsi soldats et généraux
>   Me comblaient à tout heure,
> D'amour de gloire et de butin,
> Tintin, tintin, tintin, r'lin tintin,
> D'amour de gloire et de butin,
> Soldats, voilà Catin!

The song was not altogether her song, however, for she had wept for none—wept not at all: she had never shed tears in her life. A dashing, dauntless, vivacious life, just in its youth, loving plunder, and mischief, and mirth; caring for nothing; and always ready with a laugh, a song, a slang repartee, or a shot from the dainty pistols thrust in her sash, that a general of division had given her, whichever best suited the moment.

Her mother a camp-follower, her father nobody knew who, a spoilt child of the Army from her birth, with a heart as bronzed as her cheek, and her respect for the laws of meum and tuum *nil*, yet with odd stray nature-sown instincts here and there, of a devil-may-care nobility, and of a wild grace that nothing could kill—Cigarette was the pet of the Army of Africa, and was as lawless as most of her patrons.

She would eat a succulent duck, thinking it all the spicier because it had been a soldier's 'loot'; she would wear the gold plunder off dead Arabs' dress, and never have a pang of conscience with it; she would dance all night long, when she had a chance, like a little Bacchante; she would shoot a man, if need be, with all the nonchalance in the world. She had had a thousand lovers, from handsome Marquises of the Guides to tawny black-browed scoundrels in the Zouaves, and she had never loved anything, except the roll of the *pas de charge*, and the sight of her own arch defiant face, with its scarlet lips and its short jetty hair, when she saw it by chance in some burnished cuirass, that served her for a mirror. She was more like a handsome saucy boy than anything else under the sun, and yet there was that in the pretty, impudent little Friend of the Flag that was feminine with it all—generous and graceful amidst all her boldness, and her licence, her revelries, and the unsettled life she led in the barracks and the camps, under the shadow of the eagles.

Away she went, now singing—

> Mais je ris en sage,
> Bon!
> La farira dondaine,
> Gai!
> La farira dondée!

down the crooked windings, and over the ruined gardens of the old Moorish quarter of the Kasbah, the hilts of the tiny pistols glancing in the sun, and the fierce fire of the burning sunlight pouring down unheeded on the brave bright hawk eyes that had never, since they first opened to the world, drooped or dimmed for the rays of the sun, or the gaze of a lover, for the menace of death, or the presence of war.

Of course, she was a little Amazon; of course, she was a little Guerilla; of course, she did not know what a blush meant; of course, her thoughts were as slang and as riotous as her mutinous mischief was in its act: but she was '*bon soldat*', as she was given to say, with a toss of her curly head, and she had some of the virtues of soldiers. Soldiers had been about her ever since she first remembered having a wooden casserole for a cradle, and sucking down red wine through a pipe-stem. Soldiers had been her books, her teachers, her models, her guardians, and, later on, her lovers, all the days of her life. She had had no guiding-star except the eagles on the standards; she had had no cradle-song except the rataplan and the réveillé; she had had no sense of duty taught her, except to face fire boldly, never to betray a comrade, and to worship but two deities, 'La Gloire' and 'La France'.

Yet there were tales told in the barrack-yards and under canvas of the little Amie du Drapeau, that had a gentler side. Of how softly she would touch the wounded; of how deftly she would cure them. Of how carelessly she would dash through under a raking fire, to take a draught of water to a dying man. Of how she had sat by an old Grenadier's death-couch, to sing to him, refusing to stir, though it was a fête at Châlons, and she loved fêtes as only a French girl can. Of how she had ridden twenty leagues on a saddleless Arab horse, to fetch the surgeon of the Spahis to a Bedouin perishing in the desert of shot wounds. Of how she had sent every sou of her money to her mother, so long as that mother lived—a brutal, drunk, vile-tongued old woman, who had beaten her oftentimes, as the sole maternal attention, when she was but an infant. These things were told of Cigarette, and with a perfect truth. She was '*mauvais sujet, mais bon soldat*', as she classified herself. Her own sex would have seen no good in her; but her comrades-at-arms could, and did. Of a surety, she missed virtues that women prize; but, not less of a surety, had she caught some that they miss.

Singing her refrain, on she dashed now, swift as a greyhound, light as a hare, glancing here and glancing there as she bounded over the pictur-esque desolation of the Kasbah. It was just noon, and there were few could brave the noon-heat as she did. It was very still; there was only from a little distance the roll of the French kettle-drums where the drummers of the African regiments were practising. 'Holà! le v'la!' cried Cigarette to herself, as her falcon-eyes darted right and left; and, like a chamois, she leaped down over the great masses of Turkish ruins, cleared the channel of a dry water-course, and alighted just in front of a Chasseur d'Afrique, who was sitting alone on a broken fragment of white marble, relic of some Moorish mosque, whose delicate columns, crowned with wind-sown grasses, rose behind him, against the deep intense blue of the cloudless sky.

He was sitting thoughtfully enough, almost wearily, tracing figures in the dry sand of the soil with the point of his scabbard; yet he had all the look about him of a brilliant French soldier, of one who, moreover, had seen hot and stern service. He was bronzed, but scarcely looked so after the red, brown, and black of the Zouaves and the Turco, for his skin was naturally very fair, the features delicate, the eyes very soft—for which Monsieur Tata had growled contemptuously, 'a woman's face'—a long, silken chestnut beard swept over his chest; and his figure, as he leaned there in the blue and scarlet and gold of the Chasseurs' uniform, with his spurred heel thrust into the sand, and his arm resting on his knee, was, as Cigarette's critical eye told her, the figure of a superb cavalry rider, light, supple, long of limb, wide of chest, with every sinew and nerve firm-knit as links of steel. She glanced at his hands, which were very white, despite the sun of Algiers, and the labours that fall to a private of Chasseurs.

'*Beau lion!*' she thought, 'and noble, whatever he is.'

But the best of blood was not new to her in the ranks of the Algerian regiments; she had known so many of them—those gilded butterflies of the Chaussée d'Antin, those lordly spendthrifts of the *vieille roche*, who had served in the battalions of the *demie-cavalerie*, or the squadrons of the French Horse, to be thrust nameless and unhonoured into a sand-hole hastily dug with the bayonets in the hot hush of an African night.

She woke him unceremoniously from his reverie, with a challenge to wine.

'Ah-ha, mon Roumi![1] Tata Leroux says you are English; by the faith, he must be right, or you would never sit musing there like an owl in the sunlight! Take a draught of my burgundy; bright as rubies. I never sell bad wines—not I!—I know better than to drink them myself.'

He started and rose; and, before he took the *bidon*,[2] bowed to her, raising his cap with a grave courteous obeisance; a bow that had used to be noted in throne-rooms for its perfection of grace.

'Ah, ma belle, is it you?' he said, wearily. 'You do me much honour.'

Cigarette gave a little petulant twist to the tap of her wine-barrel. She was not used to that style of salutation. She half liked it, half resented it. It made her wish, with an impatient scorn for the wish, that she knew how to read, and had not her hair cut short like a boy's—a weakness the little vivandière had never been visited with before.

'Morbleu!' she said, pettishly. 'You are too fine for us, *mon brave*. In what country, I should wonder, does one learn such dainty ceremony as that?'

'Where should one learn courtesies if not in France?' he answered, wearily. He had danced with this girl-soldier the night before at a *guinguette* ball, seeing her for the first time, for it was almost the first time he had been in the city since the night when he had thrown the dice; and lost two Napoléons and the Bedouins to Claude de Chanrellon; but his thoughts were far from her in this moment.

'Ouf! you have learnt carte and tierce with your tongue!' cried Cigarette, provoked to receive no more compliment than that. From generals and staff-officers as from drummers and trumpeters she was accustomed to flattery and wooing, luscious as sugared chocolate, and ardent as flirtation, with a barrack flavour about it, commonly is; she would, as often as not, to be sure, finish it with the butt-end of her pistol, or the butt-end of some bit of stinging sarcasm, but still for all that she liked it, and resented its omission. 'They say you are English, but I don't believe it; you speak too soft, and you sound the double Ls too well. A Spaniard, eh?'

'Do you find me so devout a Catholic that you think so?'

She laughed. 'A Greek, then?'

'Still worse. Have you seen me cheat at cards?'

'An Austrian? You waltz like a White-Coat?'

He shook his head.

---

[1] Soldier.   [2] Little wooden drinking-cup.

She stamped her little foot into the ground—a foot fit for a model, with its shapely military boot; spurred, too, for Cigarette rode like a circus-rider.

'*Bécasse!* say what you are, then, at once.'

'A soldier of France. Can you wish me more?'

For the first time her eyes flashed and softened—her one love was the tricolour.

'True!' she said, simply. 'But you were not always a soldier of France? You joined, they say, twelve years ago? What were you before then?'

She here cast herself down in front of him, and, with her elbows on the sand, and her chin on her hands, watched him with all the frank curiosity and unmoved nonchalance imaginable, as she launched the question point-blank.

'Before!' he said, slowly. 'Well—a fool.'

'You belonged to the majority, then!' said Cigarette, with a piquance made a thousand times more piquant by the camp slang she spoke in. 'You should not have had to come into the ranks, *mon ami*; majorities—specially that majority—have very smooth sailing generally!'

He looked at her more closely, though she wearied him.

'Where have you got your ironies, Cigarette? You are so young.'

She shrugged her shoulders.

'Bah! one is never young, and always young in camps. Young? Pardieu! When I was four, I could swear like a grenadier, plunder like a préfet, lie like a priest, and drink like a bohemian.'

Yet—with all that—and it was the truth, the brow was so open under the close rings of the curls, the skin so clear under the sun-tan, the mouth so rich and so arch in its youth!

'Why did you come into the service?' she went on, before he had a chance to answer her. 'You were born in the Noblesse—bah! I know an aristocrat at a glance! *Ceux qui ont pris la peine de naître!*—don't you like *Figaro*? My Spahis played it last winter, and I was Figaro myself. Now many of those aristocrats come; shoals of them, but it is always for something. They all come for something; most of them have been ruined by the *lionnes*, a hundred million of francs gone in a quarter! Ah-bah! what blind bats the best of you are! They have gambled, or bet, or got into hot water, or fought too many duels, or caused a court scandal, or something; all the aristocrats that come to Africa are ruined. What ruined you, Monsieur l'Aristocrat?'

'Aristocrat? I am none. I am a Corporal of the Chasseurs.'

'Diable! I have known a Duke a Corporal! What ruined you?'

'What ruins most men, I imagine—folly.'

'Folly sure enough!' retorted Cigarette, with scornful acquiescence. She had no patience with him. He danced so deliciously, he looked so superb, and he would give her nothing but these absent answers. 'Wisdom don't bring men who look as you look into the ranks of the volunteers for Africa. Besides, you are too handsome to be a sage!'

He langhed a little.

'I never was one, that's certain. And you are too pretty to be a cynic.'

'A what?' she did not know the word. 'Is that a good cigar you have? Give me one. Do women smoke in your old country?'

'Oh yes—many of them.'

'Where is it, then?'

'I have no country—now.'

'But the one you had?'

'I have forgotten I ever had one.'

'Did it treat you ill, then?'

'Not at all.'

'Had you anything you cared for in it?'

'Well—yes.'

'What was it? A woman?'

'No—a horse.'

He stooped his head a little as he said it, and traced more figures slowly in the sand.

'Ah!'

She drew a short, quick breath. She understood that; she would only have laughed at him had it been a woman: Cigarette was more veracious than complimentary in her estimate of her own sex.

'There was a man in the Guides I knew,' she went on softly, 'loved a horse like that—he would have died for Cossack—but he was a terrible gambler, terrible. Not but what I like play myself. Well, one day he played and played till he was mad, and everything was gone; and then in his rage he staked the only thing he had left. Staked and lost the horse! He never said a word; but he just slipped a pistol in his pocket, went to the stable, kissed Cossack once—twice—thrice—and shot himself through the heart.'

'Poor fellow!' murmured the Chasseur d'Afrique, in his chesnut beard.

Cigarette was watching him with all the keenness of her falcon eyes; 'he has gambled away a good deal too,' she thought. 'It is always the same old story with them.'

'Your cigars are good, *mon lion*,' she said, impatiently, as she sprang up, her lithe elastic figure in the bright vivandière uniform standing out in full relief against the pearly grey of the ruined pillars, the vivid green of the rank vegetation, and the intense light of the noon. 'Your cigars are good, but it is more than your company is! *Ma cantche!* If you had been as dull as this last night I would not have danced a single turn with you in the cancan!'

And with a bound to which indignation lent wings like a swallow's, the Friend of the Flag, insulted and amazed at the apathy with which her advances to friendship had been received, dashed off at her topmost speed, singing all the louder out of bravado. 'To have nothing more to say to me after dancing with me all night!' thought Cigarette, with fierce wrath at such contumely, the first neglect the pet of the Spahis had ever experienced.

She was incensed, too, that she had been degraded into that momentary wish that she knew how to read, and looked less like a boy— just because a Chasseur with white hands and silent ways had made her a grave bow! She was more incensed still because she could not get at his history, and felt, despite herself, a reluctance to bribe him for it with those cajoleries whose potency she had boasted to Tata Leroux. '*Gare à lui!*' muttered the soldier-coquette passionately, in her little white teeth, so small and so pearly, though they had gripped a bridle tight before then, when each hand was filled with a pistol. '*Gare à lui!* If he offend *me* there are five hundred swords that will thrust civility into him, five hundred shots that will teach him the cost of daring to provoke Cigarette!'

En route through the town her wayward way took the pretty brunette Friend of the Flag as many devious meanderings as a bird takes in a summer's-day flight when it stops here for a berry, there for a grass-seed, here to dip its beak into cherries, there to dart after a dragon-fly, here to shake its wings in a brook, there to poise on a lily-bell.

She loitered in a thousand places, for Cigarette knew everybody; she chattered with a group of Turcos, she emptied her barrel for some Zouaves, she ate sweetmeats with a lot of negro boys, she boxed a little drummer's ear for slurring over the 'r'lin tintin' at his practice, she drank a demi-tasse with some officers at a café, she had ten minutes' pistol-shooting, where she beat hollow a young dandy of the Guides who had come to look at Algiers for a week, and made even points with one of the first shots of the '*Cavalerie à pied*', as the Algerian antithesis runs. Finally she paused before the open French window of a snow-white villa, half-buried in tamarisk and orange and pomegranate, with the deep-hued flowers glaring in the sun, and a hedge of wild cactus fencing it in; through the cactus she made her way as easily as a rabbit burrows; it would have been an impossibility to Cigarette to enter by any ordinary means; and balancing herself lightly on the sill for a second, stood looking in at the chamber.

'Ho, M. le Marquis! the Zouaves have drunk all my wine up; fill me my keg with yours for once—the very best burgundy, mind. I'm half afraid your cellar will hurt my reputation.'

The chamber was very handsome, hung and furnished in the very best Paris fashion, and all glittering with amber and ormolu and velvets; in it half a dozen men—officers of the cavalry—were sitting over their noon breakfast, and playing a lansquenet at the same time. The table was crowded with dishes of every sort, and wines of every vintage, and the fragrance of their bouquet, the clouds of smoke, and the heavy scent of the orange-blossoms without, mingled together in an intense perfume. He whom she addressed, M. le Marquis de Châteauroy, laughed, and looked up.

'Ah, is it thee, my pretty brunette? Take what thou wantest out of the ice-pails.'

'*Premier cru?*' asked Cigarette, with the dubious air and caution of a connoisseur.

'Comet!' said M. le Marquis, amused with the precautions taken with his cellar, one of the finest in Algiers. 'Come in and have some breakfast, *ma belle*. Only pay the toll.'

Where he sat between the window and the table he caught her in his arms and drew her pretty face down; Cigarette, with the laugh of a saucy child, whisked her cigar out of her mouth, and blew a great cloud of smoke in his eyes. She had no particular fancy for him, though she had for his wines. Shouts of mirth from the other men completed the Marquis's discomfiture, as she swayed away from him, and went over to the other side of the table, emptying some bottles unceremoniously into her wine-keg; iced, ruby, perfumy claret that she could not have bought anywhere for the barracks.

'Holà!' cried the Marquis. 'Thou art not generally so coy with thy kisses, petite.'

Cigarette tossed her head.

'I don't like bad clarets after good! I've just been with your Corporal, "Bel-à-faire-peur"; you are no beauty after him, M. le Colonel.'

Châteauroy's face darkened. He was a colossal-limbed man, whose bone was iron, and whose muscles were like oak-fibres; he had a dark keen head like an eagle's, the brow narrow but very high, looking higher because the close-cut hair was worn off the temples; thin lips hidden by heavy curling moustaches, and a skin burnt black by long African service. Still he was fairly handsome enough not to have muttered so heavy an oath as he did at the vivandière's jest.

'Sacre-bleu! I wish my Corporal were shot! one can never hear the last of him.'

Cigarette darted a quick glance at him. 'Oh ho, jealous, mon brave!' thought her quick wits. 'And why, I wonder?'

'You haven't a finer soldier in your Chasseurs, mon cher; don't wish him shot for the good of the service,' said the Viscount de Chanrellon, who had now a command of his own in the Light Cavalry of Algiers. 'Pardieu! if I had to choose whether I'd be backed by "Bel-à-faire-peur", or by six other men in a skirmish, I'd choose him and risk the odds.'

Châteauroy tossed off his burgundy with a contemptuous impatience.

'Diable! That is the *galamatias* one always hears about this fellow—as if he were a second Roland, or a revivified Bayard! *I* see nothing particular in him, except that he's too fine a gentleman for the ranks.'

'Fine? Ah!' laughed Cigarette. 'He made me a bow this morning like a court chamberlain—and his beard is like carded silk—and he has such woman's hands, mon Dieu! But he is a *croc-mitaine* too.'

'Rather!' laughed Claude de Chanrellon, as magnificent a soldier himself as ever crossed swords. 'I said he would eat fire the very minute he played that queer game at dice with me years ago. I wish I had him instead of you, Châteauroy. Like lightning in a charge, and yet the very man for a dangerous bit of secret service that wants the softness of a panther. We all let our tongues go too much, but he says so little—just a word here, a word there—when one's wanted—no more; and he's the devil's own to fight.'

The Marquis heard the praise of his Corporal, knitting his heavy brows. It was evident the private was no favourite with him.

'The fellow rides well enough,' he said, with an affectation of carelessness; 'there—for what I see—is the end of his marvels. I wish you had him, Claude, with all my soul.'

'Oh-hè!' cried Chanrellon, wiping the Rhenish off his tawny moustaches, 'he should have been a captain by this, if I had. Morbleu! he is a splendid sabreur—kills as many men to his own sword as I could myself, when it comes to a hand in hand fight; breaks horses in like magic; rides them like the wind, has a hawk's eye over open country; obeys like clockwork; what more can you want?'

'Obeys! yes,' said the Colonel of Chasseurs, with a snarl. 'He'd obey without a word, if you ordered him to walk up to a cannon's mouth and be blown from it; but he gives you such a d——d languid *grand seigneur* glance as he listens that one would think he commanded the regiment.'

'But he's very popular with your men, too?'

'Monsieur, the worst quality a Corporal can have. His idea of maintaining discipline is to treat them to cognac, and give them tobacco.'

'Pardieu! not a bad way either with our French fire-eaters. *Il connaît son monde; ce brave*. Your squadrons would go to the devil after him.'

The Colonel gave a grim laugh.

'I dare say nobody knows the way better.'

Cigarette, flirting with the other officers, drinking champagne by great glassfuls, eating bonbons from one, sipping another's soup, pulling the limbs of a succulent ortolan to pieces with a relish, and devouring truffles with all the zest of a bon-vivant, did not lose a word, and catching the inflection of Châteauroy's voice, settled with her own thoughts that 'Bel-à-faire-peur' had not a fair field or a smooth course with his Colonel. The weathercock heart of the little 'Friend of the Flag' veered round, with her sex's common custom, to the side that was the weakest.

'Dieu de Dieu, M. le Colonel!' she cried, while she ate M. le Colonel's *foie gras* with as little ceremony and as much enjoyment as would be expected from a young plunderer accustomed to think a meal all the better spiced by being stolen 'by the rules of war',—'whatever else your handsome Corporal is, he is an aristocrat. Ah, ah, I know the aristocrats —I do! Their touch is so gentle, and their speech is so soft, and they have no slang of the camp, and yet they are such *diablotins* to fight and eat steel, and die laughing all so quiet and nonchalant. Give me the aristocrats—the real thing, you know. Not the ginger cakes, just gilt, that are ashamed of being honest bread—but the old blood like Bel-à-faire-peur.'

The Colonel laughed, but restlessly; the little ingrate had aimed at a sore point in him. He was of the First Empire Nobility, and he was weak enough, though a fierce, dauntless, iron-nerved soldier, to be discontented with the great fact that his father had been a hero of the Army of Italy and scarce inferior in genius to Massena, because impatient of the minor one that, before strapping on a knapsack to have his first taste of war under Custine, the Marshal had been but a postilion at a posting inn in the heart of the Nivernais.

'Ah, my brunette!' he answered, with a rough laugh, have you taken my popular Corporal for your lover? You should give your old friends warning first, or he may chance to get an ugly spit on a sabre.'

The Amie du Drapeau tossed off her sixth glass of champagne. She felt for the first time in her life a flush of hot blood on her brown clear cheek, well used as she was to such jests and such lovers as these.

'Ma foi!' she said, coolly. 'He would be more likely to spit than be spitted if it came to a duel. I should like to see him in a duel; there is not a prettier sight in the world when both men have science. As for fighting for me! Morbleu! I will thank nobody to have the impudence to do it, unless I order them out. Coqueline got shot for me, you remember;—he was a pretty fellow, Coqueline, and they killed him so clumsily, that they disfigured him terribly—it was quite a pity. I said then I would have no more handsome men fight about me. *You* may, if you like, M. le Faucon Noir.'

Which title she gave with a saucy laugh, hitting with a chocolate bonbon the black African-burnt visage of the omnipotent chief she had the audacity to attack. High or low, they were all the same to Cigarette. She would have 'slanged' the Emperor himself with the self-same coolness, and the Army had given her a passport of immunity so wide, that it would have fared ill with any one who had ever attempted to bring the vivandière to book for her uttermost mischief.

'By the way!' she went on, quick as thought, with her reckless devil-may-care gaiety. 'One thing!—Your Corporal will demoralize the Army of Africa, m'sieu?'

'Eh? He shall have an ounce of cold lead before he does. What in?'

'He will demoralize it,' said Cigarette, with a sagacious shake of her head. 'If they follow his example, we shan't have a Chasseur, or a Spahis, or a Piou-piou, or a Sapeur worth anything——'

'Sacré! What does he do?' The Colonel's strong teeth bit savagely through his cigar; he would have given much to have been able to find a single thing of insubordination or laxity of duty in a soldier who irritated and annoyed him, but who obeyed him implicitly, and was one of the most brilliant 'fire-eaters' of his regiment.

'He won't only demoralize the Army,' pursued Cigarette, with vivacious eloquence, 'but if his example is followed, he'll ruin the Préfets, close the Bureaux, destroy the Exchequer, beggar all the officials, make African life as tame as milk-and-water, and rob *you*, M. le Colonel, of your very highest and dearest privilege!'

'Sacre-bleu!' cried her hearers, as their hands instinctively sought their swords, 'what does he do?'

Cigarette looked at them out of her arch black lashes.

'Why, *he never thieves from the Arabs!* If the fashion come in, adieu to our occupation. Courtmartial him, Colonel!'

With which sally Cigarette thrust her curls back off her temples, and launched herself into lansquenet with all the ardour of a gambler and the vivacity of a child, her eyes flashing, her cheeks flushing, her little teeth set, her whole soul in the whirl of the game, made all the more riotous by the peels of laughter from her comrades, and the wines that were washed down like water. Cigarette was a terrible little gamester, and had gaming made very easy to her, for it was the creed of the Army that her losses never counted, but her gains were paid to her often double or treble. Indeed, so well did she play, and so well did the Goddess of Hazard favour her, that she might have grown a millionaire on the fruits of her dice and her cards, but for this fact, that whatever the little Friend of the Flag had in her hands one hour, was given away the next, to the first wounded soldier, or ailing veteran, or needy Arab woman that required the charity.

As much gold was showered on her as on Isabel of the Jockey Club; but Cigarette was never the richer for it. 'Bah!' she would say, when they told her of her heedlessness, 'money is like a mill, no good standing still. Let it turn, turn, turn, as fast as ever it can, and the more bread will come from it for the people to eat.'

The vivandière was by instinct a fine political economist.

Meanwhile, where she had left him among the stones of the ruined mosque, the Chasseur, whom they nicknamed Bel-à-faire-peur, in a double sense, because of his 'woman's face', as Tata Leroux termed it, and because of the terror his sword had become through North Africa, sat motionless with his right arm resting on his knee, and his spurred heel thrust into the sand, the sun shining down unheeded in its fierce burning glare on the chesnut masses of his beard, and the bright glitter of his uniform.

He was a dashing cavalry soldier, who had had a dozen wounds cut over his body by the Bedouin swords, in many and hot skirmishes; who had waited through sultry African nights for the lion's tread, and had fought the desert-king and conquered; who had ridden a thousand miles over the great sand waste, and the boundless arid plains, and slept under the stars with the saddle beneath his head, and his rifle in his hand, all through the night; who had served, and served well, in fierce, arduous, unremitting work, in trying campaigns and in close discipline; who had blent the *verve*, the brilliance, the daring, the eat-drink-and-enjoy-for-to-morrow-we-die of the French Chasseur, with something that was very different, and much more tranquil.

Yet, though as bold a man as any enrolled in the French Service, he sat alone here in the shadow of the column, thoughtful, motionless, lost in silence.

In his left hand was a *Galignani* six months old, and his eyes rested on a line in the obituary:

'On the 10th ult., at Royallieu, suddenly, the Right Hon. Denzil, Viscount Royallieu; aged 90.'

CHAPTER II

Cigarette en Bacchante

Vanitas vanitatum! The dust of death lies over the fallen altars of
Bubastis, where once all Egypt came down the flood of glowing Nile, and
Herodotus mused under the shadowy foliage, looking on the lake-like
rings of water. The Temple of the Sun, where the beauty of Asenath
beguiled the Israelite to forget his sale into bondage and banishment, lies
in shapeless hillocks, over which canter the mules of dragomen, and
chatter the tongues of tourists. Where the Lutetian Palace of Julian saw
the Legions rush, with torches and with wine-bowls, to salute their
darling as Augustus, the sledge-hammer and the stucco of the
Haussmann fiat bear desolation in their wake. Levantine dice are rattled,
where Hypatia's voice was heard. Bills of exchange are trafficked in,
where Cleopatra wandered under the palm aisles of her rose gardens.
Drummers roll their caserne-calls, where Drusus fell and Sulla laid down
dominion.

And here—in the land of Hannibal, in the conquest of Scipio, in the
Phoenicia, whose loveliness used to flash in the burning, sea-mirrored
sun, while her fleets went eastward and westward for the honey of Athens
and the gold of Spain—here Cigarette danced the cancan!

An *auberge* of the *barrière* swung its sign of the *As de Pique*, where
feathery palms once had waved above mosques of snowy gleam, with
marble domes and jewelled arabesques, and the hush of prayer under
columned aisles. '*Débits de vin, liqueurs, et tabac*,' was written, where once
verses of the Koran had been blazoned by reverent hands along porphyry
cornices and capitals of jasper. A Café Chantant reared its impudent little
roof, where once, far back in the dead cycles, Phoenician warriors had
watched the galleys of the gold-haired favourite of the gods bear down to
smite her against whom the one unpardonable sin of rivalry to Rome was
quoted.

The riot of a Paris guinguette was heard, where once the tent of
Belisarius might have been spread above the majestic head that towered
in youth above the tempestuous seas of Gothic armies, as when, silvered
with age, it rose as a rock against the on-sweeping flood of Bulgarian
hordes. The grisette charms of little tobacconists, milliners, flower-girls,
lemonade-sellers, bonbon sellers, and filles de joie flaunted themselves in

the gaslight, where the lustrous sorceress eyes of Antonina might have glanced over the Afric Sea, while her wanton's heart, so strangely filled with leonine courage and shameless licence, heroism and brutality, cruelty and self-devotion, swelled under the purples of her delicate vest, at the glory of the man she at once dishonoured and adored.

Vanitas vanitatum! Under the thirsty soil, under the ill-paved streets, under the arid turf, the Legions lay dead, with the Carthaginians they had borne down under the mighty pressure of their phalanx; and the Byzantine ranks were dust side by side with the soldiers of Gelimer. And here, above the graves of two thousand centuries, the little light feet of Cigarette danced joyously in that triumph of the Living, who never remember that they also are dancing onward to the tomb!

It was a low-roofed, white-plastered, gaudily decked, smoke-dried mimicry of the guinguettes beyond Paris. The long room, that was an imitation of the Salle de Mars on a Lilliputian scale, had some bunches of lights flaring here and there, and had its walls adorned with laurel-wreaths, stripes of tricoloured paint, vividly coloured medallions of the Second Empire, and a little pink gauze flourished about it, that flashed into brightness under the jets of flame—trumpery, yet trumpery, which, thanks to the instinct of the French esprit, harmonized, and did not vulgarize; a gift French instinct alone possesses. The floor was bare and well polished; the air full of tobacco-smoke, wine fumes, brandy odours, and an overpowering scent of oil, garlic, and *pot au feu*. Riotous music pealed through it, that even in its clamour kept a certain silvery ring, a certain rhythmical cadence. Pipes were smoked, barrack slang, camp slang, barrière slang, temple slang, were chattered volubly. Theresa's songs were sung by bright-eyed, sallow-cheeked Parisiennes, and chorused by the lusty lungs of Zouaves and Turcos. Good humour prevailed, though of a wild sort; the mad gallop of the Rigolboche had just flown round the room, like lightning, to the crash and the tumult of the most headlong music that ever set spurred heels stamping and grisettes' heels flying: and now, where the crowds of soldiers and women stood back to leave her a clear place, Cigarette was dancing alone.

She had danced the cancan; she had danced since sunset; she had danced till she had tired out cavalry-men, who could go days and nights in the saddle without a sense of fatigue, and made Spahis cry quarter, who never gave it by any chance in the battlefield; and she was dancing now like a little Bacchante, as fresh as if she had just sprung up from a long summer day's rest. Dancing as she would dance only now and then, when caprice took her, and her wayward vivacity was at its height, on the green space before a tent full of general officers, on the bare floor of a barrack-room, under the canvas of a fête-day's booth, or as here, in the *salle* of a café.

Marshals had more than once essayed to bribe the famous little Friend of the Flag to dance for them, and had failed: but, for a set of soldiers, war-worn, dust-covered, weary with toil and stiff with wounds, she would do it, till they forgot their ills, and got as intoxicated with it as with champagne. For her *gros bébées*, if they were really in want of it, she would do anything. She would flout a star-covered General, box the ears of a brilliant Aide, send killing missiles of slang at a dandy of a regiment de famille, and refuse point-blank a Russian Grand Duke; but to '*mes enfans*', as she was given to calling the rough tigers and grisly veterans of the Army of Africa, Cigarette was never capricious, however mischievously she would rally, or contemptuously would rate them, when they deserved it.

And she was dancing for them now.

Her soft short curls all fluttering, her cheeks all bright with a scarlet flush, her eyes as black as night and full of fire, her gay little uniform, with its scarlet and purple, making her look like a fuchsia bell tossed by the wind to and fro, ever so lightly, on its delicate swaying stem, Cigarette danced with the wild grace of an Almeh, of a Bayadère, of a Nautch girl, as untutored and instinctive in her as its song to a bird, as its swiftness to a chamois. To see Cigarette was like drinking light fiery wines, whose intoxication was gay as mischief, and sparkling as themselves. All the warmth of Africa, all the wit of France, all the bohemianism of the Flag, all the caprices of her sex, were in that bewitching dancing. Flashing, fluttering, circling, whirling, glancing like a sabre's gleam, tossing like a flower's head, bounding like an antelope, launching like an arrow, darting like a falcon, skimming like a swallow; then for an instant resting as indolently, as languidly, as voluptuously, as a water-lily rests on the water's breast—Cigarette *en Bacchante* no man could resist.

When once she abandoned herself to the afflatus of that dance-delirium, she did with her beholders what she would. The famous Cachucha, that made the reverend Cardinals of Spain fling off their pontifical vestments, and surrender themselves to the witchery of the castanets and the gleam of the white twinkling feet, was never more irresistible, more enchanting, more full of wild, soft, bizarre, delicious grace. It was a poem of motion and colour, an ode to Venus and Bacchus.

All her heart was in it—that heart of a girl and a soldier, of a hawk and a kitten, of a bohemian and an epicurean, of a lascar and a child, which beat so brightly and so boldly under the dainty gold aiglettes with which she laced her dashing little uniform.

In the Chambrée of Zéphyrs, among the Douars of Spahis, on sandy soil under African stars, above the heaped plunder brought in from a razzia, in the yellow light of candles fastened to bayonets stuck in the earth at a bivouac, on the broad deal table of a barrack-room full of black-browed *conscrits indigènes*, amidst the thundering echoes of the Marseillaise des Bataillons, shouted from the brawny chests of Zouaves, Cigarette had danced, danced, danced, till her whole vivacious life seemed pressed into one hour, and all the mirth and mischief of her little brigand's soul seemed to have found their utterance in those tiny, slender, spurred, and restless feet, that never looked to touch the earth which they lit on lightly as a bird alights, only to leave it afresh, with wider swifter bound, with ceaseless airy flight.

So she danced now, in the cabaret of the As de Pique. She had a famous group of spectators, not one of whom knew how to hold himself back from springing in to seize her in his arms, and whirl with her down the floor. But it had been often told them by experience, that, unless she beckoned one out, a blow of her clenched hand and a cessation of her impromptu *pas de seul*, would be the immediate result. Her spectators were renowned *croque-mitaines*; men whose names rang like trumpets in the ear of Kabyle and Marabout; men who had fought under the noble colours of the day of Mazagran, or had cherished or emulated its traditions; men who had the salient features of all the varied species that make up the soldiers of Africa.

There was Ben Arslan, with his crimson burnous wrapped round his towering stature, from whom Moor and Jew fled as before a pestilence, the fiercest, deadliest, most voluptuous of all the Spahis; brutalized in his drink, merciless in his loves; all an Arab when once back in the desert, with a blow of a scabbard his only payment for forage, and a thrust of his sabre his only apology to husbands, but to the Service a slave, and in the combat a lion.

There was Beau-Bruno, a dandy of Turcos, whose snowy turban and olive beauty bewitched half the women of Algeria, who himself affected to neglect his conquests, with a supreme contempt for those indulgences, but who would have been led out and shot rather than forego the personal adornings, for which his adjutant and his *capitaine du bureau* growled unceasing wrath at him with every day that shone.

There was Pouffer-de-Rire, a little Tringlo,[1] the wittiest, gayest, happiest, sunniest-tempered droll in all the army, who would sing the camp-songs so joyously through a burning march, that the whole of the battalions would break into one refrain as with one throat, and press on

---

[1] Soldier of the commissariat and of the baggage-trains.

laughing, shouting, running, heedless of thirst, or heat, or famine, and as full of monkey-like jests as any gamins.

There was En-ta-maboull,[1] so nicknamed from his love for that unceremonious slang phrase—a Zouave who had the history of a Gil Blas, and the talent of a Crichton, the morals of an Abruzzi brigand, and the wit of a Falstaff; aquiline-nosed, eagle-eyed, black-skinned as an African, with adventures enough in his life to outvie Munchausen, with a purse always *pleine de vide*,[2] as the camp sentence runs; who thrust his men through the body as coolly as others kill wasps; who roasted a shepherd over the camp-fire for contumacy in concealing Bedouin whereabouts; yet who would pawn his last shirt at the bazaar to help a comrade in debt, and had once substituted himself for, and received fifty blows on the loins in the stead of, his sworn friend, whom he loved with that love of David for Jonathan, which, in Caserne life, is readier found than in Club life.

There was Pattes-du-Tigre, a small wiry supple-limbed fire-eater, with a skin like a coal and eyes that sparkled like the live coal's flame, a veteran of the Joyeux, who could discipline his roughs as a sheepdog his lambs, and who had one curt martial law for his detachment, brief as Draco's, and trimmed to suit either an attack on the enemy or the chastisement of an *indiscipliné*, lying in one simple word—'Fusillez.'

There was Barbe-Grise, a grisly *ancien* of Zéphyrs, who held the highest repute of any in his battalion for rushing on to a foe with a foot speed that could equal the canter of an Arab's horse; for having stood alone once the brunt of thirty Bedouins' attack, and ended by beating them back, though a dozen spearheads were launched into his body, and his *pantalons garances* were filled with his own blood; and for framing a matchless system of night plunder that swept the country bare as a table-rock in an hour, and made the colons surrender every hidden treasure, from a pot of gold to a hen's eggs, from a caldron of couscoussou to a tom-cat.

There was Alcide Echaufourées, also a Zéphyr, who had his nickname from the marvellous changes of costume with which he would pursue his erratic expedition, and deceive the very Arabs themselves into believing him a born Mussulman; a very handsome fellow, the Lauzun of his battalion, the Brummel of his caserne; *coquette* with his kepi on one side of his graceful head, and his moustaches soft as a lady's hair, whose paradise was a score of dangerous intrigues, and whose seventh heaven was a duel with an infuriated husband; incorrigibly lazy, but with the Italian laziness, as of the panther who sleeps in the sun, and with such

---

[1] Est-ce-que-tu es fou? in ordinary French.   [2] Penniless.

episodes of romance, mischief, love, and devilry in his twenty-five years of existence as would leave behind them all the invention of Dumas, *père ou fils*.

All these and many more like them were the spectators of Cigarette's ballet, applauding with the wild hurrah of the desert, with the clashing of spurs, with the thunder of feet, with the demoniac shrieks of irrepressible adoration and delight.

And every now and then her bright eyes would flash over the ring of familiar faces, and glance from them with an impatient disappointment as she danced; her *gros bébées* were not enough for her. She wanted a Chasseur with white hands and a grave smile to be amongst them; and she shook back her curls, and flushed angrily as she noted his absence, and went on with the pirouettes, the circling flights, the wild resistless abandonment of her inspirations, till she was like a little desert-hawk that is intoxicated with the scent of prey borne down upon the wind, and wheeling like a mad thing in the transparent æther and the hot sun-glow.

L'As de Pique was the especial estaminet of the *chasses-marais*. He was in the house; she knew it; had she not seen him drinking with some others, or rather paying for all but taking little himself, just as she entered? He was in the house, this mysterious Bel-à-faire-peur—and was not here to see her dance! Not here to see the darling of the Douars; the pride of every Chacal, Zéphyr, and Chasseur in Africa; the Amie du Drapeau who was adored by every one, from Chefs de Bataillons to *fantassins*, and toasted by every drinker, from Algiers to Oran, in the Champagne of Messieurs les Généraux as in the Cric of the Loustics round a camp-fire!

He was not there; he was leaning over the little wooden ledge of a narrow window in an inner room, from which, one by one, some Spahis and some troopers of his own *tribu*,[1] with whom he had just been drinking such burgundies and brandies as the place could give, had sloped away one by one under the irresistible attraction of the vivandière. An attraction, however, that had not seduced them till all the bottles were emptied, bottles more in number and higher in cost than were prudent in a Corporal who had but his pay, and that scant enough, to keep himself, and who had known what it was to find a roll of white bread and a cup of coffee a luxury beyond all reach, and to have to *faire la lessive*[2] up to the last thing in his haversack to buy a toss of thin wine when he was dying of thirst, or a slice of melon when he was parching with African fever.

---

[1] Squadron.   [2] Sell his whole effects.

But prudence had at no time been his speciality, and the reckless life of Algeria was not one to teach it, with its frank brotherly fellowship that bound the soldiers of each battalion, or each squadron, so closely in a fraternity of which every member took as freely as he gave; its gay, careless carpe diem camp-philosophy, the unconscious philosophy of men who enjoyed heart and soul if they had a chance, because they knew they might be shot dead before another day broke; and its swift and vivid changes that made tirailleurs and troopers one hour rich as a king in loot, in wine, in dark-eyed captives at the sacking of a tribe, to be the next day famished, scorched, dragging their weary limbs, or urging their sinking horses through endless sand and burning heat, glad to sell a cartouche, if they dared so break regimental orders, or to rifle a henroost, if they came near one, to get a mouthful of food, changing everything in their haversack for a sup of dirty water, and driven to pay with the thrust of a sabre for a lock of wretched grass to keep their beasts alive through the sickliness of a sirocco.

All these taught no caution to any nature normally without it; and the chief thing that his regiment had loved, in him whom they named Bel-à-faire-peur, from the first day that he had bound his red waist-sash about his loins, and the officers of the bureau had looked over the new volunteer, murmuring admiringly in their teeth '*Ce gaillard ira loin!*' had been that all he had was given, free as the winds, to any who asked or needed.

The all was slender enough. Unless he live by the ingenuity of his own manufactures, or by thieving or intimidating the people of the country, a French soldier has but barren fare and a hard struggle with hunger and poverty; and it was the one murmur against him, when he was lowest in the ranks, that he would never follow the fashion, in wringing out by force or threat the possessions of the native population. The one reproach, that made his fellow-lascars[1] impatient and suspicious of him, was that he refused any share in those rough arguments of blows and lunges with which they were accustomed to persuade every victim they came nigh to yield them up all such treasures of food, or drink, or riches, from sheep's liver and couscoussou, to Morocco carpets and skins of brandy and coins hid in the sand, that the Arabs might be so unhappy as to own in their reach. That the fattest pullet of the poorest Bedouin was as sacred to him as the banquet of his own Chef d'Escadron, let him be ever so famished after the longest day's march, was an eccentricity, and an insult to the usages of the corps, for which not even his daring and his popularity could wholly procure him pardon.

---

[1] Soldiers.

But this defect in him was counterbalanced by the lavishness with which his *décompte*[1] was lent, given, or spent in the very moment of its receipt. If a man of his *tribu* wanted anything, he knew that Bel-à-faire-peur would offer his last sou to aid him, or, if money were all gone, would sell the last trifle he possessed to the Riz-pain-sels,[2] to get enough to assist his comrade. It was a virtue which went far to vouch for all others in the view of his lawless, open-handed brethren of the Chambrée[3] and the Camp, and made them forgive him many moments, when the mood of silence and the habit of solitude, not uncommon with him, would otherwise have incensed a fraternity with whom '*tu fais suisse!*'[4] is the deadliest charge, and the sentence of excommunication against any who dare to provoke it.

One of those moods was on him now.

He had had a drinking-bout with the men who had left him, and had laughed as gaily and as carelessly, if not as riotously, as any of them at the wild mirth, the unbridled licence, the amatory recitations, and the Bacchic odes in their lawless *sapir*, that had ushered the night in while his wines unlocked the tongues and flowed down the throats of the fierce Arab-Spahis and the French cavalry-men. But now he leant out of the pent-up casement, with his arms folded on the sill, and a short pipe in his teeth, thoughtful and solitary after the orgy, whose heavy fumes and clouds of smoke still hung heavily on the air within.

The window looked on a little, dull, close courtyard, where the yellow leaves of a withered gourd trailed drearily over the grey uneven stones. The clamour of the applause and the ring of the music from the dancing-hall echoed with a whirling din in his ear, and made, in sharper stranger contrast, the quiet of the narrow court with its strip of starry sky above its four high walls.

He leaned there musing and grave, hearing little of the noise about him; there was always noise of some sort in the clangour and tumult of barrack or bivouac life, and he had grown to heed it no more than he heeded the roar of desert beasts about him, when he slept in the desert or the hills; but, looking dreamily out at the little shadowy square, with the sear gourd leaves and the rough misshapen stones. His present and his future were neither much brighter than the gloomy walled-in den on which he gazed.

Twelve years before, when he had been ordered into the *champ de manœuvre* for the first time, to see of what mettle he was made, the

---

[1] Pay.   [2] Working-soldiers of the administration.
[3] Sleeping-room in a barrack.   [4] You live alone, or apart.

instructor had watched him with amazed eyes, muttering to himself, '*Tiens! ce n'est pas un "bleu"*[1]—*ceci!* What a rider! Dieu de Dieu! he knows more than we can teach. He has served before now—served in some Emperor's picked guard!'

And when he had passed from the exercising-ground to the campaign, the Army had found in him one of the most splendid of its many splendid soldiers; and in the *folios matricules*[2] there was no page of achievements, of exploits, of services, of dangers, that showed a more brilliant array of military deserts than his. Yet, for many years, he had been passed by unnoticed: he had now not even the cross on his chest, and he had only slowly and with infinite difficulty been promoted so far as he stood now— a Corporal in the Chasseurs d'Afrique—a step only just accorded him because wounds innumerable, and distinctions without number in countless skirmishes, had made it impossible to cast him wholly aside any longer.

The cause lay in the implacable enmity of one man—his Chief.

Far-sundered as they were by position, and rarely as they could come in actual contact, that merciless weight of animosity from the great man to his soldier had laid on the other like iron, and clogged him from all advancement. His thoughts were of it now. Only to-day, at an inspection, the accidentally-broken saddle-girth of a boy-conscript had furnished pretext for a furious reprimand, a volley of insolent opprobrium hurled at himself, under which he had had to sit mute in his saddle, with no other sign that he was human beneath the outrage than the blood that would, despite himself, flush the pale bronze of his forehead. His thoughts were on it now.

'There are many losses that are bitter enough,' he mused, 'but there is not one so bitter as the loss of the right *to resent*!'

A whirlwind of laughter, so loud that it drowned the music of the shrill violins and thundering drums, echoed through the rooms and shook him from his reverie.

'They are *bons enfants*,' he thought, with a half smile, as he listened; 'they are more honest in their mirth as in their wrath, than we ever were in that old world of mine.'

Amidst the shouts, the crash, the tumult, the gay ringing voice of Cigarette rose distinct. She had apparently paused in her dancing to exchange one of those passes of arms which were her speciality, in the Sabir that she, a child of the regiments of Africa, had known as her mother tongue.

---

[1] Raw recruit.   [2] Daily register of the troopers' conduct.

'*Il fait suisse?*'[1] she cried, disdainfully. '*Paf! et tu as bu de sa gourde, chenapan?*'

The grumbled assent of the accused was inaudible.

'*Ingrat!*' pursued the scornful, triumphant voice of the vivandière; 'you would *bazarder*[2] your mother's grave-clothes! You would eat your children, *en fricassée!* You would sell your father's bones for a draught of *tord-boyaud!*[3] *Va t'en, chien!*'

The screams of mirth redoubled; Cigarette's style of withering eloquence was suited to all her auditors' tastes, and under the chorus of laughs at his cost, her infuriated adversary plucked up courage and roared forth a defiance.

'*Ma cantche!* white hands and a brunette's face are fine things for a soldier. He kills women, he kills women with his lady's grace! *Grand' chose ça!*'

'He does not pull their ears to make them give him their *style*,[4] and beat them with a *matraque*[5] if they don't fry his eggs fast enough, as you do, Barbe-Grise,' retorted the contemptuous tones of the champion of the absent. 'White hands, morbleu! Well, his hands are not always in other people's pockets as yours are, *sacripant!*'

This forcible *tu quoque* recrimination is in high relish in the Caserne; the screams of mirth redoubled; Barbe-Grise was a redoubtable authority whom the wildest daredevil in his brigade dared not contradict, and he was getting the worst of it under the lash of Cigarette's tongue, to the infinite glee of the whole ballroom.

'Damn!—his hands cannot work as mine can!' growled her opponent.

'Oh, ho!' cried the little lady, with supreme disdain; 'they don't twist cocks' throats and skin rabbits they have thieved, perhaps, like yours, but they would wring your neck before breakfast to get an appetite, if they could touch such *canaille*.'

'*Canaille?*' thundered the insulted Barbe-Grise. '*Ma cantche!* if you were but a man!'

'What would you do to me, *brigand?*' screamed Cigarette, in fits of laughter; 'give me fifty blows of *matraque*, as your officers gave you last week for stealing his *jambon*[6] from the *blanc-bec?*'[7]

A growl like a lion from the badgered Barbe-Grise shook the walls; she had cast her mischievous stroke at him on a very sore point, the unhappy young conscript's rifle having been first dexterously thieved from him, and then as dexterously sold to an Arab.

---

[1] You call him a misanthrope? and you have been drinking at his expense, you rascal?
[2] Pawn.   [3] Brandy.
[4] Money.   [5] Stick.   [6] Gun.   [7] Newly joined soldier.

'Sacre-bleu!' he roared; 'you are in love, *au grand galop*, with this *Vanqueur des belles*—this *loustic aristocrat!*'

The only answer to this unbearable insult was a louder tumult of laughter; a crash, a splash, and a volley of oaths from Barbe-Grise. Cigarette had launched a bottle of vin ordinaire at him, blinded his eyes, and drenched his beard with the red torrent and the shower of glass shivers, and was back again dancing like a little Bacchante, and singing at the top of her sweet lark-like voice—

> Turcos! Lignards!
> Bon Zigs! Truffards!
> Autour des couscoussou,
> Sont tous mes chers zou-zous!
> Roumis
> Spahis
> Même les Arbis,
> Joyeux
> Et Bleus,
> Même les Recrues,
> Ont pour moi
> Quand on boit
> L'air des rois
> L'air des rois!
> A mon cœur le chemin
> N'est qu' par le vin!
> Le bidon qu'on savoure
> Est le titre à m'amour!

With which doggerel declaration of her own mercenary and cosmopolitan sentiments chanted in Sabir slang, the little Friend of the Flag resumed her wildest bounds and her most airy fantasias. At the sound of the animated altercation, not knowing but what one of his own troopers might be the delinquent, he who leaned our of the little casement moved forward to the doorway of the dancing-room; he did not guess that it was himself whom she had defended against the onslaught of the Zéphyr, Barbe-Grise.

His height rose far above the French soldiers, and above most even of the lofty-statured Spahis, and her rapid glance flashed over him at once. 'Did he hear?' she wondered; the scarlet flush of exercise and excitement deepened on her clear brown cheek that had never blushed at the coarsest jests or the broadest love-words of the barrack-life that had been about her ever since her eyes first opened in their infancy to laugh at the sun-gleam on a Cuirassier's corslet among the bag-gage-waggons that her mother followed. She thought he had not heard;

his face was grave, a little weary, and his gaze, as it fell on her, was abstracted.

'Oh hé! *beau Roumi!*' thought Cigarette, with a flush of hot wrath superseding her momentary and most rare embarrassment. 'You are looking at me and not thinking of me? We will soon change that!'

Such an insult she had never been subjected to, from the first day when she had danced for sweetmeats on the top of a great drum when she was three years old, in the middle of a circular camp of Tirailleurs. It sent fresh nerve into her lithe limbs, it made her eyes flash like so much fire, it gave her a million-fold more grace, more abandon, more heedlessness, more piqued and reckless *désinvolture*. She stamped her tiny spurred foot petulantly.

'Plus vite! Plus vite!' she cried, and as the musician obeyed her, she whirled, she spun, she bounded, she seemed to live in air, while her soft curls blew off her brow, and her white teeth glanced, and her cheeks glowed with a carmine glow, and the little gold aiglettes broke across her chest with the beating of her heart, that throbbed like a bird's heart when it is wild with the first breath of Spring.

She had pitted herself against him; and she won—so far.

The vivacity, the impetuosity, the antelope elegance, the voluptuous repose that now and then broke the ceaseless, sparkling movement of her dancing, caught his eyes and fixed them on her: it was bewitching, and it bewitched him for the moment; he watched her as in other days he had watched the fantastic witcheries of eastern almè, and the ballet charms of opera-dancers.

This young bohemian of the barrack danced in the dusky glare and the tavern fumes of the As de Pique to a set of soldiers in their shirt-sleeves with their short black pipes in their mouths, with as matchless a grace as ever the first ballerina of Europe danced before sovereigns and dukes on the boards of Paris, Vienna, or London. It was the eastern *bamboula* of the Harems, to which was added all the elastic joyaunce, all the gay brilliancy, of the blood of France.

Suddenly she lifted both her hands above her head.

'*A moi, Roumis!*'

It was the signal well known, the signal of permission to join in that wild vertigo for which every one of her spectators was panting; their pipes were flung away, their kepis tossed off their heads, the music clashed louder and faster, and more fiery with every sound, the chorus of the Marseillaise des Bataillons thundered from a hundred voices—they danced as only men can dance who serve under the French flag, and live under the African sun. Two, only, still looked on; the Chasseur d'Afrique, and a veteran of the 10th company, lamed for life at Mazagran.

'*En ta maboull? Tu ne danses pas—toi?*' muttered the veteran Zéphyr to his silent companion.

The Chasseur turned and smiled a little.

'I prefer a *bamboula* whose music is the cannon, *bon père*.'

'Bravo! Yet she is pretty enough to tempt you?'

'Yes; too pretty to be unsexed by such a life.'

His thoughts went to a woman he had loved well, a young Arab, with eyes like the softness of dark waters, who had fallen to him once in a razzia as his share of spoil, and for whom he had denied himself cards, or wine, or tobacco, or an hour at the Café, or anything that alleviated the privation and severity of his lot as '*simple soldat*', which he had been then, that she might have such few and slender comforts as he could give her from his miserable pay. She was dead. Her death had been the darkest passage in his life in Africa—but the flute-like music of her voice seemed to come on his ear now. This girl-soldier had little charm for him after the sweet silent tender grace of his lost Zelme.

He turned and touched on the shoulder a Chasseur who had paused a moment to get breath in the headlong whirl:

'Come, we are to be with the Djied by dawn!'

The trooper obeyed instantly; they were ordered to visit and remain with a Bedouin camp some thirty miles away on the naked plateau; a camp professedly submissive, but not so much so but that the Bureau deemed it well to profit themselves by the services of the Corporal, whose knowledge of Arabic, whose friendship with the tribes, and whose superior intelligence in all such missions rendered him peculiarly fitted for errands that required diplomacy and address as well as daring and fire.

He went thoughtfully out of the noisy reeking ballroom into the warm lustre of the Algerian night; as he went, Cigarette, who had been nearer than he knew, flashed full in his eyes the fury of her own sparkling ones, while with a contemptuous laugh she struck him across the lips with the cigar she hurled at him.

'Unsexed? Pouf! If you have a woman's face may I not have a man's soul? It is only a fair exchange. I am no kitten, *bon zig*; take care of my talons!'

The words were spoken with the fierceness of Africa; she had too much in her of the spirit of the Zéphyrs and the Chacals with whom her youth had been spent from her cradle up, not to be dangerous when roused; she was off at a bound, and in the midst of the mad whirl again, before he could attempt to soften or efface the words she had overheard, and the last thing he saw of her was in a cloud of Zouaves and Spahis with the wild tintamarre of the music shaking riotous echoes from the rafters.

But when he had passed out of sight, Cigarette shook herself free from the dancers with petulant impatience. She was not to be allured by flattery or drawn by entreaty back amongst them; she set her delicate pearly teeth tight, and vowed with a reckless contemptuous impetuous oath that she was tired, that she was sick of them, that she was no strolling player to caper for them with a tambourine; and with that declaration made her way out alone into the little open court under the stars, so cool, so still after the heat, and riot, and turbulence within.

There she dropped on a broad stone step, and leant her head on her hand.

'Unsexed! unsexed! What did he mean?' she thought, while for the first time, with a vague sense of his meaning, tears welled hot and bitter into her sunny eyes while the pained colour burned in her face. Those tears were the first that she had ever known, and they were cruel ones, though they lasted but a little time; there was too much fire in the young bohemian of the Army not to scorch them as they rose. She stamped her foot on the stones passionately, and her teeth were set like a little terrier's as she muttered:

'Unsexed! Unsexed! Bah, M'sieu l'Aristocrat! If you think so, you shall find your thought right; you shall find Cigarette can hate as men hate, and take her revenge as soldiers take theirs.'

CHAPTER III

Under the Houses of Hair

It was just sunset.

The far-off summits of the Djurjura were tinted with the intense glare the distant pines and cypresses cut sharply against the rose-warmed radiance of the sky. On the slopes of the hills white cupolas and terraced gardens, where the Algerine haouach still showed the taste and luxury of Algerine corsairs, rose up among their wild olive shadows on the groves of the lentiscus. In the deep gorges that were channelled between the riven rocks, the luxuriance of African vegetation ran riot, the feathery crests of tossing reeds, the long floating leaves of plants filling the dry water-courses of vanished streams; the broad foliage of the wild fig, and the glowing dainty blossoms of the oleander, wherever a trace of brook, or pool, or rivulet let it put forth its beautiful coronal, growing one in another in the narrow valleys, and the curving passes, wherever broken earth or rock gave shelter from the blaze and heat of the North African day.

Further inland the bare sear stretches of brown plain were studded with the dwarf palm, the vast shadowless plateaux were desolate as the great desert itself far beyond; and the sun, as it burned on them a moment in the glory of its last glow, found them naked and grand by the sheer force of immensity and desolation, but dreary and endless, and broken into refts and chasms as though to make fairer by their own barren solitude the laughing luxuriance of the sea-face of the Sahel.

A moment, and the lustre of the light flung its own magic brilliancy over the Algerine water-line, and then shone full on the heights of El Biar and Bouzariah, and on the lofty, delicate form of the Italian pines that here and there, Sicilian-like, threw out their graceful heads against the amber sun-glow and the deep azure of the heavens. Then swiftly, suddenly, the sun sank; twilight passed like a grey gliding shade, an instant, over earth and sea; and night, the balmy, sultry, star-studded night of Africa, fell over the thirsty leafage longing for its dews, the closed flowers that slumbered at its touch, the seared and blackened plains to which its coolness could bring no herbage, the massive hills that seemed to lie so calmly in its rest.

Camped on one of the bare stretches above the Mustapha Road was a circle of Arab tents. The circle was irregularly kept, and the Krümas were scattered at will; here a low one of canvas, there one of goatskin; here a white towering canopy of teleze, there a low striped little nest of shelter; and loftier than all, the stately *beit el shar* of the Sheik, with his standard struck into the earth in front of it, with its heavy folds hanging listlessly in the sultry breathless air.

The encampment stretched far over the level arid earth, and there was more than one tent where the shadowing folds of the banner marked the abode of some noble Djied. Disorder reigned supreme, in all the desert freedom; horses and mules, goats and camels, tethered, strayed among the conical houses of hair, browsing off the littered straw or the tossed-down hay; and caldrons seethed and hissed over wood fires, whose lurid light was flung on the eagle features, and the white haiks of the wanderers who watched the boiling of their mess, or fed the embers with dry sticks. Round other fires, having finished the eating of their couscoussou,

the Bedouins lay full length, enjoying the solemn silence which they love
so little to break, and smoking their long pipes, while through the
shadows about them glided the lofty figures of their brethren, with the
folds of their sweeping burnous floating in the gloom. It was a picture,
Rembrandt in colour, Oriental in composition, with the darkness sur-
rounding it stretching out into endless distance that led to the mystic
silence of the great desert, and above, the intense blue of the gorgeous
night, with the stars burning through white transparent mists of slowly
drifting clouds.

In the central tent tall and crimson striped, with its mighty standard
reared in front, and its opening free to the night, sat the Khalifa, the head
of the tribe, with a circle of Arabs about him. He was thrown on his
cushions, rich enough for a seraglio, while the rest squatted on the
morocco carpet that covered the bare ground, and that was strewn with
round brass Moorish trays, and little cups emptied of their coffee. The
sides of the tent were hung with guns and swords, lavishly adorned, and
in the middle stood a tall Turkish candle-branch in fretted-work, whose
light struggled with the white flood of the moon, and the ruddy, fitful
glare from a wood fire without.

Beneath its light, which fell full on him, flung down upon another pile
of cushions facing the open front of the tent, was a guest whom the
Khalifa delighted to honour. Only a Corporal of Chasseurs, and once a
foe, yet one with whom the Arab found the brotherhood of brave men,
and on whom he lavished, in all he could, the hospitalities and honours of
the desert.

The story of their friendship ran thus:

The tribe was now allied with France, or, at least, had accepted French
sovereignty, and pledged itself to neutrality in the hostilities still rife;
but a few years before, far in the interior and leagued with the Kabailes,
it had been one of the fiercest and most dangerous among the enemies
of France. At that time the Khalifa and the Chasseur met in many a
skirmish; hot desperate struggles, where men fought horse to horse, hand
to hand; midnight frays, when in the heart of lonely ravines, Arab
ambuscades fell on squadrons of French cavalry; terrible chases through
the heat of torrid suns, when the glittering ranks of the charging troops
swept down after the Bedouins' flight; fiery combats, when the desert
sand and the smoke of musketry circled in clouds above the close-locked
struggle, and the Leopard of France and the Lion of Sahara wrestled in a
death-grip.

In these, through four or five seasons of warfare, the Sheik and the
Chasseur had encountered each other, till each had grown to look for

the other's face as soon as the standards of the Bedouins flashed in the sunshine opposite the guidons of the Imperial forces; till each had watched and noted the other's unmatched prowess, and borne away the wounds of the other's home strokes, with the admiration of a bold soldier for a bold rival's dauntlessness and skill; till each had learned to long for an hour, hitherto always prevented by waves of battle that had swept them too soon asunder, when they should meet in a duello once for all, and try their strength together till one bore off victory and one succumbed to death.

At last it came to pass that after a lengthened term of this chivalrous antagonism the tribe were sorely pressed by the French troops, and could no longer mass its fearless front to face them, but had to flee southward to the desert, and encumbered by its flocks and its women, was hardly driven and greatly decimated. Now among those women was one whom the Sheik held above all earthly things except his honour in war, a beautiful anteloped-eyed creature, lithe and graceful as a palm, and the daughter of a pure Arab race, on whom he could not endure for any other sight than his own to look, and whom he guarded in his tent as the chief pearl of all his treasures; herds, flocks, arms, even his horses, all, save the honour of his tribe, he would have surrendered rather than surrender Djelma. It was a passion with him; a passion that not even the iron of his temper and the dignity of his austere calm could abate or conceal; and the rumour of it and of the beauty of its object reached the French camp, till an impatient curiosity was roused about her, and a raid that should bear her off became the favourite speculation round the picket-fires at night, and the scorching noons when the men lay stripped to their waist, panting like tired dogs, under the hot withering breath that stole to them from sweeping over the yellow seas of sand.

Their heated fancies had pictured this treasure of the great Djied as something beyond all that her sex had ever given them, and to snare her in some unwary moment was the chief thought of Zéphyr and Spahis when they went out on a scouting or foraging party. But it was easier said than done; the eyes of no Frank ever lit on her, and when he was most closely driven the Khalifa Ilderim abandoned his cattle and sheep, but with the females of the tribe still safely guarded, fell more and more backward and southward, drawing the French on and on further and further across the plains in the sickliest times of hottest drought.

Reinforcements could swell the Imperial ranks as swiftly as they were thinned, but with the Arabs a man once fallen was a man the less to their numbers for ever, and the lightning-like pursuit began to tell terribly on them: their herds had fallen into their pursuers' hands, and famine menaced them. Nevertheless, they were fierce in attack as tigers, rapid in swoop as vultures, and fought flying in such fashion that the cavalry lost more in this fruitless, worthless work, than they would have done in a second Hohenlinden or Austerlitz.

Moreover, the heat was intense, water was bad and very rare, dysentery came with the scorch and the toil of this endless charge; the chief in command, M. le Marquis de Châteauroy, swore heavily as he saw many of his best men dropping off like sheep in a murrain, and he offered two hundred Napoléons to whosoever should bring either the dead Sheik's head or the living beauty of Djelma.

One day the Chasseurs had pitched their camp where a few barren withered trees gave a semblance of shelter, and a little thread of brackish water oozed through the yellow earth.

It was high noon; the African sun was at its fiercest; far as the eye could reach there was only one boundless, burning, unendurable glitter of parching sand and cloudless sky, brazen beneath, brazen above, till the desert and the heavens touched, and blent in one tawny fiery glow in the measureless distance. The men lay under canvas, dead beat, half naked, without the power to do anything except to fight like thirst-maddened dogs for a draught at the shallow stream that they and their breathless horses soon drained dry.

Even Raoul de Châteauroy, though his frame was like an Arab's, and knit into Arab endurance, was stretched like a great bloodhound, chained by the sultry oppression. He was ruthless, inflexible, a tyrant to the core, and sharp and swift as steel in his rigour, but he was a fine soldier, and never spared himself any of the hardships that his regiment had to endure under him.

Suddenly the noon lethargy of the camp was broken; a trumpet-call rang through the stillness; against the amber transparency of the horizon line the outlines of half a dozen horsemen were seen looming nearer and nearer with every moment; they were some Spahis who had been out '*sondant le terrain aux environs*'.[1] The mighty frame of Châteauroy, almost as unclothed as an athlete, started from its slumberous panting rest; his eyes lightened hungrily; he muttered a fiery oath, 'Mort de Dieu!—they have the woman!'

They had the woman. She had been netted near a water-spring, to which she had wandered, too loosely guarded, and too far from the Bedouin encampment. The delight of the haughty Sidi's eyes was borne off to the tents of his foes, and the Colonel's face flushed darkly, with an

---

[1] Sweeping the country for food.

eager, lustful warmth, as he looked upon his captive. Rumour had not out-boasted the Arab girl's beauty; it was lustrous as ever was that when, far yonder to the eastward, under the curled palms of Nile, the sorceress of the Cæsars swept through her rose-strewn palace-chambers. Only Djelma was as innocent as the gazelle, whose grace she resembled, and loved her lord with a great love.

Of her suffering her captor took no more heed than if she were a young bird dying of shot-wounds; but, with one triumphant admiring glance at her, he wrote a message in Arabic, to send to the Khalifa, ere her loss were discovered—a message more cruel than iron. He hesitated a second, where he lay at the opening of his tent, whom he should send with it. His men were almost all half dead with the sun-blaze. His glance chanced to light in the distance on a soldier to whom he bore no love—causelessly, but bitterly all the same. He had him summoned, and eyed him with a curious amusement: Châteauroy treated his squadrons with much the same *sans-façon* familiarity and brutality that a chief of filibusters uses to his.

'So! you heed the heat so little, you give up your turn of water to a drummer, they say?'

The Chasseur gave the salute with a calm deference. A faint flush passed over the sun-bronze of his forehead. He had thought the Sydney-like sacrifice had been unobserved.

'The drummer was but a child, *mon Commandant.*'

'Be so good as to give us no more of those melodramatic acts!' said M. le Marquis, contemptuously. 'You are too fond of trafficking in those showy fooleries. You bribe your comrades for their favouritism too openly. Ventre-bleu! I forbid it—do you hear?'

'I hear, *mon Colonel.*'

The assent was perfectly tranquil and respectful. He was too good a soldier not to render perfect obedience, and keep perfect silence, under any goad of provocation to break both.

'Obey, then!' said Châteauroy, savagely. 'Well, since you love heat so well, you shall take a flag of truce and my scroll to the Sidi Ilderim. But tell me first—what do you think of this capture?'

'It is not my place to give opinions, M. le Colonel.'

'Pardieu! it is your place when I bid you. Speak, or I will have the *matraque* cut the words out of you!'

'I may speak frankly?'

'Ten thousand curses—yes!'

'Then, I think that those who make war on women are no longer fit to fight with men.'

For a moment the long, sinewy, massive form of Châteauroy started from the skins on which he lay full length, like a lion starting from its lair. His veins swelled like black cords; under the mighty muscle of his bare chest, his heart beat visibly, in the fury of his wrath.

'By God! I have a mind to have you shot like a dog!'

The Chasseur looked at him carelessly, composedly, but with a serene deference still, as due from a soldier to his chief.

'You have threatened it before, M. le Colonel. It may be as well to do it, or the army may think you capricious.'

Raoul de Châteauroy crushed a blasphemous oath through his clenched teeth, and laughed a certain short, stern, sardonic laugh, which his men dreaded more than his wrath.

'No; I will send you instead to the Khalifa. He often saves me the trouble of killing my own curs. Take a flag of truce and this paper, and never draw rein till you reach him, if your beast drop dead at the end.'

The Chasseur saluted, took the paper, bowed with a certain languid, easy grace, that camp life never cured him of, and went. He knew that the man who should take the news of his treasure's loss to the Emir Ilderim would, a thousand to one, perish by every torture desert cruelty could frame, despite the cover of the white banner.

Châteauroy looked after him, as he and his horse passed from the French camp in the full burning tide of noon.

'If the Arabs kill him,' he thought, 'I will forgive Ilderim five seasons of rebellion.'

The Chasseur, as he had been bidden, never drew rein across the scorching plateau. He rode to what he knew was like enough to be death, and death by many a torment, as though he rode to a midnight love-tryst. His horse was of Arab breed—young, fleet, and able to endure extraordinary pressure, both of spur and of heat. He swept on, far and fast, through the sickly lurid glitter of the day, over the loose sand, that flew in puffs around him, as the hoofs struck it flying right and left. At last, ere he reached the Bedouin tents, that were still but slender black points against the horizon, he saw the Sheik and a party of horsemen returning from a foraging quest, and in ignorance as yet of the abduction of Djelma. He galloped straight to them, and halted across their line of march, with the folds of the little white flag fluttering in the sun. The Bedouins drew bridle, and Ilderim advanced alone. He was a magnificent man, of middle age, with the noblest type of the eagle-eyed, aquiline desert beauty. He was a superb specimen of his race, without the lean, withered, rapacious, vulture look, which often mars it. His white haick floated round limbs fit for a Colossus; and under the snowy folds of his turban, the olive-bronze of his bold forehead, the sweep of his jet-black beard, and the piercing luminance of his eyes, had a grand and kingly majesty.

A glance of recognition flashed from him on the lascar, who had so often crossed swords with him; and he waved back the scroll with dignified courtesy.

'Read it me.'

It was read. Bitterly, blackly, shameful, the few brutal words were. They netted him as an eagle is netted in a shepherd's trap.

The moment that he gave a sign of advancing against his ravishers, the captive's life would pay the penalty: if he merely remained in arms, without direct attack, she would be made the Marquis's mistress, and abandoned later to the army. The only terms on which he could have her restored were instant submission to the imperial rule, and personal homage of himself and all his Djouad to the Marquis as the representative of France; homage in which they should confess themselves dogs and the sons of dogs.

So ran the message of peace.

The Chasseur read on to the end, calmly. Then he lifted his gaze, and looked at the Emir—he expected fifty swords to be buried in his heart.

As he gazed, he thought no more of his own doom: he thought only of the revelation before him, of what passion and what agony could be— things unknown in the world where the chief portion of his life had passed. He was a war-hardened campaigner, trained in the ruthless school of African hostilities, who had seen every shape of mental and physical suffering, when men were left to perish of gun-wounds, as the rush of the charge swept on; when writhing horses died by the score of famine and of thirst; when the firebrand was hurled among sleeping encampments, and defenceless women were torn from their rest by the unsparing hands of pitiless soldiers. But the torture, which shook for a second the steel-knit frame of this Arab, passed all that he had dreamed as possible; it was mute, and held in bonds of iron, for the sake of the desert pride of a great ruler's majesty; but it spoke more than any eloquence ever spoke yet on earth.

With a wild shrill yell, the Bedouins whirled their naked sabres above their heads, and rushed down on the bearer of this shame to their chief and their tribe. The Chasseur did not seek to defend himself. He sat motionless. He thought the vengeance just.

The Sheik raised his sword, and signed them back, as he pointed to the white folds of the flag. Then his voice rolled out like thunder over the stillness of the plains:

'But that you trust yourself to my honour, I would rend you limb from limb. Go back to the tiger who rules you, and tell him that—as Allah liveth—I will fall on him and smite him, as he hath never been smitten. Dead or living, I will have back my own. If he take her life, I will have ten thousand lives to answer it: if he deal her dishonour, I will light such a holy war through the length and breadth of the land, that his nation shall be driven backward, like choked dogs into the sea, and perish from the face of the earth for evermore. And this I swear by the Law and the Prophet!'

The menace rolled out, imperious as a monarch's, thrilling through the desert hush. The Chasseur bent his head as the words closed. His own teeth were tightly clenched, and his face was dark.

'Emir, listen to one word,' he said, briefly. 'Shame has been done to me as to you. Had I been told what words I bore, they had never been brought by my hand. You know me. You have had the marks of my steel, as I have had the marks of yours. Trust me in this, Sidi. I pledge you my honour, that, before the sun sets, she shall be given back to you un-harmed; or I will return here myself, and your tribe shall slay me in what fashion they will. So alone can she be saved uninjured. Answer—will you have faith in me?'

The desert chief looked at him long; sitting motionless as a statue on his stallion, with the fierce gleam of his eyes fixed on the eyes of the man who so long had been his foe in contests whose chivalry equalled their daring. The Chasseur never wavered once under the set, piercing, ruth-less gaze.

Then the Emir pointed to the sun, that was now at its zenith:

'You are a great warrior: such men do not lie. Go, and, if she be borne to me before the sun is halfway sunk towards the west, all the branches of the tribes of Ilderim shall be as your brethren, and bend as steel to your bidding. If not—as God is mighty—not one man in all your host shall live to tell the tale!'

The Chasseur bowed his head to his horse's mane; then, without a word, wheeled round, and sped back across the plain.

When he reached his own cavalry camp, he went straightway to his chief. What passed between them none ever knew. The interview was brief: it was possibly as stormy. Pregnant and decisive it assuredly was; and the squadrons of Africa marvelled that the man who dared beard Raoul de Châteauroy in his lair, came forth with his life. Whatever the spell he used, the result was a marvel.

At the very moment that the sun touched the lower half of the western heavens, the Sheik Ilderim, where he sat in his saddle, with all his tribe stretching behind him, full armed, to sweep down like falcons on the spoilers, if the hour passed with the pledge unredeemed, saw the form of the Chasseur reappear between his sight and the glare of the skies; nor did he ride alone. That night the Pearl of the Desert lay once more in the mighty sinuous arms of the great Emir.

But, with the dawn, his vengeance fell in terrible fashion on the sleeping camp of the Franks; and from that hour dated the passionate, savage, unconcealed hate of Raoul de Châteauroy to the most daring soldier of all his fiery Horse, known in his troop as 'Bel-à-faire-peur'.

It was in the tent of Ilderim now that he reclined, looking outward at the night where flames were leaping ruddily under a large caldron, and far beyond was the dark immensity of the star-studded sky. The light of the moon strayed in and fell on the chesnut waves of his beard, out of which the long amber stem of an Arab pipe glittered like a golden line; and the delicate feminine cast of his profile, which, with the fairness of the skin (fair despite a warm hue of bronze) and the long slumberous softness of the hazel eyes, were in as marked a contrast of race with the eagle outline of the Bedouins around as Frank and Arbi ever showed.

From the hour of the restoration of his treasure the Sheik had been true to his oath; his tribe in all its branches had held the French lascar in closest brotherhood; wherever they were he was honoured and welcomed; was he in war, their swords were drawn for him; was he in need, their houses of hair were spread for him; had he want of flight, the swiftest and most precious of their horses was at his service; had he thirst, they would have died themselves, wringing out the last drop from the water-skin for him. Through him their alliance, or more justly to speak their neutrality, was secured to France, and the Bedouin Chief loved him with a great, silent, noble love that was fast rooted in the granite of his nature. Between them there was a brotherhood that beat down the antagonism of race, and was stronger than the instinctive hate of the oppressed for all who came under the abhorred standard of the usurpers. He like the Arabs and they liked him; a grave courtesy, a preference for the fewest words and least demonstration possible, a marked opinion that silence was golden, and that speech was at best only silver-washed metal, an instinctive dread of all discovery of emotion, and a limitless power of resisting and suppressing suffering, were qualities the nomads of the desert and the *lion* of the Chasseurs d'Afrique had in common; as they had in unison a wild passion for war, a dauntless zest in danger, and a love for the hottest heat of fiercest battle.

Silence reigned in the tent, beyond whose first division, screened by a heavy curtain of goat's hair, the beautiful young Djelmar played with her only son, a child of three or four summers; the Sheik lay mute, the Djouad and Marabouts around never spoke in his presence unless their lord bade them, and the Chasseur was stretched motionless, his elbow resting on a cushion of Morocco fabric, and his eyes looking outward at the restless, changing movement of the firelit, starlit camp.

After the noise, the mirth, the riotous songs, and the gay elastic good-humour of his French comrades, the silence and the calm of the Emir's 'house of hair' were welcome to him. He never spoke much himself; of a truth, his gentle immutable laconism was the only charge that his Chambrée ever brought against him. That a man could be so brief in words, while yet so soft in manner, seemed a thing out of all nature to the vivacious Frenchmen; that unchanging stillness and serenity in one who was such a reckless, resistless *croc-mitaine*, swift as fire in the field, was an enigma that the Cavalerie and the Demi-cavalerie of Algeria never solved. His corps would have gone after him to the devil, as Claude de Chanrellon had averred; but they would sometimes wax a little impatient that he would never grow communicative or thread many phrases together even over the best wine which ever warmed the hearts of its drinkers or loosened all rein from their lips.

'I wish I had come straight to you, Sidi, when I first set foot in Africa,' he said at last, while the fragrant smoke uncurled from under the droop of his long pendent moustaches.

'Truly it had been well,' answered the Khalifa, who would have given the best stallions in his stud to have had this Frank with him in warfare and in peace; 'there is no life like our life.'

'Faith! I think not,' murmured the Chasseur, rather to himself than the Bedouin. 'The desert keeps you and your horse, and you can let all the rest of the world "slide".'

'But we are murderers and pillagers, say your nations,' resumed the Emir, with the shadow of a sardonic smile flickering an instant over the sternness and composure of his feature. 'To rifle a caravan is a crime, though to steal a continent is glory.'

Bel-à-faire-peur laughed slightly.

'Do not tempt me to rebel against my adopted flag.'

The Sheik looked at him in silence; the French soldiers had spent twelve years in the ceaseless exertions of an amused inquisitiveness to discover the antecedents of their volunteer; the Arabs, with their loftier instincts of courtesy, had never hinted to him a question of whence or why he had come upon African soil.

'I never thought at all in those days, else, had I thought twice, I should not have gone to your enemies,' he answered, as he lazily watched the

Bedouins without squat on their heels round the huge brass bowls of couscoussou which they kneaded into round lumps and pitched between their open bearded lips in their customary form of supper. 'Not but what our *Roumis* are brave fellows enough; better comrades no man could want.'

The Khalifa took the long pipe from his mouth and spoke; his slow sonorous accents falling melodiously on the silence in the *lingua sapir* of the France-Arab tongue.

'Your comrades are gallant men; they are *lascars kébirs*,[1] and fearless foes, against such my voice is never lifted, however my sword may cross with them. But the locust-swarms that devour the land are the money-eaters, the petty despots, the bribe-takers, the men who wring gold out of infamy, who traffic in tyrannies, who plunder under official seals, who curse Algiers with avarice, with fraud, with routine, with the hell-spawn of civilization. It is the "Bureaucratie", as your tongue phrases it, that is the spoiler and the oppressor of the soil. But, Inshallah! we endure only for a while. A little, and the shame of the invader's tread will be washed out in blood. Allah is great! We can wait.'

And with Moslem patience that the fiery gloom of his burning eyes belied, the Djied stretched himself once more into immovable and silent rest.

The Chasseur answered nothing; his sympathies were heartfelt with the Arabs, his allegiance and his esprit du corps were with the Service in which he was enrolled. He could not defend French usurpation; but neither could he condemn the Flag that had now become his Flag, and in which he had grown to feel much of national honour, to take much of national pride.

'They will never really win again, I am afraid,' he thought, as his eyes followed the wraith-like flash of the white burnous, as the Bedouins glided to and fro in the chiar'oscuro of the encampment, now in the flicker of the flame, now in the silvered lustre of the moon. 'It is the conflict of the races, as the cant runs, and their day is done. It is a bolder, freer, simpler type than anything we get in the world yonder. Shall we ever drift back to it in the future, I wonder?'

The speculation did not stay with him long; Semitic, Latin, or Teuton race was very much the same to him, and intellectual subtleties had not much attraction at any time for the most brilliant soldier in the French cavalry; he preferred the ring of the trumpets, the glitter of the sun's play along the line of steel as his regiment formed in line on the eve of a life-

---

[1] Great warriors.

and-death struggle, the wild breathless sweep of a midnight gallop over the brown swelling plateau under the light of the stars, or, in some brief interval of indolence and razzia-won wealth, the gleam of fair eyes and the flush of sparkling sherbet when some passionate darkling glance beamed on him from some Arab mistress whose scarlet lips murmured to him through the drowsy hush of an Algerine night the sense if not the song of Pelagia:

> Life is so short at best!
> Take while thou canst thy rest,
> Sleeping by me!

His thoughts drifted back over many varied scenes and changing memories of his service in Algiers, as he lay there at the entrance of the Sheik's tent, with the night of looming shadow, and reddening firelight, and picturesque movement before him. Hours of reckless headlong delight, when men grew drunk with bloodshed as with wine; hours of horrible, unsuccoured suffering, when the desert thirst had burned in his throat, and the jagged lances been broken off at the hilt in his flesh, while above head the carrion birds wheeled, waiting their meal; hours of unceasing, unsparing slaughter, when the word was given to slay and yield no mercy, where in the great, vaulted, cavernous gloom of rent rocks, the doomed were hemmed as close as sheep in shambles. Hours, in the warm flush of an African dawn, when the arbiter of the duel was the sole judge allowed or comprehended by the tigers of the tricolour, and to aim a dead shot or to receive one was the only alternative left, as the challenging eyes of 'Zéphyr' or 'Chasse-Marais' flashed death across the *barrière*, in a combat where only one might live, though the root of the quarrel had been nothing more than a toss too much of brandy, a puff of tobacco-smoke construed into insult, or a fille de joie's maliciously cast firebrand of taunt or laugh. Hours of severe discipline, of relentless routine, of bitter deprivation, of campaigns hard as steel in the endurance they needed, in the miseries they entailed; of military subjection, stern and unbending, a yoke of iron that a personal and pitiless tyranny weighted with persecution that was scarce less than hatred; of an implicit obedience that required every instinct of liberty, every habit of early life, every impulse of pride, and manhood, and freedom, to be choked down like crimes, and buried as though they had never been. Hours again, that repaid these in full, when the long line of Horse swept out to the attack, with the sun on the points of their weapons; when the wheeling clouds of Arab riders poured like the clouds of the simoom on a thinned, devoted troop that rallied and fought as hawks fight herons, and saved the day as the sky was flushed with that day's decline; when some soft-eyed captive, with limbs of free mountain grace, and the warm veins flushing under the clear olive of her cheeks, was first wild as a young fettered falcon, and then, like the falcon, quickly learned to tremble at a touch, and grow tame under a caress, and love nothing so well as the hand that had captured her. Hours of all the chanceful fortunes of a soldier's life, in hill-wars and desert raids, passed in memory through his thoughts now, where he was stretched, looking dreamily through the film of his chibouque smoke at the city of tents, and the couchant forms of camels, and the tall, white, slowly-moving shapes of the lawless marauders of the sand plains.

'Is my life worth much more under the French Flag than it was under the English?' thought the Chasseur, with a certain careless, indifferent irony on himself, natural to him. 'There I killed time—here I kill men. Which is the better pursuit, I wonder? The world would rather economize the first commodity than the last, I believe. Perhaps, it don't make an over good use of either.'

His thoughts did not stay long with that theme. He was no moralist and no philosopher, though he practised, without ever knowing it, a philosophy of the highest and simplest kind with every day that found him in the ranks of the Algerian army; and had found thought grow on him, in a grave if a desultory fashion, many a time when he had ridden alone through defiles that, for aught he knew, might harbour death with every step; or sat the only wakeful watcher beside a bivouac-fire, while his comrades slept around him, and the roar of angry beasts rolled upward from the ravines; or paced to and fro in solitude on patrol duty, with a yawning mountain pass, or a limitless night-veiled plain before him in the light of the moon. He was more silent and more meditative than seemed in keeping with a wild *lion* of the Chasseurs, whose daring out-dared all the fire-eaters, and whose negligent courage had become a password all over Africa, till 'quel p'tit verre a bu Bel-à-faire-peur?' (alias, 'what special exploit has he done today?') became the question put after every skirmish or expedition. But he was much more of a soldier than a thinker at any time, and, instead of following out the problem of the world's uses of its two raw materials, time and men, he found a subject more congenial in the discussion of stable science with the Emir.

To him the austere chief would unbend; with him the thin, compressed lips of the Arab would grow eloquent with an impressive oratory; for him all the bonds of hospitality would grow closer and warmer. Ilderim might be a pillager, with a sure swoop, and a merciless steel, as the officials of imperial government wrote him out; of a truth, caravanserais had felt the tear of his talons, and battalions staggered under the blows of his beak; but he had two desert virtues that are obsolete in the civilized world: he had gratitude, and he had sincerity. Of course he was but a nomad, a barbarian, a robber, and a ruler of robbers; of course he was but a half-savage Ishmaelite, or he would long have abandoned them.

The night was someway spent when the talk of wild-pigeon-blue mares and sorrel stallions closed between the Djied and his guest; and the French soldier, who had been sent hither from the Bureau Arabe with another of his comrades, took his way through the now stilled camp where the cattle were sleeping, and the fires were burning out, and the banner-folds hung motionless in the lustre of the stars, to the black and white tent prepared for him. A spacious one, close to the chief's, and given such luxury in the shape of ornamented weapons, thick carpets, and soft cushions, as the tribe's resources, drawn from many a raid on travellers far south, could bring together to testify their hospitality.

As he opened the folds and entered, his fellow-soldier, who was lying on his back, with his heels much higher than his head, and a short pipe in his teeth, tumbled himself up with a rapid summersault, and stood bolt upright, giving the salute; a short, sturdy little man, with a skin burnt like a coffee-berry, that was in odd contrast with his light dancing blue eyes, and his close-matted curls of yellow hair.

'Beg pardon, sir! I was half asleep!'

The Chasseur laughed a little.

'*Don't* talk English; somebody will hear you one day.'

'What's the odds if they do, sir?' responded the other. 'It relieves one's feelin's a little. All of 'em know I'm English, but never a one of 'em knew what you are. The name you was enrolled by won't really tell 'em nothing. They guess it ain't yours. That cute little chap, Tata, he says to me yesterday, "you're always a treatin' of your *galonné* like as if he was a Prince". "Dammee!" says I, "I'd like to see the Prince as would hold a candle to him." "You're right there," says the little 'un. "There ain't his equal for takin' off a beggar's head with a back sweep."'

The Corporal laughed a little again, as he tossed himself down on the carpet.

'Well, it's something to have one virtue! But have a care what those chatterboxes get out of you.'

'Lord, sir. Ain't I been a takin' care these ten years? It comes quite natural now. I couldn't keep my tongue still; that wouldn't be in anyways possible. So I've let it run on oiled wheels on a thousand rum tracks and doublins. I've told 'em such a lot of amazin' stories about where we kem from, that they've got half a million different styles to choose out of. Some thinks as how you're a Polish nob, what got into hot water with the Russians; some as how you're a Italian Prince what was cleaned out like Parma and them was; some as how you're a Austrian Archduke that have cut your country because you was in love with the Empress, and had a duel about her that scandalized the whole empire; some as how you're a exiled Spanish grandee a-come to learn tactics and that like, that you may go back, and pitch O'Donnell into the middle of next week whenever you see a chance to cut in and try conclusions with him. Bless you, sir! you may let *me* alone for bamboozlin' of anybody!'

The Corporal laughed again, as he began to unharness himself. There was in him a certain mingling of insouciance and melancholy, each of which alternately predominated; the former his by nature, the latter born of circumstance.

'If you can outwit our friends the Zéphyrs, and the Loustics, and the Indigènes, you have reached a height of diplomacy indeed! I would not engage to do it myself. Take my word for it, ingenuity is always dangerous—silence is always safe.'

'That may be, sir,' responded the Chasseur, in the sturdy English with which his bright blue eyes danced a fitting nationality. 'No doubt it's uncommon good for them as can bring their minds to it—just like water instead o' wine—but it's very tryin', like the teetotalism. You might as well tell a Newfoundland not to love a splash as me not to love a chatter. I'd cut my tongue out sooner than say never a word that you don't wish— but say *somethin'* I must, or die for it.'

With which the speaker, known to Algerian fame by the sobriquet of *Crache-au-nez-d'la-Mort*, from the hairbreadth escapes and reckless razzias from which he had come out without a scratch, dropped on his knees, and began to take off the trappings of his fellow-soldier, with as reverential a service as though he were a Lord of the Bedchamber serving a Louis Quatorze. The other motioned him gently away.

'No, no. I have told you a thousand times we are comrades and equals now.'

'And I've told you a thousand times, sir, that we *aren't*, and never will be, and don't oughtn't to be,' replied the soldier, doggedly, drawing off the spurred and dust-covered boots. 'A gentleman's a gentleman, let alone what straits he fall into.'

'But ceases to be one as soon as he takes a service he cannot requite, or claims a superiority he does not possess. We have been fellow-soldiers for twelve years——'

'So we have, sir; but we are what we always was, and always will be—
one a gentleman, t'other a scamp. If you think so be as I've done a good
thing side by side with you now and then in the fightin', give me my own
way and let me wait on you when I can. I can't do much on it when those
other fellows' eyes is on us; but here I can and I will—beggin' your
pardon—so there's an end of it. One may speak plain in this place with
nothing but them Arabs about; and all the Army know well enough, sir,
that if it weren't for that black devil, Châteauroy, you'd have had your
officer's commission and your troop too long before now——'

'Oh no. There are scores of men in the ranks merit promotion better far
than I do. And, leave the Colonel's name alone. He is our chief, whatever
else he be.'

The words were calm and careless, but they carried a weight with them
that was not to be disputed; 'Crache-au-nez-d'la-Mort' hung his head a
little and went on unharnessing his Corporal in silence, contenting himself with muttering in his throat that it was true for all that, and the whole
regiment knew it.

'*You* are happy enough in Algeria—eh?' asked the one he served, as he
stretched himself on the skins and carpets, and drank down a sherbet that
his self-attached attendant had made with a skill learned from a pretty
cantinière who had given him the lesson in return for a slashing blow with
which he had struck down two 'Riz-pain-sels', who as the best paid men
in the army had tried to cheat her in the price of her cognac.

'I, sir? Never was so happy in my life, sir. I'd be discontented indeed if
I wasn't. Always some spicy bit of fighting. If there aren't a fantasia, as
they call it, in the field, there's always somebody to pot in a small way;
and if you're lying by in barracks there's always a scrimmage hot as
pepper to be got up with fellows that love the row just as well as you do.
It's life, that's where it is; it ain't rustin'.'

'Then you prefer the French service?'

'Right and away, sir. You see this is how it is,' and the redoubtable
yellow-haired 'Crache-au-nez-d'la-Mort' paused in the vigorous cleans-
ing and brushing he was bestowing on his Corporal's uniform, and stood
at ease in his shirt and trousers with his eloquence noway impeded by the
brûle-gueule that was always between his teeth. 'Over there in England,
you know, sir, pipeclay is the deuce-and-all, you've always got to have the
stock on, and look as stiff as a stake, or it's all up with you; you're that
tormented about little things that you get riled and kick the traces before
the great 'uns come to try you. There's a lot of lads would be game as
game could be in battle, ay, and good lads to boot, doing their duty right
as a trivet when it came to anything like war, that are clean druv' out

of the service in time o' peace, along with all them petty persecutions that worry a man's skin like mosquito-bites. Now here they know that, and Lord! what soldiers they do make thro' knowing of it! It's tight enough and stern enough in big things; martial law sharp enough, and obedience to the letter all through the campaigning; but that don't grate on a fellow; if he's worth his salt he's sure to understand that he must move like clockwork in a fight, and that he's to go to hell at double-quick-march, and mute as a mouse, if his officers see fit to send him. *That*'s all right, but they don't fidget you here about the little fal-lals; you may stick your pipe in your mouth, you may have your lark, you may do as you like, you may spend your *décompte* how you choose, you may settle your little duel as you will, you may shout and sing and jump and riot on the march, so long as you *march on*; you may lounge about half-dressed in any style as suits you best, so long as you're up to time when the trumpets sound for you; and that's what a man likes. He's ready to be a machine when the machine's wanted in working trim, but when it's run off the line and the steam all let off, he do like to oil his own wheels, and lie a bit in the sun at his fancy. There aren't better stuff to make soldiers out of nowhere than Englishmen, God bless 'em! but they're badgered, they're horribly badgered, and that's why the Service don't take over there, let alone the way the country grudge 'em every bit of pay. In England you go in the ranks—well, they all just tell you you're a blackguard, and there's the lash, and you'd better behave yourself or you'll get it hot and hot; they take for granted you're a bad lot or you wouldn't be there, and in course you're riled and go to the bad according, seeing that it's what's expected of you. Here, contrariwise, you come in the ranks and get a welcome, and feel that it just rests with yourself whether you won't be a fine fellow or not; and just along of feelin' that you're pricked to show the best metal you're made on, and not to let nobody else beat you out of the race like. Ah! it makes a wonderful difference to a fellow— a wonderful difference—whether the service he's come into look at him as a scamp that never will be nothin' *but* a scamp, or as a rascal that's maybe got in him, all rascal though he is, the pluck to turn into a hero. It makes a wonderful difference, this 'ere, whether you're looked at as stuff that's only fit to be shovelled into the sand after a battle, or as stuff that'll belike churn into a great man. And it's just that difference, sir, that France has found out, and England hasn't—God bless her all the same!'

With which the soldier whom England had turned adrift, and France had won in her stead, concluded his long oration by dropping on his knees to refill his corporal's chibouque.

'A army's just a machine, sir, in course,' he concluded, as he rammed in the Turkish tobacco. 'But then it's a live machine for all that; and each little bit of it feels for itself like the joints in an eel's body. Now, if only one of them little bits smarts, the whole crittur goes wrong—there's the mischief.'

Bel-à-faire-peur listened thoughtfully to his comrade where he lay flung full length on the skins.

'I dare say you are right enough. I knew nothing of my men when—when I was in England; we none of us did; but I can very well believe what you say. Yet—fine fellows though they are here, they are terrible blackguards!'

'In course they are, sir; they wouldn't be such larky company unless they was. But what I say is that they're scamps who're told they may be great men if they like; not scamps who're told that because they're once gone to the devil they must always keep there. It makes all the difference in life.'

'Yes—it makes all the difference in life, whether hope is left, or—left out!'

The words were murmured with a half smile that had a dash of infinite sadness in it; the other looked at him quickly with a shadow of keen pain passing over the bright, frank, laughing features of his sunburnt face; he knew that the brief words held the whole history of a life.

'Won't there *never* be no hope, sir?' he whispered, while his voice trembled a little under the long fierce 'Zéphyr' sweep of his yellow moustaches.

The Chasseur rallied himself with a slight, careless laugh; the laugh with which he had met before now the onslaught of charges ferocious as those of the magnificent day of Mazagran.

'Whom for? Both of us? Oh yes, very likely we shall achieve fame, and die sous-officiers or gardes-champêtres! A splendid destiny.'

'No, sir,' said the other, with the hesitation still in the quiver of his voice. 'You know I meant, no hope of your ever being again——'

He stopped; he scarcely knew how to phrase the thoughts he was thinking.

The other moved with a certain impatience.

'How often must I tell you to forget that I was ever anything except a soldier of France?—forget as I have forgotten it?'

The audacious, irrepressible 'Crache-au-nez-d'la-Mort', whom nothing could daunt and nothing could awe, looked penitent and ashamed as a chidden spaniel.

'I know, sir. I have tried many a year, but I thought, perhaps, as how his Lordship's death——'

'No life and no death can make any difference to me, except the death that some day an Arbico's lunge will give me; and that is a long time coming.'

'Ah, for God's sake, Mr Cecil, don't talk like this!'

The Chasseur gave a short, sharp shiver, and started at the name, as if a bullet had struck him.

'Never say that again!'

Rake, Algerian-christened *Crache-au-nez-d'la-Mort*, stammered a contrite apology.

'I never have done, sir, not for never a year, but it wrung it out of me like—you talking of wanting death in that way——'

'Oh, I don't want death!' laughed the other, with a low, indifferent laughter, that had in it a singular tone of sadness all the while. 'I am of our friends the Spahis' opinion—that life is very pleasant with a handsome well-chosen harem, and a good horse to one's saddle. Unhappily, harems are too expensive for Roumis! Yet I am not sure that I am not better amused in the Chasseurs than I was in the Household—specially when we are at war. I suppose we must be wild animals at the core, or we should never find such an infinite zest in the death-grapple. Good night!'

He stretched his long, slender, symmetrical limbs out on the skins that made his bed, and closed his eyes, with the chibouque still in his mouth, and its amber bowl resting on the carpet, which the friendship and honour of Sidi-Ilderim had strewn over the bare turf, on which the house of hair was raised. He was accustomed to sleep as soldiers sleep, in all the din of a camp, or with the roar of savage brutes echoing from the hills around, with his saddle beneath his head, under a slab of rock, or with the knowledge that at every instant the alarm might be given, the drums roll out over the night, and the enemy be down like lightning on the bivouac. But now a name—long unspoken to him—had recalled years he had buried far and for ever from the first day that he had worn the *képi d'ordonnance* of the Army of Algeria, and been enrolled amongst its wild and brilliant soldiers.

Now, long after his comrade had slept soundly, and the light in the single bronze Turkish candle-branch had flickered, and died away, the Chasseur d'Afrique lay wakeful, looking outward through the folds of the tent at the dark and silent camp of the Arabs, and letting his memory drift backward to a time that had grown to be to him as a dream—a time when another world than the world of Africa had known him as Bertie Cecil.

CHAPTER IV

Cigarette en Bienfaitrice

'Oh hé! We are a queer lot; a very queer lot. Sweepings of Europe,' said Claude de Chanrellon, dashing some vermout off his golden moustaches, where he lay full length on three chairs outside the café in the Place du Gouvernement, where the lamps were just alit, and shining through the burnished moonlight of an Algerian evening, and the many-coloured, many-raced, picturesque, and polyglot population of the town were all fluttering out with the sunset, like so many gay-coloured moths.

'Hein! Diamonds are found in the *chiffonnier's* sweepings,' growled a General of Division, who was the most terrible martinet in the whole of the French service, but who loved '*mes enfans d'enfer*', as he was wont to term his men, with a great love, and who would never hear another disparage them, however he might order them blows of the *matraque*, or exile them to Beylick himself.

'You are poetic, mon Général,' said Claude de Chanrellon; 'but you are true. We are a furnace in which Blackguardism is burnt into Daredevilry, and turned out as Heroism. A fine manufacture that, and one at which France has no equal.'

'But our manufactures keep the original hallmark, and show that the devil made them if the drill have moulded them!' urged a Colonel of Tirailleurs Indigènes.

Chanrellon laughed, knocking the ash off a huge cigar.

'Pardieu! We do our original maker credit, then; nothing good in this world without a dash of diablerie. Scruples are the wet blankets, proprieties are the blank walls, principles are the quickset-hedges of life, but devilry is its champagne!'

'Ventre-bleu!' growled the General. 'We have a right to praise the blackguards; without them our conscripts would be very poor trash. The conscript fights because he has to fight, the blackguard fights because he loves to fight. A great difference that.'

The Colonel of Tirailleurs lifted his eyes; a slight pale effeminate dark-eyed Parisian, who looked scarcely stronger than a hothouse flower, yet who, as many an African chronicle could tell, was swift as fire, keen as steel, unerring as a leopard's leap, untiring as an Indian on trail, once in the field with his Indigènes.

'In proportion as one loves powder, one has been a scoundrel, mon Général,' he murmured; 'what the catalogue of your crimes must be!'

The tough old campaigner laughed grimly; he took it as a high compliment.

'Sapristi! The cardinal virtues don't send anybody, I guess, into African service. And yet, pardieu, I don't know. What fellows I have known! I have had men among my Zéphyrs—and they were the wildest *pratiques* too—that would have ruled the world! I have had more wit, more address, more genius, more devotion, in some headlong scamp of a *loustic* than all the courts and cabinets would furnish. Such lives, such lives too, morbleu!'

And he drained his absinthe thoughtfully, musing on the marvellous vicissitudes of war, and on the patrician blood, the wasted wit, the Beaumarchais talent, the Mirabeau power, the adventures like a page of fairy tale, the brains whose strength could have guided a sceptre, which he had found and known, hidden under the rough uniform of a Zéphyr, buried beneath the canvas shirts of a Roumi, lost for ever in the wild lawless escapades of rebellious *pratiques*,[1] who closed their days in the stifling darkness of the dungeons of Beylick, or in some obscure skirmish, some midnight vidette, where an Arab flissa severed the cord of the warped life, and the death was unhonoured by even a line in the Gazettes du Jour.

'Faith!' laughed Chanrellon, regardless of the General's observation, 'if we all published our memoirs, the world would have a droll book. Dumas and Terrail would be beat out of the field. The real recruiting-sergeants that send us to the ranks would be soon found to be——'

'Women,' growled the General.

'Cards,' sighed the Colonel.

'Absinthe,' muttered another.

'Mussetism in a garret.'

'Politics *un peu trop fort*.'

'A comedy that was hissed.'

'Carbonarist vows when one was a fool.'

'The spleen.'

'The dice.'

---

[1] Insubordinates.

'The roulette.'

'The natural desire of humanity to kill and to get killed!'

'Morbleu!' cried Chanrellon, as the voices closed, 'all those mischiefs beat the drum, and send volunteers to the ranks, sure enough; but the General named the worst. Look at that little Cora; the Minister of War should give her the Cross. She sends us ten times more fire-eaters than the Conscription does. Five fine fellows—of the *vieille roche* too—joined today, because she has stripped them of everything, and they have nothing for it but the service. She is invaluable, Cora.'

'And there is not much to look at in her either,' objected a Captain, who commanded Turcos. 'I saw her when our detachment went to show in Paris. A baby face, innocent as a cherub—a soft voice—a shape that looks as slight and as breakable as the stem of my glass—there is the end!'

The Colonel of Tirailleurs laughed scornfully but gently; he had been a great *lion* of the fashionable world before he came out to his Indigènes.

'The end of Cora! The end of her is—"*l'Enfer!*" My good Alcide—that "baby face" has ruined more of us than would make up a battalion. She is so quiet, so tender; smiles like an angel, glides like a fawn; is a little sad too, the innocent dove; looks at you with eyes as clear as water, and paf! before you know where you are, she has pillaged with both hands, and you wake one fine morning bankrupt!'

'Why do you let her do it?' growled the *vieille moustache*, who had served under Junot when a little lad, and had scant knowledge of the ways and wiles of the syrens of the Rue Bréda.

'Ah-bah!' said the Colonel, with a shrug of his shoulders; 'it is the thing to be ruined by Cora. There is Bébée-je-m'enfous; there is Blonde-Miou-Miou; there is the Cerisette; there is Neroli; there is Loto—any one of them is equally good style with Cora; but to be at all in the fashion, one must have been talked of with one of the six.'

'Diantre!' sighed Claude de Chanrellon, stretching his handsome limbs, with a sigh of recollection; for Paris had been a Paradise Lost to him for many seasons, and he had had of late years but one solitary glimpse of it. 'It was Cœur d'Acier who was the rage in my time. She ate me up—that woman—in three months. I had not a hundred francs left: she stripped me as bare as a pigeon. Her passion was emeralds *en cabochon* just then. Well, emeralds *en cabochon* made an end of me, and sent me out here. Cœur d'Acier was a wonderful woman!—and the chief wonder of her was, that she was as ugly as sin.'

'Ugly?'

'Ugly as sin! But she had the knack of making herself more charming than Venus. How she did it nobody knew; but men left the prettiest creatures for her: and she ruined us, I think, at the rate of a score a month.'

'Like Loto,' chimed in the Tirailleur. 'Loto has not a shred of beauty. She is a big, angular, raw-boned Normande, with a rough voice, and a villanous patois; but to be well with Loto is to have achieved distinction at once. She will have nothing under the third order of nobility; and Prince Paul shot the Duc de Var about her the other day. She is a great creature, Loto: nobody knows her secret.'

'*L'audace*, mon ami; *toujours de l'audace*,' said Chanrellon, with a twist of his superb moustaches. 'It is the finest quality out; nothing so sure to win. Hallo! there is *le beau caporal* listening. Ah! Bel-à-faire-peur, you fell, too, among the Lotos and the Cœurs d'Acier once, I will warrant.'

The Chasseur, who was passing, paused and smiled a little as he saluted.

'Cœurs d'Acier are to be found in all ranks of the sex, monsieur, I fancy?'

'Bah! you beg the question. Did not a woman send you out here—eh?'

'No, monsieur, only chance.'

'A fig for your chance! Women are the mischief that casts us adrift to chance.'

'Monsieur, we cast ourselves sometimes.'

'Dieu de Dieu! I doubt that. We should go straight enough if it were not for them.'

The Chasseur smiled again.

'M. le Viscomte thinks we are sure to be right then, if, for the key to every black story, we ask, "Who was she?"'

'Of course I do. Well! who was she? We are all quoting our tempters tonight. Give us your story, *mon brave*!'

'Monsieur, you have it in the *folios matricules*, as well as my sword could write it.'

'Good, good!' muttered the listening General. The soldier-like answer pleased him, and he looked attentively at the giver of it.

Chanrellon's brown eyes flashed a bright response.

'And your sword writes in a brave man's fashion—writes what France loves to read. But before you wore your sword here? Tell us of that. It was a romance—wasn't it?'

'If it were, I have folded down the page, monsieur.'

'Open it, then! Come—what brought you out amongst us? You had gamed *au roi dépouillé*—that was it? Out with it!'

'Monsieur, direct obedience is a soldier's duty; but I never heard that inquisitive annoyance was an officer's privilege?'

The words were calm, cold, a little languid, and a little haughty. The manner of old habit, the instinct of buried pride, spoke in them, and disregarded the barrier between a private of Chasseurs who was but a sous-officier, and a Colonel Commandant who was also a noble of France.

Involuntarily, all the men sitting round the little tables, outside the café, turned and looked at him. The boldness of speech and the quietude of tone drew all their eyes in curiosity upon him.

Chanrellon flushed scarlet over his frank brow, and an instant's passion gleamed out of his eyes: the next, he threw his three chairs down with a crash, as he shook his mighty frame like an Alpine dog, and bowed with a French grace, with a campaigner's frankness.

'A right rebuke!—fairly given, and well deserved. I thank you for the lesson.'

The Chasseur looked surprised and moved; in truth, he was more touched than he showed. Under the rule of Châteauroy, consideration or courtesy had been things long unshown to him. Involuntarily, forgetful of rank, he stretched his hand out, on the impulse of soldier to soldier, of gentleman to gentleman. Then, as the bitter remembrance of the difference of rank and station between them flashed on his memory, he was raising it, proudly but deferentially, in the salute of a subordinate to his superior, when Chanrellon's grasp closed on it readily. The victim of Cœur d'Acier was of as gallant a temper as ever blent the reckless *condottière* with the thoroughbred noble.

The Chasseur coloured slightly, as he remembered that he had forgotten alike his own position and their relative stations.

'I beg your pardon, Monsieur le Vicomte,' he said, simply, as he gave the salute with ceremonious grace, and passed onward rapidly, as though he wished to forget and to have forgotten the momentary self-oblivion of which he had been guilty.

'Dieu!' muttered Chanrellon, as he looked after him, and struck his hand on the marble-topped table till the glasses shook. 'I would give a year's pay to know that fine fellow's history. He is a gentleman—every inch of him.'

'And a good soldier, which is better,' growled the General of Brigade, who had begun life in his time driving an ox-plough over the heavy tillage of Alsace.

'A private of Châteauroy's—eh?' asked the Tirailleur, lifting his eyeglass to watch the Chasseur as he went.

'Pardieu—yes—more's the pity,' said Chanrellon, who spoke his thoughts as hastily as a hand-grenade scatters its powder. 'The Black Hawk hates him—God knows why—and he is kept down in consequence, as if he were the idlest lout or the most incorrigible rebel in the service. Look at what he has done. All the Bureaux will tell you there is not a finer Roumi in Africa—not even among our Schaouacks! Since he joined, there has not been a hot and heavy thing with the Arabs that he has not had his share in. There has not been a campaign in Oran or Kabaila that he has not gone out with. His limbs are slashed all over with Bedouin steel. He rode once twenty leagues to deliver despatches with a spear-head in his side, and fell, in a dead faint, out of his saddle just as he gave them up to the commandant's own hands. He saved the day, two years ago, at Granaila. We should have been cut to pieces, as sure as destiny, if he had not collected a handful of broken Chasseurs together, and rallied them, and rated them, and lashed them with their shame, till they dashed with him to a man into the thickest of the fight, and pierced the Arabs' centre, and gave us breathing room, till we all charged together, and beat the *Arbicos* back like a herd of jackals. There are a hundred more like stories of him—every one of them true as my sabre— and, in reward, he has just been made a *galonné!*'

'Superb!' said the General, with grim significance. '*Ce n'est pas à la France—ça!* Twelve years! In five under Napoléon, he would have been at the head of a brigade; but then'—and the veteran drank his absinthe with a regretful melancholy—'but then, Napoléon read his men himself and never read them wrong. It is a divine gift that for commanders.'

'The Black Hawk can read, too,' said Chanrellon, meditatively; it was the 'petit nom', that Châteauroy had gained long before, and by which he was best known through the army. 'No eyes are keener than his, to trace a *lascar kébir*. But, where he hates, he strikes beak and talons—pong!—till the thing drops dead—even where he strikes a bird of his own brood.'

'That is bad,' said the old General, sententiously. 'There are four people who should have no personal likes or dislikes: they are an innkeeper, a schoolmaster, a ship's skipper, and a military chief.'

With which axiom he called for some more *vert-vert*.

Meanwhile, the Chasseur went his way through the cosmopolitan groups of the great square. A little further onward, laughing, smoking, chatting, eating ices outside a Café Chantant, were a group of Englishmen—a yachting party, whose schooner lay in the harbour. He lingered a moment, and lighted a fusee, just for sake of hearing the old familiar words. As he bent his head above the vesuvian, no one saw the shadow of pain that passed over his face.

But one of them looked at him curiously and earnestly. 'The deuce,' he murmured to the man nearest him, 'who the dickens is it that French soldier's like?'

The French soldier heard, and, with the cigar in his teeth, moved away quickly. He was uneasy in the city—uneasy lest he should be recognized by any passer-by or tourist.

'I need not fear that, though,' he thought with a smile. 'Twelve years!—why, in *that* world, we used to forget the blackest ruin in ten days, and the best life amongst us ten hours after its grave was closed. Besides, I am safe enough. I am dead!'

And he pursued his onward way, with the red glow of the cigar under the chesnut splendour of his beard, and the black eyes of veiled Moresco women flashed lovingly on his tall lithe form, with the scarlet *ceinturon* swathed round his loins, and the scarlet undress fez set on his forehead, fair as a woman's still, despite the tawny glow of the Afric sun that had been on it for so long.

He was 'dead'; therein had laid all his security; thereby had 'Beauty of the Brigades' been buried beyond all discovery in 'Bel-à-faire-peur' of the 2nd Chasseurs d'Afrique. When, on the Marseilles rails, the maceration and slaughter of as terrible an accident as ever befell a train rushing through midnight darkness, at headlong speed, had left himself and the one man faithful to his fortunes unharmed by little less than a miracle, he had seen in the calamity the surest screen from discovery or pursuit.

Leaving the baggage where it was jammed among the débris, he had struck across the country with Rake for the few leagues that still lay between them and the city, and had entered Marseilles as weary foot travellers, before half the ruin on the rails had been seen by the full noon sun.

As it chanced, a trading yawl was loading in the port, to run across to Algiers that very day. The skipper was short of men, and afraid of the Lascars, who were the only sailors that he seemed likely to find, to fill up the vacant places in his small crew.

Cecil offered himself and his comrade for the passage. He had only a very few gold pieces on his person, and he was willing to work his way across, if he could.

'But you're a gentleman,' said the skipper, doubtfully eyeing him, and his velvet dress, and his black sombrero with its eagle's plume. 'I want a rare, rough, able seaman, for there'll be like to be foul weather. She looks too fair to last,' he concluded, with a glance upward at the sky.

He was a Liverpool man, master and owner of his own rakish-looking little black-hulled craft, that, rumour was wont to say, was not averse to a bit of slaving, if she found herself in far seas, with a likely run before her.

'You're a swell, that's what you are,' emphasized the skipper. 'You bean't no sort of use to me.'

'Wait a second,' answered Cecil. 'Did you ever chance to hear of a schooner called *Regina*?'

The skipper's face lighted in a moment.

'Her as was in the Biscay, July come two years?—her as druv' through the storm like a mad thing, and flew like a swallow, when everything was splittin' and founderin', and shipping seas around her?—her as was the first to bear down to the great *Wrestler*, a-lyin' there hull over in water, and took aboard all as ever she could hold o' the passengers, a-pitchin' out her own beautiful cabin fittins to have as much room for the poor wretches as ever she could? Be you a-meanin' her?'

Cecil nodded assent.

'She was my yacht, that's all; and I was without a captain through that storm. Will you think me a good enough sailor now?'

The skipper wrung his hand, till he nearly wrung it off.

'Good enough! Blast my timbers! there aren't one will beat you in any waters. Come on, sir, if so be as you wishes it; but never a stroke of work shall you do atween *my* decks. I never did think as how one of your yachting-nobs could ever be fit to lay hold of a tiller; but, hang me, if the Club make such sailors as you it's a rare 'un! Lord a mercy! why my wife was in the *Wrestler*. I've heard her tell scores of times as how she was a'most dead when that little yacht came through a swaling sea, that was all heavin' and roarin' round the wreck, and as how the swell what owned it gev' his cabin up to the womenkind, and had his swivel-guns and his handsome furniture pitched overboard, that he might be able to carry more passengers, and fed 'em, and gev' 'em champagne all round, and treated 'em like a Prince, till he ran 'em straight into Brest Harbour. But, damn me! that ever a swell like you should——'

'Let's weigh anchor,' said Bertie, quietly.

And so he crossed unnoticed to Algeria, while through Europe the tidings went, that the mutilated form, crushed between iron and wood, on the Marseilles line, was his, and that he had perished in that awful, ink-black, sultry southern night, when the rushing trains had met, as meet the thunder-clouds. The world thought him dead; as such, the journals recorded him, with the shameful outlines of imputed crime, to make the death the darker; as such, his name was forbidden to be uttered at Royallieu; as such, the Seraph mourned him with passionate loving force, refusing to the last to accredit his guilt:—and he, leaving them in their error, was drafted into the French Army under two of his Christian names, which happily had a foreign sound—Louis Victor—and laid aside for ever his identity as Bertie Cecil.

He went at once on service in the interior, and had scarcely come in any of the larger towns since he had joined. His only danger of recognition had once been when a Marshal of France, whom he had used to know well in Paris and at the Court of St James, held an inspection of the African troops.

Filing past the brilliant staff, he had ridden at only a few yards' distance from his old acquaintance, and, as he saluted, had glanced involuntarily at the face that he had seen oftentimes in the Salles des Maréchaux, and even under the roof of Royallieu. The great chief's keen blue eyes were scrutinizing the regiment, ready to note a chain loose, a belt awry, a sword specked with rust, if such a sin there were against '*les ordonnances*' in all the glittering squadrons; and swept over him, seeing in him but one among thousands—a unit in the mighty aggregate of the 'raw material' of war.

The Marshal only muttered to a General beside him, 'Why don't they all ride like that man? He has the seat of the English Guards.' But that it was in truth an officer of the English Guards, and a friend of his own, who paced past him as a private of Algerian Horse, the French leader never dreamed.

From the extremes of luxury, indolence, indulgence, pleasure, and extravagance, Cecil came to the extremes of hardship, poverty, discipline, suffering, and toil. From a life where every sense was gratified, he came to a life where every privation was endured. He had led the fashion; he came where he had to bear without a word the curses, oaths, and insults of a corporal or a sous-lieutenant. He had been used to every delicacy and delight; he came where he had to take the coarse black bread of the army as a rich repast. He had thought it too much trouble to murmur flatteries in great ladies' ears; he came where morning, noon, and night the inexorable demands of rigid rules compelled his incessant obedience, vigilance, activity, and self-denial. He had known nothing from his childhood up except an atmosphere of amusement, refinement, brilliancy, and idleness; he came where gnawing hunger, brutalized jest, ceaseless toil, coarse obscenity, agonized pain, and pandemoniac mirth, alternately filled the measure of the days.

A sharper contrast, a darker ordeal, rarely tried the steel of any man's endurance; yet, under it, he verified the truth, '*Bon sang ne peut mentir.*' No Spartan could have borne the change more mutely, more staunchly, than did the 'dandy of the Household'.

The first years were, it is true, years of intense misery to him. Misery, when all the blood glowed in him under some petty tyrant's jibe, and he had to stand immovable, holding his peace. Misery, when the hunger and thirst of long marches tortured him, and his soul sickened at the half-raw offal, and the water thick with dust, and stained with blood, which the men round him seized so ravenously. Misery, when the dreary dawn broke, only to usher in a day of mechanical manœuvres, of petty tyrannies, of barren burdensome hours in the exercise-ground, of convoy duty in the burning sun-glare, and under the heat of harness; and the weary night fell with the din and uproar, and the villanous blasphemy, and befouled merriment of the riotous Chambrée, that denied even the peace and oblivion of sleep. They were years of infinite wretchedness oftentimes, only relieved by the loyalty and devotion of the man who had followed him into his exile. But, however wretched, they never wrung a single regret or lament from Cecil. He had come out to this life; he took it as it was. As, having lost the title to command, the high breeding in him made him render implicitly the mute obedience which was the first duty of his present position, so it made him accept, from first to last, without a sign of complaint or of impatience, the altered fortunes of his career. The hardest-trained, lowest-born, longest-inured soldier in the Zephyr ranks did not bear himself with more apparent content and more absolute fortitude than did the man who had used to think it a cruelty to ride with his troop from Windsor to Wormwood Scrubs, and had never taken the trouble to load his own gun any shooting season, or to draw off his own coat any evening. He suffered acutely many times; suffered till he was heart-sick of his life; but he never sought to escape the slightest penalty or hardship, and not even Rake ever heard from him a single syllable of irritation or of self-pity.

Moreover, the war-fire woke in him.

In one shape or another, active service was almost always his lot, and hot, severe campaigning was his first introduction to military life in Algeria. The latent instinct in him—the instinct that had flashed out during his lazy fashionable calm in all moments of danger, in all days of keen sport; the instinct that had made him fling himself into the duello with the French boar, and made him mutter to Forest King, 'Kill me if you like but don't *fail* me!'—was the instinct of the born soldier. In peril, in battle, in reckless bravery, in the rush of the charge, and the excitement of the surprise, in the near presence of death, and in the chase of a foe through a hot African night when both were armed to the teeth, and one or both must fall when the grapple came—in all these that old instinct, aroused and unloosed, made him content; made him think that the life which brought them was worth the living.

There had always been in him a reckless daredevilry, which had slept under the serene effeminate insouciance of his careless temper and his pampered habits. It had full rein now, and made him, as the army affirmed, one of the most intrepid, victorious, and chivalrous lascars of its fiery ranks. Fate had flung him off his couch of down into the tempest of war, into the sternness of life spent ever on the border of the grave, ruled ever by an iron code, requiring at every step self-negation, fortitude, submission, courage, patience, the self-control which should take the uttermost provocation from those in command without even a look of reprisal, and the courageous recklessness which should meet death, and deal death, which should be as the eagle to swoop, as the lion to rend. And he was not found wanting in it.

He was too thoroughbred to attempt to claim a superiority that fortune no longer conferred on him, to seek to obtain a deference that he had no longer the position to demand. He obeyed far more implicitly than many a ruffian filibuster, who had been amongst the dregs of society from his birth. And though his quick-eyed comrades knew, before he had been amongst them five minutes, that an 'aristocrat' had taken refuge under the Flag of Mazagran, they never experienced from him one touch of the insolence that their own sous-officiers beat them with, as with the flat of the sword; and they never found in him one shadow of the arrogance that some fellow-soldier, who had swelled into a sergeant-major, or bristled into an adjutant, would strut with, like any turkey-cock.

He was too quiet, too courteous, too calmly listless; he had too easy a grace, too soft a voice, and too many gentleman-habits, for them. But when they found that he could fight like a Zouave, ride like an Arab, and bear shot-wounds or desert-thirst as though he were of bronze, it grew a delight to them to see of what granite and steel this dainty patrician was made; and they loved him with a rough, ardent, dog-like love, when they found that his last crust, in a long march, would always be divided; that the most desperate service of danger was always volunteered for by him; that no severity of personal chastisement ever made him clear himself of a false charge at a comrade's expense; and that all his *décompte* went in giving a veteran a stoup of wine, or a sick conscript a tempting meal, or a prisoner of Beylick some food through the grating, scaled too at risk of life and limb.

Cecil had all a soldier's temper in him; and the shock which had hurled him out of ease, and levity, and ultra-luxury, to stand alone before as dark and rugged a fortune as ever fronted any man, had awakened the war-fire which had only slumbered because lulled by habit, and unaroused by circumstance. He had never before been called on to exert either thought or action: the necessity for both called many latent qualities in him into play. The same nature, which had made him wish to be killed over the Grand Military course, rather than live to lose the race, made him now bear privation as calmly, and risk death as recklessly, as the hardiest and most fiery *loustic* of the African cantonments.

Bitter as the life often was, severe the suffering, and acute the deprivation, the sternest veteran scarcely took them more patiently, more silently, than the 'aristocrat', to whom a corked claret or a dusty race-day had been calamities. Cast among these wild, iron-muscled bohemians, who fought like tigers, and were as impenetrable as rhinoceri, 'race' was too strong in Cecil not to hold its own with them, whether in the quality of endurance, or the quality of daring.

'*Main de femme, mais main de fer,*' the Roumis were wont to say of their comrade, with his delicate habits, '*comme une Marquise du Faubourg*', as they would growl impatiently; and his tenacious patience, which would never give way either in the toil of the camp or the grip of the struggle.

On the surface it seemed as though never was there a life more utterly thrown away than the life of a Guardsman and a gentleman, a man of good blood, high rank, and talented gifts had he ever chosen to make anything of them, buried in the ranks of the Franco-African army, risking a nameless grave in the sand with almost every hour, associated with the roughest riffraff of Europe, liable any day to be slain by the slash of an Arab *flissa*, and rewarded for twelve years' splendid service by the distinctive badge of a corporal. Any one of the friends of his former years, seeing him thus, would have said that he might as well be thrown at once into a pit in the sand, where the dead were piled twenty deep after a skirmish, to lie and rot, or be dug up by the talons of famished beasts, whichever might chance, as live thus in the obscurity, poverty, and semi-barbarism of an Algerian private's existence.

Yet it might be doubted if any life would have done for him what this had done: it might be questioned, if, judging a career not by its social position, but by its effect on character, any other would have been so well for him or would equally have given steel and strength to the indolence and languor of his nature as this did. In his old world he would have lounged listlessly through fashionable seasons, and, in an atmosphere that encouraged his profound negligence of everything and his natural nil-admirari listlessness, would have glided from refinement to effeminacy, and from lazy grace to blasé inertia.

The severity and the dangers of the campaigns with the French army had roused the sleeping lion in him, and made him as fine a soldier as ever ranged under any flag. He had suffered, braved, resented, fought, loved, hated, endured, and even enjoyed, here in Africa, with a force and a vividness that he had never dreamed possible in his calm, passionless, insouciant world of other days. He had known what the hunger of famine, what the torment of fever, what the agony of forbidden pride, what the wild delight of combat, were. He had known what it was to long madly for a stoup of water; to lie raving, yet conscious, under the throes of gunshot wounds; to be forced to bear impassively words, for a tithe of which he could have struck across the mouth the chief who spoke them; to find in a draught of wretched wine, after days of marching, a relish that he had never found in the champagnes and burgundies of the Guards' mess; to love the dark Arab eyes, that smiled on him in his exile, as he had never loved those of any woman, and to suffer when the death-film gathered over them, as he had never thought it in him to suffer for any death or any life; to feel every nerve thrill, and every vein glow, as with fierce, exultant joy, as the musketry pealed above the plains, and his horse pressed down on to the very mouths of the rifles, and the naked sabres flashed like the play of lightnings, and over the dead body of his charger, he fought ankle deep in blood, with the Arabs circling like hawks, and their great blades whirling round him, catching the spears aimed at him with one hand, while he beat back their swords, blow for blow, with the other; he had known all these, the desert passions; and, while outwardly they left him much the same in character, they changed him vitally. They developed him into a magnificent soldier—too true a soldier not to make thoroughly his the service he had adopted, not to, oftentimes, almost forget that he had ever lived under any other flag than that tricolour, which he followed and defended now.

The quaint heroic Norman motto of his ancestors carved over the gates of Royallieu—'*Cœur Vaillant Se Fait Royaume*'—verified itself in his case. Outlawed, beggared, robbed at a stroke of every hope and prospect, he had taken his adversity boldly by the beard, and had made himself at once a country and a kingdom among the brave, fierce, reckless, loyal hearts of the men, who came from north, south, east, and west, driven by every accident, and scourged by every fate, to fill up the battalions of North Africa.

As he went now, in the warmth of the after-glow, he turned up into the Rue Babazoum, and paused before the entrance of a narrow, dark, tumbledown, picturesque shop, half like a stall of a Cairo bazaar, half like a Jew's den in a Florentine alley.

A cunning, wizen head peered out at him from the gloom.

'Ah-ha! good even, Corporal Victor!'

Cecil, at the words, crossed the sill and entered.

'Have you sold any?' he asked. There was a slight constraint and hesitation in the words, as of one who can never fairly bend his spirit to the yoke of barter.

The little, hideous, wrinkled, dwarf-like creature, a trader in curiosities, grinned with a certain gratification, in disappointing this lithe-limbed, handsome Chasseur.

'Not one. The toys don't take. Daggers now, or anything made out of spent balls, or flissas one can tell an Arab story about, go off like wildfire; but your ivory bagatelles are no sort of use, M. le Caporal.'

'Very well: no matter,' said Cecil, simply, as he paused a moment before some delicate little statuettes and carvings—miniature things, carved out of a piece of ivory, or a block of marble, the size of a horse's hoof, such as could be picked up in dry river channels, or broken off stray boulders; slender crucifixes, wreaths of foliage, branches of wild fig, figures of Arabs and Moors, dainty heads of dancing-girls, and tiny chargers fretting like Bucephalus. They were perfectly conceived and executed. He had always had a D'Orsay-like gift that way, though, in common with all his gifts, he had utterly neglected all culture of it, until, cast adrift on the world, and forced to do something to maintain himself, he had watched the skill of the French soldiers at all such expedients to gain a few coins, and had solaced many a dreary hour in barracks and under canvas with the toy-sculpture, till he had attained a singular art at it. He had commonly given Rake the office of selling them, and as commonly spent all the proceeds on all other needs save his own.

He lingered a moment, with regret in his eyes; he had scarcely a sou in his pocket, and he had wanted some money sorely that night for a comrade dying of a lung-wound—a noble fellow, a French artist who, in an evil hour of desperation, had joined the army, with a poet's temper that made its hard colourless routine unendurable, and had been shot in the chest in a night-skirmish.

'You will not buy them yourself?' he asked, at length, the colour flushing his face; he would not have pressed the question to save his own life from starving, but Léon Ramon would have no chance of a fruit or a lump of ice to cool his parched lips and still his agonized retching, unless he himself could get money to buy those luxuries that are too splendid and too merciful to be provided for a dying soldier, who knows so little of his duty to his country as to venture to die in his bed.

'Myself!' screeched the dealer, with a derisive laugh. 'Ask me to give you my whole stock next, *M. le Galonné!* These trumperies will lie on hand for a year.'

Cecil went out of the place without a word; his thoughts were with Léon Ramon, and the insolence scarce touched him. 'How shall I get him the ice?' he wondered. 'God! if I had only one of the lumps that used to float in our claret-cup!'

As he left the den, a military fairy, all gay with blue and crimson, like the fuchsia bell she most resembled, with a meerschaum in her scarlet lips and a world of wrath in her bright black eyes, dashed past him into the darkness within, and before the dealer knew or dreamt of her, tossed up the old man's little shrivelled frame like a shuttlecock, shook him till he shook like custards, flung him upward and caught him as if he were the hoop in a game of La Grace, and set him down bruised, breathless, and terrified out of his wits.

'Ah, *chénapan!*' cried Cigarette, with a volley of slang, utterly untranslatable, 'that is how you treat your betters, is it? Miser, monster, crocodile, serpent! Harpagon was an angel to you.' (She knew Harpagon because some of her Roumis chattered bits of Molière.) 'He wanted the money and you refused it? Ah—h—h! son of Satan! you live on other men's miseries! Run after him—quick, and give him this, and this, and this, and this; and say you were only in jest, and the things were worth a Sheik's ransom. Stay! you must not give him too much or he will know it is not *you*—viper! Run quick, and breathe a word about me if you dare; one whisper only, and my Spahis shall cut your throat from ear to ear. Off! or you shall have a bullet to quicken your steps; misers dance well when pistols play the minuet!'

With which exordium the little Amie du Drapeau shook her culprit at every epithet, emptied out a shower of gold and silver, just won at play, from the bosom of her uniform, forced it into the dealer's hands, hurled him out of his own door, and drew her pretty weapon with a clash from her sash.

'Run for your life!—and do just what I bid you, or a shot shall crash your skull in as sure as my name is Cigarette!'

The little old Jew flew as fast as his limbs would carry him, clutching the coins in his horny hands. He was terrified to a mortal anguish, and had not a thought of resisting or disobeying her; he knew the fame of Cigarette—as who did not? Knew that she would fire at a man as carelessly as at a cat, more carelessly in truth, for she favoured cats, saving many from going into the Zouaves' soup-caldrons, and favoured civilians not at all; and knew that at her rallying-cry all the sabres about the town would be drawn without a second's deliberation, and sheathed in anything or in anybody that had offended her, for Cigarette was, in her fashion, Generalissima of all the Regiments of Africa.

The dealer ran with all the speed of terror, and overtook Cecil, who was going slowly onward to the barracks.

'Are you serious?' he asked, in surprise at the large amount, as the little Jew panted out apologies, entreaties, and protestations of his only having been in jest, and of his fervently desiring to buy the carvings at his own price, as he knew of a great collector in Paris to whom he needed to send them.

'Serious! Indeed am I serious, M. le Caporal,' pleaded the curiosity-trader, turning his head in agonized fear to see if the vivandière's pistol was behind him. 'The things will be worth a great deal to me where I shall send them, and though they are but bagatelles, what is Paris itself but one bagatelle? Pouf! they are all children there, they will love the toys. Take the money, I pray you, take the money!'

Cecil looked at him a moment; he saw the man was in earnest, and thought but little of his repentance and trepidation, for the citizens were all afraid of slighting or annoying a soldier.

'So be it. Thank you,' he said, as he stretched out his hand and took the coins, not without a keen pang of the old pride that would not wholly be stilled, yet gladly for sake of the Chasseur dying yonder, growing delirious and wrenching the blood off his lungs, in want of one touch of the ice that was spoiled by the ton weight to keep cool the wines and the fish of M. le Marquis de Châteauroy. And he went onward to spend the gold his sculptures had brought on some mellow figs and some cool golden grapes, and some ice-chilled wines that should soothe a little of the pangs of dissolution, to his comrade, and bear him back a moment, if only in some fleeting dream, to the vine shadows, and the tossing seas of corn, and the laughing sunlit sweetness, of his own fair country by the blue Biscayan waves.

'You did it? That is well. Now, see here—one word of me, now or ever after, and there is a little present that will come to you, hot and quick, from Cigarette,' said the little Friend of the Flag, with a sententious sternness, that crushed each word deliberately through her tight-set pearly teeth. The unhappy Jew shuddered and shut his eyes as she held a bullet close to his sight, then dropped it with an ominous thud in her pistol barrel.

'Not a syllable, never a syllable,' he stammered; 'and if I had known you were in love with him, *ma belle*——'

A box on the ears sent him across his own counter.

'In love? Parbleu! I detest the fellow!' said Cigarette, with fiery scorn and as hot an oath.

'Truly? Then why give your Napoléons?——' began the bruised and stammering Israelite.

Cigarette tossed back her pretty head, that was curly and spirited and shapely as any thoroughbred spaniel's; a superb glance flashed from her eyes, a superb disdain sat on her lips.

'You are a Jew-trader; you know nothing of our code under the tri-colour. We—*nous autres soldats*—are too proud not to aid even an enemy when he is in the right, and France always arms for Justice!'

With which magnificent peroration she swept all the carvings—they were rightfully hers—off the table.

'They will light my cooking fire!' she said, contemptuously, as she vaulted lightly over the counter into the street, and pirouetted like a bit of fantoccini, that is wound up to waltz for ever, along the slope of the crowded Babazoum. All made way for her, even the mighty Spahis and the trudging Bedouin mules, for all knew that if they did not she would make it for herself, over their heads or above their prostrated bodies.

She whirled her way, like a gay-coloured top set humming down a road, through the divers motley groups, singing at the top of her sweet mirthful voice, for she was angry with herself; and, for that, sang the more loudly the most wicked and *risqué* of her slang songs, that gave the morals of a Messalina in the language of a fish-wife, and yet had an inalienable, mischievous, contagious, dauntless French grace in it withal. Finally, she whirled herself into a dark deserted Moresco archway, a little out of the town, and dropped on a stone block, as a swallow, tired of flight, drops on to a bough.

'Is that the way I revenge myself? Ah, bah! I deserve to be killed! When he called me unsexed—unsexed—unsexed!'—and with each repetition of the infamous word, so bitter because vaguely admitted to be true, with her cheeks scarlet and her eyes aflame, and her hands clenched, she flung one of the ivory wreaths on to the pavement and stamped on it with her spurred heel until the carvings were ground into powdered fragments—stamped, as though it were a living foe, and her steel-bound foot were treading out all its life with burning hate and pitiless venom.

In the act her passion exhausted itself, as the evil of such warm, impetuous, tender natures will; she was very still, and looked at the ruin she had done with regret and a touch of contrition.

'It was very pretty—and cost him weeks of labour, perhaps,' she thought.

Then she took all the rest up, one by one, and gazed at them. Things of beauty had had but little place in her lawless young life; what she thought beautiful was a regiment sweeping out in full sunlight, with its eagles, and its colours, and its kettle-drums; what she held as music was the beat of the *réveillé* and the mighty roll of the great artillery; what made her pulse throb and her heart leap was to see two fine opposing forces draw near for the onslaught and thunder of battle. Of things of grace she had no heed, though she had so much grace herself; and her life, though full of colour, pleasure, and mischief, was as rough a one in most respects as any of her comrades'. These delicate artistic carvings were a revelation to her.

Here was the slender pliant spear of the river-reed; here the rich foliage of the wild fig-tree; here the beautiful blossom of the oleander; here fruit and flower, and vine-leaf, and the pendulous ears of millet, twined together in their ivory semblance till they seemed to grow beneath her hands—and those little hands looked so brown and so powder-stained beside the pure snow-whiteness of the wreaths! She touched them reverently one by one; all the carvings had their beauty for her, but those of the flowers had far the most. She had never noted any flowers in her life before, save those she strung together for the Zéphyrs on the Jour de Mazagran. Her youth was a military ballad, rhymed vivaciously to the rhythm of the Pas de Charge; but other or softer poetry had never by any chance touched her until now. Now that in her tiny, bronzed, war-hardened palms lay the white foliage, the delicate art-trifles of this Chasseur, who bartered his talent to get a touch of ice for the burning lips of his doomed comrade.

'He is an aristocrat—he has such gifts as this—and yet he is in the ranks, has no country, is so poor that he is glad of a Jew's pittance, and must sell all this beauty to get a slice of melon for Léon Ramon!' she thought, while the silvery moon strayed in through a broken arch, and fell on an ivory coil of twisted lentiscus leaves and river grasses.

And, lost in a musing pity, Cigarette forgot her vow of vengeance.

CHAPTER V

The Ivory Squadrons

The Chambrée of the Chasseurs was bright and clean in the morning light; but in common with all Algerian barrack-rooms as unlike the barrack-rooms of the ordinary army as Cigarette, with her débonnaire devilry, smoking on a gun-waggon, was unlike a trim Normandy soubrette, sewing on a bench in the Tuileries gardens.

Disorder reigned supreme; but Disorder, although a dishevelled goddess, is very often a picturesque one, and more of an artist than her better-trained sisters; and the disorder was brightened with a thousand vivid colours and careless touches that blent in confusion to enchant a painter's eyes. The room was crammed with every sort of spoil that the adventurous pillaging temper of the troopers could forage from Arab tents, or mountain caves, or river depths, or desert beasts and birds. All things, from tiger-skins to birds'-nests, from Bedouin weapons to ostrich-eggs, from a lion's mighty coat to a tobacco-stopper chipped out of a morsel of deal, were piled together, pell-mell, or hung against the whitewashed walls, or suspended by cords from bed to bed. Everything that ingenuity and hardihood, prompted by the sharp spur of hunger, could wrest from the foe, from the country, from earth or water, from wild beasts or riven rock, were here in the midst of the soldiers' regimental pallets and regimental arms, making the Chambrée at once atelier, storehouse, workshop, and bazaar; while the men, cross-legged on their little hard couches, worked away with the zest of those who work for the few coins that alone will get them the food, the draught of wine, the hour's mirth and indulgence at the estaminet, to which they look across the long stern probation of discipline and manœuvre.

Skill, grace, talent, invention whose mother was necessity, and invention that was the unforced offshoot of natural genius, were all at work; and the hands that could send the naked steel down at a blow through turban and through brain could shape, with a woman's ingenuity, with a craftsman skill, every quaint device and dainty bijou from stone and wood, and many-coloured feathers, and mountain berries, and all odds and ends that chance might bring to hand, and that the women of Bedouin tribes or the tourists of North Africa might hereafter buy with a wondrous tale appended to them, racy and marvellous as the Sapir slang and the military imagination could weave, to enhance the toy's value, and get a few coins more on them for their manufacture.

Ignorance jostled art, and *bizarrerie* ran hand in hand with talent, in all the products of the Chasseurs' extemporised studio; but nowhere was there ever clumsiness, and everywhere was there an industry, gay, untiring, accustomed to make the best of the worst; the workers laughing, chattering, singing, in all good-fellowship, while the fingers that gave the death-thrust held the carver's chisel, and the eyes that glared blood-red in the heat of battle twinkled mischievously over the meerschaum bowl, in whose grinning form some great chief of the Bureaucratie had just been sculptured in audacious parody.

In the midst sat Rake, tattooing with an eastern skill the skin of a great lion, that a year before he had killed in single combat in the heart of Oran, having watched for the beast twelve nights in vain, high perched on a leafy crest of rock, above a watercourse. While he worked, his tongue flew far and fast over the camp slang—the slangs of all nations came easy to him—in voluble conversation with the Chasseur next him, who was making a fan out of feathers that any Peeress might have signalled with at the Opera. 'Crache-au-nez-d'la-Mort' was in high popularity with his comrades; and had said but the truth when he averred that he had never been so happy as under the tricolour. The officers pronounced him an incurably audacious '*pratique*'; he was always in mischief, and the regimental rules he broke through like a terrier through a gauze net; but they knew that when once the trumpets sounded Boot and Saddle, this yellow-haired daredevil of an English fellow would be worth a score of more orderly soldiers, and that wherever his adopted flag was carried, there would he be, first and foremost, in everything save retreat. The English service had failed to turn Rake to account; the French service made no such mistake, but knew that though this British bull-dog might set his teeth at the leash and the lash, he would hold on like grim death in a fight, and live game to the last, if well handled.

Apart, at the head of the Chambrée, sat Cecil. The banter, the songs, the laughter, the chorus of tongues, went on unslackened by his presence. He had cordial sympathies with the soldiers: with those men who had been his fellows in adversity and danger; and in whom he had found, despite all their occasional ferocity and habitual recklessness, traits and touches of the noblest instincts of humanity. His heart was with them always, as his purse, and his wine, and his bread were alike shared ever among them. He had learned to love them well—these wild wolf-dogs, whose fangs were so terrible to their foes, but whose eyes would still glisten at a kind word, and who would give a staunch fidelity unknown to tamer animals.

Living with them, one of them in all their vicissitudes, knowing all their vices, but knowing also all their virtues, owing to them many an action of generous nobility, and watching them in many an hour when their gallant self-devotion and their loyal friendships went far to redeem their lawless robberies and their ruthless crimes, he understood them thoroughly, and he could rule them more surely in their tempestuous evil, because he comprehended them so well in their mirth and in their better moods. When the grade of *sous-officier* gave him authority over them, they obeyed him implicitly because they knew that his sympathies were with them at all times, and that he would be the last to check their gaiety, or to punish their harmless indiscretions.

The warlike Roumis had always had a proud tenderness for their Bel-à-faire-peur, and a certain wondering respect for him; but they would not have adored him to a man, as they did, unless they had known that they might laugh without restraint before him, and confide any dilemma to him sure of aid, if aid were in his power.

The laughter, the work, and the clatter of conflicting tongues were at their height; Cecil sat, now listening, now losing himself in thought, while he gave the last touch to the carvings before him. They were a set of chessmen which it had taken him years to find materials for and to perfect; the white men were in ivory, the black in walnut, and were two opposing squadrons of French troops and of mounted Arabs. Beautifully carved, with every detail of costume rigid to truth, they were his master-piece, though they had only been taken up at any odd ten minutes that had happened to be unoccupied during the last three or four years. The chessmen had been about with him in so many places and under canvas so long, from the time that he chipped out their first Zouave pawn, as he lay in the broiling heat of Oran prostrate by a dry brook's stony channel, that he scarcely cared to part with them, and had refused to let Rake offer them for sale, with all the rest of the carvings. Stooping over them, he did not notice the doors open at the end of the Chambrée until a sudden silence that fell on the babble and uproar round him made him look up; then he rose and gave the salute with the rest of his discomfited and awe-stricken troopers. Châteauroy with a brilliant party had entered.

The Colonel flashed an eagle glance round.

'Fine discipline! You shall go and do this pretty work at Beylick!'

The soldiers stood like hounds that see the lash; they knew that he was like enough to carry out his threat; though they were doing no more than they had always tacit if not open permission to do. Cecil advanced, and fronted him.

'Mine is the blame, *mon Commandant*!'

He spoke simply, gently, boldly; standing with the ceremony that he never forgot to show to their chief, where the glow of African sunlight through the casement of the Chambrée fell full across his face, and his eyes met the dark glance of the 'Black Hawk' unflinchingly. He never heeded that there was a gay, varied, numerous group behind Châteauroy; visitors who were looking over the barrack; he only heeded that his soldiers were unjustly attacked and menaced.

The Marquis gave a grim significant smile, that cut like so much cord of the scourge.

'*Ça va sans dire!* Wherever there is insubordination in the regiment the blame is very certain to be yours! Corporal Gaston, if you allow your *Chambrée* to be turned into the riot of a public fair you will soon find yourself degraded from the rank you so signally contrive to disgrace.'

The words were far less than the tone they were spoken in, that gave them all the insolence of so many blows, as he swung on his heel and bent to the ladies of the party he escorted. Cecil stood mute; bearing the rebuke as it became a Corporal to bear his Commander's anger; a very keen observer might have seen that a faint flush rose over the sun-tan of his face, and that his teeth clenched under his beard, but he let no other sign escape him.

The very self-restraint irritated Châteauroy, who would have been the first to chastise the presumption of a reply, had any been attempted.

'Back to your place, sir!' he said, with a wave of his hand, as he might have waved back a cur. 'Teach your men the first formula of obedience at any rate!'

Cecil fell back in silence. With a swift warning glance at Rake, whose mouth was working, and whose forehead was hot as fire, where he clenched his lion-skin, and longed to be once free, to pull his chief down as lions pull in the death-spring, he went to his place at the further end of the chamber and stood, keeping his eyes on the chess carvings, lest the control which was so bitter to retain should be broken if he looked on at the man who had been the curse and the antagonist of his whole life in Algeria.

He saw nothing and heard almost as little of all that went on around him; there had been a flutter of cloud-like colour in his sight, a faint dreamy fragrance on the air, a sound of murmuring voices and of low laughter; he had known that some guests or friends of the Marquis's had come to view the barracks, but he never even glanced to see who or what they were. The passionate bitterness of just hatred, that he had to choke down as though it were the infamous instinct of some nameless crime, was on him.

The moments passed, the hum of the voices floated to his ear, the ladies of the party lingered by this soldier and by that, buying half the things in the chamber, filling their hands with all the quaint trifles, ordering the daggers and the flissas and the ornamented saddles and the desert skins to adorn their chateaux at home; and raining down on the troopers a shower of uncounted Napoléons until the Chasseurs, who had begun to think their trades would take them to Beylick, thought instead that they had drifted into dreams of El Dorado. He never looked up; he heard nothing, heeded nothing; he was dreamily wondering whether he should always be able so to hold his peace, and to withhold his arm, that he should never strike his tyrant down with one blow, in which all the opprobrium of years should be stamped out? A voice woke him from his reverie.

'Are those beautiful carvings yours?'

He looked up, and in the gloom of the alcove where he stood, where the sun did not stray, and two great rugs of various skins, with some conquered banners of Bedouins, hung like a black pall, he saw a woman's eyes resting on him; proud, lustrous eyes, a little haughty, very thought-ful, yet soft withal, as the deepest hue of deep waters. He bowed to her with the old grace of manner that had so amused and amazed the little vivandière.

'Yes, Madame, they are mine.'

'Ah?—what wonderful skill!'

She took the White King, an Arab Sheik on his charger, in her hand, and turned to those about her, speaking of its beauties and its workman-ship in a voice low, very melodious, ever so slightly languid, that fell on Cecil's ear like a chime of long-forgotten music. Twelve years had drifted by since he had been in the presence of a high-bred woman, and those lingering, delicate tones had the note of his dead past.

He looked at her; at the gleam of the brilliant hair, at the arch of the proud brows, at the dreaming, imperial eyes; it was a face singularly dazzling, impressive, and beautiful at all times; most so of all in the dusky shadows of the waving desert banners, and the rough, rude, barbaric life of the Caserne, where a fille de joie or a cantinnière were all of her sex that was ever seen, and those—poor wretches!—were hardened, and bronzed, and beaten, and brandy-steeped out of all likeness to the fairness of women.

'You have an exquisite art. They are for sale?' she asked him: she spoke with the careless gracious courtesy of a *grande dame* to a Corporal of Chasseurs, looking little at him, much at the ivory Kings and their mimic hosts of Zouaves and Bedouins.

'They are at your service, Madame.'

'And their price?' She had been purchasing largely of the men on all sides as she had swept down the length of the Chambrée, and she drew out some French banknotes as she spoke. Never had the bitterness of poverty smitten him as it smote him now when this young patrician offered him her gold! Old habits vanquished; he forgot who and where he now was; he bowed as in other days he had used to bow in the circle of St James's.

'Is—the honour of your acceptance, if you will deign to give that.'

He forgot that he was not as he once had been. He forgot that he stood but as a private of the French Army before an aristocrate whose name he had never heard.

She turned and looked at him, which she had never done before, so absorbed had she been in the chessmen, and so little did a Chasseur of the ranks pass into her thoughts. There was an extreme of surprise, there was something of offence, and there was still more of coldness in her glance; a proud, languid, astonished coldness of regard, though it softened slightly as she saw that he had spoken in all courtesy of intent.

She bent her graceful regal head.

'I thank you. Your very clever work can of course only be mine by purchase.'

And with that she laid aside the White King among his little troop of ivory Arabs and floated onward with her friends. Cecil's face paled slightly under the mellow tint left there by the desert sun and the desert wind; he swept the chessmen into their walnut case and thrust them out of sight under his knapsack. Then he stood motionless as a sentinel, with the great leopard skins and Bedouin banners behind him, casting a gloom that the gold points on his harness could scarcely break in its heavy shadow, and never moved till the echo of the voices, and the cloud of the draperies, and the fragrance of perfumed laces, and the brilliancy of the staff officers' uniforms had passed away, and left the soldiers alone in their Chambrée. Those careless, cold words from a woman's lips had cut him deeper than the *matraque* could have cut him, though it had bruised his loins and lashed his breast; they showed all he had lost.

'What a fool I am still!' he thought, as he made his way out of the barrack-room. 'I might have fairly forgotten by this time that I ever had the rights of a gentleman.'

So the carvings had won him one warm heart and one keen pang that day; the vivandière forgave, the aristocrat stung him, by means of those snowy, fragile, artistic toys that he had shaped in lonely nights under canvas by ruddy picket-fires, beneath the shade of wild fig-trees, and in the stir and colour of Bedouin encampments.

'I must ask to be ordered out of the city,' he thought, as he pushed his way through the crowds of soldiers and civilians. 'Here, I get bitter, restless, impatient; here, the past is always touching me on the shoulder; here, I shall soon grow to regret, and to chafe, and to look back like any pining woman. Out yonder there, with no cares to think of but my horse and my troop, I am a soldier—and nothing else: so best. I shall be nothing else as long as I live. Pardieu, though! I don't know what one wants better: it is a good life, as life goes. One must not turn compliments to great ladies, that is all—not much of a deprivation there. The chessmen are the better for that; her Maltese dog would have broken them all the first time it upset their table!'

He laughed a little as he went on smoking his *brûle-gueule*, the old carelessness, mutability, and indolent philosophies were with him still, and were still inclined to thrust away and glide from all pain as it arose. Though much of gravity and of thoughtfulness had stolen on him, much of insouciance remained; and there were times when there was not a more reckless or a more nonchalant *lion* in all the battalions than 'Bel-à-faire-peur'. Under his gentleness there was 'wild blood' in him still, and the wildness was not tamed by the fiery champagne-draught of the perilous, adventurous years he spent.

'I wonder if I shall *never* teach the Black Hawk that he may strike his beak in once too far?' he pondered, with a sudden darker, graver touch of musing; and involuntarily he stretched his arm out, and looked at the wrist, supple as Damascus steel, and at the muscles that were traced beneath the skin, as he thrust the sleeve up, clear, firm, and sinewy as any athlete's. He doubted his continence there, fast rein as he held all rebellion in, close shield as he bound to him against his own passions in the breastplate of a soldier's first duty—obedience.

He shook the thought off him as he would have shaken a snake. It had a terrible temptation—a temptation which he knew might any day overmaster him; and Cecil, who all through his life had certain inborn instincts of honour, which served him better than most codes or creeds served their professors, was resolute to follow the military religion of obedience enjoined in the Service that had received him at his needs, and to give no precedent in his own person that could be fraught with dangerous, rebellious allurement for the untamed, chafing, red-hot spirits of his comrades, for whom he knew insubordination would be ruin and death—whose one chance of reward, of success, and of a higher ambition, lay in their implicit subordination to their chiefs, and their continuous resistance of every rebellious impulse.

Cecil had always thought very little of himself.

In his most brilliant and pampered days he had always considered in his own heart that he was a graceless fellow, not worth his salt, and had occasionally wondered, in a listless sort of way, why so useless a *bagatelle à la mode* as his own life was had ever been created. He thought much the same now; but following his natural instincts, which were always the instincts of a gentleman, and of a generous temper, he did, unconsciously, make his life of much value among its present comrades.

His influence had done more to humanize the men he was associated with than any preachers or teachers could have done. The most savage and obscene brute in the ranks with him caught something gentler and better from the 'aristocrat'. His refined habits, his serene temper, his kindly forbearance, his high instinctive honour, made themselves felt imperceptibly, but surely; they knew that he was as fearless in war, as eager for danger as themselves, they knew that he was no saint, but loved the smile of women's eyes, the flush of wines, and the excitation of gaming hazards as well as they did; and hence his influence had a weight that probably a more strictly virtuous man's would have strained for, and missed, for ever. The coarsest ruffian felt ashamed to make an utter beast of himself before the calm eyes of the patrician. The most lawless *pratique* felt a lie halt on his lips when the contemptuous glance of his gentleman-comrade taught him that falsehood was poltroonery. Blasphemous tongues learnt to rein in their filthiness when this '*beau lion*' sauntered away from the picket-fire, on an icy night, to be out of hearing of their witless obscenities. More than once the weight of his arm and the slash of his sabre had called them to account in fiery fashion for their brutality to women or their thefts from the country people, till they grew aware that 'Bel-à-faire-peur' would risk having all their swords buried in him rather than stand by to see injustice done.

And throughout his corps men became unconsciously gentler, juster, with a finer sense of right and wrong, and less bestial modes of pleasure, of speech, and of habit, because he was among them. Moreover, the keen-eyed desperadoes who made up the chief sum of his comrades saw that he gave unquestioning respect to a chief who made his life a hell; and rendered unquestioning submission under affronts, tyrannies, and insults, which, as they also saw, stung him to the quick, and tortured him as no physical torture would have done—and the sight was not without a strong effect for good on them. They could tell that he suffered under these as they never suffered themselves, yet he bore them and did his duty with a self-control and patience they had never attained.

Almost insensibly they grew ashamed to be beaten by him, and strove to grow like him as far as they could. They never knew him drunk, they never heard him swear, they never found him unjust, even to a poverty-stricken *indigène*, or brutal, even to a *fille de joie*. Insensibly his presence humanized them. Of a surety, the last part Bertie dreamed of playing was that of a teacher to any mortal thing. Yet—here in Africa—it might reasonably be questioned if a second Augustine or François Xavier would ever have done half the good among the devil-may-care Roumis that was wrought by the dauntless, listless, reckless soldier, who followed instinctively the one religion which has no cant in its brave simple creed, and binds man to man in links that are true as steel—the religion of a gallant gentleman's loyalty and honour.

CHAPTER VI

Cigarette en Conseil et Cachette

'Corporal Victor, M. le Commandant desires you to present yourself at his *campagne* tonight, at ten precisely, with all your carvings; above all, with the chessmen.'

The swift sharp voice of a young officer of his regiment wakened Cecil from his musing, as he went on his way down the crowded, tortuous, stifling street. He had scarcely time to catch the sense of the words, and to halt, giving the salute, before the Chasseur's skittish little Barbary mare had galloped past him, scattering the people right and left, knocking over a sweetmeat-seller, upsetting a string of maize-laden mules, jostling a venerable marabout on to an impudent little grisette, and laming an old Moor as he tottered to his mosque, without any apology for any of the mischief, in the customary insolence, which makes 'Roumis' and 'Bureaucratie' alike execrated by the indigenous populace with a detestation that the questionable benefits of civilized importations can do very little to counter-balance in the fiery breasts of the sons of the soil.

Cecil involuntarily stood still. His face darkened. All orders that touched on the service, even where harshest and most unwelcome, he had taught himself to take without any hesitation, till he now scarcely felt the check of the steel curb; but to be ordered thus like a lackey—to take his wares thus like a hawker!

'*Ah ma cantche!* We are soldiers, not traders—aren't we? You don't like that, M. Victor? You are no pedlar—eh? And you think you would rather risk being court-martialled and shot, than take your ivory toys for the Black Hawk's talons?'

Cecil glanced up in astonishment at the divination and translation of his thoughts, to encounter the bright falcon eyes of Cigarette looking down on him from a little oval casement above, dark as pitch within, and whose embrasure, with its rim of grey stone coping, set off like a picture-frame, with a heavy background of unglazed Rembrandt shadow, the piquant head of the Friend of the Flag, with her pouting, scarlet, mocking lips, and her mischievous challenging smile, and her dainty little gold-banded foraging-cap set on curls as silken and jetty as any black Irish setter's.

'*Bon jour, ma belle!*' he answered, with a little weariness, lifting his fez to her with a certain sense of annoyance, that this young bohemian of the barracks, this child with her slang and her satire, should always be in his way like a shadow.

'*Bon jour, mon brave!*' returned Cigarette, contemptuously. 'We are not so ceremonious as all that, in Algiers! Good fellow, you should be a chamberlain, not a corporal. What fine manners, mon Dieu!'

She was incensed, and piqued, and provoked. She had been ready to forgive him because he carved so wonderfully, and sold the carvings for his comrade at the hospital; she was holding out the olive-branch after her own petulant fashion; and she thought, if he had had any grace in him, he would have responded with some such florid compliment as those for which she was accustomed to box the ears of her admirers, and would have swung himself up to the coping, to touch, or at least try to touch, those sweet, fresh, crimson lips of hers, that were like a half-opened damask rose. Modesty is apt to go to the wall in camps, and poor little Cigarette's notions of the great passion were very simple, rudimentary, and, certes, in no way coy. How should they be? She had tossed about with the army, like one of the tassels to their standards, blowing which-ever way the breath of war floated her, and had experienced, or thought she had experienced, as many *affaires* as the veriest Don Juan among them, though her heart had never been much concerned in them, but had beaten scarce a shade quicker, if a lunge in a duel, or a shot from an Indigène, had pounced off with her hero of the hour to Hades.

'Fine manners!' echoed Cecil, with a smile: 'my poor child, have you been so buffeted about that you have never been treated with common-est courtesy?'

'Whew!' cried the little lady, blowing a puff of smoke down on him. 'None of your pity for me, my *ci-devant!* Buffeted about? *Nom du diable!* do you suppose anybody ever did anything with *me* that I didn't choose? If you had as much power as I have in the army, Châteauroy would not send for you to sell your toys like a pedlar. You are a slave! I am a sovereign!'

With which she tossed back her graceful, spirited head, as though the gold band of her cap were the gold band of a diadem. She was very proud of her station in the Army of Africa, and glorified her privileges with all a child's vanity.

He listened, amused with her boastful supremacy; but the last words touched him with a certain pang just in that moment. He felt like a slave—a slave who must obey his tyrant, or go out and die like a dog.

'Well, yes,' he said, slowly, 'I am a slave, I fear. I wish a Bedouin flissa would cut my thralls in two.'

He spoke jestingly, but there was a tinge of sadness in the words that touched Cigarette's changeful temper to contrition, and filled her with the same compassion and wonder at him that she had felt when the ivory wreaths and crucifixes had laid in her hands. She knew she had been ungenerous—a crime dark as night in the sight of the little chivalrous soldier.

'*Tiens!*' she said, softly and waywardly, winding her way aright with that penetration and tact which, however unsexed in other things, Cigarette had kept thoroughly feminine. 'That was but an idle word of mine: forgive it, and forget it. You are not a slave when you fight in the *fantasias*. Morbleu! they say to see you kill a man is beautiful—so workmanlike! And you would go out and be shot tomorrow, rather than sell your honour, or stain it—eh? Bah! while you know they should cut your heart out rather than make you tell a lie, or betray a comrade, you are no slave, my *galonné*; you have the best freedom of all. Take a glass of champagne? Prut-tut! how you look! Oh, the *demoiselles* with the silver necks are not barrack drink, of course; but I drink champagne always myself. This is M. le Prince's. He knows I only take the best brands.'

With which Cigarette, leaning down from her casement, whose sill was about a foot above his head, tendered her peace-offering in a bottle of Cliquot, three of which, packed in her knapsack, she had carried off from the luncheon-table of a Russian Prince who was touring through Algiers, and who had half lost his Grand Ducal head after the bewitching, daunt-less, capricious, unattachable, unpurchasable, and coquettish little fire-eater of the Spahis, who treated him with infinitely more insolence and indifference than she would show to some battered old veteran, or some worn-out old dog, who had passed through the great Kabaila raids and battles.

'You will go to your Colonel's tonight?' she said, questioningly, as he drank the champagne, and thanked her—for he saw the spirit in which the gift was tendered—as he leaned against the half-ruined Moorish wall, with its blue and white striped awning spread over both their heads in the little street, whose crowds, chatter, thousand eyes, and incessant traffic no way troubled Cigarette, who had talked *argot* to monarchs undaunted, and who had been one of the chief sights in a hundred grand reviews ever since she had been perched on a gun-carriage at five years old, and paraded with a troop of horse artillery in the Champ de Mars, as having gone through the whole of Bugeaud's campaign, at which parade, by the way, being tendered sweetmeats by a famous General's wife, Cigarette had made the immortal reply, in lisping sabir: '*Madame, mes bonbons sont des boulets!*'

She repeated her question imperiously, as Cecil kept silent: 'You will go tonight?'

He shrugged his shoulders. He did not care to discuss his Colonel's orders with this pretty little Bacchante.

'Oh, a chief's command, you know——'

'A fico for a chief!' retorted Cigarette, impatiently. 'Why don't you say the truth? You are thinking you will disobey, and risk the rest!'

'Well, why not? I grant his right in barrack and field; but——'

He spoke rather to himself than her, and his thoughts, as he spoke, went back to the scene of the morning. He felt, with a romantic impulse that he smiled at even as it passed over him, that he would rather have half a dozen muskets fired at him in the death-sentence of a mutineer, than meet again the glance of those proud azure eyes sweep over him, in their calm indifference to a private of Chasseurs, their calm ignorance that he could be wounded or be stung.

'But?' echoed Cigarette, leaning out of her oval hole, perched in the quaint, grey, Moresco wall, particoloured with broken encaustics of varied hues. '*Chut, bon camerade!* that little word has been the undoing of the world ever since the world began. "But" is a blank cartridge, and never did anything but miss fire yet. Shoot dead, or don't aim at all, whichever you like; but never make a *coup manqué* with "but"! So you won't obey Châteauroy in this?'

He was silent again. He would not answer falsely, and he did not care to say his thoughts to her.

'"No,"' pursued Cigarette, translating his silence at her fancy, 'you say to yourself; "I am an aristocrat: I will not be ordered in this thing," you say. "I am a good soldier: I will not be sent for like a hawker," you say. "I was noble once: I will show my blood at last, if I die." Ah!—you say that?'

He laughed a little as he looked up at her.

'Not exactly that; but something as foolish, perhaps. Are you a witch, my pretty one?'

'Whoever doubted it, except you?'

She looked one, in truth, whom few men could resist, bending to him out of her owl's nest, with the flash of the sun under the blue awning brightly catching the sunny brown of her soft cheek and the cherry bloom of her lips, arched, pouting, and coquette. She set her teeth sharply, and

muttered a hot, heavy *sacré*, or even something worse, as she saw that his
eyes had not even remained on her, but were thoughtfully looking down
the chequered light and colour of the street. She was passionate, she was
vain, she was wayward, she was fierce as a little velvet leopard, as a
handsome, brilliant plumaged hawk; she had all the faults, as she had
all the virtues, of the thorough Celtic race; and, for the moment, she had
an instinct, fiery, ruthless, and full of hate, to draw the pistol out of
her belt, and teach him with a shot, crash through heart or brain, that
girls who were 'unsexed' could keep enough of the woman in them not to
be neglected with impunity, and could lose enough of it to be able to
avenge the negligence by a summary vendetta. But she was a haughty
little condottière in her fashion. She would not ask for what was not
offered her, nor give a rebuke that might be traced to mortification.
She only set her two rosebud lips in as firm a line of wrath and scorn as
ever Cæsar's or Napoléon's moulded themselves into, and spoke in the
curt, imperious, generalissimo fashion with which Cigarette before now
had rallied a demoralized troop, reeling drunk and mad, away from a
razzia.

'I am a witch? That is, I can put two and two together, and read men,
though I don't read the alphabet. Well, one reading is a good deal rarer
than the other. So you mean to disobey the Hawk tonight? I like you for
that. But listen here—did you ever hear them talk of Marquise?'

'No.'

'Parbleu!' swore the vivandière in her wrath, 'you look on at a
bamboula as if it were only a bear-cub dancing, and can only give one
"yes" and "no", as if one were a drummer-boy. Bah! are those your Paris
courtesies?'

'Forgive me, *ma belle*! I thought you called yourself our comrade, and
would have no "fine manners!" There is no knowing how to please you.'

He might have pleased her, simply and easily enough, if he had only
looked up with a shade of interest to that most picturesque picture, bright
as a pastel portrait that was hung above him in the old tumbledown
Moorish stonework. But his thoughts were with other things; and a love
scene with this fantastic young Amazon did not attract him. The warm,
ripe, mellow, little wayside-cherry hung directly in his path, with the sun
on its bloom, and the free wind tossing it merrily; but it had no charm for
him. He was musing rather on that costly, delicate, brilliant-hued, hothouse blossom, that could only be reached down by some rich man's
hand, and grew afar on heights where never winter chills, nor summer
tan, could come too rudely on it.

'Come, tell me, what is Marquise?—a kitten?' he went on, leaning his arm still on the sill of her embrasure, and willing to coax her out of her anger.

'A kitten!' echoed Cigarette, contemptuously. 'You think me a child, I suppose?'

'Surely you are not far off it?'

'Mon Dieu! why, *I* was never a child in my life,' retorted Cigarette, waxing sunny-tempered and confidential again, while she perched herself, like some gay-feathered mocking-bird on a branch, on the window-sill itself. 'When I was two, I used to be beaten, like a Turco that pawns his musket; when I was three, I used to scrape up the cigar-ends the officers dropped about, to sell them again for a bit of black bread; when I was four, I knew all about Philippe Durron's escape from Beylick, and bit my tongue through, to say nothing, when my mother flogged me with a tringlo's mule-whip because I would not tell, that she might tell again at the Bureau, and get the reward. A child?—diantre! before I was two feet high I had winged my first *Arbi*. He stole a rabbit I was roasting. Presto! how quick he dropped it when my ball broke his wrist like a twig.'

And the Friend of the Flag laughed gaily at the recollection, as at the best piece of mirth with which memory could furnish her.

'But you asked about Marquise? Well, he was what you are, a hawk among carrion crows, a gentleman in the ranks. Dieu! how handsome he was! Nobody ever knew his real name, but they thought he was of Austrian breed, and we called him Marquise because he was so womanish white in his skin and so dainty in all his ways. Just like you! Marquise could fight, fight like a hundred devils; and—pouf!—how proud he was—very much like you altogether! Now, one day something went wrong in the exercise-ground. Marquise was not to blame, but they thought he was; and an adjutant struck him—flick, flack, like that— across the face with a riding-switch. Marquise had his bayonet fixed—he belonged to the Zouaves—and before we knew what was up, crash the blade went through—through the breastbone, and out at the spine—and the adjutant fell as dead as a cat, with the blood spouting out like a fountain. "I come of a great race, that never took insult without giving back death," was all that Marquise said when they seized him, and brought him to judgement: and he would never say of what race that was. They shot him—ah, bah! discipline must be kept—and I saw him with five great wounds in his chest, and his beautiful golden hair all soiled with the sand and the powder, lying there by the open grave, that they threw him into as if he were offal: and we never knew more of him than that.'

Cigarette's radiant laugh had died, and her careless voice had sunk, over the latter words. As the little vivacious brunette told the tale of a nameless life, it took its eloquence from her, simple and brief as her speech was, and it owned a deeper pathos because the reckless young Bacchante of the As de Pique grew grave one moment while she told it. Then, grave still, she leaned her brown bright face nearer down from her oval hole in the wall.

'Now,' she whispered very low, 'if you mutiny once they will shoot you just like Marquise, and you will die just as silent, like him.'

'Well,' he answered her slowly, 'why not? Death is no great terror; I risk it every day for the sake of a common soldier's rations, why should I not chance it for the sake and in the defence of my honour?'

'Bah! men sell their honour for their daily bread all the world over!' said Cigarette, with the satire that had treble raciness from the slang in which she clothed it. 'But it is not you alone. See here—one example set on your part, and half your regiment will mutiny too. It is bitter work to obey the Black Hawk, and if you give the signal of revolt, three parts of your comrades will join you. Now what will that end in, *beau lion*—eh?'

'Tell me; you are a soldier yourself, you say.'

'Yes! I am a soldier!' said Cigarette, between her tight-set teeth, while her eyes lightened, and her voice sank down into a whisper, that had a certain terrible meaning in it, like the first dropping of the scattered opening shots in the distance before a great battle commences; 'and I have seen war, not holiday war—but war in earnest—war when men fall like hailstones, and tear like tigers, and choke like mad dogs with their throats full of blood and sand; when the gun-carriage wheels go crash over the writhing limbs, and the horses charge full gallop over the living faces, and the hoofs beat out the brains before death has stunned them senseless. Oh yes! I am a soldier, and I will tell you one thing I have seen. I have seen soldiers mutiny, a squadron of them, because they hated their chief and loved two of their sous-officiers; and I have seen the end of it all—a few hundred men, blind and drunk with despair, at bay against as many thousands, and walled in with four lines of steel and artillery, and fired on from a score of cannon-mouths—volley on volley like the thunder—till not one living man was left, and there was only a shapeless heaving moaning mass, with the black smoke over all. That is what I have seen; you will not make me see it again?'

Her face was very earnest, very eloquent, very dark and tender with thought; there was a vein of grave, even of intense feeling, that ran through the significant words to which tone and accent lent far more meaning than lay in their mere phrases; the little bohemian lost her insolence when she pleaded for her 'children', her comrades; and the mischievous pet of the camp never treated lightly what touched the France that she loved, the France that alone of all things in her careless life she held in honour and reverence.

'You will not make me see it again?' she said, once more leaning out, with her eyes that were like a brown brook sparkling deep yet bright in the sun, fixed on him. 'They would rise at your bidding, and they would be mowed down like corn. You will not?'

'Never! I give you my word.'

The promise was from his heart. He would have endured any indignity, any outrage, rather than have drawn into ruin, through him, the fiery, fearless, untutored lives of the men, who marched and slept and rode and fought, and lay in the light of the picket-fires, and swept down through the hot sandstorms on to the desert foe by his side. Cigarette stretched out her hand to him—that tiny brown hand, which, small though it was, had looked so burnt and so hard beside the delicate, fairy, ivory carvings of his workmanship—stretched it out with a frank, winning, childlike, soldierlike grace.

'*C'est ça, tu es bon soldat!*'

He bent over the hand she held to his in the courtesy natural with him to all her sex, and touched it lightly with his lips.

'Thank you, my little comrade,' he said, simply, with the graver thought still on him that her relation and her entreaty had evoked, 'you have given me a lesson that I shall not be quick to forget.'

Cigarette was the wildest little bacchanal that ever pirouetted for the delight of half a score of soldiers in their shirt-sleeves and half drunk; she was the most reckless coquette that ever made the roll-call of her lovers range from prince-marshals to ploughboy conscripts; she had flirted as far and wide as a butterfly flirts with the blossoms it flutters on to through the range of a summer-day; she took kisses, if the giver of them were handsome, as readily as a child takes sweetmeats at Mardi Gras; and of feminine honour, feminine scruples, feminine delicacy, knew nothing, save by such very dim fragmentary instincts as nature still planted in scant growth amidst the rank soil and the pestilent atmosphere of camp-life. Her eyes had never sunk, her face had never flushed, her heart had never panted, for the boldest or the wildest wooer of them all, from M. le Duc's Lauzun-esque blandishments, to Pouffer-de-Rire's or Miou-Miou's rough overtures; she had the coquetry of her nation with the audacity of a boy. Now only, for the first time, Cigarette coloured hotly at the grave, graceful, distant salute, so cold and so courteous, which was offered her in lieu of the rude and boisterous familiarities to which she was accustomed; and drew her hand away with what was, to the shame of her soldierly hardihood and her barrack tutelage, very nearly akin to an impulse of shyness.

'*Dame! Ne me donnez de la gabatine!*[1] I am not a court lady, *bon-zig!*' she cried, hastily, almost petulantly, to cover the unwonted and unwelcome weakness; while, to make good the declaration and revindicate her military renown, she balanced herself lightly on the stone ledge of her oval hole and sprang with a young wild-cat's easy vaulting leap over his head and over the heads of the people beneath, on to the ledge of the house opposite, a low-built wine-shop, whose upper story nearly touched the leaning walls of the old Moorish buildings in which she had been perched. The crowd in the street below looked up amazed and aghast at that bound from casement to casement as she flew over their heads like a blue-and-scarlet-winged bird of Oran; but they laughed as they saw who it was.

'It is Cigarette!' growled a Turco Indigène. 'Ah-ah! the devil, for a certainty, must have been her father!'

'To be sure!' cried the Friend of the Flag, looking from her elevation; 'he is a very good father, too, and I don't tease him like his sons the priests! But I have told him to take *you*, Ben Arsli, the next time you are stripping a dead body; so look out; he won't have to wait long.'

The discomfited Indigène hustled his way with many an oath through the laughing crowd as best he might, and Cigarette, with an airy pirouette on the wine-shop's roof that would have done honour to any opera boards, and was executed as carelessly, twenty feet above earth, as if she had been a pantomime-dancer all her days, let herself down by the awning, hand over hand like a little *mousse* from the harbour, jumped on to a forage-waggon that was just passing full trot down the street, and disappeared, standing on the piles of hay, and singing to the driving *tringlos*' unutterable delight the stanzas of Béranger's *Infidélités de Lisette*; her lithe slender miniature form, with its flash of gold on the breast, and its strip of rich scarlet in the fluttering sash, rising out against the blue and burning sky, the glare of the white walls, and the dusky glow and movement of the ebbing and flowing crowd.

Cecil looked after her with a certain touch of pity for her in him.

'What a gallant boy is spoilt in that little Amazon!' he thought; the quick flush of her face, the quick withdrawal of her hand he had not noticed; she had not much interest for him—scarcely any, indeed—save that he saw she was pretty, with a mignonne mischievous face, that all the

[1] 'Stuff! Don't humbug me!'

sun-tan of Africa and all the wild life of the Caserne could not harden or debase. But he was sorry a child so bright and so brave should be turned into three parts a trooper as she was, should have been tossed up on the scum and filth of the lowest barrack-life, and should be doomed in a few years' time to become the yellow, battered, foul-mouthed, vulture-eyed camp-follower that premature old age would surely render the darling of the tricolour, the pythoness of the As de Pique.

Cigarette was making scorn of her doom of Sex, dancing it down, drinking it down, laughing it down, burning it out in tobacco fumes, drowning it in tumbling cascades of wine, trampling it to dust under the cancan by her little brass-bound boots, mocking it away with her slang jests and her Theresa songs, and her devil-may-care audacities, till there was scarce a trace of it left in this prettiest and wildest little scamp of all the Army of Africa. But strive to kill it how she would, her sex would have its revenge one day and play Nemesis to her.

She was bewitching now; bewitching, though she had no witchery for him, in her youth. But when the bloom should leave her brown cheeks, and the laughter die out of her lightning glance, the womanhood she had defied would assert itself, and avenge itself, and be hideous in the sight of the men who now loved the tinkling of those little spurred feet, and shouted with applause to hear the reckless barrack-blasphemies ring their mirth from that fresh mouth which was now like a bud from a damask rose-branch, though even now it steeped itself in wine, and sullied itself with oaths and seared itself with smoke, and had never been touched from its infancy with any kiss that was innocent, not even with its mother's.

And there was a deep tinge of pity for her in Cecil's thoughts as he watched her out of sight, and then strolled across to the café opposite to finish his cigar beneath its orange-striped awning. The child had been flung upward, a little straw floating in the gutter of Paris iniquities; a little foam-bell bubbling on the sewer waters of barrack-vice; the stick had been her teacher, the baggage-waggon her cradle, the camp-dogs her playfellows, the *caserne* oaths her lullaby, the *guidons* her sole guiding-stars, the *razzia* her sole fête-day: it was little marvel that the bright, bold, insolent little Friend of the Flag had nothing left of her sex save a kitten's mischief and a coquette's archness. It said much rather for the straight fair sunlit instincts of the untaught nature, that Cigarette had gleaned, even out of such a life, two virtues that she would have held by to the death, if tried; a truthfulness that would have scorned a lie as only fit for cowards, and a loyalty that cleaved to France as a religion.

Cecil thought that a gallant boy was spoiled in this eighteen-year-old brunette of a campaigner; he might have gone further, and said that a hero was lost.

'Voilà!' said Cigarette between her little teeth.

She stood in the glittering Algerine night, brilliant with a million stars, and balmy with a million flowers, before the bronze trellised gate of the villa on the Sahel, where Châteauroy when he was not on active service— which chanced rarely, for he was one of the finest soldiers and most daring chiefs in Africa—indemnified himself with the magnificence that his private fortune enabled him to enjoy, for the unsparing exertions and the rugged privations that he always shared willingly with the lowest of his soldiers. It was the grandest trait in the man's character that he utterly scorned the effeminacy which many commanders provided for their table, their comfort, and their gratification whilst campaigning, and would commonly neither take himself nor allow to his officers any more indulgence on the march than his troopers themselves enjoyed. But his villa on the Sahel was a miniature palace; it had formerly been the harem of a great Rais, and the gardens were as enchanting as the interior was, if something florid, still as elegant as Paris art and Paris luxury could make it; for ferocious as the Black Hawk was in war, and well as he loved the chase and the slaughter, he did not disdain, when he had whetted beak and talons to satiety, to smoothe his ruffled plumage in downy nests and under caressing hands.

Tonight the windows of the pretty, low, snow-white, far-stretching building were lighted and open, and through the wilderness of cactus, myrtle, orange, citron, fuchsia, and a thousand flowers that almost buried it under their weight of leaf and blossom, a myriad of lamps were gleaming like so many glow-worms beneath the foliage, while from a cedar grove some slight way further out, the melodies and over-tures of the best military bands in Algiers came mellowed, though not broken, by the distance, and the fall of the bubbling fountains. Cigarette looked and listened, and her gay brown face grew duskily warm with wrath.

'Ah, bah!' she muttered, as she pressed her pretty lips to the lattice-work. 'The men die like murrained sheep in the hospital, and get sour bread tossed to them as if they were pigs, and are thrashed if they pawn their muskets for a stoup of drink when their throats are dry as the desert—and *you* live like a *coq en pâte!*[1] Morbleu! what fools the people are to fight, and toil, and get their limbs broken, and have their brains dashed

---

[1] In clover.

out by spent balls, that M. le Maréchal may send home a grand story with his own name flaring in letters a yard long on the placards, and M. le Colonel give his fêtes with stars and ribbons on his breast, while those who won the battle lie rotting in the sand!'

Cigarette was a resolute little democrat; she had loaded the carbines behind the barricade in an émeute in Paris before she was ten years old, and was not seldom in the perplexity of conflicting creeds when her loyalty to the tricolour and the guidons smote with a violent clash on her love for the populace and their liberty. She was given, however, usually to reconciling the dilemma with all her sex's illogical ingenuity, and so far thoroughly carried out her republicanism that she boxed a Prince's ear without ceremony when one tried to subjugate her, and never by any chance veiled the sun of her smiles to her 'children' the troopers—not even when she was tired to death after a burning march across leagues on leagues of locust-wasted country, or had spent half the night, after a skirmish, dressing wounds, soothing fever, seeking out the dying men who lay scattered on the outskirts of the field of carnage, with a magic and a sweetness, and a patience that seemed rather fitting for the gentle *Sœurs Grises* than for the wayward, mischievous, insolent young reveller of the As de Pique.

She looked a moment longer through the gilded scroll-work; then, as she had done once before, thrust her pistols well within her sash that they should not catch upon the boughs, and pushing herself through the prickly cactus hedge, impervious to anything save herself or a Barbary marmoset, twisted with marvellous ingenuity through the sharp-pointed leaves and the close barriers of spines, and launched herself with inimitable dexterity on to the other side of the cacti. Cigarette had too often played a game at spying and reconnoitring for her regiments, and played it with a cleverness that distanced even the most rusé of the Zéphyrs, not to be able to do just whatever she chose, in taking the way she liked, and lurking unseen at discretion.

She crossed the breadth of the grounds under the heavy shade of arbutus-trees with a hare's fleetness, and stood a second looking at the open windows and the terraces that lay before them, brightly lighted by the summer moon and by the lamps that sparkled among the shrubs. Then down she dropped, as quickly, as lightly, as a young setter down charging among the ferns, into a shower of rhododendrons, whose rose and lilac blossoms shut her wholly within them like a fairy enclosed in bloom. The good fairy of one life there she was assuredly, though she might be but a devil-may-care, audacious, careless little feminine Belphégor and military Asmodeus.

'Ah!' she said, quickly and sharply, with a deep-drawn breath. The single ejaculation was at once a menace, a tenderness, a whirlwind of rage, a volume of disdain, a world of pity. It was intensely French, and the whole nature of Cigarette was in it.

Yet all she saw was a small and brilliant group sauntering to and fro before the open windows, after dinner, listening to the bands, which, through dinner, had played to them, and laughing low and softly; and, at some distance from them, beneath the shade of a cedar, the figure of a Corporal of Chasseurs, calm, erect, motionless, as though he were the figure of a soldier cast in bronze. The scene was simple enough, though very picturesque; but it told, by its vivid force of contrast, a whole history to Cigarette.

'A true soldier!' she muttered, where she lay among the rhododendrons, while her eyes grew very soft, as she gave the highest word of praise that her whole range of language held. 'A true soldier! How he keeps his promise! But it must be bitter.'

She looked awhile, very wistfully, at the Chasseur, where he stood under the Lebanon boughs; then her glance swept bright as a hawk's over the terrace, and lighted with a prescient hatred on the central form of all—a woman's. There were two other great ladies there; but she passed them, and darted with unerring instinct on that proud, fair, patrician head, with its haughty stag-like carriage and the crown of its golden hair.

Cigarette had seen *grandes dames* by the thousand, though never very close; seen them in Paris, when they came to look on at a grand review; seen them in their court attire, when the Guides had filled the Carrousel on some palace ball night, and lined the Cour des Princes, and she had bewitched the officers of the guard into letting her pass in to see the pageantry. But she had never felt for those *grandes dames* anything save a considerably contemptuous indifference. She had looked on them pretty much as a war-worn powder-tried veteran looks on the curled dandy of some fashionable home-staying corps. She had never realised the difference betwixt them and herself, save in so far as she thought them useless butterflies, worth nothing at all, and laughed as she triumphantly remembered how she could shoot a man like any Tirailleur, and break in a colt like any rough-rider.

Now, for the first time, the sight of one of those aristocrats smote her with a keen hot sting of heart-burning jealousy. Now, for the first time, the little Friend of the Flag looked at all the nameless graces of rank with an envy that her sunny, gladsome, generous nature had never before been touched with—with a sudden perception, quick as thought, bitter as gall, wounding, and swift, and poignant, of what this womanhood, that he had said she herself had lost, might be in its highest and purest shape.

'Unsexed—he said I was unsexed,' she mused, while her teeth clenched on the ruby fulness of her lips, and her heart swelled, half with impotent rage, half with unconfessed pain. For the first time, looking on this imperial foreign beauty, sweeping so slowly and so idly along there in the Algerian starlight, she understood all that he had missed, all that he had meant, when he had used that single word, for which she had vowed on him her vengeance and the vengeance of the Army of Africa.

'If those are the women that he knew before he came here, I do not wonder that he never cared to watch even my *bamboula*,' was the latent, unacknowledged thought that was so cruel to her: the consciousness— which forced itself in on her, while her eyes jealously followed the perfect grace of the one in whom instinct had found her rival—that, while she had been so proud of her recklessness, and her devilry, and her trooper's slang, and her deadly skill as a shot, she had only been something very worthless, something very lightly held by those who liked her for a ribald jest, and a guinguette dance, and a Spahis' supper of headlong riot and drunken mirth.

The mood did not last. She was too brave, too fiery, too dauntless, too untamed. The dusky angry flush upon her face grew deeper, and the passion gathered more stormily in her eyes, while she felt the pistol-butts in her sash, and laughed low to herself, where she lay stretched under her flowery nest.

'Bah! she would faint, I dare say, at the mere sight of these,' she thought, with her old disdain, 'and would stand fire no more than a gazelle! They are only made for summer-day weather, those dainty, gorgeous, silver pheasants. A breath of war, a touch of tempest, would soon beat them down—crash!—with all their proud crests drooping!'

Like many another, Cigarette underrated what she had no knowledge of, and depreciated an antagonist the measure of whose fence she had no power to gauge.

Crouched there among the rhododendrons, she lay as still as a mouse, moving nearer and nearer, though none would have told that so much as a lizard even stirred under the blossoms, until her ear, quick and unerring as an Indian's, could detect the sense of the words spoken by that group, which so aroused all the hot ire of her warrior's soul and her democrat's impatience. Châteauroy himself was bending his fine dark head towards the patrician on whom her instinct of sex had fastened her hatred.

'You expressed your wish to see my Corporal's little sculptures again, Madame,' he was murmuring now, as Cigarette got close enough under her flower shadows to catch the sense of the words. 'To hear was to obey with me. He waits your commands yonder.'

'Mille tonneres! It was *you*, was it, brought him here?' muttered the Friend of the Flag to herself, with the passion in her burning more hotly against that 'silver pheasant', whose delicate train was sweeping the white marbles of Châteauroy's terraces, and whose reply, 'with fashion, not with feeling, softly freighted', she lost, though she could guess what it had been, when a lacquey crossed the lawn, and summoned the Chasseur from his waiting-place beneath the cedars.

Cecil obeyed, passed up the terrace stairs, and stood before his Colonel, giving the salute; the shade of some acacias still fell across him, whilst the party he fronted were in all the glow of a full Algerian moon, and of the thousand lamps among the belt of flowers and trees. Cigarette gave another sharp deep-drawn breath, and lay as mute and motionless as she had done before then among the rushes of some dried brook's bed, scanning a hostile Kabyl camp, when the fate of a handful of French troops had rested on her surety and her caution.

Châteauroy spoke with a carelessness of a man to a dog, turning to his Corporal.

'Victor, Madame la Princesse honours you with the desire to see your toys again. Spread them out.'

The savage authority of his general speech was softened for sake of his guest's presence, but there was a covert tone in the words that made Cigarette murmur to herself:

'If he forget his promise, I will forgive him!'

Cecil had not forgotten it; neither had he forgotten the lesson that this fair *aristocrate* had read him in the morning. He saluted his chief again, set the chess-box down upon the ledge of the marble balustrade and stood silent, without once glancing at the fair and haughty face that was more brilliant still in the African starlight than it had been in the noon sun of the Chasseurs' *Chambrée*. Courtesy was forbidden him as insult from a Corporal to a nobly born beauty; he no more quarrelled with the decree than with other inevitable consequences, inevitable degradations, that followed on his entrance as a private under the French flag. He had been used to the impassable demarcations of Caste, he did not dispute them more now that he was without, than he had done when within, their magic pale.

The carvings were passed from hand to hand as the Marquis's six or eight guests, listlessly willing to be amused in the warmth of the evening after their dinner, occupied themselves with the ivory chess armies, cut with a skill and a finish worthy a Roman studio. Praise enough was awarded to the art, but none of them remembered the artist who stood apart, grave, calm, with a certain serene dignity that could not be degraded because others chose to treat him as the station he filled gave them fit right to do.

Only one glanced at him with a touch of wondering pity, softening her pride; she who had rejected the gift of those mimic squadrons.

'You were surely a sculptor, once?' she asked him, with that graceful distant kindness which she might have shown some Arab outcast.

'Never, Madame.'

'Indeed! Then who taught you such exquisite art?'

'It cannot claim to be called art, Madame.'

She looked at him with an increased interest: the accent of his voice told her that this man, whatever he might be now, had once been a gentleman.

'Oh yes; it is perfect of its kind. Who was your master in it?'

'A common teacher, Madame—Necessity.'

There was a very sweet gleam of compassion in the lustre of her dark dreaming eyes.

'Does necessity often teach so well?'

'In the ranks of our army, Madame, I think it does;—often indeed much better.'

Châteauroy had stood by and heard, with as much impatience as he cared to show before guests whose rank was precious to the man who had still weakness enough to be ashamed that his father's brave and famous life had first been cradled under the thatch roof of a little posting-house.

'Victor knows that neither he nor his men have any right to waste their time on such trash,' he said, carelessly; 'but the truth is, they love the canteen so well that they will do anything to add enough to their pay to buy brandy.'

She whom he had called Madame la Princesse looked with a doubting surprise at the sculptor of the white Arab King she held.

'That man does not carve for brandy,' she thought.

'It must be a solace to many a weary hour in the barracks to be able to produce such beautiful trifles as these,' she said aloud. 'Surely you encourage such pursuits, Monsieur?'

'Not I,' said Châteauroy, with a dash of his camp tone that he could not withhold. 'There are but two arts or virtues for a trooper to my taste—fighting and obedience.'

'You should be in the Russian service, M. de Châteauroy,' said the lady, with a smile, that, slight as it was, made the Marquis's eyes flash fire.

'Almost I wish I had been,' he answered her; 'men are made to keep their grades there, and privates who think themselves fine gentlemen receive the lash they merit.'

'How he hates his Corporal!' thought Miladi, while she laid aside the White King once more.

'Nay,' interposed Châteauroy, recovering his momentary self-abandonment, 'since you like the bagatelles, do me honour enough to keep them.'

'Oh no, I offered your soldier his own price for them this morning, and he refused any.'

Châteauroy swung round.

'*Ah, sacripant!* you dared refuse your bits of ivory when you were honoured by an offer for them.'

Cecil stood silent; his eyes met his chief's steadily; Châteauroy had seen that look when his Chasseur had bearded him in the solitude of his tent, and demanded back the Pearl of the Desert.

The Princess glanced at both; then she stooped her elegant head slightly to the Marquis.

'Do not blame your Corporal unjustly through me, I pray you. He refused any price, but he offered them to me very gracefully as a gift, though of course it was not possible that I should accept them so.'

'The man is the most insolent *larron* in the service,' muttered her host, as he motioned Cecil back off the terrace. 'Get you gone, sir, and leave your toys here, or I will have them broken up by a hammer.'

The words were low, that they should not offend the ears of the great ladies who were his listeners, but they were coarsely savage in their whispered command, and the Princess heard them.

'He has brought his Chasseur here only to humiliate him,' thought Miladi with the same thought that flashed through the mind of the little Friend of the Flag where she hid among her rhododendrons. Now the dainty aristocrat was very proud, but she was not so proud but that justice was stronger in her than pride, and a noble generous temper mellowed the somewhat too cold and languid negligence of one of the fairest and haughtiest women that ever adorned a court. She was too generous not to rescue any one who suffered through her the slightest injustice, not to interfere when through her any misconception lighted on another; she told with her sex's rapid perception and sympathy that the man, whom Châteauroy addressed with the brutal insolence of a bully to his disobedient dog, had once been a gentleman, though he now held but the rank of a sous-officier in the Algerian Cavalry, and she saw that he suffered all the more keenly under an outrage he had no power to resist because of that enforced serenity, that dignity of silence and of patience, with which he stood before his tyrant.

'Wait,' she said, moving a little towards them, while she let her eyes rest on the carver of the sculptures with a grave compassion, though she addressed his chief. 'You wholly mistake me. I laid no blame whatever on your Corporal. Let him take the chessmen back with him; I would on no account rob him of them. I can well understand that he does not care to part with such masterpieces of his art; and that he would not appraise them by their worth in gold only shows that he is a true artist, as doubtless also he is a true soldier.'

The words were spoken with a gracious courtesy, the clear cold tone of her habitual manner just marking in them still the difference of caste between her and the man for whom she interceded, as she would equally have interceded for a dog who should have been threatened with the lash because he had displeased her. That very tone struck a sharper blow to Cecil than the insolence of his commander had power to deal him. His face flushed a little; he lifted his cap to her with a grave reverence, and moved away:

'I thank you, Madame. Keep them, if you will so far honour me.'

The words reached only her ear; in another instant he had passed away down the terrace steps, obedient to his chief's dismissal.

'Ah! have no kind scruples in keeping them, Madame,' Châteauroy laughed to her, as she still held in her hand, doubtfully, the White Sheik of the chess Arabs; 'I will see that Bel-à-faire-peur, as they call him, does not suffer by losing these trumperies, which, I believe, old Zist-et-Zest, a veteran of ours and a wonderful carver, had really far more to do with producing than he. You must not let your gracious pity be moved by such fellows as these troopers of mine; they are the most ingenious rascals in the world, and know as well how to produce a dramatic effect in your presence as they do how to drink and to swear when they are out of it.'

'Very possibly,' she said, with an indolent indifference; 'but that man was no actor, and I never saw a gentleman if he have not been one.'

'Like enough,' answered the Marquis. 'I believe many "gentlemen" come in our ranks who have fled their native countries, and broken all laws from the Decalogue to the Code Napoléon. So long as they fight well, we don't ask their past criminalities. We cannot afford to throw away a good *sabreur* because he has made his own land too hot to hold him.'

'Of what country is your Corporal, then?'

'I have not an idea. I imagine his past must have been something very black indeed, for the slightest trace of it has never, that I know of, been allowed to let slip from him. He encourages the men in every insubordination, buys their favour with every sort of stage trick, thinks himself the finest gentleman in the whole brigades of Africa, and ought to have been shot long ago if he had had his real deserts.'

She let her glance dwell on him with a contemplation that was half contemptuous amusement, half unexpressed dissent.

'I wonder he has not been, since *you* have the ruling of his fate,' she said, with a slight smile lingering about the proud rich softness of her lips.

'So do I.'

There was a gaunt, grim, stern significance in the three monosyllables that escaped him unconsciously; it made her turn and look at him more closely.

'How has he offended you?' she asked.

Châteauroy laughed off the question.

'In a thousand ways, Madame. Chiefly because I received my regimental training under one who followed the traditions of the Armies of Egypt and the Rhine, and have, I confess, little tolerance, in consequence, of a rebel who plays the martyr, and a soldier who is too effeminate an idler to do anything except attitudinize in interesting situations to awaken sympathy.'

She listened with something of distaste upon her face where she still leaned against the marble balustrade toying with the ivory Bedouins.

'I am not much interested in military discussions,' she said, coldly, 'but I imagine—if you will pardon me for saying so—that you do your Corporal some little injustice here. I should not fancy he "affects" anything, to judge from the very good tone of his manners. For the rest, I shall not keep the chessmen without making him fitting payment for them; since he declines money, you will tell me what form that had better take to be of real and welcome service to a Chasseur d'Afrique.'

Châteauroy, more incensed than he chose or dared to show, bowed courteously, but with a grim ironic smile.

'If you really insist, give him a Napoléon or two whenever you see him; he will be very happy to take it and spend it *au cabaret*, though he played the aristocrat to-day. But you are too good to him; he is one of the very worst of my *pratiques*, and you are as cruel to me in refusing to deign to accept my trooper's worthless bagatelles at my hands.'

She bent her superb head silently, whether in acquiescence or rejection he could not well resolve with himself, and turned to the staff-officers, among them the heir of a princely semi-royal French House, who surrounded her, and sorely begrudged the moments she had given to those miniature carvings and the private soldier who had wrought them. She was no coquette; she was of too imperial a nature, had too lofty a pride, and was too difficult to charm or to enchain; but those meditative, brilliant, serene eyes had a terrible gift of wakening without ever seeking love, and of drawing without ever recompensing homage.

Couched down among her rose-hued covert, Cigarette had watched and heard, her teeth set tightly, her breath coming and going swiftly, her hand clenched close on the butts of her pistols, fiery curses, with all the infinite variety in cursing of a barrack *répertoire*, chasing one another in hot fast mutterings off those bright lips, that should have known nothing except a child's careless and innocent song.

'*Comme elle est belle! comme elle est belle!*' she whispered every now and then to herself, with a new, bitter, ferocious meaning in the whisper that had, with all its hate, something pathetic too. She had never looked at a beautiful high-bred woman before, holding them in gay satirical disdain as mere *papillons rouants* who could not prime a revolver and fire it off to save their own lives, if ever such need arose; a depth of ignorance that was, to the vivandière's view, the *ne plus ultra* of crassitude and impotence. But now she studied one through all the fine, quickened, unerring instincts of jealousy; and there is no instinct in the world that gives such thorough appreciation of the very rival it reviles. She saw the courtly negligence, the regal grace, the fair brilliant loveliness, the delicious serene languor, of a pure '*aristocrate*' for the very first time to note them, and they made her heart sick with a new and deadly sense; they moved her much as the white delicate carvings of the lotus-lilies and the lentiscus-leaves had done; they, like the carvings, showed her all she had missed. She dropped her head suddenly like a wounded bird, and the racy vindictive camp-oaths died off her lips. She thought of herself as she had danced that mad bacchic *bamboula* amidst the crowd of shouting, stamping, drunken, half-infuriated soldiery, and for the moment she hated herself more even than she hated that patrician yonder.

'I know what he meant *now!*' she pondered, and her spirited, sparkling, brunette face was dark and weary, like a brown sun-lightened brook over whose radiance the heavy shadow of some broad-spread eagle's wings hovers, hiding the sun.

She looked once, twice, thrice, more enquiringly, envyingly, thirstily; then, as the band under the cedars rolled out their music afresh, and light laughter echoed to her from the terrace, she turned and wound herself back under the cover of the shrubs, not joyously and mischievously as she had come, but almost as slowly, almost as sadly, as a hare that the greyhounds have coursed drags itself through the grasses and ferns.

Once through the cactus hedge her old spirit returned; she shook herself angrily with petulant self-scorn; she swore a little, and felt that the fierce familiar words did her good like brandy poured down her throat; she tossed her head like a colt that rebels against the gall of the curb; then, fleet as a fawn, she dashed down the moonlit road at topmost speed. 'Diantre! she can't do what I do!' she thought.

And she ran the faster, and sang a drinking-song of the Spahis all the louder, because still at her heart a dull pain was aching.

CHAPTER VII

Cigarette en Condottiera

Cigarette always went fast. She had a bird-like way of skimming her ground that took her over it with wonderful swiftness, all the tassels, and ribbon-knots, and sashes with which her uniform was rendered so gay and so distinctive fluttering behind her, and her little military boots, with the bright spurs twinkling, flying over the earth too lightly for a speck of dust, though it lay thick as August suns could parch it, to rest upon her. Thus she went now, along the lovely moonlight, singing her drinking-song so fast and so loud that had it been any other than this young fire-eater of the African squadrons it might have been supposed she sang out of fear and bravado—two things, however, that never touched Cigarette; for she exulted in danger as friskily as a young salmon exults in the first fresh, crisp, tumbling crest of a sea-wave, and would have backed up the most vainglorious word she could have spoken with the cost of her life, had need been. Suddenly, as she went, she heard a shout on the still night air—very still now, that the lights, and the melodies, and the laughter of Châteauroy's villa lay far behind, and the town of Algiers was yet distant, with its lamps glittering down by the sea.

The shout was, '*A moi, Roumis! Pour la France!*' And Cigarette knew the voice, ringing melodiously and calmly still, though it gave the sound of alarm.

'Cigarette *au secour!*' she cried in answer; she had cried it many a time over the heat of battlefields, and when the wounded men in the dead of the sickly night writhed under the knife of the camp-thieves. If she had gone like the wind before, she went like the lightning now.

A few yards onward she saw a confused knot of horses and of riders struggling one with another in a cloud of white dust, silvery and hazy in the radiance of the moon.

The centre figure was Cecil's; the four others were Arabs, armed to the teeth and mad with drink, who had spent the whole day in drunken debauchery, pouring in raki down their throats until they were wild with its poisonous fire, and had darted headlong all abreast down out of the town overriding all that came in their way, and lashing their poor beasts with their sabres till the horses' flanks ran blood. Just as they neared Cecil, they had knocked aside and trampled over a worn-out old colon, of age too feeble for him to totter in time from their path. Cecil had reined up and shouted to them to pause; they, inflamed with the perilous drink, and senseless with the fury which seems to possess every Arab once started in a race neck to neck, were too blind to see, and too furious to care, that they were faced by a soldier of France, but rode down on him at once, with their curled sabres flashing round their heads. His horse stood the shock gallantly, and he sought at first only to parry their thrusts and to cut through their stallion's reins; but the latter were chain bridles, and only notched his sword as the blade struck them, and the former became too numerous and too savagely dealt to be easily played with in carte and tierce. The Arabs were dead-drunk, he saw at a glance, and had got the blood-thirst upon them; roused and burning with brandy and raki, these men were like tigers to deal with; the words he had spoken they never heard, and their horses hemmed him in powerless, whilst their steel flashed on every side; they were not of the tribe of the Khalifa.

If he struck not, and struck not surely, he saw that a few moments more of that moonlight night were all that he would live. He wished to avoid bloodshed, both because his sympathies were always with the conquered tribes, and because he knew that every one of these quarrels and combats between the vanquisher and the vanquished served further to widen the breach, already broad enough, between them. But it was no longer a matter of choice with him, as his shoulder was grazed by a thrust which, but for a swerve of his horse, would have pierced to his lungs; and the four riders, yelling like madmen, forced the animal back on its haunches and assaulted him with breathless violence. He swept his own arm back, and brought his sabre down straight through the sword-arm of the foremost; the limb was cleft through as if the stroke of an axe had severed it, and, thrice infuriated, the Arabs closed in on him. The points of their weapons were piercing his harness when, sharp and swift, one on another, three shots hissed past him; the nearest of his assailants fell stone dead, and the others, wounded and startled, loosed their hold, and tore off down the lonely road, while the dead man's horse, shaking his burden from him out of the stirrups, followed them at a headlong gallop through a cloud of dust.

'That was a pretty cut through the arm; better had it been through the throat. Never do things by halves, ami Victor,' said Cigarette; carelessly, as she thrust her pistols back into her sash, and looked, with the tranquil appreciation of a connoisseur, on the brown, brawny, naked limb, where it lay severed on the sand, with the hilt of the weapon still hanging in the sinewy fingers. Cecil threw himself from his saddle and gazed at her in bewildered amazement; he had thought those sure, cool, death-dealing shots had come from some Spahis or Chasseur.

'I owe you my life!' he said, rapidly. 'But—good God!—you have shot the fellow dead——'

Cigarette shrugged her shoulders with a contemptuous glance at the Bedouin's corpse.

'To be sure—I am not a bungler.'

'Happily for me, or I had been where he lies now. But wait—let me look; there may be breath in him yet.'

Cigarette laughed, offended and scornful, as with the offence and scorn of one whose first science was impeached.

'*Pas si bête!* Look and welcome; but if you find any life in that Arbi, make a laugh of it before all the army tomorrow.'

She was at her fiercest. A thousand new emotions had been roused in her that night, bringing pain with them, that she bitterly resented; and, moreover, this child of the Army of Africa caught fire at the flame of battle with instant contagion, and had seen slaughter around her from her first infancy.

Cecil, disregarding her protest, stooped and raised the fallen Bedouin. He saw at a glance that she was right; the lean, dark, lustful face was set in the rigidity of death; the bullet had passed straight through the temples.

'Did you never see a dead man before?' demanded Cigarette, impatiently, as he lingered; even in this moment he had more thought of this *Arbico* than he had of her!

He laid the Arab's body gently down, and looked at her with a glance that, rightly or wrongly, she thought had a rebuke in it.

'Very many. But—it is never a pleasant sight. And they were in drink; they did not know what they did.'

'Pardieu! What divine pity! Good powder and ball were sore wasted, it seems; you would have preferred to lie there yourself, it appears. I beg your pardon for interfering with the preference.'

Her eyes were flashing, her lips very scornful and wrathful. This was his gratitude!

'Wait, wait,' said Cecil, rapidly, laying his hand on her shoulder, as she flung herself away. 'My dear child, do not think me ungrateful. I know well enough I should be a dead man myself had it not been for your gallant assistance. Believe me, I thank you from my heart.'

'But you think me "unsexed" all the same! I see, *beau lion*!'

The word had rankled in her; she could launch it now with telling reprisal.

He smiled; but he saw that this phrase, which she had overheard, had not alone incensed, but had wounded her.

'Well, a little, perhaps,' he said, gently. 'How should it be otherwise? And, for that matter, I have seen many a great lady look on and laugh her soft cruel laughter while the pheasants were falling by hundreds, or the stags being torn by the hounds. They called it "sport"; but there was not much difference—in the mercy of it, at least—from your war. And they had not a tithe of your courage.'

The answer failed to conciliate her; there was an accent of compassion in it that ill-suited her pride, and a lack of admiration that was not less new and unwelcome.

'It *was* well for you I *was* unsexed enough to be able to send an ounce of lead into a drunkard!' she pursued, with immeasurable disdain. 'If I had been like that dainty aristocrat down there—pardieu! it had been worse for you. I should have screamed, and fainted, and left you to be killed whilst I made a *tableau*. Oh-hé, that is to be "feminine", is it not?'

'Where did you see that lady?' he asked, in some surprise.

'Oh! I was there!' answered Cigarette, with a toss of her head southward to where the villa lay. 'I went to see how you would keep your promise.'

'Well, you saw I kept it.'

She gave her little teeth a sharp click like the click of a trigger.

'Yes. And I would have forgiven you if you had broken it.'

'Would you? I should not have forgiven myself.'

'Ah! you are just like Marquise. And you will end like him.'

'Very probably.'

She knitted her pretty brows, standing there in his path, with the pistols thrust in her sash, and her hands resting lightly on her hips as a good workman rests after a neatly finished job, and her dainty fez set half on one side on her brown tangled curls, while upon them the intense lustre of the moonlight streamed, and in the dust, well-nigh at their feet, lay the gaunt white-robed form of the dead Arab, with the olive saturnine face turned upward to the stars.

'Why did you give those chessmen to that silver pheasant?' she asked him, abruptly.

'Silver pheasant?'

'Yes. See how she sweeps—sweeps—sweeps so languid, so brilliant, so useless—bah! Why did you give them?'

'She admired them. It was not much to give.'

'Diantre! You would not have given them to a daughter of the people.'

'Why not?'

'Why not? Oh-hé! Because her hands would be hard, and brown, and coarse, not fit for those ivory puppets; but Miladi's are white like the ivory, and cannot soil it. She will handle them so gracefully, for five minutes; and then buy a new toy, and let her lapdog break yours!'

'Like enough.' He said it with his habitual gentle temper, but there was a shadow of pain in the words. The chessmen had become in some sort like living things to him, through long association; he had parted from them not without regret, though, for the moment, courtesy and generosity of instinct had overcome it; and he knew that it was but too true how, in all likelihood these trifles of his art, that had brought him many a solace and been his companion through many a lonely hour, would be forgotten by the morrow, where he had bestowed them, and at best put aside in a cabinet to lie unnoticed among bronzes or porcelain, or be set on some boudoir-table to be idled with in the mimic warfare that would serve to cover some listless flirtation.

Cigarette, quick to sting, but as quick to repent using her sting, saw the regret in him; with the rapid uncalculating liberality of an utterly unselfish and intensely impulsive nature, she hastened to make amends by saying what was like gall on her tongue in the utterance.

'*Tiens!*' she said, quickly. 'Perhaps she will value them more than that. I know nothing of the aristocrats—not I! When you were gone, she championed you against the Black Hawk. She told him that if you had not been a gentleman before you came into the ranks, she never had seen one. *Ma cantche!* she spoke well if you had but heard her.'

'She did!'

She saw his glance brighten as it turned on her in a surprised gratification.

'Well! What is there so wonderful?'

Cigarette asked it with a certain petulance and doggedness, taking a namesake out of her breast-pocket, biting its end off, and striking a fusee. A word from this aristocrat was more welcome to him than a bullet that had saved his life!

Her generosity had gone very far, and, like most generosity, got nothing for its pains.

He was silent a few moments, tracing lines in the dust with the point of his scabbard. Cigarette, with the cigar in her mouth, stamped her foot impatiently.

'Corporal Victor! are you going to dream there all night? What is to be done with this dog of an Arbico?'

She was angered by him; she was in the mood to make herself seem all the rougher, fiercer, naughtier, and more callous. She had shot the man—pouf! what of that? She had shot men before, as all Africa knew. She would defend a half-fledged bird, a terrified sheep, a worn-out old cur; but a man! Men were the normal and natural food for pistols and rifles, she considered. A state of society in which firearms had been unknown was a thing Cigarette had never heard of, and in which she would have contumeliously disbelieved if she had been told of it.

Cecil looked up from his musing; he thought what a pity it was this pretty graceful French kitten was such a bloodthirsty young panther at heart.

'I scarcely know what to do,' he answered her, doubtfully. 'Put him across my saddle, poor wretch, I suppose: the fray must be reported.'

'Leave that to me,' said Cigarette, decidedly and with a certain haughty patronage. '*I* shot him; I will see the thing gets told right. It might be awkward for you; they are growing so squeamish about the *Roumis* killing the natives. Draw him to one side there, and leave him. The crows will finish his affair.'

The coolness with which this handsome child disposed of the fate of what, a moment or two before, had been a sentient, breathing, vigorous frame, sent a chill through her hearer, though he had been seasoned by a decade of slaughter.

'No,' he said, briefly. 'Suspicion might fall on some innocent passer-by. Besides—he shall have decent burial.'

'Burial for an Arbi—faugh!' cried Cigarette, in derision. 'Parbleu, M. Bel-à-faire-peur, I have seen hundreds of *our* best lascars lie rotting on the plains with the birds' beaks at their eyes and the jackals' fangs in their flesh. What was good enough for them is surely good enough for him. You are an eccentric fellow—you——'

He laughed a little.

'Time was when I should have begged you not to call me any such "bad form!" Eccentric! I am not genius enough for that.'

'Eh?'—she did not understand him. 'Well, you want that carrion poked into the earth, instead of lying atop of it. I don't see much difference myself. I would like to be in the sun as long as I could, I think, dead or alive. Ah! how odd it is to think one will be dead some day—never wake for the réveillé—never hear the cannon or the caissons roll by—never stir when the trumpets sound the charge, but lie there dead—dead—dead—while the squadrons thunder above one's grave! Droll, eh?'

A momentary pathos softened her voice (which could melt and change into a wonderful music), where she stood in the glistening moonlight. That the time would ever come when her glad laughter would be hushed, when her young heart would beat no more, when the bright, abundant, passionate blood would bound no longer through her veins, when all the vivacious, vivid, sensuous charms of living would be ended for her for ever, was a thing that she could no better bring home to her than a bird that sings in the light of the sun could be made to know that the time would come when its little melodious throat would be frozen in death, and give song never more.

The tone touched him; made him think less of her as a daredevil boy, as a reckless child-soldier, and more of her as what she was, than he had done before: he touched her almost caressingly.

'*Pauvre enfant!* I hope that day will be very distant from you. And yet—how bravely you risked death for me just now!'

Cigarette, though accustomed to the lawless loves of the camp, flushed ever so slightly at the mere caress of his hand.

'*Chut!* I risked nothing!' she said, rapidly. 'As for death—when it comes, it comes. Every soldier carries it in his wallet, and it may jump out on him any minute. I would rather die young than grow old. Pardi! age is nothing else but death that is *conscious*.'

'Where do you get your wisdom, little one?'

'Wisdom? Bah! living is learning. Some people go through life with their eyes shut, and then grumble there is nothing to see in it! Well—you want that Arbi buried? What a fancy! Look you, then; stay by him, since you are so fond of him, and I will go and send some men to you with a stretcher to carry him down to the town. As for reporting, leave that to me; I shall tell them *I* left you on guard. That will square things, if you are late at the barrack.'

'But that will give you so much trouble, Cigarette.'

'Trouble? Morbleu! Do you think I am like that silver pheasant yonder? Lend me your horse, and I shall be in the town in ten minutes!'

She vaulted, as she spoke, into the saddle; he laid his hand on the bridle, and stopped her.

'Wait! I have not thanked you half enough, my brave little champion. How am I to show you my gratitude?'

For a moment the bright, brown, changeful face, that could look so fiercely scornful, so sunnily radiant, so tempestuously passionate, and so tenderly childlike, in almost the same moment, grew warm as the warm suns that had given their fire to her veins; she glanced at him almost shyly, while the moonlight slept lustrously in the dark softness of her eyes; there was an intense allurement in her in that moment—the allurement of a woman's loveliness, bitterly as she disdained a woman's charms. It might have told him, more plainly than words, how best he could reward her for the shot that had saved him; yet, though a man on whom such beguilement usually worked only too easily and too often, it did not now touch him. He was grateful to her; but, despite himself, he was cold to her; despite himself, the life which that little hand that he held had taken so lightly made it the hand of a comrade to be grasped in alliance, but never the hand of a mistress to steal to his lips and to lie in his breast.

Her rapid and unerring instinct made her feel that keenly and instantly; she had seen too much passion not to know when it was absent. The warmth passed off her face, her teeth clenched, she shook the bridle out of his hold.

'Take gratitude to Miladi there! She will value fine words; I set no count on them. I did no more for you than I have done scores of times for my Spahis. Ask them how many I have shot with my own hand!'

In another instant she was away like a sirocco, a whirlwind of dust that rose in the moonlight marking her flight as she rode full gallop down to Algiers.

'A kitten with the tigress in her,' thought Cecil, as he seated himself on a broken pile of stone to keep his vigil over the dead Arab. It was not that he was callous to the generous nature of the little Friend of the Flag, or that he was insensible either of the courage that beat so dauntlessly in her pulses, or of the piquant picturesque grace that accompanied even her wildest actions; but she had nothing of her sex's charm for him. He thought of her rather as a young soldier than as a young girl. She amused him as a wayward, bright, mischievous, audacious boy might have done; but she had no other interest for him. He had given her little attention; a waltz, a cigar, a passing jest, were all he had bestowed on the little *lionne* of the Spahis corps; and the deepest sentiment she had ever awakened in him was an involuntary pity—pity for this flower which blossomed on the polluted field of war, and under the poison-dropping branches of lawless crime. A flower bright-hued, sun-fed, glancing with the dews of youth now, when it had just unclosed, in all its earliest beauty but already soiled and tainted by the bed from which it sprang, and doomed to be swept away with time, scentless and loveless, down the rapid noxious current of that broad black stream of vice on which it now floated so heedlessly.

Even now, his thoughts drifted from her almost before the sound of the horse's hoofs had died where he sat on a loose pile of stones, with the lifeless limbs of the Arab at his feet.

'Who was it in my old life that she is like?' he was musing. It was the deep-blue, dreaming, haughty eyes of 'Miladi' that he was bringing back to memory, not the brown mignon face that had been so late close to his in the light of the moon.

Meanwhile, on his good grey Cigarette rode like a true Chasseur herself. She was used to the saddle, and would ride a wild desert colt without stirrup or bridle, balancing her supple form now on one foot now on the other on the animal's naked back, while they flew at full speed, with a skill and address that would have distanced the best heroines of manège and hippodrome. Not so fantastically, but full as speedily, she dashed down into the city, scattering all she met with right and left, till she rode straight up to the barracks of the Chasseurs d'Afrique. At the entrance, as she reined up, she saw the very person she wanted, and signed him to her as carelessly as if he were a conscript, instead of that powerful officer, François Vireflau, captain and adjutant.

'Holà!' she cried, as she signalled him; Cigarette was privileged all through the army, and would have given the *langue verte* to the Emperor himself, had she met him. 'Adjutant Vireflau, I come to tell you a good story for your *folios matricules*. There is your Corporal there—*le beau Victor*—has been attacked by four drunken dogs of Arbicos, dead drunk and four against one. He fought them superbly; but he would only parry, not thrust, because he knows how strict the rules are about dealing with the scoundrels—even when they are murdering you, parbleu! He had behaved splendidly. *I* tell you so. And he was so patient with these dogs that he would not have killed one of them. But I did; shot one straight through the brain—a beautiful thing—and he lies on the Oran road now. Victor would not leave him, for fear some passer-by should be thought guilty of a murder; so I came on to tell you, and ask you to send some men up for the jackal's body. Ah! he is a fine soldier, that Bel-à-faire-peur of yours. Why don't you give him a step—two steps—three steps? Diantre! It is not like France to leave him a corporal!'

Vireflau listened attentively—a short, lean, black-visaged campaigner, who yet relaxed into a grim half-smile as the vivandière addressed him with that air as of a generalissimo addressing a subordinate, which always characterized Cigarette the more strongly the higher the grade of her companion or opponent.

'Always eloquent, pretty one!' he growled. 'Are you sure he did not begin the fray?'

'*Ma cantche!* Don't I tell you the four Arabs were like four devils? They knocked down an old colon, and Bel-à-faire-peur tried to prevent their doing more mischief, and they set on him like so many wildcats. He kept his temper wonderfully; he always tries to preserve order; you can't say so much of your riffraff, Captain Vireflau, commonly! Here! this is his horse. Send some men to him; and mind the thing is reported fairly, and to his credit, tomorrow.'

With which command, given as with the air of a commander-in-chief, in its hauteur and its nonchalance, Cigarette vaulted off the charger, flung the bridle to a soldier, and was away and out of sight before François Vireflau had time to consider whether he should laugh at her caprices, as all the army did, or resent her insolence to his dignity. But he was a good-natured man, and, what was better, a just one; and Cigarette had judged rightly that the tale she had told would weigh well with him to the credit side of his Corporal; and would not reach his Colonel in any warped version that could give pretext for any fresh exercise of tyranny over 'Bel-à-faire-peur' under the title of 'discipline'.

'Dieu de Dieu!' thought his champion, as she made her way through the gay-lit streets. 'I swore to have my vengeance on him. It is a droll vengeance—to save his life, and plead his cause with Vireflau! No matter! one could not look on and let a set of Arbicos kill a good *lascar* of France, and the thing that is just must be said, let it go as it will against one's grain. Public Welfare before Private Pique!'

A grand and misty generality which consoled Cigarette for an abandonment of her sworn revenge, which she felt was a weakness utterly unworthy of her, and too much like that inconsequent weathercock, that useless insignificant part of creation, those objects of her supreme derision and contempt, those frivolous trifles which she wondered the good God had ever troubled himself to make—namely, '*Les Femmes.*'

'Holà, Cigarette!' cried the Zouave Tata, leaning out of a little casement of the As de Pique as she passed it. '*A la bonne heure, ma belle!* Come in; we have the devil's own fun here——'

'No doubt!' retorted the Friend of the Flag. 'It would be odd if the master-fiddler would not fiddle for his own!'

Through the window, and over the sturdy shoulders in their canvas shirt of the hero Tata, the room was visible, full of smoke, through which the lights glimmered like the sun in a fog, reeking with bad wines, crowded with laughing bearded faces, and the battered beauty of women revellers, while on the table, singing with a voice Mario himself could not have rivalled for exquisite sweetness, was a slender Zouave, gesticulating with the most marvellous pantomime, while his melodious tones rolled out the obscenest and wittiest ballad that ever was carolled in a guinguette.

'Come in, my pretty one!' entreated Tata, stretching out his brawny arms. 'You will die of laughing if you hear Gris-Gris tonight. Such a song!'

'A pretty song, yes, for a pigsty!' said Cigarette, with a glance into the chamber, as she shook his hand off her, and went on down the street. A night or two before a new song from Gris-Gris, the best tenor in the whole army, would have been paradise to her, and she would have vaulted through the window at a single bound into the pandemonium. Now, she did not know why, she found no charm in it.

And she went quietly home to her little straw-bed in her garret, and curled herself up like a kitten to sleep; but for the first time in her young life sleep did not come readily to her, and when it did come, for the first time found a restless sigh upon her laughing mouth, as she murmured, dreaming: '*Comme elle est belle! Comme elle est belle!*'

CHAPTER VIII

The Mistress of the White King

'Fighting in the Kabaila, life was well enough; but here!' thought Cecil as, earlier awake than those of his Chambrée, he stood looking down the lengthy narrow room where the men lay asleep along the bare floor.

Tired as overworked cattle, and crouched or stretched like worn-out homeless dogs, they had never wakened as he had noiselessly harnessed himself, and he looked at them with that interest in other lives that had come to him through adversity; for if misfortune had given him strength, it had also given him sympathy.

They were of marvellously various types—those sleepers brought under one roof by fates the most diverse. Close beside a huge and sinewy brute of an Auvergnat, whose coarse bestial features and massive bull's head were fitter for a galley-slave than a soldier, were the symmetrical limbs and the oval delicate face of a man from the Valley of the Rhône. Beneath a canopy of flapping tawny wild-beast skins, the spoils of his own hand, was flung the naked torso of one of the splendid peasants of the Sables d'Olonne; one steeped so long in blood and wine and alcohol, that he had forgotten the blue bright waves that broke on the western shores of his boyhood's home, save when he muttered thirstily in his dreams of the cool sea, as he was muttering now. Next him, curled, dog-like, with its round black head meeting its feet, was a wiry frame on which every muscle was traced like network, and the skin burnt black as jet under twenty years of African sun. The midnight streets of Paris had seen its birth, the thieves' quarter had been its nest; it had no history, it had almost no humanity; it was a perfect machine for slaughter, no more—who had ever tried to make it more?

Farther on lay, sleeping fitfully, a boy of scarcely more than seventeen, with rounded cheek and fair white limbs like a child's, whose uncovered chest was delicate as a girl's, and through whose long brown lashes tears in his slumber were stealing as his rosy mouth murmured, '*Mère! Mère! Pauvre mère!*' He was a young conscript taken from the glad vine-country of the Loire, and from the little dwelling up in the rock beside the sunny brimming river, and half-buried under its grape-leaves and coils, that was dearer to him than is the palace to its heir. There were many others beside these; and Cecil looked at them with those weary speculative meditative fancies which, very alien to his temperament, stole on him occasionally in the privations and loneliness of his existence here—loneliness in the midst of numbers, the most painful of all solitude.

Life was bearable enough to him in the activity of campaigning, in the excitement of warfare; there were times even when it yielded him absolute enjoyment, and brought him interests more genuine and vivid than any he had known in his former world. But, in the monotony and the confinement of the barrack routine, his days were often intolerable to him. Morning after morning he rose to the same weary round of duty, the same series of petty irritations, of physical privations, of irksome repetitions, to take a toss of black rough coffee, and begin the day knowing it would bring with it endless annoyances without one gleam of hope. Rose to spend hours on the exercise-ground in the glare of a burning sun, railed at if a trooper's accoutrement were awry, or an insubordinate scoundrel had pawned his regulation-shirt; to be incessantly witness of tyrannies and cruelties he was powerless to prevent, and which he continually saw undo all he had done, and render men desperate whom he had spent months in endeavouring to make contented; to have as the only diversions for his few instants of leisure loathsome pleasures that disgusted the senses they were meant to indulge, and that brought him to scenes of low debauchery from which all the old fastidious instincts of his delicate luxurious taste recoiled: with such a life as this he often wondered regretfully why, out of the many Arab swords that had crossed his own, none had gone straight to his heart; why out of the many wounds that had kept him hovering on the confines of the grave, none had ever brought him the end and the oblivion of death.

Had he been subject to all the miseries and personal hardships of his present career, but had only owned the power to command, to pardon, to lead, and to direct, as Alan Bertie before him had done with his Irregular Cavalry in the Indian plains, such a thought would never have crossed him; he was far too thorough a soldier not *then* to have been not only satisfied, but happy. What made his life in the barracks of Algiers so bitter were the impotency, the subjection, the compelled obedience to a bidding that he knew often capricious and unjust as it was cruel, which were so unendurable to his natural pride, yet to which he had hitherto rendered undeviating adhesion and submission, less for his own sake than for that of the men around him, who, he knew, would back him in revolt to the death, and be dealt with, for such loyalty to him, in the fashion that the vivandière's words had pictured with such terrible force and truth.

'Is it worth while to go on with it? Would it not be the wiser way to draw my own sabre across my throat?' he thought, as the brutalized companionship in which his life was spent struck on him all the more darkly because, the night before, a woman's voice and a woman's face had recalled memories buried for twelve long years.

But, after so long a stand-up fight with fate, so long a victory over the temptation to let himself drift out in an opium-sleep from the world that had grown so dark to him, it was not in him to give under now. In his own way he had found a duty to do here, though he would have laughed at any one who should have used the word 'duty' in connection with him. In his own way, amidst these wild spirits, who would have been blown from the guns' mouths to serve him, he had made good the 'Cœur Vaillant Se Fait Royaume' of his House. And he was, moreover, by this time, a French soldier at heart and in habit, in almost all things, though the English gentleman was not dead in him under the harness of a Chasseur d'Afrique.

This morning he roused the men of his Chambrée with that kindly gentleness which had gone so far in its novelty to attach their liking; went through the customary routine of his post with that exactitude and punctuality of which he was always careful to set the example; made his breakfast off some wretched onion soup and a roll of black bread; rode fifty miles in the blazing heat of the African day at the head of a score of his *chasses-marais* on convoy duty, bringing in escort a long string of maize-waggons from the region of the Kabaila, which, without such guard, might have been swooped down on and borne off by some predatory tribe; and returned, jaded, weary, parched with thirst, scorched through with heat, and covered with white dust, to be kept waiting in his saddle, by his Colonel's orders, outside the barrack for three-quarters of an hour, whether to receive a command or a censure he was left in ignorance.

When the three-quarters had passed, he was told M. le Commandant had gone long ago, and did not require him!

Cecil said nothing.

Yet he reeled slightly as he threw himself out of saddle; a nausea and a giddiness had come on him. To have passed nigh an hour motionless in his stirrups, with the skies like brass above him, whilst he was already worn with riding from sunrise well-nigh to sunset, with little to appease hunger and less to slake thirst, made him, despite himself, stagger dizzily under a certain sense of blindness and exhaustion as he dismounted.

The Chasseur who had brought him the message caught his arm eagerly.

'Are you hurt, *mon Caporal*?'

Cecil shook his head. The speaker was one known in the regiment as Petit Picpon, who had begun life as a *gamin* of Paris, and now bade fair to make one of the most brilliant of the soldiers of Africa. Petit Picpon had but one drawback to his military career—he was always in insubordination; the old *gamin* daredevilry was not dead in him, and never would die; and Petit Picpon accordingly was perpetually a hero in the field and a ragamuffin in the times of peace. Of course he was always arrayed against authority, and now, being fond of his *galonné* with that curious dog-like deathless attachment that these natures, all reckless, wanton, destructive, and mischievous though they be, so commonly bestow, he muttered a terrible curse under his fiercely curled moustaches.

'If the Black Hawk were nailed up in the sun like a kite on a barn-door, I would drive twenty nails through his throat!'

Cecil turned rapidly on him.

'Silence, sir! or I must report you. Another speech like that, and you shall have a turn at Beylick.'

It went to his heart to rebuke the poor fellow for an outburst of indignation which had its root in regard for himself, but he knew that to encourage it by so much even as by an expression of gratitude for the affection borne him would be to sow further and deeper the poison-seeds of that inclination to mutiny, and that rebellious hatred against their chief, already only planted too strongly in the squadrons under Châteauroy's command.

Petit Picpon looked as crestfallen as one of his fraternity could; he knew well enough that what he had said could get him twenty blows of the *matraque*, if his Corporal chose to give him up to judgement; but he had too much of the Parisian in him still not to have his say, though he should be shot for it.

'Send me to Beylick if you like, Corporal,' he said, sturdily; 'I was in wrath for you—not for myself. Diantre!'

Cecil was infinitely more touched than he dared, for sake of discipline, for sake of the speaker himself, to show; but his glance dwelt on Petit Picpon with a look that the quick, black, monkey-like eyes of the rebel were swift to read.

'I know,' he said, gravely. 'I do not misjudge you; but, at the same time, my name must never serve as a pretext for insubordination. Such men as care to pleasure me will best do so in making my duty light by their own self-control and obedience to the rules of their service.'

He led his horse away, and Petit Picpon went on on an errand he had been sent to do in the streets for one of the officers. Picpon was unusually thoughtful and sober in deportment for him, since he was usually given to making his progress along a road, taken unobserved by those in command over him, '*faisant roue*', with hands and heels in the dexterous somersaults of his early days.

Now he went along without any unprofessional antics, biting the tip of a smoked-out cigar which he had picked up off the pavement in sheer instinct, retained from the old times when he had used to rush in, the foremost of *la queue*, into the forsaken theatres of Bouffes or of Variétés in search for those odds and ends which the departed audience might have left behind them;—one of the favourite modes of seeking a livelihood with the Parisian night-birds.

'Damn! I will give it up, then,' resolved Picpon, half aloud, valorously.

Now Picpon had come forth on evil thoughts intent.

His officer—a careless and extravagant man, the richest man in the regiment—had given him a rather small velvet bag, sealed, with directions to take it to a certain notorious beauty of Algiers, whose handsome Moresco eyes smiled—or at least he believed so—exclusively for the time on the sender. Picpon was very quick, intelligent, and much liked by his superiors, so that he was often employed on errands; and the tricks he played in the execution thereof were so adroitly done that they were never detected. Picpon had chuckled to himself over this mission. It was but the work of an instant for the lithe nimble fingers of the ex-gamin to undo the bag without touching the seal, to see that it contained a hundred Napoléons with a note, to slip the gold into the folds of his *ceinturon*, to fill up the sack with date-stones to make it assume its original form so that none could have imagined it had been touched, and to proceed with it thus to the Moorish *lionne*'s dwelling. The negro who always opened her door would take it in; Picpon would hint to him to be careful, as it contained some rare and rich sweetmeats; negro nature, he well knew, would impel him to search for the bonbons; and the bag, under his clumsy treatment, would bear plain marks of having been tampered with, and, as the African had a most thievish reputation, he would never be believed if he swore himself guiltless. *Voilà!* here was a neat trick! If it had a drawback, it was that it was too simple, too little *risqué*. A child might do it.

Still—a hundred Naps.! What fat geese, what flagons of brandy, what dozens of wine, what rich soups, what handsome moukieras, what tavern banquets they would bring! Picpon had chuckled again as he arranged the little bag so carefully, with its date-stones, and pictured the rage of the beautiful Moor when she should discover the contents, and order the stick to her negro. Ah! that was what Picpon called fun!

To appreciate the full force of such fun, it is necessary to have also appreciated the *gamin*. To understand the legitimate aspect such a theft bore, it is necessary to have also understood the unrecordable codes that govern the genus *pratique*, into which the genus *gamin*, when at maturity, develops.

Picpon was quite in love with his joke; it was only a good joke in his sight; and, indeed, men need to live as hardly as an African soldier lives, to estimate the full temptation that gold can have when you have come to look on a cat as very good eating, and to have nothing to gnaw but a bit of old shoe-leather through the whole of the long hours of a burning day of fatigue-duty; and to estimate, as well, the full width and depth of the renunciation that made him mutter now so valorously, 'Dieu! I will give it up, then!'

Picpon did not know himself as he said it. Yet he turned down into a lonely narrow lane, under marble walls, overtopped with fig and palm from some fine gardens, undid the bag for the second time, whisked out the date-stones and threw them over the wall, so that they should be out of his reach if he repented, put back the Napoléons, closed the little sack, ran as hard as he could scamper to his destination, delivered his charge into the fair lady's own hands, and relieved his feelings by a score of somersaults along the pavement as fast ever he could go.

'*Ma cantche!*' he thought, as he stood on his head, with his legs at an acute angle in the air, a position very favoured by him for moments of reflection—he said his brain worked better upside down. '*Ma cantche!* what a weakness, what a weakness! What remorse to have yielded to it! Beneath you, Picpon—utterly beneath you. Just because that *ci-devant* says such follies please him in us!'

Picpon (then in his *gamin* stage) had been enrolled in the Chasseurs at the same time with the '*ci-devant*', as they called Bertie, and, following his *gamin* nature, had exhausted all his resources of impudence, maliciousness, and power of tormenting, on the 'aristocrat'; somewhat disappointed, however, that the utmost ingenuities of his insolence and even his malignity never succeeded in breaking the 'aristocrat's' silence and contemptuous forbearance from all reprisal. For the first two years the hell-on-earth which life with a Franco-Arab regiment seemed to Cecil, was a hundredfold embittered by the brutalized jests and mosquito-like torments of this little odious chimpanzee of Paris.

One day, however, it chanced that a detachment of Chasseurs, of which Cecil was one, was cut to pieces by such an overwhelming mass of Arabs, that scarce a dozen of them could force their way through the Bedouins with life; he was amongst those few, and a flight at full speed was the sole chance of regaining their encampment. Just as he had shaken his bridle free of the Arabs' clutch, and had mowed himself a clear path through their ranks, he caught sight of his young enemy, Picpon, on the ground, with a lance broken off in his ribs, guarding his head, with bleeding hands, as the horses trampled over him. To make a dash at the boy, though to linger a moment was to risk certain death, to send his steel through an Arab who came in his way, to lean down and catch hold of the lad's sash, to swing him up into his saddle and throw him across it in front of him, and to charge afresh through the storm of musket-balls, and ride on thus burdened, was the work of ten seconds with 'Bel-à-faire-peur'. And he brought the boy safe over a stretch of six leagues in a flight for life, though the imp no more deserved the compassion than a scorpion that has spent all its noxious day stinging at every point of uncovered flesh would merit tenderness from the hand it had poisoned.

When he was swung down from the saddle and laid in front of a vidette fire, sheltered from the bitter north wind that was then blowing cruelly, the bright, black, ape-like eyes of the Parisian *diablotin* opened with a strange gleam in them:

'*Picpon s'en souviendra,*' he murmured.

And Picpon had kept his word; he had remembered often; he remembered now, standing on his head and thinking of his hundred Napoléons surrendered because thieving and stealing in the regiment gave pain to that oddly prejudiced *ci-devant*. This was the sort of loyalty that the Franco-Arabs rendered; this was the sort of influence that the English Guardsman exercised amongst his Roumis.

Meantime, while Picpon made a human cone of himself, to the admiration of the polyglot crowd of the Algerine street, Cecil, having watered, fed, and littered down his tired horse, made his way to a little café he commonly frequented, and spent the few sous he could afford on an iced draught of lemon-flavoured drink. Eat he could not; over-fatigue had given him a nausea for food, and the last hour, motionless in the intense glow of the afternoon sun, had brought that racking pain through his temples which assailed him rarely now, but which in his first years in Africa had given him many hours of agony. He could not stay in the café; it was the hour of dinner for many, and the odours joined with the noise were insupportable to him.

A few doors farther in the street, which was chiefly of Jewish and Moslem shops, there was a quaint place, kept by an old Moor, who had some of the rarest and most beautiful treasures of Algerian workmanship in his long, dark, silent chambers. With this old man Cecil had something of a friendship; he had protected him one day from the mockery and outrage of some drunken Indigènes, and the Moor, warmly grateful, was ever ready to give him a cup of coffee and a huble-bubble in the stillness of his dwelling. Its resort was sometimes welcome to him as the one spot, quiet and noiseless, to which he could escape out of the continuous turmoil of street and of barrack, and he went thither now. He found the old man sitting cross-legged behind his counter; a noble-looking aged Mussulman, with a long beard like white silk, with cashmeres and broidered stuffs of peerless texture hanging above his head, and all around him things of silver, of gold, of ivory, of amber, of feathers, of bronze, of emeralds, of ruby, of beryl, whose rich colours glowed through the darkness.

'No coffee, no sherbet, thanks, good father,' said Cecil, in answer to the Moor's hospitable entreaties. 'Give me only licence to sit in the quiet here. I am very tired.'

'Sit and be welcome, my son,' said Ben Arsli. 'Whom should this roof shelter in honour, if not thee? Musjid shall bring thee the supreme solace.'

The supreme solace was a narghilé, and its great bowl of rose-water was soon set down by the little Moorish lad at Cecil's side. Whether fatigue really weighted his eyes with slumber, or whether the soothing sedative of the pipe had its influence, he had not sat long in the perfect stillness of the Moor's shop before the narrow view of the street under the awning without was lost to him, the lustre and confusion of shadowy hues swam awhile before his eyes, the throbbing pain in his temples grew duller, and he slept—the heavy, dreamless sleep of intense exhaustion.

Ben Arsli glanced at him, and bade Musjid be very quiet. Half an hour or more passed; none had entered the place. The grave old Moslem was half slumbering himself, when there came a delicate odour of perfumed laces, a delicate rustle of silk swept the floor; a lady's voice asked the price of an ostrich-egg, superbly mounted in gold. Ben Arsli opened his eyes— the Chasseur slept on: the newcomer was one of those great ladies who now and then winter in Algeria.

Her carriage waited without: she was alone, making purchase of those innumerable splendid trifles with which Algiers is rife, while she drove through the town in the cooler hour before the sun sank into the western sea.

The Moor rose instantly, with profound salaams, before her, and began to spread before her the richest treasures of his stock. Under plea of the light, he remained near the entrance with her; money was dear to him, and must not be lost, but he would make it if he could without awakening the tired soldier. Marvellous caskets of mother-of-pearl; carpets soft as down, with every brilliant hue melting one within another; coffee equipages, of inimitable metal work; silver statuettes, exquisitely chased and wrought; feather-fans, and screens of every beauty of device, were spread before her, and many of them were bought by her with that unerring grace of taste and lavishness of expenditure which were her characteristics, but which are far from always found in unison; and throughout her survey, Ben Arsli had kept her near the entrance, and Cecil had slept on unaroused by the low tones of their voices.

A roll of notes had passed from her hand to the Moslem's, and she was about to pass out to her carriage, when a lamp which hung at the farther end caught her fancy. It was very singular, a mingling of coloured glass, silver, gold, and ivory being wrought in with much beauty in its formation.

'Is that for sale?' she enquired.

As he answered in the affirmative, she moved up the shop, and, her eyes being lifted to the lamp, had drawn close to Cecil before she saw him. When she did so, she paused near, in astonishment:

'Is that soldier asleep?'

'He is, Madame,' softly answered the old man, in his slow, studied French. 'He comes here to rest sometimes out of the noise; he was very tired today, and I think ill, would he have confessed it.'

'Indeed!' Her eyes fell on him with compassion; he had fallen into an attitude of much grace, and of utter exhaustion; his head was uncovered and rested on one arm, so that the face was turned upward. With a woman's rapid comprehensive glance she saw the dark shadow like a bruise under his closed aching eyes, she saw the weary pain upon his forehead, she saw the whiteness of his hands, the slenderness of his wrists, the softness of his hair; she saw, as she had seen before, that whatever he might be now, in some past time he had been a man of gentle blood, of courtly bearing.

'He is a Chasseur d'Afrique?' she asked the Moslem.

'Yes, Madame, I think——he must have been something very different some day.'

She did not answer; she stood with her thoughtful eyes gazing on the worn-out soldier.

'He saved me once, Madame, at much risk to himself, from the savagery of some Turcos,' the old man went on. 'Of course he is always welcome under my roof. The companionship he has must be bitter to him, I fancy; they do say he would have had his officer's grade, and the cross, too, long before now, if it were not for his Colonel's hatred.'

'Ah! I have seen him before now; he carves in ivory. I suppose he has a good sale for those things with you?'

The Moor looked up in amazement.

'In ivory, Madame?—*he?* Allah-il-Allah! I never heard of it. It is strange——'

'Very strange. Doubtless you would have given him a good price for them?'

'Surely I would; any price he should have wished. Do I not owe him my life?'

At that moment little Musjid let fall a valuable coffee-tray, inlaid with amber; his master, with muttered apology, hastened to the scene of accident; the noise startled Cecil, and his eyes unclosed to all the dreamy fantastic colours of the place, and met those bent on him in musing pity— saw that lustrous, haughty, delicate head bending slightly down through the many-coloured shadows.

He thought he was dreaming, yet on instinct he rose, staggering slightly, for sharp pain was still darting through his head and temples.

'Madame! pardon me! Was I sleeping?'

'You were, and rest again. You look ill?' she said, gently; and there was, for a moment, less of that accent in her voice which, the night before, had marked so distinctly, so pointedly, the line of demarcation between a Princess of Spain and a soldier of Africa.

'I thank you, I ail nothing.'

He had no sense that he did, in the presence of that face which had the beauty of his old life; under the charm of that voice which had the music of his buried years.

'I fear that is scarcely true?' she answered him. 'You look in pain; though as a soldier, perhaps, you will not own it?'

'A headache from the sun—no more, Madame.'

He was careful not again to forget the social gulf which yawned between them.

'That is quite bad enough! Your service must be severe?'

'In Africa, Miladi, one cannot expect indulgence.'

'I suppose not. You have served long?'

'Twelve years, Madame.'

'And your name?'

'Louis Victor.' She fancied there was a slight abruptness in the reply, as though he were about to add some other name, and checked himself.

She entered it in the little book from which she had taken her banknotes.

'I may be able to serve you,' she said, as she wrote. 'I will speak of you to the Marshal; and when I return to Paris, I may have an opportunity to bring your name before the Emperor. He is as rapid as his uncle to reward military merit; but he has not his uncle's opportunities for personal observation of his soldiers.'

The colour flushed his forehead.

'You do me much honour,' he said, rapidly, 'but if you would gratify me, Madame, do not seek to do anything of the kind.'

'And why? Do you not even desire the cross?'

'I desire nothing, except to be forgotten.'

'You seek what others dread, then!'

'It may be so. At any rate, if you would serve me, Madame, never say what can bring me into notice.'

She regarded him with much surprise, with some slight sense of annoyance; she had bent far in tendering her influence at the French Court to a private soldier, and his rejection of it seemed as ungracious as it was inexplicable.

At that moment the Moor joined them.

'Miladi has told me, Monsieur Victor, that you are a first-rate carver of ivories. How is it you have never let me benefit by your art?'

'My things are not worth a sou,' muttered Cecil, hurriedly.

'You do them great injustice, and yourself also,' said the *grande dame*, more coldly than she had before spoken. 'Your carvings are singularly perfect, and should bring you considerable returns.'

'Why have you never shown them to me at least?' pursued Ben Arsli—'why not have given me my option?'

The blood flushed Cecil's face again; he turned to the Princess.

'I withheld them, Madame, not because he would have underpriced, but overpriced them. He rates a trifling act of mine of long ago so unduly.'

She bent her head in silence; yet a more grateful comprehension of his motive she could not have given than her glance alone gave.

Ben Arsli stroked his great beard; more moved than his Moslem dignity would show.

'Always so!' he muttered, 'always so! My son, in some life before this, was not generosity your ruin?'

'Miladi was about to purchase that lamp?' asked Cecil, avoiding the question. 'Her Highness will not find anything like it in all Algiers.'

The lamp was taken down, and the conversation turned from himself.

'May I bear it to your carriage, Madame?' he asked, as she moved to leave, having made it her own, while her footman carried out the smaller articles she had bought to the equipage. She bowed in silence; she was very proud, she was not wholly satisfied with herself for having conversed thus with a Chasseur d'Afrique in a Moor's bazaar. Still, she vaguely felt pity for this man; she equally vaguely desired to serve him.

'Wait, Monsieur Victor!' she said, as he closed the door of her carriage. 'I accepted your chessmen last night, but you are very certain that it is impossible I can retain them on such terms.'

A shadow darkened his face.

'Let your dogs break them, then, Madame. They shall not come back to me.'

'You mistake, I did not mean that I would send them back. I simply desire to offer you some equivalent for them. There must be something that you wish for?—something which would be acceptable to you in the life you lead?'

'I have already named the only thing I desire.'

He had been solicitous to remember and sustain the enormous difference in their social degrees; but at the offer of her gifts, of her patronage, of her recompense, the pride of his old life rose up to meet her own.

'To be forgotten? A sad wish! Nay, surely life in a regiment of Africa cannot be so cloudless that it can create in you no other?'

'It is not. I have another.'

'Then tell it to me; it shall be gratified.'

'It is to enjoy a luxury long ago lost for ever. It is—to be allowed to give the slight courtesy of a gentleman without being tendered the wage of a servant.'

She understood him; she was moved, too, by the inflexion of his voice. She was not so cold, not so negligent, as the world called her.

'I had passed my word to grant it; I cannot retract,' she answered him, after a pause. 'I will press nothing more on you. But—as an obligation to me—can you find no way in which a rouleau of gold would benefit your men?'

'No way that I can take it for them. But, if you care indeed to do them a charity, a little wine, a little fruit, a few flowers (for there are those among them who love flowers), sent to the hospital, will bring many benedictions on your name, Madame. They lie in infinite misery there!'

'I will remember,' she said, simply, while a thoughtful sadness passed over her brilliant face. 'Adieu! M. le Caporal; and, if you should think better of your choice, and will allow your name to be mentioned by me to His Majesty, send me word through my people. There is my card.'

The carriage whirled away down the crooked street; he stood under the tawny awning of the Moorish house, with the thin glazed card in his hand. On it was printed:

Mme. la Princesse Corona d'Amägué,

Hôtel Corona, Paris.

In the corner was written, 'Villa Aïoussa, Algiers'. He thrust it in the folds of his sash, and turned within.

'Do you know her?' he asked Ben Arsli.

The old man shook his head.

'She is the most beautiful of thy many fair Frankish women. I never saw her till today. She seemed to have an interest in thee, my son. But listen here. Touching these ivory toys—if thou dost not bring henceforth to me all the work in them that thou doest, thou shalt never come here more to meet the light of her eyes.'

Cecil smiled and pressed the Moslem's hand.

'I kept them away because you would have given me a hundred piastres for what had not been worth one. As for her eyes, they are stars that shine on another world than an African trooper's. So best!'

Yet they were stars of which he thought more, as he wended his way back to the barracks, than of the splendid constellations of the Algerian evening that shone with all the lustre of the day, but with a soft enchanted light which transfigured sea and earth and sky as never did the day's full glow, as he returned to the mechanical duties, to the thankless services, to the distasteful meal, to the riotous mirth, to the coarse comradeship, which seemed to him to-night more bitter than they had ever done since his very identity, his very existence, had been killed and buried past recall, past resurrection, under the *képi d'ordonnance* of a Chasseur d'Afrique.

Meantime the Princesse Corona drove homeward—homeward to where a temporary home had been made by her in the most elegant of the many snow-white villas that stud the sides of the Sahel and face the bright bow of the sunlit bay; a villa with balconies, and awnings, and cool silent chambers, and rich glowing gardens, and a broad low roof half hidden in bay and orange and myrtle and basilica, and the liquid sound of waters bubbling beneath a riotous luxuriance of blossom.

Madame la Princesse passed from her carriage to her own morning-room, and sank down on a couch a little listless and weary with her search among the treasures of the Algerine bazaars. It was purposeless work, after all. Had she not bronzes, and porcelains, and bric-à-brac, and *objets d'art* in profusion in her Roman villa, her Parisian hotel, her great grim palace in the far Asturias?

'Not one of those things do I want—not one shall I look at twice. The money would have been better at the soldiers' hospital,' she thought, while her eyes dwelt on a chess-table near her—a table on which the mimic hosts of Chasseurs and Arabs were ranged in opposed squadrons.

She took the White King in her hand and gazed at it with a certain interest.

'That man has been noble once,' she thought. 'What a fate!—what a cruel fate!'

It touched her to great pity; although proud with too intense a pride, her nature was exceedingly generous, and, when once moved, deeply compassionate. The unerring glance of a woman habituated to the first society of Europe had told her that the accent, the bearing, the tone, the features of this soldier, who only asked of life 'oblivion', were those of one originally of gentle blood; and the dignity and patience of his acceptance of the indignities which his present rank entailed on him had not escaped her, any more than the delicate beauty of his face as she had seen it, weary, pale, and shadowed with pain, in the unconscious revelation of sleep.

'How bitter his life must be!' she mused. 'When Philip comes, perhaps he will know of some way to aid him. And yet—who can serve a man who only desires to be forgotten?'

Then, with a certain impatient sense of some absurd discrepancy, of some unseemly occupation, in *her* thus dwelling on the wishes and the burdens of a *sous-officier* of Light Cavalry, she laughed a little and put the White Chief back once more in his place. Yet even as she set the king amongst his mimic forces, the very carvings themselves seemed to retain their artist in her memory.

There was about them an indescribable elegance, an exceeding grace and beauty, which spoke of a knowledge of art and of a refinement of taste far beyond those of a mere military amateur in the one who had produced them.

'What could bring a man of that talent, with that address, into the ranks?' she mused. 'Persons of good family, of once fine position, come here, they say, and live and die unrecognized under the Imperial flag. It is usually some dishonour that drives them out of their own worlds; it may be so with him. Yet he does not look like one whom shame has touched; he is proud still—prouder than he knows. More likely it is the old old story—a high name and a narrow fortune—the ruin of thousands! He is French, I suppose; a French aristocrat who has played *au roi dépouillé*, most probably, and buried himself and his history for ever beneath those two names that tell one nothing—Louis Victor. Well, it is no matter of mine. Very possibly he is a mere adventurer with a good manner. This army here is a pot-pourri, they say, of all the varied scoundrelisms of Europe!'

She left the chess-table and went onward to the dressing and bath and bed-chambers, which opened in one suite from her boudoir, and resigned herself to the hands of her attendants for her dinner-toilette.

The Moslem had said aright of her beauty; and now, as her splendid hair was unloosened and gathered up afresh with a crescent-shaped comb of gold that was not brighter than the tresses themselves, the brilliant, haughty, thoughtful face was of a truth, as he had said, the fairest that had ever come from the Frankish shores to the hot African seaboard. Many beside the old Moslem had thought it 'the fairest that e'er the sun shone on', and held one grave lustrous glance of the blue imperial eyes above aught else on earth. Many had loved her—all without return. Yet, although only twenty years had passed over her proud head, the Princess Corona d'Amägué had been wedded and been widowed.

Wedded, with no other sentiment than that of a certain pity and a certain honour for the man whose noble Spanish name she took. Widowed, by a death that was the seal of her marriage-sacrament, and left her his wife only in name and law.

The marriage had left no chain upon her; it had only made her mistress of wide wealth, of that villa on the Sicilian Sea, of that light spacious palace-dwelling in Paris that bore her name, of that vast majestic old castle throned on brown Estremaduran crags, and looking down on mighty woods of cork and chesnut, and flashing streams of falling water hurling through the gorges. The death had left no regret upon her; it had only given her for a while a graver shadow over the brilliancy of her youth and of her beauty, and given her for always—or for so long, at least, as she so chose to use it—a plea for that indifference to men's worship of her which their sex called heartlessness, which her own sex thought an ultra-refined coquetry, and which was in real truth neither the one nor the other, but simply the negligence of a woman very difficult to touch, and, as it had seemed, impossible to charm.

None knew quite aright the history of that marriage. Some were wont to whisper it 'ambition'; and, when that whisper came round to her, her splendid lips would curl with as splendid a scorn.

'Do they not know that scarce any marriage can mate *us* equally?' she would ask; for she came of a great Line that thought few royal branches on equality with it; and she cherished as things of strictest creed and fact the legends that gave to her race, with its amber hair and its eyes of sapphire blue, the blood of King Arthur in their veins.

Of a surety it was not ambition that had allied her, on his deathbed, with Beltran Corona d'Amägué; but what it was the world could never tell precisely. The world would not have believed if it had heard the truth—the truth that it had been, in a different fashion, a gleam of something of that same compassion which now made her merciful to a private trooper of Africa which had wedded her to the dead Spanish Prince: compassion which, with many another rich and generous thing, lay beneath her coldness and her pride as the golden stamen lies folded within the white virginal chill cup of the lily.

She had never felt a touch of even passing preference to any one out of the many who had sought her high-born beauty; she was too proud to be easily moved to such selection, and she was far too habituated to homage to be wrought upon by it ever so slightly. She was of a noble, sun-lit, gracious-nature; she had been always happy, always obeyed, always caressed, always adored; it had rendered her immeasurably contemptuous of flattery; it had rendered her a little contemptuous of pain. She had never had aught to regret; it was not possible that she could realize what regret was.

Hence men called and found her very cold; yet those of her own kin whom she loved knew that the heart of a summer rose was not warmer, nor sweeter, nor richer than her. And first amongst these was her half-brother, twenty years her senior—at once her guardian and her slave—who thought her perfect, and would no more have crossed her will than he would have set his foot on her beautiful imperial head. Corona d'Amägué had been his friend; the only one for whom he had ever sought to break her unvarying indifference to her lovers, but for whom even he had pleaded vainly until one autumn season, when they had stayed together at a great archducal castle in South Austria.

In one of the forest-glades, awaiting the *fanfare* of the hunt, she rejected, for the third time, the passionate supplication of the superb noble who ranked with the D'Ossuna and the Medina-Sidonia. He rode from her in great bitterness, in grief that no way moved her—she was importuned with these entreaties to weariness. An hour after, he was brought past her, wounded and senseless; he had saved her brother from imminent death at his own cost, and the tusks of the mighty Styrian boar had plunged through and through his frame as they had met in the narrow woodland glade.

'He will be a cripple—a paralysed cripple—for life!' said the one whose life had been rescued by this devotion to her; and his lips shook a little under his golden beard as he spoke.

She looked at him; she loved him well, and no homage to herself could have moved her as this sacrifice for her brother had done.

'You think he will live?' she asked.

'They say it is sure. He may live on to old age. But how? My God! what a death in life! And all for my sake, in my stead!'

She was silent several moments; then she raised her face, a little paler than it had been, but with a passionless resolve set on it.

'Philip, *we* do not leave our debts unpaid. Go; tell him I will be his wife.'

'His wife—now! Venetia!——'

'Go!' she said, briefly. 'Tell him what I say.'

'But what a sacrifice! In your beauty, in your youth——'

'He did not count cost. Are we less generous? Go—tell him.'

He was told; and was repaid. Such a light of unutterable joy burnt through the misty agony of his eyes as never, it seemed to those who saw, had beamed before in mortal eyes. He did not once hesitate at the acceptance of her self-surrender; he only pleaded that the marriage ceremony should pass between them that night.

There were notaries and many priests in the great ducal household; all was done as he desired. She consented without wavering; she had passed her word, she would not have withdrawn it if its redemption had been a thousand times more bitter. The honour of her house was dearer to her than any individual happiness. This man, for them, had lost peace, health, joy, strength, every hope of life; to dedicate her own life to him, as he had vainly prayed her when in the full glow and vigour of his manhood, was the only means by which their vast debt to him could be paid. To thus pay it was the instant choice of her high code of honour, and of a generosity that would not be outrun. Moreover, she pitied him unspeakably, though her heart had no tenderness for him; she had dismissed him with cold disdain, and he had gone from her to save the only creature that she loved, and was stretched a stricken, broken, helpless wreck, with endless years of pain and weariness before him!

So—at midnight, in the great dim magnificence of the state chamber where he lay, and with the low, soft changing of the chapel choir from afar echoing through the incensed air, she bent her haughty head down over his couch, and the marriage benediction was spoken over them.

His voice was faint and broken, but it had the thrill of a passionate triumph in it. When the last words were uttered, he lay awhile, exhausted, silent, only looking ever upward at her with his dark, dreamy eyes, in which the old love glanced so strangely through the blindness of pain. Then he smiled as the last echo of the choral melodies died softly on the silence.

'That is joy enough! Ah! have no fear. With the dawn you will be free once more. Did you think that I could have taken your sacrifice? I knew well, let them say as they would, that I should not live the night through. But, lest existence should linger to curse me, to chain you, I rent the linen bands off my wounds an hour ago. All their science will not put back the life *now*! My limbs are dead, and the cold steals up! Ah, love! ah, love! You never thought how men can suffer! But have no grief for me. I am happy. Bend your head down, and lay your lips on mine once. You are my own!—death is sweeter than life!'

And before sunrise he died.

Some shadow from that fatal and tragic midnight marriage rested on her still. Though she was blameless, some vague remorse ever haunted her; though she had been so wholly guiltless of it, this death for her sake ever seemed in some sort of her bringing. Men thought her only colder, only prouder; but they erred. She was one of those women who, beneath the courtly negligence of a chill manner, are capable of infinite tenderness, infinite nobility, and infinite self-reproach.

A great French painter once, in Rome, looking on her from a distance, shaded his eyes with his hand, as if her beauty, like the sun, dazzled him.

'Exquisite—superb!' he muttered; and he was a man whose own ideals were so matchless that living women rarely could ring out his praise. 'She is nearly perfect, your Princess Corona!'

'Nearly!' cried a Roman sculptor. 'What, in Heaven's name, can she want?'

'Only one thing!'

'And that is——?'

'*To have loved.*'

Wherewith he turned into the Greco.

He had found the one flaw—and it was still there. What he had missed in her, still was wanting.

CHAPTER IX

The little Leopard of France

'*V'la ce que c'est la gloire—au grabat!*'

The contemptuous sentence was crushed through Cigarette's tight-pressed bright-red lips with an irony sadder than tears. She was sitting on the edge of a *grabat*, hard as wood, comfortless as a truss of straw, and looking down the long hospital-room with its endless rows of beds and its hot sun shining blindingly on its glaring whitewashed walls.

She was well known and well loved there. When her little brilliant-hued figure fluttered, like some scarlet bird of Africa, down the dreary length of those chambers of misery, bloodless lips close-clenched in torture would stir with a smile, would move with a word of welcome. No tender-voiced dove-eyed Sister of Orders of Mercy, gliding grey and soft, and like a living psalm of consolation, beside those couches of misery, bore with them the infinite inexpressible charm that the Friend of the Flag brought to the sufferers. The Sisters were good, were gentle, were valued as they merited by the greatest blackguard prostrate there;

but they never smiled, they never took the dying heart of a man back with one glance to the days of his childhood, they never gave a wild snatch of song, like a bird's on a spring-blossoming bough, that thrilled through half-dead senses with a thousand voices from a thousand buried hours.

'But the Little One,' as said a gaunt grey-bearded Zéphyr once, where he lay with the death-chill stealing slowly up his jagged, torn frame—'the Little One—do you see—she is youth, she is life; she is all we have lost. That is her charm! The Sisters are good women, they are very good; but they only *pity* us. The little one, she *loves* us. That is the difference, do you see?'

It was all the difference—a wide difference; she loved them all, with the warmth and fire of her young heart, for sake of France and of their common Flag. And though she was but a wild wayward mischievous *gamin*, a *gamin* all over though in a girl's form, men would tell in camp and hospital, with great tears coursing down their brown scarred cheeks, how her touch would fall as softly as a snowflake on their heated foreheads, how her watch would be kept by them through long nights of torment, how her gifts of golden trinkets would be sold or pawned as soon as received to buy them ice or wine, and how in their delirium the sweet fresh voice of the Child of the Regiment would soothe them, singing above their wretched beds some carol or chant of their own native province, which it always seemed that she must know by magic. For, were it Basque or Breton, were it a sea-lay of Vendée or a mountain-song of the Orientales, were it a mere ringing rhyme for the mules of Alsace, or a wild bold romanesque from the country of Berri, Cigarette knew each and all, and never erred by any chance, but ever sung to every soldier the rhythm familiar from his infancy, the melody of his mother's cradle-song and of his first love's lips. And there had been times when those songs suddenly breaking through the darkness of night, suddenly lulling the fiery anguish of wounds, had made the men who one hour before had been like mad dogs, like goaded tigers, men full of the lusts of slaughter and the lusts of the senses, and chained powerless and blaspheming to a bed of agony, tremble and shudder at themselves, and turn their faces to the wall and weep like children, and fall asleep, at length, with wondering dreams of God.

'*V'la ce que c'est la gloire—au grabat!*' said Cigarette, now grinding her pretty teeth. She was in her most revolutionary and reckless mood, drumming the rataplan with her spurred heels, and sitting smoking on the corner of old Miou-Matou's mattress. Miou-Matou, who had acquired that title among the *joyeux* for his scientific powers of making a tom-cat

into a stew so divine that you could not tell it from rabbit, being laid up
with a ball in his hip, a spear-head between his shoulders, a rib or so
broken, and one or two other little trifling casualties.

Miou-Matou, who looked very like an old grizzly bear, laughed in the
depths of his great hairy chest.

'Dream of glory, and end on a *grabat!* Just so, just so. And yet one has
pleasures—to sweep off an Arbico's neck nice and clean—swish!' And he
described a circle with his lean brawny arm with as infinite a relish as a
dilettante grown blind would listen thirstily to the description of an
exquisite bit of Faïence or Della Quercia work.

'Pleasures! My God! Infinite, endless misery!' murmured a man on her
right hand. He was not thirty years of age, with a delicate, dark, beautiful
head that might have passed as model to a painter for a St John. He was
dying fast of the most terrible form of pulmonary maladies.

Cigarette flashed her bright falcon glance over him.

'Well! is it not misery that is glory?'

'We think that it is when we are children. God help us!' murmured the
man who lay dying of lung-disease.

'Ouf! Then we think rightly! Glory! Is it the cross, the star, the bâton?
No! He who wins those[1] only reins his horse up on a hill, out of shot
range, and watches through his glass how his troops surge up, wave on
wave, in the great sea of blood. It is misery that is glory—the misery that
toils with bleeding feet under burning suns without complaint; that
lies half dead through the long night with but one care, to keep the torn
flag free from the conqueror's touch; that bears the rain of blows in
punishment rather than break silence and buy release by betrayal of a
comrade's trust; that is beaten like the mule, and galled like the horse,
and starved like the camel, and housed like the dog, and yet does the
thing which is right and the thing which is brave despite all; that suffers,
and endures, and pours out his blood like water to the thirsty sands whose
thirst is never stilled, and goes up in the morning sun to the combat as
though death were the paradise that the Arbicos dream, knowing the
while that no paradise waits save the crash of the hoof through the
throbbing brain, or the roll of the gun-carriage over the writhing limb.
That is glory. The misery that is heroism because France needs it, because
a soldier's honour wills it. *That* is glory. It is here today in the Hospital, as

[1] Having received ardent reproaches from Field-Officers and Commanders of Divisions for
the injustice done to military chiefs by this sentence, I beg to assure them the injustice is
Cigarette's—not mine. I should be very sorry, for an instant, to seem to depreciate the 'genius
of command', without whose guiding will an army is but a rabble; or to underrate that noblest
courage which accepts the burden of arduous responsibilities, and of duties as bitter in anxiety
as they are precious in honour.

it never is in the Cour des Princes where the glittering host of the Marshals gather!'

Her voice rang clear as a clarion; the warm blood burnt in her bright cheeks; the swift, fiery, pathetic eloquence of her nation moved her, and moved strangely the hearts of her hearers; for though she could neither read nor write, there was in Cigarette the germ of that power which the world mistily calls Genius.

There were men lying in that sick-chamber brutalized, crime-stained, ignorant as the bullocks of the plains, and, like them, reared and driven for the slaughter; yet there was not one among them to whom some ray of light failed to come from those words, through whom some thrill failed to pass as they heard them. Out yonder in the free air, in the barrack-court, or on the plains, the Little One would rate them furiously, mock them mercilessly, rally them with the flat of a sabre if they were mutinous, and lash them with the most pitiless ironies if they were grumbling; but here, in the hospital, the Little One loved them, and they knew it, and that love gave a flute-like music to the passion of her voice.

Then she laughed, and drummed the rataplan again with her brass heel.

'All the same; one is not in paradise *au grabat*, eh, Père Matou?' she said, curtly.

She was half impatient of her own momentary lapse into enthusiasm, and she knew the temper of her 'children' as accurately as a bugler knows the notes of the réveillé—knew that they loved to laugh even with the death-rattle in their throats, and, with their hearts half breaking over a comrade's corpse, would cry in burlesque mirth, '*Ah, le bon-zig! Il a avalé sa cartouche!*'

'Paradise!' growled Père Matou. 'Ouf! Who wants that? If one had a few *bidons* of brandy, now——'

'Brandy? Oh-hé. You are to be much more of aristocrats now than that!' cried Cigarette, with an immeasurable satire curling on her rosy, piquante lips. 'The Silver Pheasants have taken to patronise you. *Ma cantche!* if I were you, I would not touch a glass nor eat a fig; you will not if you have the spirit of a rabbit. You! Fed like dogs with the leavings of her table—pardieu! that is not for soldiers of France!'

'Eh? What dost thou say?' growled Miou-Matou, peering up under his grey shaggy brows.

'Only that a *grande dame* has sent you champagne. That is all. Sapristi! how easy it is to play the saint and Samaritan with two words to one's *maître d'hôtel*, and a *rouleau* of gold that one never misses! The rich—they can buy all things, you see, even heaven, so cheap!'

With which withering satire Cigarette left Père Matou in the conviction that he must be already dead and amongst the angels if people began to talk of champagne to him, and flitting down between the long rows of beds with the old disabled veterans who tended them, skimmed her way, like a bird as she was, into another great chamber, filled, like the first, with suffering in all stages and at all years, from the boy-conscript, tossing in African fever, to the white-haired campaigner of a hundred wounds.

Cigarette was as caustic as a Voltaire this morning. Coming through the entrance of the hospital, she had casually heard that Madame la Princesse Corona d'Amägué had made a gift of singular munificence and mercy to the invalid soldiers—a gift of wine, of fruit, of flowers, that would brighten their long dreary hours for many weeks. Who Madame la Princesse might be she knew nothing; but the title was enough, she was a silver pheasant—bah! And Cigarette hated the aristocrats—when they were of the sex feminine.

'An aristocrat in adversity is an eagle,' she would say; 'but an aristocrat in prosperity is a peacock.'

Which was the reason why she flouted glittering young nobles with all the insolence imaginable, but took the part of 'Marquise', of 'Bel-à-faire-peur', and of such wanderers like them, who had buried their sixteen quarterings under the black shield of the Enfans Perdus of Africa.[1]

With a word here and a touch there—tender, soft, and bright, since, however ironic her mood, she never brought anything except sunshine to those who lay in such sore need of it, beholding the sun in the heavens only through the narrow chink of a hospital window;—at last she reached the bed she came most specially to visit: a bed on which was stretched the emaciated form of a man once beautiful as a Greek dream of a God.

The dews of a great agony stood on his forehead; his teeth were tight clenched on lips white and parched; and his immense eyes, with the heavy circles round them, were fastened on vacancy with the yearning misery that gleams in the eyes of a Spanish bull when it is struck again and again by the matador, and yet cannot die.

She bent over him softly.

'*Tiens, Monsieur Léon!* I have brought you some ice.'

His weary eyes turned on her gratefully; he sought to speak, but the effort brought the spasm on his lungs afresh; it shook him with horrible violence from head to foot, and the foam on his auburn beard was red with blood.

---

[1] The 'écusson noir' on the uniform of the 'Zéphyrs'.

There was no one by to watch him; he was sure to die; a week sooner or later—what mattered it? He was useless as a soldier; good only to be thrown into a pit, with some quicklime to hasten destruction and do the work of the slower earthworms.

Cigarette said not a word, but she took out of some vine-leaves a cold hard lump of ice, and held it to him; the delicious coolness and freshness in that parching noontide heat stilled the convulsion; his eyes thanked her, though his lips could not; he lay panting, exhausted, but relieved; and she—thoughtfully for her—slid herself down on the floor and began singing, low and sweetly, as a fairy might sing on the raft of a water-lily leaf.

She sang *gaudrioles*, to be sure, Béranger's songs and odes of the camp; for she knew of no hymn but the *Marseillaise*, and her chants were all chants like the *Laus Veneris*. But the voice that gave them was pure as the voice of a thrush in the spring, and the cadence of its music was so silvery-sweet that it soothed like a spell all the fever-racked brains, all the pain-tortured spirits.

'Ah! that is good,' murmured the dying man. 'It is like the brooks—like the birds—like the winds in the leaves.'

He was but half conscious; but the lulling of that gliding voice brought him peace. And Cigarette sang on, only moving to reach him some fresh touch of ice, while time travelled fast, and the first afternoon shadows crept across the bare floor. Every now and then, dimly through the openings of the windows, came a distant roll of drums, a burst of military music, an echo of the laughter of a crowd; and then her head went up eagerly, an impatient shade swept across her expressive face.

It was a fête-day in Algiers; there were flags and banners fluttering from the houses, there were Arab races and Arab manœuvres, there was a review of troops for some foreign general, there were all the mirth and the mischief that she loved, and that never went on without her, and she knew well enough that from mouth to mouth there was sure to be asking, '*Mais où donc est Cigarette?*'—Cigarette, who was the Generalissima of Africa!

But still she never moved; though all her vivacious life was longing to be out and in their midst, on the back of a desert horse, on the head of a huge drum, perched on the iron support of a high-hung lantern, standing on a cannon while the Horse Artillery swept full gallop, firing down a volley of *argot* on the hot homage of a hundred lovers, drinking creamy liqueurs and filling her pockets with bonbons from handsome subalterns and aides-de-camp, doing as she had done ever since she could remember her first rataplan. But she never moved. She knew that in the general gala these sick-beds would be left more deserted and less soothed than ever. She knew, too, that it was for the sake of this man, lying dying here from the lunge of a Bedouin lance through his lungs, that the ivory wreaths and crosses and statuettes had been sold.

And Cigarette had done more than this ere now many a time for her 'children'.

The day stole on; Léon Ramon lay very quiet; the ice for his chest and the song for his ear gave him that semi-oblivion, dreamy and comparatively painless, which was the only mercy which could come to him. All the chamber was unusually still; on three of the beds the sheet had been drawn over the face of the sleepers, who had sunk to a last sleep since the morning rose. The shadows lengthened, the hours followed one another; Cigarette sang on to herself with few pauses: whenever she did so pause to lay soaked linen on the soldier's hot forehead, or to tend him gently in those paroxysms that wrenched the clotted blood from off his lungs, there was a light on her face that did not come from the golden heat of the African sun.

Such a light those who know well the Children of France may have seen, in battle or in insurrection, grow beautiful upon the young face of conscript or of boy-insurgent as he lifted a dying comrade, or pushed to the front to be slain in another's stead; the face that a moment before had been keen for the slaughter as the eyes of a kite, and recklessly gay as the saucy refrain the lips carolled.

A step sounded on the bare boards; she looked up, and the wounded man raised his weary lids with a gleam of gladness under them; Cecil bent above his couch.

'Dear Léon! how is it with you?'

His voice was softened to infinite tenderness; Léon Ramon had been for many a year his comrade and his friend; an artist of Paris, a man of marvellous genius, of high idealic creeds, who, in a fatal moment of rash despair, had flung his talents, his broken fortunes, his pure and noble spirit, into the fiery furnace of the hell of military Africa; and now lay dying here, a common soldier, forgotten as though he were already in his grave.

'The review is just over. I got ten minutes to spare, and came to you the instant I could,' pursued Cecil. 'See here what I bring you! You, with your artist's soul, will feel yourself all but well when you look on these!'

He spoke with a hopefulness he could never feel, for he knew that the life of Léon Ramon was doomed; and as the other strove to gain breath enough to answer him, he gently motioned him to silence, and placed on his bed some peaches bedded deep in moss and circled round with stephanotis, with magnolia, with roses, with other rarer flowers still.

The face of the artist-soldier lightened with a longing joy; his lips quivered.

'Ah, God! they have the fragrance of my France!'

Cecil said nothing, but moved them nearer in to the clasp of his eager hands. Cigarette he did not see.

There were some moments of silence, while the dark eyes of the dying man thirstily dwelt on the beauty of the flowers, and his dry ashen lips seemed to drink in their perfumes as those athirst drink in water.

'They are beautiful,' he said, faintly, at length. 'They have our youth in them! How came you by them, dear friend?'

'They are not due to me,' answered Cecil, hurriedly. 'Madame la Princesse Corona sends them to you. She has sent great gifts to the hospital—wines, fruits, a profusion of flowers, such as those. Through her, these miserable chambers will bloom for a while like a garden; and the best wines of Europe will slake your thirst in lieu of that miserable *tisane*.'

'It is very kind,' murmured Léon Ramon, languidly; life was too feeble in him to leave him vivid pleasure in aught. 'But I am ungrateful. La Cigarette here—she has been so good, so tender, so pitiful. For once I have *almost* not missed you!'

Cigarette, thus alluded to, sprang to her feet with her head tossed back, and all her cynicism back again; a hot petulant colour was on her cheeks, the light had passed from her face, she struck her white teeth together. She had thought 'Bel-à-faire-peur' chained to his regiment in the field of manœuvre, or she would never have come thither to tend his friend. She had felt happy in her self-sacrifice; she had grown into a gentle, pensive, merciful mood, singing here by the side of the dying soldier, and now the first thing she heard was of the charities of Madame la Princesse!

That was all her reward! Cigarette received the recompense that usually comes to generous natures which have strung themselves to some self-surrender that costs them dear!

Cecil looked at her surprised, and smiled.

'*Ma belle*, is it you? That is, indeed, good. You were the good angel of my life the other night, and today come to bring consolation to my friend——'

'"Good angel!" *Chut*, M. Victor! One does not know those *mots sucrés* in Algiers. There is nothing of the angel about me, I hope. Your friend, too! Ouf! Do you think I have never been used to taking care of my comrades in hospital before *you* played the sick-nurse here?'

She spoke with all her brusque petulance in arms again; she hated that he should imagine she had sacrificed her fête-day to Léon Ramon, because the artist-trooper was dear to *him*; she hated him to suppose that she had waited there all the hours through on the chance that *he* would find her at her post, and admire her for her charity. Cigarette was far too proud and disdainful a young soldier to seek either his presence or his praise.

He smiled again; he did not understand the caprices of her changeful moods, and he did not feel that interest in her which would have made him divine the threads of their vagaries.

'I did not think to offend you, my Little One,' he said, gently. 'I meant only to thank you for your goodness to Ramon in my absence.'

Cigarette shrugged her shoulders.

'There was no goodness, and there need be no thanks. Ask Père Matou how often I have sat with *him* hours through.'

'But on a fête-day! And you who love pleasure, and grace it so well——'

'Ouf! I have had so much of it,' said the Little One, contemptuously. 'It is so tame to *me*. Clouds of dust, scurry of horses, fanfare of trumpets, thunder of drums, and all for nothing! Bah! I have been in a dozen battles—I—and I am not likely to care much for a sham fight.'

'Nay, she is unjust to herself,' murmured Léon Ramon. 'She gave up the fête to do this mercy—it has been a great one. She is more generous than she will ever allow. Here, Cigarette, look at these scarlet rosebuds; they are like your bright cheeks. Will you have them? I have nothing else to give.'

'Rosebuds?' echoed Cigarette, with supreme scorn. 'Rosebuds for me? I know no rose but the red of the tricolour; and I could not tell a weed from a flower. Besides, I told Miou-Matou just now, if my children do as *I* tell them, they will not take a leaf or a peach-stone from this *grande dame*—how does she call herself?—Madame Corona d'Amägué!'

Cecil looked up quickly: 'Why not?'

Cigarette flashed on him her brilliant brown eyes with a fire that amazed him.

'Because we are soldiers, not paupers!'

'Surely; but——'

'And it is not for the silver pheasants, who have done nothing to deserve their life but lain in nests of cotton wool, and eaten grain that others sow and shell for them, and spread their shining plumage in a sun that never clouds above their heads, to insult, with the insolence of their "pity" and their "charity", the Heroes of France, who perish, as they have lived, for their Country and their Flag!'

It was a superb peroration! If the hapless flowers lying there had been a cartel of outrage to the concrete majesty of the French Army, the Army's champion could not have spoken with more impassioned force and scorn.

Cecil laughed slightly; but he answered, with a certain annoyance:

'There is no "insolence" here; no question of it. Madame la Princesse desired to offer some gift to the soldiers of Algiers; I suggested to her that to increase the scant comforts of the hospital, and gladden the weary eyes of sick men with beauties that the Executive never dreams of bestowing, would be the most merciful and acceptable mode of exercising her kindness. If blame there be in the matter, it is mine.'

In defending the generosity of what he knew to be a genuine and sincere wish to gratify his comrades, he betrayed what he did not intend to have revealed, namely, the conversation that had passed between himself and the Spanish Princess. Cigarette caught at the inference with the quickness of her lightning-like thought.

'Oh-hé! So it is *she*!'

There was a whole world of emphasis, scorn, meaning, wrath, comprehension, and irony in the four monosyllables; the dying man looked at her with languid wonder.

'She? Who? What story goes with these roses?'

'None,' said Cecil, with the same inflection of annoyance in his voice; to have his passing encounter with this beautiful patrician pass into a barrack *canard*, through the unsparing jests of the soldiery around him, was a prospect very unwelcome to him. 'None whatever. A generous thoughtfulness for our common necessities as soldiers——'

'Ouf!' interrupted Cigarette, before his phrase was one-third finished. 'The stalled mare will not go with the wild coursers; an aristocrat may live with us, but he will always cling to his old Order. This is the story that runs with the roses. Miladi was languidly insolent over some ivory chessmen, and Corporal Victor thought it divine, because languor and insolence are the twin gods of the noblesse, parbleu! Miladi, knowing no gods but those two, worships them, and sends to the soldiers of France, as the sort of sacrifice her gods love, fruits and wines that, day after day, are set on her table, to be touched, if tested at all, with a butterfly's sip; and Corporal Victor finds this a charity sublime—to give what costs nothing, and scatter a few crumbs out from the profusion of a life of waste and indulgence! And I say, that if my children are of my fashion of thinking, they will choke like dogs dying of thirst rather than slake their throats with alms cast to them as if they were beggars!'

With which fiery and bitter enunciation of her views on the gifts of the Princess Corona d'Amägué Cigarette struck light to her *brûle-gueule*, and thrusting it between her lips, with her hands in the folds of her scarlet waist-sash, went off with the light swift step natural to her, exaggerated into the carriage she had learned of the Zouaves, laughing her good-morrows noisily to this and that trooper as she passed their couches, and not dropping her voice even as she passed the place where the dead lay, but singing, as loud as she could, the most impudent drinking-song out of the taverns of the Spahis that ever celebrated wine, women, and war in the lawlessness of the lingua Sabir.

Her wrath was hot, and her heart heavy within her. She had given up her whole fête-day to wait on the anguish and to soothe the solitude of his friend lying dying there; and her reward had been to hear him speak of this aristocrat's donations, that cost her nothing but the trouble of a few words of command to her household, as though they were the saintly charities of some angel from heaven!

'Diantre!' she muttered, as her hand wandered to the ever-beloved forms of the pistols within her sash. 'Echaffaurées or Achmet, or any one of them, would throw a draught of wine in his face, and lay him dead for me, with a pass or two, ten minutes after. Why don't I bid them? I have a mind——'

In that moment she could have shot him dead herself without a second's thought. Storm and sunlight swept, one after another, with electrical rapidity at all times through her vivid changeful temper; and here she had been wounded and been stung in the very hour in which she had subdued her national love of mirth, and her childlike passion for show, and her impatience of all confinement, and her hatred of all things mournful, to the attainment of this self-negation!

Moreover, there mingled with it the fierce and intolerant heat of the passionate and scarce-conscious jealousy of an utterly untamed nature, and of Gallic blood, quick and hot as the steaming springs of the Geyser.

'You have vexed her, Victor,' said Léon Ramon, as she was lost to sight through the doors of the great desolate chamber.

'I hope not; I do not know how,' answered Cecil. 'It is impossible to follow the windings of her wayward caprices. A child—a soldier—a dancer—a brigand—a spoilt beauty—a mischievous gamin—how is one to treat such a little fagot of opposites?'

The other smiled.

'Ah! you do not know the Little One yet. She is worth a study. I painted her years ago—"La Vivandière à Sept Ans." There was not a picture in the Salon that winter that was sought like it. I had travelled in Algeria then; I had not entered the Army. The first thing I saw of Cigarette was this: She was seven years old; she had been beaten black and blue; she had had two of her tiny teeth knocked out. The men were furious, she was a pet with them; and she would not say who had done it, though she knew twenty swords would have beaten him flat as a fritter if she had given his name. I got her to sit to me some days after. I pleased her with her own picture. I asked her to tell me why she would not say who had ill-treated her. She put her head on one side like a robin, and told me, in a whisper: "It was one of my comrades—because I would not steal for him. I would not have the army know—it would demoralize them! If a French soldier ever does a cowardly thing, another French soldier must not betray it." That was Cigarette—at seven years. The esprit du corps was stronger than her own wrongs. What do you say to that nature?'

'That it is superb!—that it might be moulded to anything. The pity is——'

'Ah, *tais-toi!*' said the artist-trooper, half wearily, half laughingly. 'Spare me the old world-worn, threadbare formulas. Because the flax and the colza blossom for use, and the garden-flowers grow trained and pruned, must there be no bud that opens for mere love of the sun, and swings free in the wind in its fearless fair fashion? Believe me, dear Victor, it is the lives which follow no previous rule that do the most good and give the most harvest.'

'Surely. Only for this child—a woman—in her future——'

'Her future? Well, she will die, I dare say, some bright day or another, at the head of a regiment, with some desperate battle turned by the valour of her charge, and the sight of the torn tricolour upheld in her little hands. That is what Cigarette hopes for—why not? There will always be a million of commonplace women ready to keep up the decorous traditions of their sex, and sit in safety over their needles by the side of their hearths. One little lioness here and there in a generation cannot do overmuch harm.'

Cecil was silent. He would not cross the words of the wounded man by saying what might bring a train of less pleasant thoughts—saying what, in truth, was in his mind, that the future which he had meant for the little Friend of the Flag was not that of any glorious death by combat, but that of a life (unless no bullet early cut its silver cord in twain) when youth should have fled, and have carried for ever with it her numberless graces, and left in its stead that ribaldry-stained, drink-defiled, hardened, battered, joyless, cruel, terrible thing which is unsightly and repugnant to even the lowest amongst men, which is as the lees of the drunk wine, as the ashes of the burnt-out fires, as the discord of the broken and earth-clogged lyre.

Cigarette was charming now—a fairy-story set into living motion—a fantastic little firework out of an extravaganza, with the impudence of a boy-harlequin and the witching kittenhood of a girl's beauty. But when this youth that made it all fair should have passed (and youth passes soon when thus adrift on the world), when there should be left in its stead only shamelessness, hardihood, vice, weariness—those who found the prettiest jest in her now would be the first to cast aside, with an oath, the charred wrecked rocket-stick of a life from which no golden careless stream of many-coloured fires of coquette caprices would rise and enchant them then.

'Who is it that sent these?' asked Léon Ramon, later on, as his hands still wandered among the flowers: for the moment he was at peace; the ice and the hours of quietude had calmed him.

Cecil told him again.

'What does Cigarette know of her?' he pursued.

'Nothing, except, I believe, she knew that Madame Corona accepted my chess-carvings.'

'Ah!' he said, with a smile, 'I thought the Little One was jealous, Victor.'

'Jealous? Pshaw! Of whom?'

'Of any one you admire—specially of this *grande dame*.'

'Absurd!' said Cecil, with a sense of annoyance. 'Cigarette is far too bold a little trooper to have any thoughts of those follies; and as for this *grande dame*, as you call her, I shall, in every likelihood, never see her again—unless when the word is given to "Carry Swords", or "Lances", at the general salute, where she reins her horse beside M. le Maréchal's at a review, as I have seen her this morning.'

The keen ear of the sick man caught the inflection of an impatience, of a mortification, in the tone that the speaker himself was unconscious of. He guessed the truth—that Cecil had never felt more restless under the shadow of the Eagles than he had done when he had carried his lance up in the salute as he passed with his regiment the flagstaff where the aristocracy of Algiers had been gathered about the Marshal and his staff, and the azure eyes of Madame la Princesse had glanced carelessly and critically over the long line of grey horses of those Chasseurs d'Afrique amongst whom he rode a *bas-officier*.

'Cigarette is right,' said Ramon, with a slight smile again. 'Your heart is with your old Order. You are "*aristocrat au bout des ongles*".'

'Indeed I am not, *mon ami*; I am a mere trooper.'

'Now! Well, keep your history as you have always done, if you will. What my friend was matters nothing; I know well what he is, and how true a friend. As for Miladi, she will be best out of your path, Victor. Women! God!—they are so fatal!'

'Does not our folly make their fatality?'

'Not always; not often. The madness may be ours, but they sow it. Ah! do they not know how to rouse and enrage it; how to fan, to burn, to lull, to pierce, to slake, to inflame, to entice, to sting? Heavens! so well they know—that their beauty must come, one thinks, out of hell itself!'

His great eyes gleamed like fire, his hollow chest panted for breath, the sweat stood out on his temples. Cecil sought to soothe him, but his words rushed on with the impetuous course of the passionate memories that arose in him.

'Do you know what brought me here? No! As little as I know what brought you, though we have been close comrades all these years. Well, it was *she!* I was an artist. I had no money, I had few friends; but I had youth, I had ambition, I had, I think, genius, till she killed it. I loved my art with a great love, and I was happy. Even in Paris one can be so happy without wealth, whilst one is young. The mirth of the Barrière—the grotesques of the Halles—the wooden booths on New Year's Day— the bright midnight crowds under the gaslights—the bursts of music from the gay cafés—the grey little nuns flitting through the snow—the Mardi Gras and its old-world fooleries—the summer Sundays under the leaves while we laughed like children—the silent dreams through the length of the Louvre—dreams that went home with us and made our garret bright with their visions! One was happy in them—happy, happy!'

His eyes were still fastened on the blank white wall before him while he spoke, as though the things that his words sketched so faintly were painted in all their vivid colours on the dull blank surface. And so in truth they were, as remembrance pictured all the thousand perished hours of his youth.

'Happy—until she looked at me,' he pursued, while his voice flew in feverish haste over the words. 'Why would she not let me be? She had them all in her golden nets; nobles, and princes, and poets, and soldiers, she swept them in far and wide. She had her empire; why must she seek out a man who had but his art and his youth, and steal *those?* Women are so insatiate, look you; though they held all the world, they would not rest if one mote in the air swam in sunshine free of them! It was the first year I touched triumph that I saw her. They began for the first time to speak of me; it was the little painting of Cigarette, as a child of the Army, that did it. Ah, God! I thought myself already so famous! Well, she sent for me to take her picture, and I went. I went and I painted her as Cleopatra— by her wish. Ah! it was a face for Cleopatra—the eyes that burn your youth dead, the lips that kiss your honour blind! A face—my God! how beautiful! She had set herself to gain my soul; and as the picture grew, and grew, and grew, so my life grew into hers till I lived only by her breath. Why did she want my life? She had so many! She had rich lives, great lives, grand lives at her bidding; and yet she knew no rest till she had leaned down from her cruel height and had seized mine, that had nothing on earth but the joys of the sun and the dew, and the falling of night, and the dawning of day, that are given to the birds of the fields.'

His chest heaved with the spasms that with each throe seemed to tear his frame asunder; still he conquered them, and his words went on, his eyes fastened on the burning white glare of the wall as though all the beauty of this woman glowed afresh there to his sight.

'She was great; no matter her name, she lives still. She was vile; ay, but not in my sight till too late. Why is it that men never love so well as where they love their own ruin? that the heart which is pure never makes ours beat upon it with the rapture sin gives? Through month on month my picture grew, and my passion grew with it, fanned by her hand. She knew that never would a man paint her beauty like one who gave his soul for the price of success. I had my paradise; I was drunk; and I painted as never the colours of mortals painted a woman. I think even she was content; even she, who in her superb arrogance thought she was matchless and deathless. Then came my reward; when the picture was done, her fancy had changed! A light scorn, a careless laugh, a touch of her fan on my cheek; could I not understand? Was I still such a child? Must I be broken more harshly in to learn to give place? That was all! and at last her lacquais pushed me back with his wand from her gates! What would you? I had not known what a great lady's illicit caprices meant; I was still but a boy! She had killed me; she had struck my genius dead; she had made earth my hell—what of that? She had her beauty eternal in the picture she needed, and the whole city rang with her loveliness as they looked on my work. I have never painted again. I came here. What of that? An artist the less then, the world did not care; a life the less soon, she will not care either!'

Then, as the words ended, a great wave of blood beat back his breath and burst from the pent-up torture of his striving lungs, and stained red the dark and silken masses of his beard. His comrade had seen the haemorrhage many times, yet now he knew, as he had never known before, that this was death.

As he held him upward in his arms, and shouted loud for help, the great luminous eyes of the French soldier looked up at him through their mist with the deep, fond gratitude that beams in the eyes of a dog as it drops down to die, knowing one touch and one voice to the last.

'*You* do not forsake,' he murmured, brokenly, while his voice ebbed faintly away as the stream of his life flowed faster and faster out. 'It is over now, so best! If only I could have seen France once more. France——'

He stretched his arms outward as he spoke with the vain longing of a hopeless love. Then a deep sigh quivered through his lips; his hand strove to close on the hand of his comrade, and his head fell, resting on the flushed blossoms of the rosebuds of Provence.

He was dead.

*   *   *

An hour later Cecil left the hospital, seeing and hearing nothing of the gay riot of the town about him, though the folds of many-coloured silk and bunting fluttered across the narrow Moorish streets, and the whole of the populace was swarming through them with the vivacious enjoyment of Paris mingling with the stately picturesque life of Arab habit and custom. He was well used to pain of every sort; his bread had long been the bread of bitterness, and the waters of his draught been of gall. Yet this stroke, though looked for, fell heavily and cut far.

Yonder, in the dead-room, there lay a broken, useless mass of flesh and bone that in the sight of the Bureau Arabe was only a worn-out machine that had paid its due toll to the wars of the Second Empire, and was now valueless; only fit to be cast in to rot, unmourned, in the devouring African soil. But to him that lifeless, useless mass was dear still; was the wreck of the bravest, tenderest, and best-beloved friend that he had found in his adversity.

In Léon Ramon he had found a man whom he had loved, and who had loved him. They had suffered much, and much endured together; their very dissimilarities had seemed to draw them nearer to each other. The gentle impassiveness of the Englishman had been like rest to the ardent impetuosity of the French soldier; the passionate and poetic temperament of the artist-trooper had revealed to Cecil a thousand views of thought and of feeling which had never before then dawned on him. And now that the one lay dead, a heavy, weary sense of loneliness rested on the other. They died around him every day; the fearless, fiery blood of France watered in ceaseless streams the arid, harvestless fields of northern Africa; death was so common, that the fall of a comrade was no more noted by them than the fall of a loose stone that their horse's foot shook down a precipice. Yet this death was very bitter to him; he wondered with a dull sense of aching impatience why no Bedouin bullet, no Arab sabre, had ever found his own life out, and cut his thralls asunder.

The evening had just followed on the glow of the day—evening, more lustrous even than ever, for the houses were all a-glitter with endless lines of coloured lamps and strings of sparkling illuminations, a very sea of bright-hued fire. The noise, the mirth, the sudden swell of music, the pleasure-seeking crowds, all that were about him, served only to make more desolate and more oppressive by their contrast his memories of that life, once gracious, and gifted, and content with the dower of its youth, ruined by a woman, and now slaughtered here, for no avail and with no honour, by a lance-thrust in a midnight skirmish, which had been unrecorded even in the few lines of the gazette that chronicled the war-news of Algeria.

Passing one of the cafés, a favourite resort of the officers of his own regiment, he saw Cigarette. A sheaf of blue, and white, and scarlet lights flashed with tongues of golden flame over her head, and a great tricolour flag with the brass eagle above it was hanging in the still, hot air from the balcony from which she leaned. Her tunic-skirt was full of bon-bons and of crackers that she was flinging down among the crowd whilst she sang, stopping every now and then to exchange some passage of *gaulois* wit with them that made her hearers scream with laughter, whilst behind her was a throng of young officers drinking champagne, eating ices, and smoking, echoing her songs and her satires with enthusiastic voices and stamps of their spurred boot-heels. As he glanced upward, she looked literally in a blaze of luminance, and the wild, mellow tones of her voice ringing out in the '*Rien n'est sacré pour un Sapeur*,' sounded like a mockery of that dying-bed beside which they had both so late stood together.

'She has the playfulness of the young leopard, and the cruelty,' he thought; with a sense of disgust, forgetting that she did not know what he knew, and that if Cigarette had waited to laugh until death had passed by she would have never laughed all her life through in the battalions of Africa.

She saw him, as he went beneath her balcony; and she sung all the louder, she flung her sweetmeat missiles with the reckless force of a Roman Carnivalist, she launched bolts of tenfold more audacious raillery at the delighted mob below. Cigarette was '*bon soldat*'; when she was wounded, she wound her scarf round the nerve that ached, and only laughed the gayer.

And he did her that injustice which the best amongst us are apt to do to those whom we do not feel interest enough in to study with that closeness which can alone give comprehension of the intricate and complex rebus, so faintly sketched, so marvellously involved, of human nature.

He thought her a little leopard, in her vivacious play and her inborn bloodthirstiness.

Well, the little leopard of France played recklessly enough that evening. Algiers was *en fête*, and Cigarette was sparkling over the whole of the town like a humming-bird or a firefly—here, and there, and everywhere, in a thousand places at once, as it seemed; staying long with none, making music and mirth with all. Waltzing like a thing possessed, pelting her lovers with a tempest storm of dragées, standing on the head of a gigantic Spahis *en tableau* amidst a shower of fireworks, improvising slang songs worthy of Jean Vadé and his Poissardes, and chorused by a hundred lusty lungs that yelled the burden in riotous glee as furiously as they were accustomed to shout 'En avant!' in assault and in charge, Cigarette made amends to herself at night for her vain self-sacrifice of the fête-day.

She had her wound; yes, it throbbed still now and then, and stung like a bee in the warm core of a rose. But she was young, she was gay, she was a little philosopher, above all she was French, and in the real French blood happiness runs so richly that it will hardly be utterly chilled until the veins freeze in the coldness of death. She enjoyed—enjoyed all the more fiercely, perhaps, because a certain desperate bitterness mingled with the abandonment of her Queen Mab-like revelries. Until now Cigarette had been as absolutely heedless and without a care as any young bird taking its first summer circles downward through the intoxication of the sunny air. It was not without fiery resistance and scornful revolt that the madcap Figlia del Reggimento would be prevailed on to admit that any shadow could have power to rest on her.

She played through more than half the night, the agile bounding graceful play of the young leopard to which he had likened her, and with a quick punishment from her velvet-sheathed talons if any durst offend her. Then when the dawn was nigh, leopard-like, the Little One sought her den.

She was most commonly under canvas; but when she was in the town it was at one with the proud independence of her nature that she rejected all offers made her, and would have her own nook to live in, even though she were not there one hour out of the twenty-four.

'Le Château de Cigarette' was a standing jest of the Army; for none was ever allowed to follow her thither, or to behold the interior of her fortress, and one over-venturous Spahi scaling the ramparts had been rewarded with so hot a deluge of lentiles-soup from a boiling casserole poured on his head from above, that he had beaten a hasty and ignominious retreat, which was more than a whole tribe of the most warlike of his countrymen could ever have made him do.

'Le Château de Cigarette' was neither more nor less than a couple of garrets, high in the air, in an old Moorish house, in an old Moorish court, decayed, silent, poverty-struck, with the wild pumpkin thrusting its leaves through the broken fretwork, and the green lizard shooting over the broad pavements, once brilliant in mosaic, that the robes of the princes of Islam had swept, now carpeted deep with the dry white drifted dust, and only crossed by the tottering feet of aged Jews or the laden steps of Algerine women.

Up a long winding rickety stair Cigarette approached her castle, which was very near the sky indeed. 'I like the blue,' said the châtelaine, laconically, 'and the pigeons fly close by my window.' And through it, too, she might have added, for though no human thing might invade her château, the pigeons circling in the sunrise light always knew well there were rice and crumbs spread for them in that eyelet-hole of a casement.

Cigarette threaded her agile way up the dark ladder-like shaft, and opened her door. There was a dim oil-wick burning; the garret was large, and as clean as a palace could be; its occupants were various, and all sound asleep except one, who, rough, and hard, and small, and three-legged, limped up to her and rubbed a little bullet head against her lovingly.

'Bouffarick—p'tit Bouffarick!' returned Cigarette, caressingly, in a whisper; and Bouffarick, content, limped back to a nest of hay, being a little wiry dog that had lost a leg in one of the most famous battles of Oran, and lain in its dead master's breast through three days and nights on the field. Cigarette, shading the lamp with one hand, glanced round on her family.

They had all histories—histories in the French Army, which was the only history she considered of any import to the universe. There was a raven perched high, by name Vole-qui-Veut: he was a noted character amongst the Zouaves, and had made many a campaign riding on his owner's bayonet; he loved a combat, and was specially famed for scream-ing '*Tue! tue! tue!*' all over a battlefield; he was very grey now, and the Zouave's bones had long bleached on the edge of the desert.

There was a tame rat who was a *vieille moustache*, and who had lived many years in a Lignard's pocket, and munched waifs and strays of the military rations, until the enormous crime being discovered that it was taught to sit up and dress its whiskers to the heinous air of the *Marseillaise*, the Lignard got the *matraque*, and the rat was condemned to be killed, had not Cigarette dashed in to the rescue and carried the long-tailed revolutionist off in safety.

There was a big white cat curled in a ball, who had been the darling of a Tringlo, and had travelled all over North Africa on the top of his mule's back, seven seasons through; in the eighth the Tringlo was picked off by a flying shot, and an Indigène was about to skin the shrieking Boule Blanche for the soup-pot, when a bullet broke his wrist, making him drop the cat with a yell of pain, and the Friend of the Flag, catching it up, laughed in his face: 'A lead comfit instead of slaughter-soup, my friend!'

There was little Bouffarick and three other brother-dogs of equal celebrity, one, in especial, who had been brought from Châlons, in defiance of the regulations, inside the drum of his regiment, and had been wounded a dozen times, always seeking the hottest heat of the skirmish. And there was, besides these, sleeping serenely on a straw palliasse, a very old man with a snowy beard and a head fit for Gerome to give to an Abraham.

A very old man—one who had been a conscript in the bands of Young France, and marched from his Pyrennean village to the battle tramp of the Marseillaise, and charged with the Enfans de Paris across the plains of Gemappes; who had known the passage of the Alps, and lifted the long curls from the dead brow of Désaix at Marengo, and seen in the sultry noonday dust of a glorious summer the Guard march into Paris, while the people laughed and wept with joy, surging like the mighty sea around one pale frail form, so young by years, so absolute by genius.

A very old man; long broken with poverty, with pain, with bereavement, with extreme old age; and, by a long course of cruel accidents, alone, here in Africa, without one left of the friends of his youth, or of the children of his name, and deprived even of the charities due from his country to his services—alone save for the little Friend of the Flag, who, for four years, had kept him on the proceeds of her wine trade in this Moorish attic, tending him herself when in the town, taking heed that he should want for nothing when she was campaigning.

'I will have a care of him,' she had said, curtly, when she had found him in great misery and learned his history from others; and she had had the care accordingly, maintaining him at her own cost in the Moorish building, and paying a good Jewess of the quarter to tend him when she was not herself in Algiers.

The old man was almost dead, mentally, though in bodily strength still well able to know the physical comforts of food, and rest, and attendance; he was in his second childhood, in his ninetieth year, and was unconscious of the debt he owed her; even, with a curious caprice of decrepitude, he disliked her, and noticed nothing except the raven when it shrieked its '*Tue! tue! tue!*' But to Cigarette he was as sacred as a god; had he not fought beneath the glance, and gazed upon the face, of the First Consul?

She bent over him now, saw that he slept, busied herself noiselessly in brewing a little tin pot full of coffee and hot milk, set it over the lamp to keep it warm, and placed it beside him ready for his morning meal, with a roll of white wheat bread; then, with a glance round to see that her other dependents wanted for nothing, went to her own garret adjoining, and with the lattice fastened back, that the first rays of sunrise and the first white flash of her friends the pigeons' gleaming wings might awaken her, threw herself on her straw and slept with all the graceful careless rest of the childhood which though in one sense she had never known, yet in another had never forsaken her.

She hid as her lawless courage would not have stooped to hide a sin, had she chosen to commit one, this compassion which she, the young condottiera of Algeria, showed with so tender a charity to the soldier of Bonaparte. To him, moreover, her fiery imperious voice was gentle as the dove, her wayward dominant will was pliant as the reed, her contemptuous sceptic spirit was reverent as a child's before an altar. In her sight the survivor of the Army of Italy was sacred; sacred the eyes which, when full of light, had seen the sun glitter on the squadrons of the Hussars of Murat, the Dragoons of Kellerman, the Cuirassiers of Milhaud; sacred the hands which when nervous with youth, had borne the standard of the Republic victorious against the gathered Teuton host in the Thermopylæ of Champagne; sacred the ears which, when quick to hear, had heard the thunder of Arcola, of Lodi, of Rivoli, and, above even the tempest of war, the clear still voice of Napoléon; sacred the lips which, when their beard was dark in the fulness of manhood, had quivered, as with a woman's weeping, at the farewell, in the spring night, in the moonlit Cour des Adieux.

Cigarette had a religion of her own; and followed it more closely than most disciples follow other creeds.

CHAPTER X

'Miladi aux beaux Yeux bleus'

Early that morning, when the snowy cloud of pigeons were circling down to take their daily alms from Cigarette, where her bright brown face looked out from the lattice-hole, Cecil, with some of the roughriders of his regiment, was sent far into the interior to bring in a string of colts, bought of a friendly desert tribe, and destined to be shipped to France for the Imperial Haras. The mission took two days; early on the third they returned with the string of wild young horses, whom it had taken not a little exertion and address to conduct successfully through the country into Algiers.

He was usually kept in incessant activity, because those in command over him had quickly discovered the immeasurable value of a *bas-officier* who was certain to enforce and obtain implicit obedience, and certain to execute any command given him with perfect address and surety, yet who, at the same time, was adored by his men, and had acquired a most singularly advantageous influence over them. But of this he was always glad: throughout his twelve years' service under the Emperor's flag, he had only found those moments in which he was unemployed intolerable; he would willingly have been in saddle from dawn till midnight.

Châteauroy was himself present when the colts were taken into the stable-yard; and himself inquired, without the medium of any third person, the whole details of the sale and of the transit. It was impossible, with all his inclination, to find any fault either with the execution of the errand or with the brief respectful answers by which his Corporal replied to his rapid and imperious cross-questionings. There were a great number of men within hearing, many of them the most daring and rebellious *pratiques* of the regiment; and Cecil would have let the coarsest upbraidings scourge him, rather than put the temptation to mutiny in their way which one insubordinate or even not strictly deferential word from him would have given. Hence the inspection passed off peaceably; as the Marquis turned on his heel, however, he paused a moment.

'Victor!'

'*Mon Commandant?*'

'I have not forgotten your insolence with those ivory toys. But Madame la Princesse herself has deigned to solicit that it shall be passed over unpunished. She cannot, of course, yield to your impertinent request to remain also unpaid for them. I charged myself with the fulfilment of her wishes. You deserve the *matraque*, but since Miladi herself is lenient enough to pardon you, you are to take this instead. Hold your hand, sir!'

Cecil put out his hand; he expected to receive a heavy blow from his commander's sabre, that possibly might break the wrist. These little trifles were common in Africa.

Instead, a *rouleau* of Napoléons was laid on his open palm. Châteauroy knew the gold would sting more than the blow.

For the moment Cecil had but one impulse—to dash the pieces in the giver's face. In time to restrain the impulse, he caught sight of the wild eager hatred gleaming in the eyes of Rake, of Petit Picpon, of a score of others who loved him and cursed their Colonel, and would at one signal from him have sheathed their sword in the mighty frame of the Marquis, though they should have been fired down the next moment themselves for the murder. The warning of Cigarette came to his memory; his hand clasped on the gold; he gave the salute calmly as Châteauroy swung himself away.

The troopers looked at him with longing questioning eyes; they knew enough of him by now to know the bitterness such gold, so given, had for him. Any other, even a corporal, would have been challenged with a storm of raillery, a volley of congratulation, and would have had shouted or hissed after him opprobrious accusations of '*faisant suisse*' if he had not forthwith treated his comrades royally from such largesse. With Bel-à-faire-peur they held their peace; they kept the silence which they saw that he wished to keep, as, his hour of liberty being come, he went slowly out of the great court with the handful of Napoléons thrust in the folds of his sash.

Rather unconsciously than by premeditation his steps turned through the streets that led to his old familiar haunt, the As de Pique, and dropping down on a bench under the awning, he asked for a draught of water. It was brought him at once, the hostess, a quick brown little woman from Paris, whom the lovers of Eugène Sue called Rigolette, adding of her own accord a lump of ice and a slice or two of lemon, for which she vivaciously refused payment, though generosity was by no means her cardinal virtue.

'Bel-à-faire-peur' awakened general interest through Algiers; he brought so fiery and so daring a reputation with him from the wars and raids of the interior, yet he was so calm, so grave, so gentle, so listless; it was known that he had made himself the terror of Kabyle and Bedouin, yet here in the city he thanked the negro boy who took him a glass of lemonade at an estaminet, and sharply rebuked one of his men for knocking down an old colon with a burden of gourds and of melons; such a Roumi as this the good people of the Franco-African capital held as a perfect gift of the gods, and not understanding one whit, nevertheless fully appreciated.

He did not look at the newspapers she offered him; but sat gazing out from the tawny awning, like the sail of a Neapolitan felucca, down the chequered shadows and the many-coloured masses of the little crooked, rambling, semi-barbaric alley. He was thinking of the Napoléons in his sash and of the promise he had pledged to Cigarette. That he would keep it he was resolved. The few impressive vivid words of the young vivandière had painted before him like a picture the horrors of mutiny and its hopelessness; rather than that, through him, these should befall the men who had become his brethren-in-arms, he felt ready to let the Black Hawk do his worst on his own life. Yet a weariness, a bitterness, he had never known in the excitement of active service came on him, brought by this sting of insult from the fair hand of an *aristocrate*.

There was absolutely no hope possible in his future. The uttermost that could ever come to him would be a grade something higher in the army that now enrolled him; the gift of the cross, or a post in the bureau. Algerine warfare was not like the campaigns of the armies of Italy or the Rhine, and there was no Napoléon here to discern with unerring omniscience a leader's genius under the képi of a common trooper. Though he should show the qualities of a Massena or a Kléber, the chances were a million to one that he would never get even so much as a lieutenancy; and the raids on the decimated tribes, the obscure skirmishes of the interior, though terrible in slaughter and venturesome enough, were not the fields on which great military successes were won and great military honours acquired. The French fought for a barren strip of brown plateau that, gained, would be of little use or profit to them; he thought that he did much the same, that his future was much like those arid sand-plains, those thirsty verdureless stretches of burnt earth—very little worth the reaching.

The heavy folds of a Bedouin's haick brushing the papers off the bench, broke the thread of his musings. As he stooped for them, he saw that one was an English journal some weeks old. His own name caught his eye—the named buried so utterly, whose utterance in the Sheik's tent had struck him like a dagger's thrust. The flickering light and darkness, as the awning waved to and fro, made the lines move dizzily upwards and downwards as he read—read the short paragraph touching the fortunes of the race that had disowned him:

THE ROYALLIEU SUCCESSION.—We regret to learn that the Right Hon. Viscount Royallieu, who so lately succeeded to the family title on his father's death, has expired at Mentone, whither his health had induced him to go some months previous. The late Lord was unmarried. His next brother was, it will be remembered, many years ago, killed on a southern railway. The title, therefore, now falls to the third and only remaining son, the Hon. Berkeley Cecil, who, having lately inherited considerable properties from a distant relative, will, we believe, revive all the old glories of this Peerage, which have, from a variety of causes, lost somewhat of their ancient brilliancy.

Cecil sat quite still, as he had sat looking down on the record of his father's death, when Cigarette had rallied him with her gay challenge among the Moresco ruins. His face flushed hotly under the warm golden hue of the desert bronze, then lost all its colour as suddenly, till it was as pale as any of the ivory he carved. The letters of the paper reeled and wavered and grew misty before his eyes; he lost all sense of the noisy changing polyglot crowd thronging past him; he, a common soldier in the Algerian Cavalry, knew that, by every law of birthright, he was now a Peer of England.

His first thought was for the dead man. True, there had been little amity, little intimacy, between them; a negligent friendliness whenever they had met had been all that they had ever reached. But in their childhood they had been carelessly kind to one another, and the memory of the boy who had once played beside him down the old galleries and under the old forests, of the man who had now died yonder where the southern seaboard lay across the warm blue Mediterranean, was alone on him for the moment. His thoughts had gone back, with a pang, almost ere he had read the opening lines, to autumn mornings in his youngest years when the leaves had been flushed with their earliest red, and the brown still pools had been alive with water-birds, and the dogs had dropped down charging among the flags and rushes, and his brother's boyish face had laughed on him from the wilderness of willows, and his brother's boyish hands had taught him to handle his first cartridge, and to fire his first shot. The many years of indifference and estrangement were forgotten, the few years of childhood's confidence and comradeship alone remembered, as he saw the words that brought him in his exile the story of his brethren's fate and of his race's fortunes. His head sank, his face was still colourless, he sat motionless with the printed sheet in his hand. Once his eyes flashed, his breath came fast and uneven; he rose with a sudden impulse, with a proud bold instinct of birth and freedom. Let him stand here in what grade he would, with the badge of a Corporal of the Army of Africa on his arm, this inheritance that had come to him was his; he bore the name and the title of his house as surely as any had ever borne it since the first of the Norman owners of Royallieu had followed the Bastard's banner.

The vagabond throngs, Moorish, Frank, negro, colon, paused as they pushed their way over the uneven road, and stared at him vacantly where he stood. There was something in his attitude, in his look, which swept over them seeing none of them, in the eager lifting of his head, in the excited fire in his eyes, that arrested all, from the dullest muleteer plodding on with his string of patient beasts, to the most volatile French girl laughing on her way with a group of *fantassins*. He did not note them, hear them, think of them; the whole of the Algerine scene had faded out as if it had no place before him; he had forgot that he was a cavalry soldier of the Empire; he saw nothing but the green wealth of the old home woods far away in England; he remembered nothing save that he, and he alone, was the rightful Lord of Royallieu.

'*Tiens, es-tu fou, mon brave? Bois de m'avoine,*[1] *Bel-à-faire-peur!*'

The coarse good-humoured challenge, as the hand of a broad-chested black-visaged veteran of Chasseurs fell on his shoulder, and the wooden rim of a little wine-cup was thrust towards him with the proffered drink, startled him and recalled him to the consciousness of where he was. He stared one moment absently in the trooper's amazed face, then shook him off with a suddenness that tossed back the cup to the ground, and, holding the journal clenched close in his grasp, went swiftly through the masses of the people out and away, he little noted where, till he had forced his road beyond the gates, beyond the town, beyond all reach of its dust and its babble and its discord, and was alone in the further outskirts, where to the north the calm sunlit bay slept peacefully with a few scattered ships riding at anchor, and southward the luxuriance of the Sahel stretched to meet the wide and cheerless plateaux, dotted with the conical houses of hair, and desolate as though the locust-swarm had just alighted there to lay them waste.

Reaching the heights he stood still involuntarily, and looked down once more on the words that told him of his birthright; in the blinding intense light of the African day they seemed to stand out as though carved in stone, and as he read them once more a great darkness passed over his face; this heritage was his, and he could never take it up; this thing had come to him, and he must never claim it. He was Viscount Royallieu as surely as any of his fathers had been so before him, and he was dead for ever in the world's belief; he must live, and grow old, and perish by shot or steel, by sickness or by age, with his name and his rights buried, and his years passed as a private soldier of France.

The momentary glow which had come to him with the sudden resurrection of hope and of pride faded utterly as he slowly read and re-read the lines of the journal on the broken terraces of the hillside, where the great fig-trees spread their fantastic shadows, and through a rocky channel a russet stream of shallow waters threaded its downward path under the reeds, and no living thing was near him save some quiet browsing herds far off, and their Arab shepherd-lad that an artist might have sketched as Ishmael. What his future might have been rose before his thoughts; what it must be rose also, bitterly, blackly, drearily in contrast. A noble without even a name; a chief of his race without even the power

---

[1] Brandy.

to claim kinship with that race; owner by law of three thousand broad English acres, yet an exile without freedom to set foot on his native land; by heritage one amongst the aristocracy of England, by circumstance, now and for ever, till an Arab bullet should cut in twain his thread of life, a soldier of the African legions, bound to obey the commonest and coarsest boor that had risen to a rank above him: this was what he knew himself to be, and knew that he must continue to be without one appeal against it, without once stretching out his hand towards his right of birth and station.

There was a passionate revolt, a bitter heart-sickness on him; all the old freedom and peace and luxury and pleasure of the life he had left so long allured him with a terrible temptation; the honours of the rank that he should now have filled were not what he remembered; what he longed for with an agonised desire was to stand once more stainless among his equals, to reach once more the liberty of unchallenged unfettered life, to return once more to those who held him but as a dishonoured memory, as one whom violent death had well snatched from the shame of a criminal career.

'But who would believe me now?' he thought. 'Besides, this makes no difference. If three words spoken would reinstate me, I could not speak them at that cost. The beginning perhaps was folly, but for sheer justice sake there is no drawing back now. Let him enjoy it; God knows *I* do not grudge him it.'

Yet though it was true to the very core that no envy and no evil lay in his heart against the younger brother to whose lot had fallen all good gifts of men and fate, there was almost unbearable anguish on him in this hour in which he learned the inheritance that had come to him, and remembered that he could never take again even so much of it as lay in the name of his fathers. When he had given his memory up to slander and oblivion and the shadow of a great shame, when he had let his life die out from the world that had known him, and buried it beneath the rough, weather-stained, blood-soaked cloth of a private soldier's uniform, he had not counted the cost then nor foreseen the cost hereafter. It had fallen on him very heavily now.

Where he stood under some sheltered columns of a long-ruined mosque whose shafts were bound together by a thousand withes and wreaths of the rich fantastic Sahel foliage, an exceeding weariness of longing was upon him—longing for all that he had forfeited, for all that was his own, yet never could be claimed as his.

The day was intensely still; there was not a sound except when here and there the movement of a lizard under the dry grasses gave a low crackling rustle. He wondered almost which was the dream and which the truth; that old life that he had once led, and that looked now so far away and so unreal, or this which had been about him for so many years in the camps and the bivouacs, the barracks and the battlefields. He wondered almost which he himself was—an English Peer on whom the title of his line had fallen, or a Corporal of Chasseurs who must take his chief's insults as patiently as a cur takes the blows of its master; that he was *both*, seemed to him, as he stood there with the glisten of the sea before and the swelling slopes of the hillside above, a vague distorted nightmare.

Hours might have passed, or only moments, he could not have told; his eyes looked blankly out at the sun-glow, his hand instinctively clenched on the journal whose stray lines had told him in an Algerine *trattoria* that he had inherited what he never could enjoy.

'Are *they* content, I wonder?' he thought, gazing down that fiery blaze of shadowless light; 'do they ever remember?'

He thought of those for whose sakes he had become what he was.

The distant mellow ringing notes of a trumpet-call floated to his ear from the town at his feet; it was sounding the '*rentrée en caserne*'. Old instinct, long habit, made him start and shake his harness together and listen. The trumpet-blast winding cheerily from afar off recalled him to the truth, summoned him sharply back from vain regrets to the facts of daily life. It woke him as it wakes a sleeping charger; it roused him as it rouses a wounded trooper.

He stood hearkening to the familiar music till it had died away, spirited, yet still lingering; full of fire, yet fading softly down the wind. He listened till the last echo ceased; then he tore the paper that he held in strips, and let it float away, drifting down the yellow current of the reedy river-channel; and he half drew from its scabbard the sabre whose blade had been notched and dinted and stained in many midnight skirmishes and many headlong charges under the desert suns, and looked at it as though a friend's eye gazed at him in the gleam of the trusty steel. And his soldier-like philosophy, his campaigner's carelessness, his habitual easy negligence that had sometimes been weak as water and sometimes heroic as martyrdom, came back to him with a deeper shadow on it, that was grave with a calm, resolute, silent courage.

'So best after all, perhaps,' he said, half aloud, in the solitude of the ruined and abandoned mosque. 'He cannot well come to shipwreck with such a fair wind and such a smooth sea. And I—I am just as well here. To ride with the Chasseurs is more exciting than to ride with the Pytchley; and the rules of the Chambrée are scarce more tedious than the rules of a Court. Nature turned me out for a soldier, though Fashion spoiled me for one. I can make a good campaigner—I should never make anything else.'

And he let his sword drop back again into the scabbard, and quarrelled no more with fate.

His hand touched the thirty gold pieces in his sash.

He started, as the recollection of the forgotten insult came back on him. He stood awhile in thought; then he took his resolve.

A half-hour of quick movement, for he had become used to the heat as an Arab, and heeded it as little, brought him before the entrance-gates of the Villa Aïoussa. A native of Soudan, in a rich dress, who had the office of porter, asked him politely his errand. Every *indigène* learns by hard experience to be courteous to a French soldier. Cecil simply asked, in answer, if Madame la Princesse were visible. The negro returned, cautiously, that she was at home, but doubted her being accessible. 'You come from M. le Marquis?' he enquired.

'No; on my own errand.'

'You!' Not all the native African awe of a *Roumi* could restrain the contemptuous amaze in the word.

'I. Ask if Corporal Victor, of the Chasseurs, can be permitted a moment's interview with your mistress. I come by permission,' he added, as the native hesitated between his fear of a Roumi and his sense of the appalling unfittingness of a private soldier seeking an audience of a Spanish Princess. The message was passed about between several of the household; at last a servant of higher authority appeared:

'Madame permitted Corporal Victor to be taken to her presence. Would he follow?'

He uncovered his head and entered, passing through several passages and chambers, richly hung and furnished; for the villa had been the 'campagne' of an illustrious French personage, who had offered it to the Princess Corona when, for some slight delicacy of health, the air of Algeria was advocated. A singular sensation came on him, half of familiarity, half of strangeness, as he advanced along them; for twelve years he had seen nothing but the bare walls of barrack-rooms, the goat-skin of douars, and the canvas of his own camp-tent. To come once more, after so long an interval, amidst the old things of luxury and grace that had been so long unseen, wrought curiously on him. He could not fairly disentangle past and present. For the moment, as his feet fell once more on soft carpets, and his eyes glanced over gold and silver, malachite and bronze, white silk and violet damasks, he almost thought the Algerian years were a disordered dream of the night.

His spur caught in the yielding carpet, and his sabre clashed slightly against it; as the *rentrée au caserne* had done an hour before, the sound recalled the actual present to him. He was but a French soldier, who went on sufferance into the presence of a great lady. All the rest was dead and buried.

Some half-dozen apartments, large and small, were crossed; then into that presence he was ushered. The room was deeply shaded, and fragrant with the odours of the innumerable flowers of the Sahel soil; there was that about it which struck on him as some air, long unheard but once intimately familiar, on the ear will revive innumerable memories; like the '*vieil air languissant et funèbre*', for which Gerard de Nerval was willing to give 'all Rossini and Weber'. She was at some distance from him, with the trailing draperies of eastern fabrics falling about her in a rich, unbroken, shadowy cloud of melting colour, through which, here and there, broke threads of gold; involuntarily he paused on the threshold looking at her. Some faint, far-off remembrance stirred in him, but deep down in the closed grave of his past; some vague intangible association of forgotten days, forgotten thoughts, drifted before him as it had drifted before him when first in the Chambrée of his barracks he had beheld the Venetia Corona.

She moved forward as her servant announced him; she saw him pause there like one spell-bound, and thought it the hesitation of one who felt sensitively his own low grade in life. She came towards him with the silent sweeping grace that gave her the carriage of an empress; her voice fell on his ear with the accent of a woman immeasurably proud, but too proud not to bend softly and graciously to those who were so far beneath her that without such aid from her they could never have addressed or have approached her.

'You have come, I trust, to withdraw your prohibition? Nothing will give me greater pleasure than to bring his Majesty's notice to one of the best soldiers his Army holds.'

There was that in the words, gently as they were spoken, that recalled him suddenly to himself; they had that negligent, courteous pity she would have shown to some colon begging at her gates! He forgot—forgot utterly—that he was only an African trooper. He only remembered that he had once been a gentleman, that—if a life of honour and of self-negation can make any so—he was one still. He advanced and bowed with the old serene elegance that his bow had once been famed for; and she, well used to be even over-critical in such trifles, thought, 'That man has once lived in courts!'

'Pardon me, Madame, I do not come to trespass so far upon your benignity,' he answered, as he bent before her. 'I come to express, rather, my regret that you should have made one signal error.'

'Error!'—a haughty surprise glanced from her eyes as they swept over him. Such a word had never been used to her in the whole course of her brilliant and pampered life of sovereignty and indulgence.

'One common enough, Madame, in your Order. The error to suppose that under the rough cloth of a private trooper's uniform there cannot possibly be such aristocratic monopolies as nerves to wound.'

'I do not comprehend you.' She spoke very coldly; she repented her profoundly of her concession in admitting a Chasseur d'Afrique to her presence.

'Possibly not. Mine was the folly to dream that you would ever do so. I should not have intruded on you now, but for this reason: the humiliation you were pleased to pass on me I could neither refuse nor resent to the dealer of it. Had I done so, men who are only too loyal to me would have resented with me, and been thrashed or been shot, as payment. I was compelled to accept it, and to wait until I could return your gift to you. I have no right to complain that you pained me with it, since one who occupies my position ought, I presume, to consider remembrance, even by an outrage, an honour done to him by the Princess Corona.'

As he said the last words he laid on a table that stood near him the gold of Châteauroy's insult. She had listened with a bewildered wonder, held in check by the haughtier impulse of offence, that a man in this grade could venture thus to address, thus to arraign her. His words were totally incomprehensible to her, though, by the grave rebuke of his manner, she saw that they were fully meant, and, as he considered, fully authorized by some wrong done to him. As he laid the gold pieces down upon her table, an idea of the truth came to her.

'I know nothing of what you complain of; I sent you no money. What is it you would imply?' she asked him, looking up from where she leaned back in the low couch into whose depths she had sunk as he had spoken.

'You did not send me these? Not as payment for the chess service?'

'Assuredly not. After what you said the other day, I should have scarcely been so ill-bred and so heedless of inflicting pain. Who used my name thus?'

His face lightened with a pleasure and a relief that changed it wonderfully; that brighter look of gladness had been a stranger to it for so many years.

'You give me infinite happiness, Madame. You little dream how bitter such slights are where one has lost the power to resent them! It was M. de Châteauroy, who this morning——'

'Dared to tell you I sent you those coins?'

The serenity of a courtly woman of the world was unbroken, but her blue and brilliant eyes darkened and gleamed beneath the sweep of their lashes.

'Perhaps I can scarcely say so much. He gave them, and he implied that he gave them from you. The words he spoke were these.'

He told her them as they had been uttered, adding no more; she saw the construction they had been intended to bear, and that which they had borne naturally to his ear; she listened earnestly to the end. Then she turned to him with the exquisite softness of grace which, when she was moved to it, contrasted so vividly with the haughty and almost chill languor of her habitual manner.

'Believe me, I regret deeply that you should have been wounded by this most coarse indignity; I grieve sincerely that through myself in any way it should have been brought upon you. As for the perpetrator of it, M. de Châteauroy will be received here no more; and it shall be my care that he learns not only how I resent his unpardonable use of my name, but how I esteem his cruel outrage to a defender of his own Flag. You did exceedingly well and wisely to acquaint me; in your treatment of it as an affront that I was without warrant to offer you, you showed the just indignation of a soldier, and—of what I am very sure that you are—a gentleman.'

He bowed low before her.

'Madame, you have made me the debtor of my enemy's outrage. Those words from you are more than sufficient compensation for it.'

'A poor one, I fear! Your Colonel is your enemy, then? And wherefore?'

He paused a moment:

'Why at first I scarcely know. We are antagonistic, I suppose.'

'But is it usual for officers of his high grade to show such malice to their soldiers?'

'Most unusual. In this service especially so; although officers rising from the ranks themselves are more apt to contract prejudices and ill-feeling against, as they are to feel favouritism to, their men, than where they enter the regiment in a superior grade at once. At least, that is the opinion I myself have formed, studying the working of the different systems.'

'You know the English service, then?'

'I know something of it.'

'And still, though thinking this, you prefer the French?'

'I distinctly prefer it, as one that knows how to make fine soldiers, and how to reward them; as one in which a brave man will be valued, and a worn-out veteran will not be left to die like a horse at a knacker's.'

'A brave man valued, and yet you are a corporal!' thought Miladi, as he pursued:

'Since I am here, Madame, let me thank you, in the Army's name, for your infinite goodness in acting so munificently on my slight hint. Your generosity has made many happy hearts in the hospital.'

'Generosity! Oh, do not call it by any such name! What did it cost me? We are terribly selfish here. I am indebted to you that for once you made me remember those who suffered.'

She spoke with a certain impulse of candour and of self-accusation that broke with great sweetness the somewhat careless coldness of her general manner; it was like a gleam of light that showed all the depth and the warmth that in truth lay beneath that imperial languor of habit. It broke further the ice of distance that severed the *grande dame* from the cavalry soldier.

Insensibly to himself, the knowledge that he had, in fact, the right to stand before her as an equal gave him the bearing of one who exercised that right, and her rapid perception had felt before now that this *Roumi* of Africa was as true a gentleman as any that had ever thronged about her in palaces. Her own life had been an uninterrupted course of luxury, prosperity, serenity, and power: the adversity which she could not but perceive had weighed on his had a strange interest to her. She had heard of many calamities, and aided many; but they had always been far sundered from her, they had never touched her: in this man's presence they seemed to grow very close, terribly real. She led him on to speak of his comrades, of his daily life, of his harassing routine of duties in peace, and of his various experiences in war. He told her, too, of Léon Ramon's history; and as she listened, he saw a mist arise and dim the brilliancy of those eyes that men complained would never soften. The very fidelity with which he sketched to her the bitter sufferings and the rough nobility that were momentarily borne and seen in that great military family of which he had become a son by adoption, interested her by its very unlikeness to anything in her own world.

His voice had still its old sweetness, his manner still its old grace; and added to these were a grave earnestness and a natural eloquence that the darkness of his own fortunes and the sympathies with others that pain had awakened had brought to him. He wholly forgot their respective stations; he only remembered that for the first time for so many years he had the charm of converse with a woman of high breeding, of inexpressible beauty, and of keen and delicate intuition. He wholly forgot how time passed, and she did not seek to remind him; indeed, she but little noted it herself.

At last the conversation turned back to his Chief.

'You seem to be aware of some motive for your commandant's dislike?' she asked him. 'Tell me to what you attribute it?'

'It is a long tale, Madame.'

'No matter, I would hear it.'

'I fear it will only weary you?'

'Do not fear that. Tell it me?'

He obeyed, and told to her the story of the Emir and of the Pearl of the Desert; and Venetia Corona listened, as she had listened to him throughout, with an interest that she rarely vouchsafed to the recitals and the witticisms of her own circle. He gave to the narrative a soldierly simplicity, and a picturesque colouring that lent a new interest to her; and she was of that nature which, however it may be led to conceal feeling from pride and from hatred, never fails to awaken to indignant sympathy at wrong.

'This barbarian is your chief?' she said, as the tale closed. 'His enmity is your honour! I can well credit that he will never pardon your having stood between him and his crime.'

'He has never pardoned it yet, of a surety.'

'I will not tell you it was a noble action,' she said, with a smile sweet as the morning, a smile that few saw light on them. 'It came too naturally to a man of honour for you to care for the epithet. Yet it was a great one, a most generous one. But I have not heard one thing: what argument did you use to obtain her release?'

'No one has ever heard it,' he answered her, while his voice sank low. 'I will trust you with it; it will not pass elsewhere. I told him enough of— of my own past life to show him that I knew what his had been, and that I knew moreover, though they were dead to me now, men in that greater world of Europe who would believe my statement if I wrote them his outrage on the Emir, and would avenge it for the reputation of the Empire. And unless he released the Emir's wife, I swore to him that I would so write, though he had me shot on the morrow; and he knew I should keep my word.'

She was silent some moments, looking on him with a musing gaze, in which some pity and more honour for him were blended.

'You told him your past. Will you confess it to me?'

'I cannot, Madame.'

'And why?'

'Because I am dead! Because, in your presence, it becomes more bitter to me to remember that I ever lived.'

'You speak strangely. Cannot your life have a resurrection?'

'Never, Madame. For a brief hour you have given it one—in dreams. It will have no other.'

'But surely there may be ways—such a story as you have told me brought to the Emperor's knowledge, you would see your enemy disgraced, yourself honoured?'

'Possibly, Madame. But it is out of the question that it should ever be so brought. As I am now, so I desire to live and die.'

'You voluntarily condemn yourself to this?'

'I have voluntarily chosen it. I am well sure that the silence I entreat will be kept by you?'

'Assuredly; unless by your wish it be broken. Yet, I await my brother's arrival here; he is a soldier himself; I shall hope that he will persuade you to think differently of your future. At any rate, both his and my own influence will always be exerted for you, if you will avail yourself of it.'

'You do me much honour, Madame. All I will ever ask of you is to return those coins to my Colonel, and to forget that your gentleness has made *me* forget, for one merciful half-hour, the sufferance on which alone a private trooper can present himself here.'

He swept the ground with his kepi as though it were the plumed hat of a Marshal, and backed slowly from her presence, as he had many a time long before backed out of a throne-room.

As he went, his eyes caught the armies of the ivory chessmen; they stood under glass, and had not been broken by her lapdog.

Miladi, left alone there in her luxurious morning-room, sat awhile lost in thought. He attracted her; he interested her; he aroused her sympathy and her wonder as the men of her own world had failed to do—aroused them despite the pride which made her impatient of lending so much attention to a mere Chasseur d'Afrique. His knowledge of the fact that he was in reality the representative of his race, although the power to declare himself so had been for ever abandoned and lost, had given him in her presence that day a certain melancholy, and a certain grave dignity, that would have shown a far more superficial observer than she was that he had come of a great race, and had memories that were of a very different hue to the coarse and hard life which he led now. She had seen much of the world, and was naturally far more penetrative and more correct in judgement than are most women. She discovered the ring of true gold in his words, and the carriage of pure breeding in his actions. He interested her; more than it pleased her that he should. A man so utterly beneath her!—doubtless brought into the grade to which he had fallen by every kind of error, of improvidence, of folly, of probably worse than folly!

It was too absurd that she, so difficult to interest, so inaccessible, so fastidious, so satiated with all that was brilliant and celebrated, should find herself seriously spending her thoughts, her pity, and her speculation on an adventurer of the African Army! She laughed a little at herself as she stretched out her hand for a new volume of French poems dedicated to her by their accomplished writer, who was a Parisian diplomatist.

'One would imagine I was just out of a convent, and weaving a marvellous romance from a mystery and a *tristesse*, because the first soldier I notice in Algeria has a gentleman's voice and is ill-treated by his officers!' she thought, with a smile, while she opened the poems which had that day arrived, radiant in the creamy vellum, the white velvet, and the gold of a dedication copy, with the coronet of the Corona d'Amägué on their binding. The poems were sparkling with all the grace of airy *vers de société* and elegant silvery harmonies; but they served ill to chain her attention, for whilst she read her eyes wandered at intervals to the chess battalions.

'Such a man as that buried in the ranks of this brutalized army!' she mused. 'What fatal chance could bring him here? Misfortune, not misconduct, surely. I wonder if Lyon could learn! He shall try.'

'Your Chasseur has the air of a Prince, my love,' said a voice behind her.

'Equivocal compliment! A much better air than most Princes,' said Madame Corona, glancing up with a slight shrug of her shoulders, as her guest and travelling companion, the Marquise de Rénardière, entered.

'Indeed! I saw him as he passed out; and he saluted me as if he had been a Marshal. Why did he come?'

Venetia Corona pointed to the Napoléons, and told the story; rather listlessly and briefly.

'Ah! The man has been a gentleman, I dare say. So many of them come to our army. I remember General Villefleur's telling me—he commanded here awhile—that the ranks of the Zéphyrs and Zouaves were full of well-born men, utterly good-for-nothing, the handsomest scoundrels possible, who had every gift and every grace, and yet come to no better end than a pistol-shot in a ditch or a mortal thrust from Bedouin steel. I dare say your Corporal is one of them.'

'It may be so.'

'But you doubt it, I imagine.'

'I am not sure that I do. But this person is certainly unlike a man to whom disgrace has ever attached.'

'You think your protégé, then, has become what he is through adversity, I suppose? Very interesting!'

'I really can tell you nothing of his antecedents. Through his skill at sculpture, and my notice of it, considerable indignity has been brought upon him; and a soldier can feel, it seems, though it is very absurd that he should! That is all my concern with the matter, except that I have to teach his commander not to play with my name in his barrack-yard.'

She spoke with that negligence which always sounded very cold, though the words were so gently spoken. Her best and most familiar friends always knew when, with that courtly chillness, she had signed them their line of demarcation.

And the Marquise de Rénardière said no more, but talked of the Ambassador's poems.

CHAPTER XI

'Le Bon-Zig'

Meanwhile, the subject of their first discourse returned to the Chambrée.

He had encouraged the men to pursue those various industries and ingenuities which, though they are affectedly considered against 'discipline', formed, as he knew well, the best preservative from real insubordination, and the best instrument in humanizing and ameliorating the condition of his comrades. The habit of application alone was something gained; and if it kept them only for a while from the haunts of those coarsest debaucheries, which are the only possible form in which the soldier can pursue the forbidden licence of vice, it was better than that leisure should be spent in that joyless bestiality which made Cecil, once used to every refinement of luxury and indulgence, sicken with a pitying wonder for those who found in it the only shape they knew of 'pleasure'.

He had seen from the first, in many men of his *tribu*, capabilities that might be turned to endless uses; in the conscript drawn from the populace of the provinces there was almost always a knowledge of self-help, and often of some trade, coupled with habits of diligence; in the soldier made from the street-Arab of Paris there was always inconceivable intelligence, rapidity of wit, and plastic vivacity; in the adventurers come, like himself, from higher grades of society, and burying a broken career under the shelter of the tricolour, there were continually gifts and acquirements, and even genius, that had run to seed and brought forth no fruit. Of all these France always avails herself in a great degree; but, as far as Cecil's influence extended, they were developed much more than usual. As his own character gradually changed under the force of fate, the desire for some interest in life grew on him (every man, save one absolutely brainless and self-engrossed, feels this sooner or later); and that interest he found, or rather created, in his regiment. All that he could do to contribute to its efficiency in the field he did; all that he could do to further its internal excellence he did likewise.

Coarseness perceptibly abated, and violence became much rarer in that portion of his corps with which he had immediately to do; the men gradually acquired from him a better, a higher tone; they learned to do duties inglorious and distasteful as well as they did those which led them to the danger and the excitation that they loved; and, having their good faith and sympathy, heart and soul, with him, he met, in these lawless leopards of African France, with loyalty, courage, generosity, and self-abnegation far surpassing those which he had ever met with in the polished civilization of his early experience.

For their sakes, he spent many of his free hours in the Chambrée. Many a man, seeing him, there came and worked at some ingenious design, instead of going off to burn his brains out with brandy, if he had sous enough to buy any, or to do some dexterous bit of thieving on a native, if he had not. Many a time, knowing him to be there, sufficed to restrain the talk around from lewdness and from ribaldry, and turn it into channels at once less loathsome and more mirthful, because they felt that obscenity and vulgarity were alike jarring on his ear, although he had never more than tacitly shown that they were so. A precisian would have been covered with their contumely and ridicule; a saint would have been driven out from their midst with every missile merciless tongues and merciless hands could pelt with; a martinet would have been cursed aloud, and cheated, flouted, rebelled against, on every possible occasion. But the man who was 'one of them' entirely, whilst yet simply and thoroughly a gentleman, had great influence—an influence exclusively for good.

The Chambrée was empty when he returned; the men were scattered over the town in one of their scant pauses of liberty; there was only the dog of the regiment, Flick-Flack, a snow-white poodle, asleep in the heat, on a sack, who, without waking, moved his tail in a sign of gratification as Cecil stroked him, and sat down near, betaking himself to the work he had in hand.

It was a stone for the grave of Léon Ramon. There was no other to remember the dead Chasseur; no other beside himself, save an old woman sitting spinning at her wheel under the low-sloping shingle roof of a cottage by the western Biscayan sea, who, as she spun, and as the thread flew, looked with anxious aged eyes over the purple waves where she had seen his father—the son of her youth—go down beneath the waters, and murmured ever and again, '*Il r'viendra! il r'viendra!*'

But the thread of her flax would be spun out, and the thread of her waning life be broken, ere ever the soldier for whom she watched would go back to her and to Languedoc.

For life is brutal; and to none so brutal as to the aged who remember so well, yet are forgotten as though already they were amidst the dead.

Cecil's hand pressed the graver along the letters, but his thoughts wandered far from the place where he was. Alone there, in the great sun-scorched barrack-room, the news that he had read, the presence he had quitted, seemed alike a dream.

He had never known fully all that he had lost until he had stood before the beauty of this woman, in whose deep imperial eyes the light of other years seemed to lie, the memories of other worlds seemed to slumber.

Those blue, proud, fathomless eyes! Why had they looked on him? He had grown content with his fate; he had been satisfied to live and to fall a soldier of France; he had set a seal on that far-off life of his earlier time, and had grown to forget that it had ever been. Why had chance flung him in her way, that, with one careless haughty glance, one smile of courteous pity, she should have undone in a moment all the work of a half-score years, and shattered in a day the serenity which it had cost him such weary self-contest, such hard-fought victory, to attain?

She had come to pain, to weaken, to disturb, to influence him, to shadow his peace, to wring his pride, to unman his resolve, as women do mostly with men. Was life not hard enough here already, that she must make it more bitter yet to bear?

He had been content, with a soldier's contentment, in danger and in duty; and she must waken the old coiled serpent of restless stinging regret which he had thought lulled to rest for ever!

'If I had my heritage?' he thought; and the chisel fell from his hands as he looked down the length of the barrack-room with the blue glare of the African sky through the casement.

Then he smiled at his own folly, in dreaming idly thus of things that might have been.

'I will see her no more,' he said to himself. 'If I do not take care, I shall end by thinking myself a martyr—the last refuge and consolation of emasculate vanity, of impotent egotism!'

For though his whole existence was a sacrifice, it never occurred to him that there was anything whatever great in its acceptation, or unjust in its endurance. He thought too little of his life's value, or of its deserts, ever to consider by any chance that it had been harshly dealt with, or unmeritedly visited.

At that instant Petit Picpon's keen, pale, Parisian face peered through the door, his great black eyes, that at times had so pathetic a melancholy, and at others such a monkeyish mirth and malice, were sparkling excitedly and gleefully.

'Mon Caporal!'

'You, Picpon? What is it?'

'*Mon Caporal*, there is great news. *La danse commence là-bas.*'[1]

'Ah! Are you sure?'

'Sure, *mon Caporal*. The *Arbicos* want a *fantasia à la clarinette*.[2] We are not to know just yet: we are to have the *ordre de route* tomorrow. I overheard our officers say so. They think we shall have brisk work. And for that they will not punish the *vieille lame*.'

'Punish! Is there fresh disobedience? In my squadron; in my absence?'

He rose instinctively, buckling on his sword which he had put aside.

'Not in your *tribu*, mon Caporal,' said Picpon, quickly. 'It is not much, either. Only the *bon-zig* Rac.'

'Rake! What has he been doing?'

There was infinite anxiety and vexation in his voice. Rake had recently been changed into another squadron of the regiment, to his great loss and regret; for not only did he miss the man's bright face and familiar voice from the Chambrée, but he had much disquietude on the score of his safety, for Rake was an incorrigible *pratique*, had only been kept from scrapes and mischief by Cecil's influence, and even despite that had been often in hot water, and once even had been drafted for a year or so of chastisement amongst the 'Zéphyrs', a mode of punishment which, but for its separation of him from his idol, would have given unmitigated delight to the audacious offender.

'Very little, *mon Caporal!*' said Picpon, eagerly. 'A mere nothing—a bagatelle! Run a Spahi through the stomach, that is all. I don't think the man is so much as dead, even!'

---

[1] There is fighting broken out yonder. [2] A skirmish to the music of musketry.

'I hope not, indeed. When will you cease this brawling amongst your-selves? A soldier's blade should never be turned upon men of his own army. How did it happen?'

'*Pour si peu de chose, mon Caporal.* A woman! They quarrelled about a little fruit-seller. The *homard*[1] was in fault. *Crache-au-nez-d'la-Mort* was there before him; and was preferred by the girl; and women should be allowed something to do with choosing their lovers, that I think, though it is true they often take the worst man. They quarrelled; the *homard* drew first; and then, *pouf et passe!* quick as thought, Rac lunged through him. He has always a most beautiful stroke. Le Capitaine Argentier was passing, and made a fuss; else nothing would have been done. They have put him under arrest, but I heard them say they would let him free tonight because we should march at dawn.'

'I will go and see him at once.'

'Wait, *mon Caporal*; I have something to tell you,' said Picpon, quickly. 'The *zig* has a motive in what he does. Rac wanted to get the *trou*.[2] He has done more than one bit of mischief only for that.'

'Only for what? He cannot be in love with the *trou*?'

'It serves his turn,' said Picpon, mysteriously. 'Did you never guess why, *mon Caporal*? Well, I have. *Crache-au-nez-d'la-Mort* is a *risquetout*.[3] The officers know it; the bureaux know it. He would have mounted, mounted, mounted, and been a Captain long before now, if he had not been a *pratique*.'

'I know that; so would many of you.'

'Ah, *mon Caporal*; but that is just what Rac does not choose. In the books his page beats every man's, except yours. They have talked of him many times for the cross and for promotion; but whenever they do—*cric-crac!* he goes off to a bit of mischief, and gets himself punished. Any *rabiat*,[4] long or short, serves his purpose. They think him too wild to take out of the ranks. You remember, *mon Caporal*, that splendid thing that he did five years ago at Sabasasta? Well, you know they spoke of promoting him for it, and he would have run up all the grades like a squirrel, and died a *Kébir*,[5] I dare say. What did he do to prevent it? Why, went that escapade into Oran disguised as a Dervish, and got the *trou* instead.'

'To prevent it? Not purposely?'

'Purposely, *mon Caporal*,' said Petit Picpon, with a sapient nod that spoke volumes. 'He always does something when he thinks promotion is coming—something to get himself out of its way, do you see? And the reason is this: 'tis a good *zig*, and loves you, and will not be put over your

---

[1] Spahi.   [2] Prison.   [3] A fine, fearless soldier.
[4] Term of punishment.   [5] General.

head. "Me rise afore him?" said the *zig* to me once. "I'll have the *As de pique* [1] on my collar fifty times over first! He's a Prince, and I'm a mongrel got in a gutter! I owe him more'n I'll ever pay, and I'll kill the Kébir himself afore I'll insult him that way." So say little to him about the Spahi, *mon Caporal*. He loves you well, does your Rac.'

'Well indeed! Good God! what nobility!'

Picpon glanced at him; then, with the tact of his nation, glided away and busied himself teaching Flick-Flack to shoulder and present arms, the weapon being a long chibouque-stick.

'After all, Diderot was in the right when he told Rousseau which side of the question to take,' mused Cecil, as he crossed the barrack-yard a few minutes later to visit the incarcerated *pratique*. 'On my life, civilization develops comfort, but I do believe it kills nobility. Individuality dies in it, and egotism grows strong and specious. Why is it that in a polished life a man, whilst becoming incapable of sinking to crime, almost always becomes also incapable of rising to greatness? Why is it that misery, tumult, privation, bloodshed, famine, beget, in such a life as this, such countless things of heroism, of endurance, of self-sacrifice—things worthy of demigods—in men who quarrel with the wolves for a wild-boar's carcase, for a sheep's offal?'

A question which perplexes, very wearily, thinkers who have more time, more subtlety, and more logic to bring to its unravelment than Bertie had either leisure or inclination to do.

'Is this true, Rake—that you intentionally commit these freaks of misconduct to escape promotion?' he asked of the man when he stood alone with him in his place of confinement.

Rake flushed a little.

'Mischief's bred in me, sir; it must come out. It's just bottled up in me like ale; if I didn't take the cork out now and then, I should fly apieces!'

'But many a time when you have been close on the reward of your splendid gallantry in the field, you have frustrated your own fortunes and the wishes of your superiors by wantonly proving yourself unfit for the higher grade they were going to raise you to; why do you do that?'

Rake fidgeted restlessly, and, to avoid the awkwardness of the question, replied, like a Parliamentary orator, by a flow of rhetoric.

'Sir, there's a many chaps like me. They can't help *nohow* bustin' out when the fit takes 'em. 'Tain't reasonable to blame 'em for it; they're just made so, like a chesnut's made to bust its pod, and a chicken to bust its shell. Well, you see, sir, France she know that, and she say to herself,

[1] A little mark in black cloth that distinguishes the uniform of the 'Incorrigibles'.

"Here are these madcaps, if I keep 'em tight in hand I shan't do nothin'
with 'em—they'll turn obstreperous and cram my convict-cells. Now, I
want soldiers, I don't want convicts. I can't let 'em stay in the Regulars,
'cause they'll be for making all the army wildfire like 'em; I'll just draft
'em by theirselves, treat 'em different, and let 'em fire away. They've got
good stuff in 'em, though too much of the curb riles 'em." Well, sir, she
do that; and aren't the Zéphyrs as fine a lot of fellows as any in the
service? Of course they are; but if they'd been in England—God bless
her, the dear old d—d obstinate soul!—they'd have been druv' crazy
along o' pipeclay and razors; *she'd* never have seed what was in 'em, her
eyes are so bunged up with routine. If a pup riot in the pack, she's
no notion but to double-thong him, and a-course, in double-quick time,
she finds herself obliged to go further and hang him. She don't ever
remember that it may be only just along of his breedin', and that he may
make a very good hound elseways let out a bit, though he'll spile the
whole pack if she *will* be a fool and try to make a steady line-hunter of
him straight agin his nature.'

Rake stopped breathless in his rhetoric, which contained more truth in
it, as also more roughness, than most rhetoric does.

'You are right. But you wander from my question,' said Cecil, gently.
'Do you avoid promotion?'

'Yes, sir, I do,' said Rake, something sulkily; for he felt he was being
driven 'up a corner'. 'I *do*. I ain't not one bit fitter for an officer than that
rioting pup I talk on is fit to lead them crack packs at home. I should be
in a strait-waistcoat if I was promoted; and as for the cross—Lord, sir, that
would get me into a world o' trouble! I should pawn it for a toss of wine
the first day out, or give it to the first *moukiera* that winked her black eye
for it! The star put on my buttons suits me a deal better; if you'll believe
me, sir, it do.'[1]

Cecil's eyes rested on him with a look that said far more than his
answer.

'Rake, I know you better than you would let me do, if you had your
way. My noble fellow! you reject advancement, and earn yourself an
unjust reputation for mutinous conduct, because you are too generous to
be given a step above mine in the regiment.'

'Who's a been a tellin' you that trash, sir?' retorted Rake, with ferocity.

'No matter who. It is no trash. It is a splendid loyalty of which I am
utterly unworthy, and it shall be my care that it is known at the bureaux,
so that henceforth your great merits may be——'

---

[1] The star on the metal buttons of the insubordinates or Zéphyrs.

'Stow that, sir!' cried Rake, vehemently. 'Stow that, *if* you please! Promoted I *won't* be—no, not if the Emperor hisself was to order it, and come across here to see it done! A pretty thing, surely! Me a officer, and you never a one—me a commandin' of you, and you a salutin' of me! By the Lord, sir! we might as well see the camp-scullions a ridin' in state, and the Marshal a scouring out the soup-pots!'

'Not at all. This Army has not a finer soldier than yourself; you have a right to the reward of your services in it. And I assure you you do me a great injustice if you think I would not as willingly go out under your orders as under those of all the Marshals of the Empire.'

The tears rushed into the hardy eyes of the redoubtable 'Crache-au-nez-d'la-Mort', though he dashed them away in a fury of eloquence.

'Sir, if you don't understand as how you've given me a power more than all the crosses in the world in sayin' of them there words, why *you* don't know *me* much either, that's all. You're a gentleman—a right on rare thing that is—and, bein' a gentleman, a course you'd be too generous and too proud like not to behave well to me, whether I was a servin' you as I've always served you, or a insultin' of you by ridin' over your head in that way as we're speakin' on. But I know my place, sir, and I know yours. If it wasn't for that 'ere Black Hawk—damn him!—I can't help it, sir, I *will* damn him, if he shoot me for it—you'd a been a Chef d'Escadron by now. There ain't the leastest doubt of it. Ask all the *zigs* what they think. Well, sir, now you know I'm a man what do as I say; if you don't let me have my own way, and if you do the littlest thing to get me a step, why, sir, I swear as I'm a livin' bein', that I'll draw on Châteauroy the first time I see him afterwards, and slit his throat as I'd slit a jackal's! There!—my oath's took!'

And Cecil saw that it would also be kept. The natural lawlessness and fiery passion inborn in Rake had of course not been cooled by the teachings of African warfare; and his hate was intense against the all-potent Chief of his regiment, as intense as the love he bore to the man whom he had followed out into exile.

Cecil tried vainly to argue with him; all his reasonings fell like hailstones on a cuirass, and made no more impression; he was resolute.

'But listen to one thing,' he urged as last. 'Can you not see how you pain me by this self-sacrifice? If I knew that you had attained a higher grade, and wore your epaulettes in this service, can you not fancy I should feel pleasure then (as I feel regret, even remorse, now) that I brought you to Africa through my own follies and misfortunes?'

'Do you, sir? There ain't the least cause for it, then,' returned Rake, sturdily. 'Lor bless you, sir, why this life's made a purpose for me! If ever a round peg went trim and neat into a round hole, it was when I came into this here Army. I never was so happy in all my days before. They're right on good fellows, and 'll back you to the death if so be as you've allays been share-and-share-alike with 'em, as a *zig* should. As a private, sir, I'm happy and I'm *safe*; as a officer, I should be kicking over the traces, and blunderin' everlastingly. However, there ain't no need to say a word more about it; I've sworn, and you've heerd me swear, sir, and you know as how I shall keep my oath if ever I'm provoked to it by bein' took notice of. I stuck that *homard* just now just by way of a lark, and only 'cause he come where he'd no business to poke his turbaned old pate; 'tain't likely as I shall stop at givin' the Hawk two inches of steel if he comes such a insult over us both as to offer a blackguard like me the epaulettes as you ought to be a wearin'!'

And Cecil knew that it was hopeless either to persuade him to his own advantage or to convince him of his disobedience in speaking thus of his supreme, before his non-commissioned, officer. He was himself, moreover, deeply moved by the man's fidelity.

He stretched his hand out:

'I wish there were more blackguards with hearts like yours. I cannot repay your love, Rake, but I can value it.'

Rake put his own hands behind his back.

'God bless you, sir, you've repaid it ten dozen times over. But you shan't do *that*, sir. I told you, long ago, I'm too much of a scamp! Some day, p'rhaps, as I said, when I've settled scores with myself, and wiped off all the bad 'uns with a clear sweep, tolerably clean. Not afore, sir!'

And Rake was too sturdily obstinate not to always carry his point.

The love that he bore to Cecil was very much such a wild, chivalric, romantic fidelity as the Cavaliers or the Gentlemen of the North bore to their Stuart idols. That his benefactor had become a soldier of Africa in no way lessened the reverent love of his loyalty, any more than theirs was lessened by the adversities of their royal masters. Like theirs, also, it had beauty in its blindness—the beauty that lies in every pure unselfishness.

Meanwhile, Picpon's news was correct.

The regiment was ordered out *à la danse*.[1] There was fresh war in the interior; and wherever there was the hottest slaughter, there the Black Hawk always flew down with his falcon-flock. When Cecil left his incorrigible *zig*, the trumpets were sounding an assembly; there were noise, tumult, eagerness, excitement, delighted zest on every side; a general order was read to the enraptured squadrons; they were to leave the town at the first streak of dawn.

[1] On the march.

There were before them death, deprivation, long days of famine, long days of drought and thirst; parching sun-baked roads; bitter chilly nights; fiery furnace-blasts of sirocco; killing, pitiless, northern winds; hunger, only sharpened by a snatch of raw meat or a handful of maize; and the probabilities, ten to one, of being thrust under the sand to rot, or left to have their skeletons picked clean by the vultures. But what of that? There were also the wild delight of combat, the freedom of lawless warfare, the joy of deep strokes thrust home, the chance of plunder, of wine-skins, of cattle, of women; above all, that lust for slaughter which burns so deep down in the hidden souls of men, and gives them such brotherhood with wolf, and vulture, and tiger, when once its flame bursts forth.

That evening, at the Villa Aïoussa there gathered a courtly assembly, of much higher rank than Algiers can commonly afford, because many of station as lofty as her own had been drawn thither to follow her to what the Princess Corona called her banishment—an endurable banishment enough under those azure skies, in that clear elastic air, and with that charming 'bonbonnière' in which to dwell, yet still a banishment to the reigning beauty of Paris, to one who had the habits and the commands of a wholly undisputed sovereignty in the royal splendour of her womanhood.

There was a variety of distractions to prevent ennui; there were half a dozen clever Paris actors playing the airiest of vaudevilles in the Bijou theatre, beyond the drawing-rooms; there were some celebrated Italian singers whom an Imperial Prince had brought over in his yacht; there was the best music; there was wit as well as homage whispered in her ear. Yet she was not altogether amused; she was a little touched with ennui.

'Those men are very stupid. They have not half the talent of that soldier!' she thought once, turning from a Peer of France, an Austrian Archduke, and a Russian diplomatist. And she smiled a little, furling her fan and musing on the horror that the triad of fashionable conquerors near her would feel if they knew that she thought them duller than an African *lascar*!

But they only told her things of which she had been long weary, specially of her own beauty; he had told her of things totally unknown to her, things real, terrible, vivid, strong, sorrowful—strong as life, sorrowful as death.

'Châteauroy and his Chasseurs have an *order de route*,' a voice was saying, that moment, behind her chair.

'Indeed?' said another. 'The Black Hawk is never so happy as when unhooded. When do they go?'

'Tomorrow. At dawn.'

'There is always fighting here, I suppose?'

'Oh yes. The losses in men are immense; only the journals would get a *communiqué*, or worse, if they ventured to say so in France. How delicious La Doche is! She comes in again with the next scene.'

The Princess Corona listened; and her attention wandered further from the Archduke, the Peer, and the diplomatist, as from the vaudeville. She did not find Madame Doche very charming; and she was absorbed for a time looking at the miniatures on her fan.

At the same moment, through the lighted streets of Algiers, Cigarette, like a union of fairy and of fury, was flying with the news. Cigarette had seen the flame of war at its height, and had danced in the midst of its whitest heat, as young children dance to see the fires leap red in the black winter's night. Cigarette loved the battle, the charge, the wild music of bugles, the thunder-tramp of battalions, the sirocco-sweep of light squadrons, the mad *tarantala* of triumph when the slaughter was done, the grand swoop of the Eagles down unto the carnage, the wild hurrah of France.

She loved them with all her heart and soul; and she flew now through the starlit sultry night, crying, 'La guerre! La guerre! La guerre!' and chanting to the enraptured soldiery a *Marseillaise* of her own improvization, all slang, and doggrel, and barrack-grammar; but fire-giving as a torch, and rousing as a bugle in the way she sang it, waving the tricolour high above her head:

> Fantasia,
> Deo Gratia!
> En avant!
> On t'attend!
> Au cor et à cri
> Suivez, mes Spahis!
> On s'élance à la danse,
> Pour la gloire de la France.
> Fusillons,
> Bataillons!
> Et marchons
> Au guidons!
> Va, loustic,
> Et du cric
> Vides ton verre,
> À la guerre!
> C'est l'Amie du Drapeau
> Qui s'appelle son troupeau!
> Faisons pouff à l'Emir,
> Faisons style à venir,

De l'avoine la moisson,
Portera belle boisson,
Le Zéphyr au douar
F'ra retentir son cor,
Chasse-marais cont' fleurettes
S'emparant des fillettes,
Et sous l'Aigle mes Roumis,
Vont gorger les Arbis,
À la musique si nette
De la haute clarinette!
      Razzia,
      Grazia,
      Est ici,
      Mes Spahis,
A l'amour! Aux beaux jours
Rataplan des tambours,
Nous appelle, 'R'lin tintin,
Vite au rire, au butin!'
      Vive la gloire!
      Vive le boire!
Vive le vin rosé du sang!
Vive le feu volage des rangs!
Vive tout ça qui va nous faire
Paradis au fond d'enfer,
Par la Guerre, par la Guerre!
En avant! Allons! Buvons!
En avant! Allons! Mourrons!

VOLUME III

CHAPTER I

Zarâila

The African day was at its noon.

From the first break of dawn the battle had raged; now, at mid-day, it was at its height. Far in the interior, almost on the edge of the great desert, in that terrible season when air that is flame by day is ice by night, and when the scorch of a blazing sun may be followed in an hour by the blinding fury of a snowstorm, the slaughter had gone on hour through hour under a shadowless sky, blue as steel, hard as a sheet of brass. The Arabs had surprised the French encampment where it lay in the centre of an arid plain that was called Zarâila. Hovering like a cloud of hawks on the entrance of the Sahara, massed together for one mighty if futile effort, with all their ancient war-lust, and with a new despair, the tribes who refused the yoke of the alien empire were once again in arms, were once again combined in defence of those limitless kingdoms of drifting sand, of that beloved belt of bare and desolate land so useless to the conqueror, so dear to the nomad. When they had been, as it had been thought, beaten back into the desert wilderness, when, without water and without cattle, it had been calculated that they would, of sheer necessity, bow themselves in submission, or perish of famine and of thirst, they had recovered their ardour, their strength, their resistance, their power to harass without ceasing, if they could never arrest, the enemy. They had cast the torch of war afresh into the land, and here, southward, the flame burned bitterly, and with a merciless tongue devoured the lives of men, licking them up as a forest fire the dry leaves and the touchwood.

Circling, sweeping, silently, swiftly, with that rapid spring, that marvellous whirlwind of force, that is of Africa, and of Africa alone, the tribes had rushed down in the darkness of night, lightly as a kite rushes through the gloom of the dawn. For once the vigilance of the invader served him nought; for once the Frankish camp was surprised off its guard. Whilst the air was still chilly with the breath of the night, whilst the first gleam of morning had barely broken through the mists of the east, whilst the picket-fires burned through the dusky gloom, and the sentinels paced slowly to and fro, hearing nothing worse than the stealthy tread of the jackal or the muffled flight of a night-bird, afar in the south a great dark cloud had risen, darker than the brooding shadows of the earth and sky.

The cloud swept onward, like a mass of cirri, in those shadows shrouded. Fleet as though wind-driven, dense as though thunder-charged, it moved over the plains. As it grew nearer and nearer, it grew greyer, a changing mass of white and black that fused, in the obscurity, into a shadow-colour; a dense array of men and horses flitting noiselessly like spirits, and as though guided alone by one rein and moved alone by one breath and one will; not a bit champed, not a linen-fold loosened, not a shiver of steel was heard; as silently as the winds of the desert sweep up northward over the plains, so they rode now, host upon host of the warriors of the soil.

The outlying videttes, the advanced sentinels, had scrutinized so long through the night every wavering shade of cloud and moving form of buffalo in the dim distance, that their sleepless eyes, strained and aching, failed to distinguish this moving mass that was so like the brown plains and starless sky that it could scarce be told from them. The night, too, was bitter; northern cold cut hardly chillier than this that parted the blaze of one hot day from the blaze of another. The sea-winds were blowing cruelly keen, and men who at noon gladly stripped to their shirts shivered now where they lay under canvas.

Awake while his comrades slept around him, Cecil was stretched half unharnessed. The foraging duty of the past twenty-four hours had been work harassing and heavy, inglorious and full of fatigue. The country round was bare as a table-rock; the watercourses poor, choked with dust and stones, unfed as yet by the rains or snows of the approaching winter. The horses suffered sorely, the men scarce less. The hay for the former was scant and bad; the rations for the latter often cut off by flying skirmishers of the foe. The campaign, so far as it had gone, had been fruitless, yet had cost largely in human life. The men died rapidly of dysentery, disease, and the chills of the nights, and had severe losses in countless obscure skirmishes that served no end except to water the African soil with blood.

True, France would fill the gaps up as fast as they occurred, and the *Moniteur* would only allude to the present operations when it could give a flourishing line descriptive of the Arabs being driven back decimated to the borders of the Sahara. But as the flourish of the *Moniteur* would never reach a thousand little wayside huts, and seaside cabins, and vine-dressers' sunny nests, where the memory of some lad who had gone forth never to return would leave a deadly shadow athwart the humble threshold, so the knowledge that they were only so many automata in the hands of government, whose loss would merely be noted that it might be efficiently supplied, was not that wine-draught of La Gloire which poured the strength and the daring of gods into the limbs of the men of Jena and of Austerlitz. Still, there was the war-lust in them, and there was the fire of France; they fought not less superbly here, where to be food for jackal and kite was their likeliest doom, than their sires had done under the eagles of the First Empire, when the Conscript hero of today was the glittering Marshal of tomorrow.

Cecil had awakened while the camp still slept. Do what he would, force himself into the fulness of this fierce and hard existence as he might, he could not burn out or banish a thing that had many a time haunted him, but never as it did now—the remembrance of a woman. He almost laughed as he lay there on a pile of rotting straw, and wrung the truth out of his own heart that he—a soldier of these exiled squadrons— was mad enough to love that woman whose deep proud eyes had dwelt with such serene pity upon him.

Yet his hand clenched on the straw as it had clenched once when the operator's knife had cut down through the bones of his breast to reach a bullet that, left in his chest, would have been death. If in the sight of men he had only stood in the rank that was his by birthright, he could have strived for—it might be that he could have roused—some answering passion in her. But that chance was lost to him for ever. Well, it was but one thing more that was added to all that he had of his own will given up. He was dead; he must be content, as the dead must be, to leave the warmth of kisses, the glow of delight, the possession of a woman's loveliness, the homage of men's honour, the gladness of successful desires, to those who still lived in the light he had quitted. He had never allowed himself the emasculating indulgence of regret; he flung it off him now.

Flick-Flack, coiled asleep in his bosom, thrilled, stirred, and growled. He rose, and, with the little dog under his arm, looked out from the canvas. He knew that the most vigilant sentry in the service had not the instinct for a foe afar off that Flick-Flack possessed. He gazed keenly southward, the poodle growling on; that cloud so dim, so distant, caught his sight. Was it a moving herd, a shifting mist, a shadow-play between the night and dawn?

For a moment longer he watched it; then what it was he knew, or felt by such strong instinct as makes knowledge; and like the blast of a clarion his alarm rang over the unarmed and slumbering camp.

An instant, and the hive of men, so still, so motionless, broke into violent movement; and from the tents the half-clothed sleepers poured, wakened, and fresh in wakening as hounds. Perfect discipline did the rest. With marvellous, with matchless swiftness and precision they harnessed and got under arms. They were but fifteen hundred or so in all—a single squadron of Chasseurs, two battalions of Zouaves, half a corps of Tirailleurs, and some Turcos; only a branch of the main body, and without artillery. But they were some of the flower of the Army of Algiers, and they roused in a second with the vivacious ferocity of the bounding tiger, with the glad eager impatience for the slaughter of the unloosed hawk. Yet, rapid in its wondrous celerity as their united action was, it was not so rapid as the downward sweep of that war-cloud that came so near, with the tossing of white draperies and the shine of countless sabres, now growing clearer and clearer out of the darkness, till, with a whirr like the noise of an eagle's wings, and a swoop like an eagle's seizure, the Arabs whirled down upon them, met a few yards in advance by the answering charge of the Light Cavalry.

There was a crash as if rock were hurled upon rock, as the Chasseurs, scarce seated in saddle, rushed forward to save the pickets, to encounter the first blind force of the attack, and to give the infantry, further in, more time for harness and defence. Out of the caverns of the night an armed multitude seemed to have suddenly poured. A moment ago, they had slept in security; now, thousands on thousands whom they could not number, whom they could but dimly even perceive, were thrown on them in immeasurable hosts, which the encircling cloud of dust served but to render vaster, ghastlier, and more majestic. The Arab line stretched out with wings that seemed to extend on and on without end; the line of the Chasseurs was not one-half its length; they were but a single squadron flung in their stirrups, scarcely clothed, knowing only that the foe was upon them, caring only that their sword-hands were hard on their weapons. With all the *élan* of France they launched themselves forward to break the rush of the desert horses; they met with a terrible sound, like falling trees, like clashing metal.

The hoofs of the rearing chargers struck each other's breasts, and these bit and tore at each other's manes, while their riders reeled down dead. Frank and Arab were blended in one inextricable mass as the charging squadrons encountered. The outer wings of the tribes were spared the shock, and swept on to meet the bayonets of Zouaves and Turcos as at their swift foot-gallop the Enfans Perdus of France threw themselves forward from the darkness. The cavalry was enveloped in the overwhelming numbers of the centre; and the flanks seemed to cover the Zouaves and Tirailleurs as some great settling mist may cover the cattle who move beneath it.

It was not a battle; it was a frightful tangling of men and brutes. No contest of modern warfare such as commences and conquers by a duel of artillery, and gives the victory to whosoever has the superiority of ordnance; but a conflict, hand to hand, breast to breast, life for life, a Homeric combat of spear and of sword even whilst the first volleys of the answering musketry pealed over the plain.

For once the Desert avenged in like that terrible inexhaustibility of supply wherewith the Empire so long had crushed them beneath the overwhelming difference of numbers. It was the Day of Mazagran once more, as the light of the morning broke, grey, silvered, beautiful, in the far, dim distance, beyond the tawny seas of reeds. Smoke and sand soon densely rose above the struggle, white, hot, blinding; but out from it the lean dark Bedouin faces, the snowy haïcks, the red burnous, the gleam of the Tunisian muskets, the flash of the silver-hilted yataghans, were seen fused in a mass with the brawny naked necks of the Zouaves, with the shine of the French bayonets, with the tossing manes and glowing nostrils of the Chasseurs' horses, with the torn, stained silk of the raised Tricolour, through which the storm of balls flew thick and fast as hail, yet whose folds were never suffered to fall, though again and again the hand that held its staff was cut away or was unloosed in death, yet ever found another to take its charge before the Flag could once have trembled in the enemy's sight.

The Chasseurs could not charge; they were hemmed in, packed between bodies of horsemen that pressed them together as between iron plates; now and then they could cut their way through, clear enough to reach their comrades of the *demie cavalerie*, but as often as they did so, so often the overwhelming numbers of the Arabs surged in on them afresh like a flood, and closed upon them, and drove them back.

Every soldier in the squadron that lived kept his life by sheer, breathless, ceaseless, hand-to-hand sword-play, hewing right and left, front and rear, without pause, as in the great tangled forests of the west men hew aside branch and brushwood ere they can force one forward step.

The gleam of the dawn spread in one golden glow of morning, and the day rose radiant over the world; they stayed not for its beauty or its peace; the carnage went on hour upon hour; men began to grow drunk with slaughter as with raki. It was sublimely grand; it was hideously hateful— this wild-beast struggle, this heaving tumult of striving lives that ever and anon stirred the vast war-cloud of smoke and broke from it as the lightning from the night. The sun laughed in its warmth over a thousand hills and streams, over the blue seas lying northward, and over the yellow sands of the south; but the touch of its heat only made the flame in their blood burn fiercer; the fulness of its light only served to show them clearer where to strike and how to slay.

It was bitter, stifling, cruel work; with their mouths choked with sand, with their throats caked with thirst, with their eyes blind with smoke; cramped as in a vice, scorched with the blaze of powder, covered with blood and with dust; while the steel was thrust through nerve and sinew, or the shot ploughed through bone and flesh. The answering fire of the Zouaves and Tirailleurs kept the Arabs further at bay, and mowed them faster down; but in the Chasseurs' quarter of the field—parted from the rest of their comrades as they had been by the rush of that broken charge with which they had sought to save the camp and arrest the foe—the worst pressure of the attack was felt, and the fiercest of the slaughter fell.

The Chef d'Escadron had been shot dead as they had first swept out to encounter the advance of the desert horsemen; one by one the officers had been cut down, singled out by the keen eyes of their enemies, and throwing themselves into the deadliest of the carnage with the impetuous self-devotion characteristic of their service. At the last there remained but a mere handful out of all the brilliant squadron that had galloped down in the grey of the dawn to meet the whirlwind of Arab fury. At their head was Cecil.

Two horses had been killed under him, and he had thrown himself afresh across unwounded chargers, whose riders had fallen in the mêlée, and at whose bridles he had caught as he shook himself free of the dead animals' stirrups. His head was uncovered; his uniform, hurriedly thrown on, had been torn aside, and his chest was bare to the red folds of his sash: he was drenched with blood, not his own, that had rained on him as he fought; and his face and his hands were black with smoke and with powder. He could not see a yard in front of him; he could not tell how the day went anywhere, save in that corner where his own troop was hemmed in. As fast as they beat the Arabs back, and forced themselves some clearer space, so fast the tribes closed in afresh. No orders reached him from the General of Brigade in command; except for the well-known war-shouts of the Zouaves that ever and again rang above the din, he could not tell whether the French battalions were not cut utterly to pieces under the immense numerical superiority of their foes. All he could see was that every officer of Chasseurs was down, and that unless he took the vacant place, and rallied them together, the few score troopers that were still left would scatter, confused and demoralized, as the best soldiers will at times when they can see no chief to follow.

He spurred the horse he had just mounted against the dense crowd opposing him, against the hard black wall of dust, and smoke, and steel, and savage faces, and lean swarthy arms, which were all that his eyes could see, and that seemed impenetrable as granite, moving and changing though it was. He thrust the grey against it, whilst he waved his sword above his head:

'*En avant, mes frères! France! France! France!*'

His voice, well known, well loved, thrilled the hearts of his comrades, and brought them together like a trumpet-call. They had gone with him many a time into the hell of battle, into the jaws of death. They surged about him now, striking, thrusting, forcing with blows of their sabres or their lances and blows of their beasts' fore-feet a passage one to another, until they were reunited once more as one troop, whilst their shrill shouts, like an oath of vengeance, echoed after him in the butchery, that has pealed victorious over so many fields from the soldiery of France. They loved him; he had called them his brethren. They were like lambs for him to lead, like tigers for him to incite.

They could scarcely see his face in that great red mist of combat, in that horrible stifling pressure on every side that jammed them as if they were in a press of iron, and gave them no power to pause, though their animals' hoofs struck the lingering life out of some half-dead comrade, or trampled over the writhing limbs of the brother-in-arms they loved dearest and best. But his voice reached them, clear and ringing in its appeal for sake of the country they never once forgot or once reviled, though in her name they were starved and beaten like rebellious hounds, though in her cause they were exiled all their manhood through under the sun of this cruel, ravenous, burning Africa. They could see him lift aloft the Eagle he had caught from the last hand that had borne it, the golden gleam of the young morning flashing like flame upon the brazen wings; and they shouted, as with one throat, '*Mazagran! Mazagran!*' As the battalion of Mazagran had died keeping the ground through the whole of the scorching day, while the fresh hordes poured down on them like ceaseless torrents snow-fed and exhaustless, so they were ready to hold the ground here, until of all their numbers there should be left not one living man.

He glanced back on them, guarding his head the while from the lances that were rained on him; and he lifted the Guidon higher and higher, till, out of the ruck and the throng, the brazen bird caught afresh the rays of the rising sun.

'*Suivez-moi!*' he shouted.

Then, like arrows launched at once from twice a hundred bows, they charged, he still slightly in advance of them, the bridle flung upon his horse's neck, his head and breast bare, one hand striking aside with his blade the steel shafts as they poured on him, the other holding high above the press the Eagle of the Bonapartes.

The effort was superb.

Dense bodies of Arabs parted them in the front from the camp where the battle raged, harassed them in the rear with flying shots and hurled lances, and forced down on them on either side, like the closing jaws of a trap. The impetuosity of their onward movement was, for the moment, irresistible; it bore headlong all before it; the desert horses recoiled, and the desert riders themselves yielded, crushed, staggered, trodden aside, struck aside, by the tremendous impetus with which the Chasseurs were thrown upon them. For the moment, the Bedouins gave way, shaken and confused, as at the head of the French they saw this man, with his hair blowing in the wind, and the sun on the fairness of his face, ride down on them thus unharmed, though a dozen spears were aimed at his naked breast, dealing strokes sure as death right and left as he went, with the light from the hot blue skies on the ensign of France that he bore.

They knew him; they had met him in many conflicts; and wherever the '*fair Frank*', as they called him came, there they knew of old the battle was hard to win; bitter to the bitterest end, whether that end were defeat, or victory costly as defeat in its achievement.

And for the moment they recoiled under the shock of that fiery on-slaught; for the moment they parted, and wavered, and oscillated beneath the impetus with which he hurled his hundred Chasseurs on them, with that light, swift, indescribable rapidity and resistlessness of attack charac-teristic of the African Cavalry.

Though a score or more, one on another, had singled him out with special and violent attack, he had gone, as yet, unwounded, save for a lance-thrust in his shoulder, of which, in the heat of the conflict, he was unconscious. The 'fighting fury' was upon him; and when once this had been lit in him, the Arabs knew of old that the fiercest vulture in the Frankish ranks never struck so surely home as this hand that his comrades called '*main de femme, mais main de fer*'.

As he spurred his horse down on them now, twenty blades glittered against him; the foremost would have cut straight down through the bone of his bared chest and killed him at a single lung, but as its steel flashed in the sun, one of his troopers threw himself against it, and parried the stroke from him by sheathing it in his own breast. The blow was mortal; and the one who had saved him reeled down off his saddle under the hoofs of the trampling chargers. '*Picpon s'en souvient*', he murmured with a smile; and as the charge swept onward, Cecil, with a great cry of horror, saw the feet of the maddened horses strike to pulp the writhing body, and saw the black wistful eyes of the Enfant de Paris look upwards to him once, with love, and fealty, and unspeakable sweetness gleaming through their darkened sight.

But to pause was impossible. Though the French horses were forced with marvellous dexterity through a bristling forest of steel, though the remnant of the once-glittering squadron was cast against them in as headlong a daring as if it had half the regiments of the Empire at its back, the charge availed little against the hosts of the desert that had rallied and swooped down afresh almost as soon as they had been, for the instant of the shock, panic-stricken. The hatred of the opposed races was aroused in all its blind ravening passion; the conquered had the conquering nation for once at their mercy, for once at tremendous disadvantage; on neither side was there aught except that one instinct for slaughter, which, once awakened, kills every other in the breast in which it burns.

The Arabs had cruel years to avenge—years of a loathed tyranny, years of starvation and oppression, years of constant flight southward, with no choice but submission or death. They had deadly memories to wash out—memories of brethren who had been killed like carrion by the invaders' shot and steel; of nomadic freedom begrudged and crushed by civilization; of young children murdered in the darkness of the caverns with the sulphurous smoke choking the innocent throats that had only breathed the golden air of a few summers; of women, well-beloved, torn from them in the hot flames of burning tents and outraged before their eyes with insult, whose end was a bayonet-thrust into their breasts, whose sin was fidelity to the vanquished.

They had vengeance to do that made every stroke seem righteous and holy in their sight; that nerved each of their bare and sinewy arms as with the strength of a thousand limbs. Right—so barren, so hopeless, so unavailing—had long been with them. Now with it was added at last the power of might; and they exercised the power with the savage ruthlessness of the desert. They closed in on every side; wheeling their swift coursers hither and thither; striking with lance and blade; hemming in, beyond escape, the doomed fragment of the Frankish squadron till there remained of them but one small nucleus, driven close together, rather as infantry will form than as cavalry usually does—a ring of horsemen, of which every one had his face to the foe; a solid circle curiously wedged one against the other, with the bodies of chargers and of men deep around them, and with the ground soaked with blood till the sand was one red morass.

Cecil held the Eagle still, and looked round on the few left to him.

'You are sons of the Old Guard: die like them.'

They answered with a pealing cry, terrible as the cry of the lion in the hush of night, but a shout that had in it assent, triumph, fealty, victory, even as they obeyed him and drew up to die, whilst in their front was the young brow of Petit Picpon turned upwards to the glare of the skies.

There was nothing for them but to draw up thus, and await their butchery, defending the Eagle to the last; looking till the last towards that 'woman's face of their leader', as they had often termed it, that was to them now as the face of Napoléon was to the soldiers who loved him.

There was a pause, brief as is the pause of the lungs to take a fuller breath. The Arabs honoured these men who, alone in the midst of the hostile force, held their ground and prepared thus to be slaughtered one by one, till of all the squadron that had ridden out in the darkness of the dawn there should be only a black, huddled, stiffened heap of dead men and of dead beasts. The chief who led them pressed them back, withholding them from the end that was so near to their hands when they should stretch that single ring of horsemen all lifeless in the dust.

'You are great warriors,' he cried, in the Sabir tongue; 'surrender, we will spare!'

Cecil looked back once more on the fragment of his troop, and raised the Eagle higher aloft where the wings should glisten in the fuller day. Half-naked. scorched, blinded, with an open gash in his shoulder where the lance had struck, and with his brow wet with the great dews of the noon-heat and the breathless toil, his eyes were clear as they flashed with the light of the sun in them; his mouth smiled as he answered:

'Have we shown ourselves cowards, that you think we shall yield?'

A *hourrah* of wild delight from the Chasseurs he led greeted and ratified the choice: '*On meurt—on ne se rend pas!*' they shouted in the words, which, even if they be but legendary, are too true to the spirit of the soldiers of France not to be as truth in their sight. Then, with their swords above their heads, they waited for the collision of the terrible attack which would fall on them upon every side, and strike all the sentient life out of them before the sun should be one point higher in the heavens. It came: with a yell as of wild beasts in their famine, the Arabs threw themselves forward, the chief himself singling out the 'fair Frank' with the violence of a lion flinging himself on a leopard. One instant longer, one flash of time, and the tribes pressing on them would have massacred them like cattle driven into the pens of slaughter. Ere it could be done, a voice like the ring of a silver trumpet echoed over the field:

'*En avant! En avant! Tue, tue, tue!*'

Above the din, the shouts, the tumult, the echoing of the distant musketry, that silvery cadence rung; down into the midst, with the Tricolour waving above her head, the bridle of her fiery mare between her teeth, the raven of the dead Zouave flying above her head, and her pistol levelled in deadly aim, rode Cigarette.

The lightning fire of the crossing swords played round her, the glitter of the lances dazzled her eyes, the reek of smoke and of carnage was round her; but she dashed down into the heart of the conflict as gaily as though she rode at a review, laughing, shouting, waving the torn colours that she grasped, with her curls blowing back in the breeze, and her bright young face set in the warrior's lust. Behind her, by scarcely a length, galloped three squadrons of Chasseurs and Spahis; trampling headlong over the corpse-strewn field, and breaking through the masses of the Arabs as though they were seas of corn.

She wheeled her mare round by Cecil's side at the moment when, with six swift passes of his blade he had warded off the Chief's blows and sent his own sword down through the chestbones of the Bedouin's mighty form.

'Well struck! The day is turned. Charge!'

She gave the order as though she were a Marshal of the Empire, the sun-blaze full on her where she sat on the rearing, fretting, half-bred grey, with the Tricolour folds above her head, and her teeth tight gripped on the chain-bridle, and her face all glowing and warm and full of the fierce fire of war—a little Amazon in scarlet, and blue, and gold; a young Jeanne d'Arc, with the crimson fez in lieu of the silvered casque, and the gay broderies of her fantastic dress instead of the breastplate of steel. And with the Flag of her idolatry, the Flag that was as her religion, floating back as she went, she spurred her mare straight against the Arabs, straight over the lifeless forms of the hundreds slain; and after her poured the fresh squadrons of cavalry, the ruby burnous of the Spahis streaming on the wind as their darling led them on to retrieve the day for France.

Not a bullet struck, not a sabre grazed her; but there, in the heat and the press of the worst of the slaughter, Cigarette rode hither and thither, to and fro, her voice ringing like a brid's song over the field, in command, in applause, in encouragement, in delight; bearing her standard aloft and untouched; dashing heedless through a storm of blows; cheering on her 'children' to the charge again and again; and all the while with the sunlight full on her radiant spirited head, and with the grim grey raven flying above her, shrieking shrilly its '*Tue, tue, tue!*' The Army believed with superstitious faith in the potent spell of that veteran bird, and the story ran that whenever he flew above a combat France was victor before the sun set. The echo of the raven's cry, and the presence of the child who, they knew, would have a thousand musket-balls fired in her fair young breast rather than live to see them defeated, made the fresh squadrons sweep in like a whirlwind, bearing down all before them.

Cigarette saved the day.

CHAPTER II

The Love of the Amazon

Before the sun had declined from his zenith the French were masters of the field, and pursued the retreat of the Arabs till for miles along the plain the line of their flight was marked with horses that had dropped dead in the strain, and with the motionless forms of their desert-riders; their cold hands clenched in the loose hot sands, and their stern faces turned upwards to the cloudless scorch of their native skies, under whose freedom they would never again ride forth to the joyous clash of the cymbals and the fierce embrace of the death-grapple.

When at length she returned, coming in with her ruthless Spahis, whose terrible passions she feared no more that Virgil's Volscian huntress feared the beasts of forest and plain, the raven still hovered above her exhausted mare, the torn flag was still in her left hand; and the bright laughter, the flash of ecstatic triumph, was still in her face as she sang the last lines of her own war-chant. The leopard nature was roused in her. She was a soldier; death had been about her from her birth; she neither feared to give nor to receive it; she was proud as ever was young Pompeius flushed with the glories of his first eastern conquests; she was happy as such elastic, sunlit, dauntless youth as hers alone can be, returning in the reddening after-glow at the head of her comrades to the camp that she had saved.

She could be cruel—women are, when roused, as many a revolution has shown; she could be heroic—she would have died a hundred deaths for France; she was vain with a vivacious child-like vanity; she was brave with a bravery beside which many a man's high courage palled. Cruelty, heroism, vanity, and bravery were all on fire, and all fed to their uttermost,

most eager, most ardent flame now that she came back at the head of her Spahis; while all who remained of the soldiers who, but for her, would have been massacred long ere then, without one spared amongst them, threw themselves forward, crowded round her, caressed, and laughed, and wept, and shouted with all the changes of their intense mercurial temperaments, kissed her boots, her sash, her mare's drooping neck, and, lifting her, with wild vivas that rent the sky, on to the shoulders of the four tallest men amongst them, bore her to the presence of the only officer of high rank who had survived the terrors of the day, a Chef de Bataillon of the Zouaves.

And he, a grave and noble-looking veteran, uncovered his head and bowed before her as courtiers bow before their queens.

'Mademoiselle, you saved the honour of France. In the name of France, I thank you.'

The tears rushed swift and hot into Cigarette's bright eyes—tears of joy, tears of pride. She was but a child still in much, and she could be moved by the name of France as other children by the name of their mothers.

'Chut! I did nothing,' she said, rapidly. 'I only rode fast.'

The frenzied hurrahs of the men who heard her drowned her words. They loved her for what she had done; they loved her better still because she set no count on it.

'The Empire will think otherwise,' said the Major of the Zouaves. 'Tell me, my Little One, how did you do this thing?'

Cigarette, balancing herself with a foot on either shoulder of her supporters, gave the salute and answered:

'Simply, mon Commandant—very simply. I was alone, riding midway between you and the main army—three leagues, say, from each. I was all alone; only Vole-qui-veut flying with me for fun. I met a colon. I knew the man. For the matter of that, I did him once a service—saved his geese and his fowls from burning, one winter's day, in their house, while he wrung his hands and looked on. Well, he was full of terror, and told me there was fighting yonder—here he meant—so I rode nearer to see. That was just upon sunrise. I dismounted, and ran up a high tree there.' And Cigarette pointed to a far-off slope crowned with the remains of a once-mighty forest. 'I got up very high. I could see miles round. I saw how things were with you. For the moment I was coming straight to you. Then I thought I should do more service if I let the main army know, and brought you a reinforcement. I rode fast. Dieu! I rode fast. My horse dropped under me twice; but I reached them at last, and I went at once to the General. He guessed at a glance how things were, and I told him to give me my Spahis and let me go. So he did. I got on a mare of his own staff, and away we came. Ma foi! it was a near thing. If we had been a minute later, it had been all up with you.'

'True indeed,' muttered the Zouave in his beard. 'A superb action, my Little One. But did you meet no Arab scouts to stop you?'

Cigarette laughed.

'Did I not? Met them by dozens. Some had a shot at me; some had a shot from me. One fellow nearly winged me; but I got through them all, somehow. Sapristi! I galloped so fast I was very hard to hit flying. Those things only require a little judgement; but some men, pardi! always are creeping when they should fly, and always are scampering when they should saunter; and then they wonder when they make *fiasco!* Bah!'

And Cigarette laughed again. Men were such bunglers—ouf!

'Mademoiselle, if all soldiers were like you,' answered the Major of Zouaves, curtly, 'to command a battalion would be paradise!'

'All soldiers would do anything I have done,' retorted Cigarette, who never took a compliment at the expense of her 'children'. 'They do not all get the opportunity, look you; *c'est tout!* Opportunity is a little angel; some catch him as he goes, some let him pass by for ever. You must be quick with him, for he is like an eel to wriggle away. If you want a good soldier, take that aristocrat of the *Chasse-Marais*—that *beau Victor*. Pouf! all his officers were down; and how splendidly he led the troop! He was going to die with them rather than surrender. Napoléon'—and Cigarette uncovered her curly head reverentially as at the name of a deity— 'Napoléon would have given him his brigade ere this. If you had seen him kill the chief!'

'He will have justice done him, never fear. And for you—the Cross shall be on your breast, Cigarette, if I live over tonight to write my dispatches.'

And the Chef de Bataillon saluted her once more, and turned away to view the carnage-strewn plain, and number the few who remained out of all those who had been wakened by the clash of the Arab arms in the grey of the earliest dawn.

Cigarette's eyes flashed like sun playing on water and her flushed cheeks grew scarlet. Since her infancy it had been her dream to have the Cross, to have the *Grande Croix* to lie above her little lion's heart; it had been the one longing, the one ambition, the one undying desire of her soul; and lo! she touched its realization!

The wild, frantic, tumultuous cheers and caresses of her soldiery, who could not triumph in her and triumph with her enough to satiate them, recalled her to the actual moment. She sprang down from her elevation, and turned on them with a rebuke. 'Ah! you are making this fuss about me while hundreds of better soldiers than I lie yonder. Let us look to them first; we will play the fool afterwards.'

And, though she had ridden fifty miles that day, if she had ridden one—though she had eaten nothing since sunrise, and had only had one draught of bad water—though she was tired, and stiff, and bruised, and parched with thirst, Cigarette dashed off as lightly as a young goat to look for the wounded and the dying men who strewed the plain far and near.

She remembered one whom she had not seen after that first moment in which she had given the word to the squadrons to charge.

It was a terrible sight—the arid plain, lying in the scarlet glow of sunset, covered with dead bodies, with mutilated limbs, with horses gasping and writhing, with men raving like mad creatures in the torture of their wounds. It was a sight which always went to her heart. She was a true soldier, and, though she could deal death pitilessly, could, when the delirium of war was over, tend and yield infinite compassion to those who were in suffering. But such scenes had been familiar to her from the earliest years when, on an infant's limbs, she had toddled over such battlefields, and wound tiny hands in the hair of some dead trooper who had given her sweetmeats the hour before, vainly trying to awaken him. And she went through all the intense misery and desolation of the scene now without shrinking, and with that fearless tender devotion to the wounded which Cigarette showed in common with other soldiers of her nation, being, like them, a young lion in the combat, but a creature unspeakably gentle and full of sympathy when the fury of the fight was over.

She had seen great slaughter often enough, but even she had not seen any struggle more close, more murderous, than this had been. The dead lay by hundreds; French and Arab locked in one another's limbs as they had fallen when the ordinary mode of warfare had failed to satiate their violence, and they had wrestled together like wolves fighting and rending each other over a disputed carcase. The bitterness and the hatred of the contest were shown in the fact that there were very few merely wounded or disabled; almost all of the numbers that strewed the plain were dead. It had been a battle-royal, and, but for her arrival with the fresh squadrons, not one among her countrymen would have lived to tell the story of this terrible duello, which had been as magnificent in heroism as any Austerlitz or Gemappes, but which would pass unhonoured, almost unnamed, amongst the futile fruitless heroisms of Algerian warfare.

'Is he killed? Is he killed?' she thought, as she bent over each knot of motionless bodies where here and there some faint stifled breath or some moan of agony told that life still lingered beneath the huddled stiffening heap. And a tightness came at her heart, an aching fear made her shrink, as she raised each hidden face that she had never known before. 'What if he be?' she said fiercely to herself. 'It is nothing to me. I hate him, the cold aristocrat. I ought to be glad if I see him lie here.'

But, despite her hatred for him, she could not banish that hot feverish hope, that cold suffocating fear, which, turn by turn, quickened and slackened the bright flow of her warm young blood as she searched amongst the slain.

'Ah! le pauvre Picpon!' she said, softly, as she reached at last the place where the young Chasseur lay, and lifted the black curls off his forehead. The hoofs of the charging cavalry had cruelly struck and trampled his frame; the back had been broken, and the body had been mashed as in a mortar under the thundering gallop of the Horse; but the face was still uninjured, and had a strange pathetic beauty, a calm and smiling courage, on it. It was ashen pale; but the great black eyes that had glistened in such malicious mirth, and sparkled in such malignant mischief during life, were open, and had a mournful pitiful serenity in their look as if from their depths the soul still gazed—that soul which had been neglected and cursed, and left to wander amongst evil ways, yet which, through all its darkness, all its ignorance, had reached, unguided, to love and to nobility.

Cigarette closed their long black lashes down on the white cheeks with soft and reverent touch; she had seen that look ere now on the upturned faces of the dead who had strewn the barricades of Paris, with the words of the Marseillaise the last upon their lips.

To her there could be no fate fairer, no glory more glorious, than this of his—to die for France. And she laid him gently down, and left him, and went on with her quest.

It was here that she had lost sight of Cecil as they had charged together, and her mare, enraged and intoxicated with noise and terror, had torn away at a full speed that had outstripped even the swiftest of her Spahis. A little further on a dog's moan caught her ear; she turned and looked across. Upright, amongst a ghastly pile of men and chargers, sat the small snowy poodle of the Chasseurs, beating the air with its little paws as it had been taught to do when it needed anything, and howling piteously as it begged.

'Flick-Flack? What is it, Flick-Flack?' she cried to him, while, with a bound, she reached the spot. The dog leaped on her, rejoicing. The dead were thick there—ten or twelve deep—French trooper and Bedouin rider flung across each other, horribly entangled with the limbs, the manes, the shattered bodies of their own horses. Among them she saw the face she sought as the dog eagerly ran back, caressing the hair of a soldier who lay underneath the weight of his grey charger, that had been killed by a musket-ball.

Cigarette grew very pale, as she had never grown when the hailstorm of shots had been pouring on her in the midst of a battle; but, with the rapid skill and strength she had acquired long before, she reached the place, lifted aside first one, then another, of the lifeless Arabs that had fallen above him, and drew out from beneath the suffocating pressure of his horse's weight the head and the frame of the Chasseur whom Flick-Flack had sought out and guarded.

For the moment she thought him dead; then, as she drew him out where the cooler breeze of the declining day could each him, a slow breath, painfully drawn, moved his chest; she saw that he was unconscious from the stifling oppression under which he had been buried since the noon; an hour more without the touch of fresher air, and life would have been extinct.

Cigarette had with her the flask of brandy that she always brought on such errands as these; she forced the end between his lips, and poured some down his throat; her hand shook slightly as she did so, a weakness the gallant little campaigner never before then had known.

It revived him in a degree; he breathed more freely, though heavily, and with difficulty still; but gradually the deadly leaden colour of his face was replaced by the hue of life, and his heart began to beat more loudly. Consciousness did not return to him; he lay motionless and senseless, with his head resting on her lap, and with Flick-Flack, in eager affection, licking his hands and his hair.

'He was as good as dead, Flick-Flack, if it had not been for you and me,' said Cigarette, while she wetted his lips with more brandy. 'Ah bah! and he would be more grateful, Flick-Flack, for a scornful scoff from Miladi!'

Still, though she thought this, she let his head lie on her lap, and, as she looked down on him, there was the glisten as of tears in the brave sunny eyes of the little Friend of the Flag.

'*Il est si beau, si beau, si beau!*' she muttered in her teeth, drawing the silk-like lock of his hair through her hands, and looking at the stricken strength, the powerless limbs, the bare chest, cut and bruised, and heaved painfully by each uneasy breath. She was of a vivid, voluptuous, artistic nature; she was thoroughly woman-like in her passions and her instincts, though she so fiercely contemned womanhood. If he had not been beautiful, she would never have looked twice at him, never once have pitied his fate.

And he was beautiful still, though his hair was heavy with dew and dust, though his face was scorched with powder, though his eyes were closed as with the leaden weight of death, and his beard was covered with the red stain of blood that had flowed from the lance-wound on his shoulder.

He was not dead; he was not even in peril of death. She knew enough of medical lore to know that it was but the insensibility of exhaustion and suffocation: and she did not care that he should waken. She drooped her head over him, moving her hand softly among the masses of his curls, and watching the quickening beatings of his heart under the bare strong nerves. Her face grew tender, and warm, and eager, and melting with a marvellous change of passionate hues. She had all the ardour of southern blood; without a wish he had wakened in her a love that grew daily and hourly, though she would not acknowledge it. She loved to see him lie there as though he were asleep, to cheat herself into the fancy that she watched his rest to wake it with a kiss on his lips. In that unconsciousness, in that abandonment, he seemed wholly her own; passion which she could not have analysed made her bend above him with a half-fierce half-dreamy delight in that solitary possession of his beauty, of his life.

The restless movements of little Flick-Flack detached a piece of twine passed round his favourite's throat; the glitter of gold arrested Cigarette's eyes. She caught what the poodle's impatient caress had broken from the string; it was a small blue enamel medallion bonbon-box, with a hole through it by which it had been slung—a tiny toy once costly, now tarnished, for it had been carried through many rough scenes and many years of hardship, had been bent by blows struck at the breast against which it rested, and was clotted now with blood. Inside it was a woman's ring, of sapphires and opals.

She looked at both close in the glow of the setting sun; then passed the string through and fastened the box afresh. It was a mere trifle, but it sufficed to banish her dream, to arouse her to contemptuous impatient bitterness with that new weakness that had for the hour broken her down to the level of this feverish folly. He was beautiful—yes! She could not bring herself to hate him; she could not help the brimming tears blinding her eyes when she looked at him stretched senseless thus. But he was wedded to his past; that toy in his breast, whatever it might be, whatever tale might cling to it, was sweeter to him than her lips would ever be. Bah! there were better men than he; why had she not let him lie and die as he might under the pile of dead?

Bah! she could have killed herself for her folly! She, who had scores of lovers, from princes to piou-pious, and never had a heart-ache for one of them, to go and care for a silent '*ci-devant*', who had never even noticed that her eyes had any brightness or her face had any charm!

'You deserve to be shot—you!' said Cigarette, fiercely abusing herself as she put his head off her lap, and rose abruptly and shouted to a Tringlo who was at some distance searching for the wounded. 'Here is a Chasse-Marais with some breath in him,' she said, curtly, as the man with his mule-cart and its sad burden of half-dead, moaning, writhing frames drew near at her summons. 'Put him in. Soldiers cost too much, training, to waste them on jackals and kites, if one can help it. Lift him up—quick!'

'He is badly hurt?' said the Tringlo.

She shrugged her shoulders.

'Oh no! I have had worse scratches myself. The horse fell on him, that was the mischief. Most of them here have swallowed the "*petite pilule d'oubli*" once and for all. I never saw a prettier thing—every Lascar has killed his own little knot of Arbicos. Look how nice and neat they lie.'

Cigarette glanced over the field with the satisfied appreciation of a connoisseur glancing over a Soltykoff or Blacas collection unimpeachable for accuracy and arrangement; and drank a toss of her brandy, and lighted her little amber pipe, and sang loudly as she did so the gayest ballad of the Langue Verte.

She was not going to have him imagine she cared for that Chasseur whom he lifted up on his little waggon with so kindly a care—not she! Cigarette was as proud in her way as was ever the Princess Venetia Corona.

Nevertheless, she kept pace with the mules, carrying little Flick-Flack, and never paused on her way, though she passed scores of dead Arabs, whose silver ornaments and silk broideries commonly after such a fantasia replenished the knapsack and adorned in profusion the uniform of the young filibuster, being gleaned by her, right and left, as her lawful harvest after the fray.

'Leave him there. I will have a look at him,' she said, at the first empty tent they reached: the camp had been the scene of as fierce a struggle as the part of the plain which the cavalry had held, and it was strewn with the slaughter of Zouaves and Tirailleurs. The Tringlo obeyed her, and went about his errand of mercy. Cigarette, left alone with the wounded man, lying insensible still on a heap of forage, ceased her song, and grew very quiet. She had a certain surgical skill, learnt as her untutored genius learned most things, with marvellous rapidity, by observation and intuition; and she had saved many a life by her knowledge and her patient attendance on the sufferers—patience that she had been famed for when she had been only six years old, and a surgeon of the Algerian regiments had affirmed that he could trust her to be as wakeful, as watchful, and as sure to obey his directions as though she were a Sœur de Charité. Now 'the little fagot of opposites', as Cecil had called her, put this skill into active use.

The tent had been a scullion's tent; the poor *marmiton* had been killed, and lay outside, with his head clean-severed by an Arab flissa; his fire had gone out, but his brass pots and pans, his jar of fresh water, and his various preparations for the General's dinner, were still there. The General was dead also; far yonder, where he had fallen in the van of his Zouaves, exposing himself with all the splendid reckless gallantry of France; and the soup stood unserved, the wild plovers were taken by Flick-Flack, the empty dishes waited for the viands which there were no hands to prepare and no mouths to eat. Cigarette glanced round, and saw all with one flash of her eyes; then she knelt down beside the heap of forage, and, for the first thing, dressed his wounds with the cold, clear water, and washed away the dust and the blood that covered his chest.

'He is too good a soldier to die; one must do it for France,' she said to herself, in a kind of self-apology. And as she did it, and bound the lance-gash close, and bathed his breast, his forehead, his hair, his beard, free from the sand, and the powder, and the gore, a thousand changes swept over her mobile face. It was one moment soft, and flushed, and tender as passion; it was the next jealous, fiery, scornful, pale, and full of impatient self-disdain.

He was nothing to her—morbleu! He was an aristocrat, and she was a child of the people. She had been besieged by Dukes, and had flouted Princes; she had borne herself in such gay liberty, such vivacious freedom, such proud and careless sovereignty—bah! what was it to her whether this man lived or died? If she saved him, he would give her a low bow as he thanked her, thinking all the while of Miladi!

And yet she went on with her work.

Cecil had been stunned by a stroke from his horse's hoof as the poor beast fell beneath and rolled over him. His wounds were slight—marvellously so for the thousand strokes that had been aimed at him; but it was difficult to rouse him from unconsciousness, and his face was white as death where he lay on the heap of dry reeds and grasses. She began to feel fear of that lengthened syncope; a chill, tight, despairing fear that she had never known in her life before. She knelt silent a moment, drawing through her hand the wet locks of his hair with the bright threads of gold gleaming in it.

Then she started up, and, leaving him, found a match, and lighted the died-out wood afresh; the fire soon blazed up, and she warmed above it the soup that had grown cold, poured into it some red wine that was near, and forced some, little by little, down his throat. It was with difficulty at first that she could pass any through his tightly-locked teeth; but by degrees she succeeded, and, only half conscious still, he drank it faster, the heat and the strength reviving him as its stimulant warmed his veins. His eyes did not unclose, but he stirred, moved his limbs, and, with some muttered words she could not hear, drew a deeper breath and turned.

'He will sleep now—he is safe,' she thought to herself while she stood watching him with a curious conflict of pity, impatience, anger, and relief at war within her.

Bah! Why was she always doing good service to this man, who only cared for the blue serene eyes of a woman who would never give him aught except pain? Why should she take such care to keep the fire of vitality alight in him, when it had been crushed out in thousands as good as he, who would have no notice save a hasty thrust into the earth, no funeral chant except the screech of the carrion-birds?

Cigarette had been too successful in her rebellion against all weakness, and was far too fiery a young warrior to find refuge or consolation in the poet's plea,

> How is it under our control to love or not to love?

To allow anything to gain ascendancy over her that she resisted, to succumb to any conqueror that was unbidden and unwelcome, was a submission beyond words degrading to the fearless soldier-code of the Friend of the Flag. And yet—there she stayed and watched him. She took some food, for she had been fasting all day; then she dropped down before the fire she had lighted, and, in one of those soft, curled, kitten-like attitudes that were characteristic of her, kept her vigil over him.

She was bruised, stiff, tired, longing like a tired child to fall asleep; her eyes felt hot as flame, her rounded supple limbs were aching, her throat was sore with long thirst and the sand that she seemed to have swallowed till no draught of water or wine would take the scorched, dry pain out of it. But, as she had given up her fête-day in the hospital, so she sat now— as patient in the self-sacrifice as she was impatient when the vivacious agility of her young frame was longing for the frenzied delights of the dance or the battle.

Yonder she knew, where her Spahis bivouacked on the hard-won field, there were riotous homage, wild applause, intoxicated triumph waiting for the Little One who had saved the day, if she chose to go out for it; and she loved to be the centre of such adoration and rejoicing with all the exultant vanity of a child and a hero in one. Here there were warmth of flames, quietness of rest, long hours for slumber, all that her burning eyes and throbbing nerves were longing for, as the sleep she would not yield to stole on her, and the racking pain of fatigue cramped her bones. But she would not go to the pleasure without, and she would not give way to the weariness that tortured her.

Cigarette could crucify self with a generous courage, all the purer because it never occurred to her that there was anything of virtue or of sacrifice in it. She was acting *en bon soldat*—that was all. Pouf! that wanted no thanks.

Silence settled over the camp; half the slain could not be buried, and the clear luminous stars rose on the ghastly plateau. All that she heard were the challenge of sentinels, the tramp of patrols. The guard visited her once: '*C'est Cigarette*,' she said, briefly, and she was left undisturbed.

She kept herself awake in the little dark tent, only lit by the glow of the fire. Dead men were just without, and in the moonlight without, as the night came on, she could see the severed throat of the scullion, and the head further off, like a round grey stone. But that was nothing to Cigarette; dead men were no more to her than dead trees are to others.

Every now and then, four or five times in an hour, she gave him whom she tended the soup or the wine that she kept warmed for him over the embers. He took it without knowledge, sunk half in lethargy, half in sleep; but it kept the life glowing in him which, without it, might have perished of cold and exhaustion as the chills and northerly wind of the evening succeeded to the heat of the day, and pierced through the canvas walls of the tent. It was very bitter; more keenly felt because of the previous burning of the sun. There was no cloak or covering to fling over him; she took off her blue cloth tunic and threw it across his chest, and, shivering despite herself, curled closer to the little fire.

She did not know why she did it—he was nothing to her—and yet she kept herself wide awake through the dark autumn night, lest he should sigh or stir and she not hear him.

'I have saved his life twice,' she thought, looking at him; 'beware of the third time, they say!'

He moved restlessly, and she went to him. His face was flushed now; his breath came rapidly and shortly; there was some fever on him. The linen was displaced from his wounds; she dipped it again in water, and laid the cooled bands on them. 'Ah, bah! If I were not unsexed enough for this, how would it be with you now?' she said in her teeth. He tossed wearily to and fro; detached words caught her ear as he muttered them:

'Let it be, let it be—he is welcome! How could I prove it at his cost? I saved him—I could do that. It was not much——'

She listened with intense anxiety to hear the other whispers ending the sentence, but they were stifled and broken.

'*Tiens!*' she murmured below her breath. 'It is for some other he has ruined himself.'

She could not catch the words that followed, They were in an unknown language to her, for she knew nothing of English, and they poured fast and obscure from his lips as he moved in feverish unrest; the wine that had saved him from exhaustion inflaming his brain in his sleep. Now and then French phrases crossed the English ones; she leaned down to seize their meaning till her cheek was against his forehead, till her lips touched his hair; and at that half caress her heart beat, her face flushed, her mouth trembled with a too vivid joy, with an impulse, half fear and half longing, that had never so moved her before.

'If I had my birthright,' he muttered in her own tongue. 'If I had it—would she look so cold then? She might love me—women used once. O God! if *she* had not looked on me, I had never known all I have lost!'

Cigarette started as if a knife had stabbed her, and sprang up from her rest beside him.

'She—she—always she!' she muttered fiercely, while her face grew duskily scarlet in the fire-glow of the tent; and she went slowly away, back to the low wood fire.

This was to be ever her reward!

Her eyes glistened and flashed with the fiery vengeful passions of her hot and jealous instincts. Cigarette had in her the violence as she had the nobility of a grand nature that has gone wholly untutored and unguided; and she had the power of southern vengeance in her, though she had also the scant and rapid impulse to forgiveness of a generous and sunlit temper. It was bitter, beyond any other bitterness that could have wounded her, for the spoilt, victorious, imperious, little empress of the Army of Algeria to feel that, though she had given his life twice back to this man, she was less to him than the tiny white dog that nestled in his breast; that she who never before had endured a slight, or known what neglect could mean, gave care, and pity, and aid, and even tenderness, to one whose only thought was for a woman who had accorded him nothing but a few chill syllables of haughty condescension!

He lay there unconscious of her presence, tossing wearily to and fro in fevered unrefreshing sleep, murmuring incoherent words of French and English strangely mingled; and Cigarette crouched on the ground, with the firelight playing all over her picturesque, child-like beauty, and her large eyes strained and savage, yet with a strange mistful pain in them, looking out at the moonlight where the headless body lay in a cold grey sea of shadow.

Yet she did not leave him.

She was too generous for that. 'What is right is right. He is a soldier of France,' she muttered, while she kept her vigil. She felt no want of sleep; a hard hateful wakefulness seemed to have banished all rest from her; she stayed there all the night through. Whenever she could ease or aid him she rose and did so, with the touch of water on his forehead, or of cooled wine to his lips, by the alteration of the linen on his wounds, or the shifting of the rough forage that made his bed. But she did it without anything of that loving lingering attendance she had given before; she never once drew out the task longer than it needed, or let her hands wander among his hair, or over his lips, as she had done before.

And he never once was conscious of it; he never once knew that she was near. He did not waken from the painful, delirious, stupified slumber that had fallen on him; he only vaguely felt that he was suffering pain; he only vaguely dreamed of what he murmured of—his past, and the beauty of the woman who had brought all the memories of that past back on him.

And this was Cigarette's reward—to hear him mutter wearily of the proud eyes and of the lost smile of another!

The dawn came at last; her constant care and the skill with which she had cooled and dressed his wounds had done him infinite service; the fever had subsided, and towards morning his incoherent words ceased, his breathing grew calmer and more tranquil; he fell asleep—sleep that was profound, dreamless, and refreshing.

She looked at him with a tempestuous shadow darkening her face that yet was soft with a tenderness that she could not banish. She hated him; she ought to have stabbed or shot him rather than have tended him thus; he neglected her, and only thought of that woman of his old Order. As a daughter of the People, as a child of the Army, as a soldier of France, she ought to have killed him rather than have caressed his hair and soothed his pain! Pshaw! She ground one in another her tiny white teeth, that were like a spaniel's.

Then gently, very gently, lest she should waken him, she took her tunic skirt with which she had covered him from the chills of the night, put more broken wood on the fading fire, and with a last lingering look at him where he slept, passed out from the tent as the sun rose in a flushed and beautiful dawn. He would never know that she had saved him thus: he never should know it, she vowed in her heart.

Cigarette was very haughty in her own wayward, careless fashion. At a word of love from him, at a kiss from his lips, at a prayer from his voice, she would have given herself to him in all the abandonment of a first passion, and have gloried in being known as his mistress. But she would have perished by a thousand deaths rather than have sought him through his pity or through his gratitude; rather than have accepted the compassion of a heart that gave its warmth to another; rather than have ever let him learn that he was any more to her than all their other countless comrades who filled up the hosts of Africa.

'He will never know,' she said to herself, as she passed through the disordered camp, and in a distant quarter coiled herself amongst the hay of a forage-waggon, and covered up in dry grass, like a bird in a nest, let her tired limbs lie and her aching eyes close in repose. She was very tired; and every now and then as she slept a quick sobbing breath shook her as she slumbered, like a worn-out fawn who has been wounded whilst it played.

CHAPTER III

The Leathern Zackrist

With the reveille and the full daybreak Cigarette woke, herself again; she gave a little petulant shake to her fairy form when she thought of what folly she had been guilty of. 'Ah bah! you deserve to be shot,' she said to herself afresh. 'One would think you were a Silver Pheasant—you grow such a little fool!'

Love was all very well, so Cigarette's philosophy had always reckoned; a chocolate bonbon, a firework, a bagatelle, a draught of champagne, to flavour an idle moment. '*Vin et Vénus*' she had always been accustomed to see worshipped together, as became their alliterative; it was a bit of fun—that was all. A passion that had pain in it had never touched the Little One; she had disdained it with lightest, airiest contumely. 'If your dragée have a bitter almond in it, eat the sugar and throw the almond away, you goose! that is simple enough, isn't it? Bah! I don't pity the people who eat the bitter almond; not I—*ce sont bien bêtes, ces gens!*' she had said once, when arguing with an officer on the absurdity of a melancholy love that possessed him, and whose sadness she rallied most unmercifully. Now, for once in her young life, the Child of France found that it was remotely possible to meet with almonds so bitter that the taste will remain and taint all things, do what philosophy may to throw its acridity aside.

With the reveille she awoke, herself again, though she had not had more than an hour's slumber—awoke, it is true, with a dull ache at her heart that was very new and bitterly unwelcome to her, but with the buoyant vivacity and the proud carelessness of her nature in arms against it, and with that gaiety of childhood inherent to her repelling, and very nearly successfully, the foreign depression that weighed on it.

Her first thought was to take care that he should never learn what she had done for him. The Princess Corona would not have more utterly disdained to solicit regard through making a claim upon gratitude than the fiery little warrior of France would have done. She went straight to the Tringlo who had known her at her mission of mercy.

'Georges, *mon brave*,' said the Little One, with that accent of authority which was as haughty as any General's, 'do you know how that Chasseur is that we brought in last night?'

'Not heard, ma belle,' said the cheery little Tringlo, who was hard pressed; for there was much to be done, and he was very busy.

'What is to be done with the wounded?'

Georges lifted his eyebrows:

'Ma belle! there are very few. There are hundreds of dead. It was a duel *à outrance* yesterday. The few there are we shall take with an escort of Spahis to headquarters.'

'Good. I will go with you. Have a heed, Georges, never to whisper that I had anything to do with saving that man I called to you about.'

'And why, my Little One?'

'Because *I* desire you!' said Cigarette, with her most imperious emphasis. 'They say he is English, and a ruined milord, pardieu! Now I would not have an Englishman think I thought his six feet of carcase worth saving for a ransom.'

The Tringlo chuckled; he was an Anglophobist. In the Chinese expedition his share of 'loot' had been robbed from him by a trick of which two English soldiers had been the concocters, and a vehement animosity against the whole British race had been the fruit of it in him.

'*Non, non, non!*' he answered her, heartily. 'I understand. Thou art very right, Cigarette. If we have ever obliged an Englishman, he thinks his obligation to us opens him a neat little door through which to cheat us. It is very dangerous to oblige the English; they always hate you for it. That is their way. They may have virtues; they may,' he added, dubiously, but with an impressive air of strictest impartiality, 'but among them is not written gratitude. Ask that man, Rac, how they treat their soldiers!' and M. Georges hurried away to his mules and his duties, thinking with loving regret of the delicious Chinese plunder of which the dogs of Albion had deprived him.

'He is safe!' thought Cigarette; of the patrol who had seen her she was not afraid—he had never noticed with whom she was when he had put his head into the scullion's tent; and she made her way towards the place where she had left him, to see how it went with this man whom she was so careful should never know that which he had owed to her.

It went well with him, thanks to her; care, and strengthening nourishment, and the skill of her tendance, had warded off all danger from his wound. The bruise and pressure from the weight of the horse had been more ominous, and he could not raise himself or even breathe without severe pain; but his fever had left him, and he had just been lifted into a mule-drawn ambulance-waggon as Cigarette reached the spot.

'How goes the day, Monsieur Victor? So you got sharp scratches, I hear? Ah! that was a splendid thing we had yesterday! When did you go down? We charged together!' she cried gaily to him; then her voice dropped suddenly, with an indescribable sweetness and change of tone. 'So!—you suffer still?' she asked, softly.

Coming close up to where he lay on the straw, she saw the exhausted languor of his regard, the heavy darkness under his eyelids, the effort with which his lips moved as the faint words came broken through them.

'Not very much, *ma belle*, I thank you. I shall be fit for harness in a day or two. Do not let them send me into hospital. I shall be perfectly— well—soon.'

Cigarette swayed herself upon the wheel and leaned towards him, touching and changing his bandages with clever hands:

'They have dressed your wound ill; whose doing is that?'

'It is nothing. I have been half cut to pieces before now; this is a mere bagatelle. It is only——'

'That it hurts you to breathe? I know! Have they given you anything to eat this morning?'

'No. Everything is in confusion. We——'

She did not stay for the conclusion of his sentence; she had darted off, quick as a swallow. She knew what she had left in her dead scullion's tent. Everything was in confusion, as he had said. Of the few hundreds that had been left after the terrific onslaught of the past day, some were employed far out, thrusting their own dead into the soil; others were removing the tents and all the equipage of the camp; others were busied with the wounded, of whom the greatest sufferers were to be borne to the nearest hospital (that nearest many leagues away over the wild and barren country); while those who were likely to be again soon ready for service were to be escorted to the headquarters of the main army. Among the latter Cecil had passionately entreated to be numbered; his prayer was granted to the man who had kept at the head of his Chasseurs and borne aloft the Tricolour through the whole of the war-tempest on which the dawn had risen, and which had barely lulled and sunk by the setting of the sun. Châteauroy was away with the other five of his squadrons; and the Zouave chef de bataillon, the only officer of any rank who had come alive through the conflict, had himself visited Bertie, and given him warm words of eulogy, and even of gratitude, that had soldierly sincerity and cordiality in them.

'Your conduct was magnificent,' he had said, as he had turned away. 'It shall be my care that it is duly reported and rewarded.'

Cigarette was but a few seconds absent; she soon bounded back like the swift little chamois she was, bringing with her a huge bowl full of red wine with bread broken in it.

'This is the best I could get,' she said; 'it is better than nothing. It will strengthen you.'

'What have you had yourself, *petite*?'

'Ah bah! Leave off thinking for others; I have breakfasted long ago,' she answered him. (She had only eaten a biscuit well-nigh as hard as a flint.) 'Take it—here, I will hold it for you.'

She perched herself on the wheel like a bird on a twig; she had a bird's power of alighting and sustaining herself on the most difficult and most airy elevation; but Cecil turned his eyes on the only soldier in the cart beside himself, one of the worst men in his regiment—a murderous, sullen, black-browed, evil wretch, fitter for the bench of the convict-galley than for the ranks of the cavalry.

'Give half to Zackrist,' he said. 'I know no hunger; and he has more need of it.'

'Zackrist! that is the man who stole your lance and accoutrements, and got you into trouble by taking them to pawn in your name, a year and more ago.'

'Well, what of that? He is not the less hungry.'

'What of that? Why, you were going to be turned into the First Battalion[1] disgraced for the affair, because you would not tell of him: if Vireflou had not found out the rights of the matter in time!'

---

[1] The battalion of the criminal outcasts of all corps, whether horse or foot; answering to the *Straf-bataillons* of the Austrian service.

'What has that to do with it?'

'This, Monsieur Victor, that you are a fool.'

'I dare say I am. But that does not make Zackrist less hungry.'

He took the bowl from her hands, and emptying a little of it into the wooden *bidon* that hung to her belt, kept that for himself, and stretching his arm across the straw, gave the bowl to Zackrist, who had watched it with the longing ravenous eyes of a starving wolf, and seized it with rabid avidity.

A smile passed over Cecil's face, amused despite the pain he suffered.

'That is one of my "sensational tricks", as M. de Châteauroy calls them. Poor Zackrist! did you see his eyes?'

'A jackal's eyes, yes!' said Cigarette, who, between her admiration for the action and her impatience at the waste of her good bread and wine, hardly knew whether to applaud or to deride him. 'What recompense do you think you will get? He will steal your things again, first chance.'

'May be. I don't think he will. But he is very hungry all the same; that is about the only question just now,' he answered her, as he drank and ate his portion, with a need of it that could willingly have made him take thrice as much, though for the sake of Zackrist he had denied his want of it.

The Zackrist himself, who could hear perfectly what was said, uttered no word; but when he had finished the contents of the bowl, lay looking at his corporal with an odd gleam in the dark sullen savage depths of his hollow eyes. He was not going to say a word of thanks; no!—none had ever heard a grateful or a decent word from him in his life; he was proud of that. He was the most foul-mouthed brute in the army, and, like *Snake* in the *School for Scandal*, thought a good action would have ruined his character for ever. Nevertheless, there came into his cunning and ferocious eyes a glisten of the same light which had been in the little *gamin*'s when first by the bivouac-fire he had murmured, '*Picpon s'en souviendra.*'

'When anybody stole from me,' muttered Cigarette, 'I shot him.'

'You would have fed him had he been starving. Do not belie yourself, Cigarette; you are too generous ever to be vindictive.'

'Pooh! Revenge is one's right.'

'I doubt that. We are none of us good enough to claim it, at any rate.'

Cigarette shrugged her shoulders in silence; then poising herself on the wheel, she sprang from thence on to the back of her little mare which she had brought up, having the reins in one of her hands and the wine-bowl in the other; the animal had not had a scratch in the battle, and was fresh and bright after the night's repose.

'I will ride with you, with my Spahis,' she said, as a young queen might have promised protection from her escort. He thanked her, and sank back among the straw, exhausted and worn out with pain and with languor; the weight that seemed to oppress his chest was almost as hard to bear as when the actual pressure of his dead charger's body had been on him.

Yet, as he had said, it was but a bagatelle beside the all but mortal wounds, the agonising neuralgia, the prostrating fever, the torture of bullet-torn nerves, and the scorching fire of inflamed sword-wounds, that had in their turn been borne by him in his twelve years of African service—things which, to men who have never suffered them, sound like the romanced horrors of an exaggerated imagination; yet things which are daily and quietly borne, by such soldiers as the soldiers of the Algerian Army, as the natural accompaniments of a military life—borne, too, in brave simple unconscious heroism by men who know well that the only reward for it will be their own self-contentment at having been true to the traditions of their regiment.

Four other troopers were placed on the straw beside him, and the mule-carts with their mournful loads rolled slowly out of camp eastward towards the quarters of the main army; the Spahis, glowing red against the sun, escorting them, with their darling in their midst, while from their deep chests they shouted war-songs in Sabir with all the wild and riotous delight that the triumph of victory and the glow of bloodshed roused in those who combined in them the fire of France and the fanaticism of Islamism—an irresistible union.

Though the nights were now cold, and before long even the advent of snow might be looked for, the days were hot and even scorching still. Cigarette and her Spahis took no heed of it; they were desert born and bred; and she was well-nigh invulnerable to heat as any little salamander. But, although they were screened as well as they could be under an improvised awning, the wounded men suffered terribly. Gnats and mosquitoes and all the winged things of the African air tormented them, and tossing on the dry hot straw they grew delirious, some falling asleep and murmuring incoherently, others lying with wide open eyes of half-sense-less straining misery. Cigarette had known well how it would be with them; she had accompanied such escorts many a time; and ever and again when they halted she dismounted and came to them, and mixed wine with some water that she had slung a barrel of to her saddle, and gave it to them, and moved their bandages, and spoke to them with a soft caressing consolation that pacified them as if by some magic. She had led them like a young lion on to the slaughter in the past day; she soothed them now with a gentleness that the gentlest daughter of the Church could not have surpassed.

The way was long; the road ill formed, leading for the most part across a sear and desolate country, with nothing to relieve its barrenness except long stretches of the great spear-headed reeds. At noon the heat was intense; the little cavalcade halted for half an hour under the shade of some black towering rocks which broke the monotony of the district, and commenced a more hilly and more picturesque portion of the country. Cigarette came to the side of the temporary ambulance in which Cecil was placed. He was asleep—sleeping for once peacefully with little trace of pain upon his features, as he had slept the previous night. She saw that his face and chest had not been touched by the stinging insect-swarm; he was doubly screened by a shirt hung above him dexterously on some bent sticks.

'Who has done that?' thought Cigarette. As she glanced round she saw:—without any linen to cover him, Zackrist had reared himself up and leaned slightly forward over against his comrade. The shirt that protected Cecil was his; and on his own bare shoulders and mighty chest the tiny armies of the flies and gnats were fastened, doing their will uninterrupted.

As he caught her glance, a sullen ruddy glow of shame shone through the black hard skin of his sunburnt visage—shame to which he had been never touched when discovered in any one of his guilty and barbarous actions.

'Dame!' he growled, savagely; 'he gave me his wine; one must do something in return. Not that I feel the insects—not I; my skin is leather, see you? they can't get through it; but his is *une peau de femme*—white and soft—bah! like tissue-paper!'

'I see, Zackrist; you are right. A French soldier can never take a kindness from an English fellow without outrunning him in generosity. Look—here is some drink for you.'

She knew too well the strange nature with which she had to deal to say a syllable of praise to him for his self-devotion, or to appear to see that, despite his boast of his leather skin, the stings of the cruel winged tribes were drawing his blood and causing him alike pain and irritation which, under that sun, and added to the torment of his gunshot-wound, were a martyrdom as great as the noblest saint ever endured.

'*Tiens—tiens!* I did him wrong,' murmured Cigarette. 'That is what they are—the children of France—even when they are at their worst, like that devil, Zackrist. Who dare say they are not the heroes of the world?'

And all through the march she gave Zackrist a double portion of her water dashed with red wine, that was so welcome and so precious to the parched and aching throats; and all through the march Cecil lay asleep, and the man who had thieved from him, the man whose soul was stained with murder and pillage and rapine, sat erect beside him, letting the insects suck his veins and pierce his flesh.

It was only when they drew near the camp of the main army that Zackrist beat off the swarm and drew his old shirt over his head. 'You do not want to say anything to him,' he muttered to Cigarette. 'I am of leather, you know; I have not felt it.'

She nodded; she understood him. Yet his shoulders and his chest were well-nigh flayed, despite the tough and horny skin of which he made his boast.

'Dieu! we are droll!' mused Cigarette. 'If we do a good thing, we hide it as if it were a bit of stolen meat, we are so afraid it should be found out; but, if they do one in the world there, they bray it at the tops of their voices from the houses' roofs, and run all down the streets screaming about it for fear it should be lost. Dieu! we are droll!'

And she dashed the spurs into her mare and galloped off at the height of her speed into camp—a very city of canvas, buzzing with the hum of life, regulated with the marvellous skill and precision of French warfare, yet with the carelessness and the picturesqueness of the desert-life pervading it.

'*C'est la Cigarette!*' ran from mouth to mouth as the bay mare with her little Amazon rider, followed by the scarlet cloud of the Spahis, all ablaze like poppies in the sun, rose in sight, thrown out against the azure of the skies.

What she had done had been told long before by an orderly, riding hard in the early night to take the news of the battle; and the whole host was on watch for its darling—the saviour of the honour of France. Like wave rushing on wave of some tempestuous ocean, the men swept out to meet her in one great surging tide of life, impetuous, passionate, idolatrous, exultant, with all the vivid ardour, all the uncontrolled emotion, of natures south-born, sun-nurtured. They broke away from their mid-day rest as from their military toil, moved as by one swift breath of fire, and flung themselves out to meet her, the chorus of a thousand voices ringing in deafening vivas to the skies. She was enveloped in that vast sea of eager furious lives, in that dizzy tumult of vociferous cries, and stretching hands, and upturned faces. As her soldiers had done the night before, so these did now—kissing her hands, her dress, her feet, sending her name in thunder through the sunlit air, lifting her from off her horse, and bearing her, in a score of stalwart arms, triumphant in their midst.

She was theirs—their own—the Child of the Army, the Little One whose voice above their dying brethren had the sweetness of an angel's song, and whose feet, in their hours of revelry, flew like the swift and dazzling flight of gold-winged orioles. And she had saved the honour of their Eagles; she had given to them and to France their god of Victory. They loved her—O God, how they loved her!—with that intense, breathless, intoxicating love of a multitude which, though it may stone tomorrow what it adores today, has yet for those on whom it has once been given thus a power no other love can know—a passion unutterably sad, deliriously strong.

That passion moved her strangely.

As she looked down upon them, she knew that not one man breathed amongst that vast tumultuous mass but would have died that moment at her word; not one mouth moved amongst that countless host but breathed her name in pride, and love, and honour.

She might be a careless young coquette, a lawless little brigand, a child of sunny caprices, an elf of dauntless mischief; but she was more than these. The divine fire of genius had touched her, and Cigarette would have perished for her country not less surely than Jeanne d'Arc. The holiness of an impersonal love, the glow of an imperishable patriotism, the melancholy of a passionate pity for the concrete and unnumbered sufferings of the people were in her, instinctive and inborn, as fragrance in the heart of flowers. And all these together moved her now, and made her young face beautiful as she looked down upon the crowding soldiery.

'It was nothing,' she answered them—'it was nothing. It was for France.'

For France! They shouted back the beloved word with tenfold joy; and the great sea of life beneath her tossed to and fro in stormy triumph, in frantic paradise of victory, ringing her name with that of France upon the air, in thunder-shouts like spears of steel smiting on shields of bronze.

But she stretched her hand out, and swept it backward to the desert-border of the south with a gesture that had awe for them.

'Hush!' she said, softly, with an accent in her voice that hushed the riot of their rejoicing homage till it lulled like the lull in a storm. 'Give me no honour whilst *they* sleep yonder. With the dead lies the glory!'

CHAPTER IV

By the Bivouac-fire

Le Roi Gaillard qui s'appelle la Guerre,
C'est mon souverain tout débonnair;
Au bouche qui rit, au main qui tue,
Au front d'airain, aux yeux de feu!
Comme il est beau ce roi si gai,
Qui fait le diable à quatre au gré,
Qui brûle, qui boit, qui foudre, qui fume,
Qui aime le vin, le sang, l'écume,
Qui jette la torche——

'Holà! nous v'la!' cried Cigarette, interrupting herself in her chant in honour of the attributes of war, as the Tringlo's mules which she was driving some three weeks after the fray of Zarâila, stopped, by sheer force of old habit, in the middle of a green plateau on the outskirts of a camp pitched in its centre, and overlooked by brown rugged scarps of rock, stunted bushes on their summits, and here and there a maritime pine clinging to their naked slopes. At sight of the food-laden little beasts, and the well-known form behind them, the Tirailleurs, the Indigènes, and the Zouaves, on whose side of the encampment she had approached, rushed towards her with frantic shouts, and wild delight, and vehement hurrahs in a tempest of vociferous welcome that might have stunned any ears less used, and startled any nerves less steeled, to military life than the Friend of the Flag's. She signed back the shouting disorderly crowd with her mule-whip as superbly as though she were a Marshal of France signing back a whole army's mutiny.

'What children you are! You push, and scramble, and tear, like a set of monkeys over a nut. Get out of my way, or I swear you shall none of you have so much as a morsel of black bread—do you hear?'

It was amusing to see how they minded her contemptuous orders; how these black-bearded fire-eaters, the terror of the country, each one of whom could have crushed her in his grasp as a wolf crushes a lamb, slunk back, silenced and obedient, before the imperious bidding of the little vivandière. They had heeded her and let her rule over them almost as much when she had been seven years old, and her curls, now so dark, had been yellow as corn in the sun.

'Ouf!' growled only one insubordinate, 'if you had been a day and night eating nothing but a bit of moist clay, *you* might be hungry too, *fanfan?*'

The humiliated supplication of the reply appeased their autocratic sovereign. She nodded her head in assent.

'I know; I know. I have gone days on a handful of barley-ears. M. le Colonel has his *marmitons*, and his *fricassées*, and his *batterie de cuisine* where he camps—oh-hé!—but we soldiers have nothing but a hunch of baked chaff. Well, we win battles on it—eh? *"Quand la panse est vide, l'épée mange vite!"* '

Which was one of the impromptu proverbs that Cigarette was wont to manufacture and bring into her discourse with an air of authority as of one who quotes from profound scholastic lore. It was received with a howl of applause and of ratification. The entrails often gnaw with bitter pangs of famine in the Army of Algiers, and they knew well how sharp an edge hunger gives to the steel.

Nevertheless, the sullen angry roar of famished men, that is so closely, so terribly, like the roar of wild beasts, did not cease.

'Where is Biribi?' they growled. 'Biribi never keeps us waiting. Those are Biribi's beasts.'

'Right,' said Cigarette, laconically, with a crack of her mule-whip on to the arm of a Zouave who was attempting to make free with her convoy and purloin a loaf off the load.

'Where is Biribi, then?' they roared in concert, a crowd of eager, wolfish, ravenous, impatient men, hungry as camp-fasting could make them, and half inclined even to tear their darling in pieces, since she kept them thus from the stores.

Cigarette uncovered her head with a certain serious grace very rare in her.

'Biribi has made a good end.'

Her assailants grew very quiet.

'Shot?' they asked, briefly. Biribi was a Tringlo well beloved in all the battalions.

Cigarette nodded, with a gesture outward to the solitary country. She was accustomed to these incidents of war; she thought of them no more than a girl of civilized life thinks of the grouse or the partridges that are killed by her lovers and brothers.

'I was out yonder, two leagues or more away. I was riding; I was on my own horse; Etoile-Filante. Well, I heard shots; of course I made for the place by my ear. Before I got up I saw what was the mischief. There were the mules in a gorge, and Biribi in front of them, fighting, mon Dieu!— fighting like the devil—with three Arbis upon him. They were trying to stop the convoys, and Biribi was beating them back with all his might. I was too far off to do much good; but I shouted and dashed down to them.

The Arbis heard, Biribi heard; he flew on to them like a tiger, that little Tringlo. It was wonderful! Two fell dead under him; the third took fright and fled. When I got up, Biribi lay above the dead brutes with a dozen wounds in him, if there were one. He looked up, and knew me. "Is it thee, Cigarette?" he asked; and he could hardly speak for the blood in his throat. "Do not wait with me; I am dead already. Drive the mules into camp as quick as thou canst; the men will be thinking me late."'

'Biribi was always *bon enfant*,' muttered the listening throng; they forgot their hunger as they heard.

'Ah, *chenapans!* he thought more of you than you deserve, you jackals! I drew him aside into a hole in the rocks out of the heat. He was dead; he was right. No man could live slashed about like that. The Arbicos had set on him as he went singing along; if he would have given up the brutes and the stores, they would not have harmed him; but that was not Biribi. I did all I could for him. Dame! it was no good. He lay very still for some minutes with his head on my lap; then he moved restlessly and tossed about. "They will think me so late—so late," he muttered; "and they are famished by this. There is that letter, too, from his mother for Petit-Pot-de-Terre; there is all that news from France; I have so much for them, and I shall be so late—so late!" All he thought was that he should be so late into camp. Well, it was all over very soon. I do not think he suffered; but he was so afraid you should not have the food. I left him in the cave, and drove the mules on as he asked. Etoile-Filante had galloped away; have you seen him home?'

There broke once more from the hearkening throng a roar that shook the echoes from the rocks; but it was not now the rage of famished longing, but the rage of the lust for vengeance, and the grief of passionate hearts blent together. Quick as the lightning flashes, their swords leaped from their scabbards and shook in the sun-lightened air.

'We will avenge him!' they shouted as with one throat, the hoarse cry rolling down the valley like a swell of thunder. If the bonds of discipline had loosed them, they would have rushed forth on the search and to the slaughter, forgetful of hunger, of heat, of sun-stroke, of self-pity, of all things save the dead Tringlo, whose only fear in death had been lest they should want and suffer through him.

Their adjutants, alarmed by the tumult, hurried to the spot, fearing a bread-riot; for the camp was far from supplies, and had been ill victualled for several days. They asked rapidly what was the matter.

'Biribi has been killed,' some soldier answered.

'Ah! and the bread not come?'

'Yes, *mon adjutant*; the bread is there, and Cigarette too.'

'There is no need for me, then,' muttered the Adjutant of Zouaves; 'the Little One will keep order.'

The Little One had before now quelled a mutiny with her pistol at the ringleader's forehead, and her brave scornful words scourging the insubordinates for their dishonour to their arms, for their treason to the tricolour; and she was equal to the occasion now. She lifted her right hand:

'We will avenge him. That is of course. The Flag of France never hangs idly when there is a brave life's loss to be reckoned for; I shall know again the cur that fled. Trust to me, and now be silent. You bawl out your oath of vengeance, oh yes! But you bawled as loud a minute ago for bread. Biribi loved you better than you deserved. You deserve nothing; you are hounds as ready to tear for offal to eat as to rend the foe of your dead friend. Bah!'

The roar of the voices sank somewhat; Cigarette had sprung aloft on a gun-carriage, and as the sun shone on her face it was brilliant with the scorn that lashed them like whips.

'Sang de Dieu?' fiercely swore a Zouave. 'Hounds, indeed! If it were any one but you! When one has had nothing but a snatch of raw bullock's meat, and a taste of coffee black with mud, for a week through, is one a hound because one hungers?'

'No,' said the orator from her elevation, and her eyes softened wonderfully. In her heart she loved them so well, these wild barbaric warriors that she censured—'no, one is not a hound because one hungers; but one is not a soldier if one complains. Well! Biribi loved you; and I am here to do his will, to do his work. He came laden; his back was loaded heavier than the mules'. To the front, all of you, as I name you! Petit-Pot-de-Terre, there is your old mother's letter. If she knew as much as I do about you, scapegrace, she would never trouble herself whether you were dead or alive! Fagotin! here is a bundle of Paris newspapers for you; they are quite new—only nine months old! Potélé! some woman has sent you a love-scrawl and some tobacco; I suppose she knew your passions all ended in smoke! Rafle! here is a little money come for you from France; it has not been stolen, so it will have no spice for you! Racoleur! here is a *poulet*[1] from some simpleton, with a knife as a souvenir; sharpen it on the Arbicos. Poupard, Loup-terrible, Jean Pagnote, Pince-Maille, Louis Magot, Jules Goupil—here! There are your letters, your papers, your commissions. Biribi forgot nothing. As if you deserved to be worked for or thought of, *sacripans*!'

---

[1] Love-billet.

With which reproach, Cigarette relieved herself of the certain pain that was left on her by the death of Biribi; she always found, that to work yourself into a passion with somebody is the very best way in the world to banish an unwelcome emotion.

The men summoned by their camp-sobriquets, which were so familiar that they had, many of them, fairly forgotten their original names, rallied around her to receive the various packets with which a Tringlo is commonly charged by friends in the towns, or relatives away in France, for the soldiers of African brigades, and which, as well as his convoy of food and his budget of news, render him so precious and so welcome an arrival at an encampment. The dead Biribi had been one of the lightest, brightest, cheeriest, and sauciest of the gay, kindly, industrious wanderers of his branch of the service; always willing to lend, always ready to help; always smoking, singing, laughing, chattering; treating his three mules as an indulgent mother her children; calling them Plick, Plack, et Plock, and thinking of Plick, Plack, et Plock far beyond himself at all times; a merry, busy, smiling, tender-hearted soul, who was always happy, trudging along the sunburnt road, and carolling in his joyous voice *chansonnettes* and *gaudrioles* to the African flocks and herds, amidst the African solitudes. If there were a man they loved, it was Biribi; Biribi, whose advent in camp had always been the signal for such laughter, such abundance, such shower of newspapers, such quantities of intelligence from that France for tidings of which the hardest-featured veteran amongst them would ask with a pang at the heart, with a thrill in the words. And they had sworn, and would keep what they had sworn in bitter intensity, to avenge him to the uttermost point of vengeance. Yet five minutes afterwards, when the provisions Plick, Plack, et Plock had brought were divided and given out, they were shouting, eating, singing, devouring, with as eager a zest, and as hearty an enjoyment, as though Biribi were amongst them, and did not lie dead two leagues away, with a dozen wounds slashed on his stiffening frame.

'What heartless brutes! Are they always like that?' muttered a gentleman-painter who, travelling through the interior to get military sketches, had obtained permission to take up quarters in the camp.

'If they were not like that, they could not live a day,' a voice answered, curtly, behind him. 'Do you know what this service is, that you venture to judge them? Men who meet death in the face every five minutes they breathe cannot afford the space for sentimentalism which those who saunter at ease and in safety can do. They laugh when we are dead, perhaps, but they are true as steel to us while we live; it is the reverse of the practice of the world!'

The tourist started, turned, and looked aghast at the man who had reproved him; it was a Chasseur d'Afrique, who, having spoken, was already some way onward, moving through the press and tumult of the camp to his own regiment's portion of it.

Cigarette, standing by to see that Plick, Plack, and Plock were properly baited on the greenest forage to be found, heard, and her eyes flashed with a deep delight.

'Dame!' she thought, 'I could not have answered better myself! He is a true soldier, that.' And she forgave Cecil all his sins to her with the quick, impetuous, generous pardon of her warm little Gallic heart.

Cigarette believed that she could hate very bitterly; indeed, her power of resentment she rated high amongst her grandest qualities. Had the little leopard been told that she could not resent to the death what offended her, she would have held herself most infamously insulted. Yet hate was, in truth, foreign to her frank, vivacious nature; its deadliness never belonged to her, if its passion might; and at a trait akin to her, at a flash of sympathetic spirit in the object of her displeasure, Cigarette changed from wrath to friendship with the true instinct of her little heart of gold. A heart which, though it had been tossed about on a sea of blood, and had never been graven with so much as one tender word or one moral principle from the teachings of any creature, was still gold, despite all, no matter the bruises and the stains and the furnace-heats that had done their best to harden it into bronze, to debase it into brass.

The camp was large, and a splendid picture of colour, movement, picturesque combination, and wonderful light and shadow, as the sun-glow died out and the fires were lighted; for the nights were now intensely cold, cold with the cutting, icy, withering *bise*, and clear above as an Antarctic night, though the days were still hot and dry as flame.

On the left were the Tirailleurs, the Zouaves, the Zéphyrs; on the right were the Cavalry and the Artillery; in the centre of all was the tent of the Chief. Everywhere, as evening fell, the red warmth of fires rose; the caldron of soup or of coffee simmered, gipsy-like, above; the men lounged around, talking, laughing, cooking, story-telling at their pleasure; after the semi-starvation of the last week, the abundance of stores that had come in with other Tringlos besides poor Biribi, caused an universal hilarity. The glitter of accoutrements, the contents of open knapsacks, the skins of animals just killed for the *marmite*, the boughs of pines broken for firewood, strewed the ground. Tethered horses, stands of arms, great drums and eagle-guidons, the looming darkness of huge cannon, the blackness, like dromedaries couched, of caissons and ambulance-waggons, the whiteness of the canvas tents, the incessant movement as the crowds of soldiery stirred, and chattered, and worked, and sang—all these, on the green level of the plain, framed in by the towering masses of the rugged rocks, made a picture of marvellous effect and beauty.

Cecil, looking at it, thought so; though the harsh and bitter misery which he knew that glittering scene enfolded, and which he had suffered so many years himself—misery of hunger, of cold, of shot-wounds, of racking bodily pains—stole from it, in his eyes, that poetry and that picturesque brilliancy which it bore to the sight of the artist and the amateur. He knew the naked terrors of war, the agony, the travail, the icy chills, the sirocco heats, the grinding routine, the pitiless chastisements of its reality; to those who do, it can no longer be a spectacle dressed in the splendid array of romance. It is a fearful tragedy and farce woven close one in another; and its sole joy is in that blood-thirst which men so lustfully share with the tiger, and yet shudder from when they have sated it.

It was this knowledge of war, in its bitter and deadly truth, which had made him give the answer that had charmed Cigarette, to the casual visitor of the encampment.

He sat now, having recovered from the effects of the day of Zaràila, within a little distance of the fire at which his men were stewing some soup in the great simmering copper bowl. They had eaten nothing for nigh a week, except some mouldy bread, with the chance of a stray cat or a shot bird to flavour it. Hunger was a common thorn in Algerian warfare, since not even the matchless *intendance* of France could regularly supply the troops across those interminable breadths of arid land, those sun-scorched plains, swept by Arab foragers.

'Beau Victor! you took their parts well,' said a voice behind him, as Cigarette vaulted over a pile of knapsacks and stood in the glow of the fire, with a little pipe in her pretty rosebud mouth, and her cap set daintily on one side of her curls.

He looked up, and smiled.

'Not so well as your own clever tongue would have done. Words are not my weapons.'

'No! You are as silent as the grave commonly; but when you do speak, you speak well,' said the vivandière-Demosthenes, condescendingly. 'I hate silence myself! Thoughts are very good grain, but if they are not whirled round, round, round, and winnowed and ground in the mill-stones of talk, they keep little, hard, useless kernels, that not a soul can digest.'

With which metaphor Cigarette blew a cloud of smoke into the night air, looking the prettiest little genre picture in the ruddy firelight that ever was painted on such a background of wavering shadow and undulating flame.

'Will your allegory hold good, *petite*?' smiled Cecil, thinking but little of his answer or of his companion, of whose service to him he remained utterly ignorant. 'I fancy speech is the chaff most generally, little better. So, they talk of you for the Cross? No soldier ever, of a surety, more greatly deserved it.'

Her eyes gleamed with a lustre like the African planets above her; her face caught all the fire, the light, the illumination of the flames flashing near her.

'I did nothing,' she said, curtly. 'Any man on the field would have done the same.'

'That is easy to say; not so easy to prove. In all great events there may be the same strength, courage, and desire to act greatly in those who follow as in the one that leads; but it is only in that one that there is also the daring to originate, the genius to seize aright the moment of action and of success.'

Cigarette was a little hero, she was, moreover, a little desperado; but she was a child in years and a woman at heart, valiant and ruthless young soldier though she might be. She coloured all over her *mignonne* face at the words of eulogy from this man whom she had told herself she hated: her eyes filled; her lips trembled.

'It was nothing,' she said, softly, under her breath. 'I would die twenty deaths for France.'

He looked at her, and for the hour understood her aright; he saw that there was the love for her country and the power of sacrifice of a Viriathus or an Arminius in this gay-plumaged and capricious little hawk of the desert.

'You have a noble nature, Cigarette,' he said, with an earnest regard at her. 'My poor child, if only——' He paused. He was thinking what it was hard to say to her—if only the accidents of her life had been different, what beauty, grace, and genius might have been developed out of the untamed, untutored, inconsequent, but glorious nature of the child-warrior.

As by a fate, unconsciously his pity embittered all the delight his praise had given, and this implied regret for her stung her as the rend of the spur a young Arab colt—stung her inwardly into cruel wrath and pain; outwardly into irony, devilry, and contemptuous retort.

'Oh-hé! Child, indeed! Was I a child the other day, my good fellow, when I saved your squadron from being cut to pieces like grass with a scythe? As for nobility? Pouf! Not much of that in me. I love France— yes. A soldier always loves his country. She is so brave, too, and so fair, and so *riante*, and so gay. Not like your Albion—if it *is* yours—who is a great *gobemouche* stuffed full of cotton, steaming with fog, clutching gold with one hand and the Bible with the other, that she may swell her money-bags, and seem a saint all the same; never laughing, never learning, always growling, always shuffling, who is like this spider—look!—a tiny body and huge hairy legs—pull her legs, the Colonies, off, and leave her little English body, all shrivelled and shrunk alone, and I should like to know what size she would be then, and how she would manage to swell and to strut?'

Wherewith Cigarette tossed the spider into the air, with all the supreme disdain she could impel into that gesture. Cigarette, though she knew not her A, B, C, D, and could not have written her name to save her own life, had a certain bright intelligence of her own that caught up political tidings, and grasped at public subjects with a skill education alone will not bestow. One way and another, she had heard most of the floating opinions of the day, and stored them up in her fertile brain as a bee stores honey into his hive by much as nature-given and unconscious an instinct as the bee's own.

Cecil listened amused.

'You little Anglophobist! You have the tongue of a Voltaire.'

'Voltaire?' questioned Cigarette. 'Voltaire. Let me see. I know that name. He was the man who championed Calas? who had a fowl in the pot for every poor wretch that passed his house? who was taken to the Panthéon by the people in the Revolution?'

'Yes. And the man whom the wise world pretends still to call without a heart or a God!'

'Chut! He fed the poor, and freed the wronged. Better than pattering Paters, that!' said Cigarette, who thought a midnight mass at Notre-Dame or a Salutation at the Madeleine a pretty *coup de théâtre* enough, but who had for all churches and creeds a serene contempt and a fierce disdain. 'Go to the grandams and the children!' she would say, with a shrug of her shoulders, to a priest, whenever one in Algiers or Paris attempted to reclaim her; and a son of the Order of Jesus, famed for persuasiveness and eloquence, had been fairly beaten once when, in the ardour of an African missionary, he had sought to argue with the little Bohemian of the Tricolour, and had had his logic rent in twain, and his rhetoric scattered like dust, under the merciless home-thrusts and the sarcastic artillery of Cigarette's replies and enquiries.

'Holà!' she cried, leaving Voltaire for what took her fancy. 'We talk of Albion—there is one of her sons. I detest your country, but, *ma foi!* I must confess she breeds uncommonly handsome men.'

She was a dilettante in handsome men; she nodded her head now to where, some yards off, at another of the camp-fires, stood, with some officers of the regiment, one of the tourists; a very tall, very fair man, with a gallant bearing, and a tawny beard that glittered to gold in the light of the flames.

Cecil's glance followed Cigarette's. With a great cry he sprang to his feet and stood entranced, gazing at the stranger. She saw the startled amaze, the longing love, the agony of recognition, in his eyes; she saw the impulse in him to spring forward, and the shuddering effort with which the impulse was controlled. He turned to her almost fiercely:

'He must not see me! Keep him away—away, for God's sake!'

He could not leave his men; he was fettered there where his squadron was camped. He went as far as he could from the flame-light into the shadow, and thrust himself amongst the tethered horses. Cigarette asked nothing; comprehended at a glance with all the tact of her nation; and sauntered forward to meet the officers of the regiment as they came up to the picket-fire with the yellow-haired English stranger. She knew how charming a picture there, with her hands lightly resting on her hips, and her bright face danced on by the ruddy fire-glow, she made: she knew she could hold thus the attention of a whole brigade. The eyes of the stranger lighted on her, and his voice laughed in mellow music to his companions and *ciceroni*.

'Your intendance is perfect; your ambulance is perfect; your camp-cookery is perfect, messieurs; and here you have even perfect beauty too! Truly, campaigning must be pleasant work in Algeria!'

Then he turned to her with compliments frank and gay, and full of a debonair grace that made her doubt he could be of Albion.

Retort was always ready to her; and she kept the circle of officers in full laughter round the vidette-fire with a shower of repartee that would have made her fortune on the stage of the Châtelet or Folies Marigny. And every now and then her glance wandered to the shadow where the horses were tethered.

Bah! why was she always doing him service? She could not have told. '*Parcequ' j' suis bien bête*,' said Cigarette mentally, with a certain fiery contempt for herself.

Still she went on—and did it.

It was a fantastic picture by the bright scarlet light of the camp-fire, with the Little One in her full glory of mirth and mischief, and her circle of officers laughing on her with admiring eyes; nearest her the towering height of the English stranger with the gleam of the flame in the waves of his leonine beard.

From the darkness, where the scores of grey horses were tethered, Cecil's eyes were riveted on it. There were none near to see him; had there been, they would have seen an agony in his eyes that no physical misery, no torture of the battle-field, had brought there. His face was bloodless, and his gaze strained through the gleam on to the fire-lit group with a passionate intensity of yearning—he was well used to pain, well used to self-control, well used to self-restraint, but for the first time in his exile the bitterness of a struggle almost vanquished him. All the old love of his youth went out to this man, so near beside him, yet so hopelessly severed from him: looking on the face of his friend, a violence of longing shook him: 'O God, if I were dead!' he thought, 'they might know then——'

He would have died gladly to have had that familiar hand once more touch his; those familiar eyes once more look on him with the generous tender trust of old.

His brain reeled, his thoughts grew blind, as he stood there amongst his horses with the stir and tumult of the bivouac about him. There was nothing simpler, nothing less strange, than that an English soldier should visit the Franco-Arab camp; but to him it seemed like the resurrection of the dead.

Whether it was a brief moment, or an hour through, that the circle stood about the great black caldron that was swinging above the flames, he could not have told: to him it was an eternity. The echo of the mellow ringing tones that he knew so well came to him from the distance, till his heart seemed breaking with but one forbidden longing, to look once more in those brave eyes that made every coward and liar quail, and say only, 'I was guiltless.'

It is bitter to know those whom we love dead; but it is more bitter to be as dead to those who, once having loved us, have sunk our memory deep beneath oblivion that is not the oblivion of the grave.

Awhile, and the group broke up and was scattered, the English traveller throwing gold pieces by the score amongst the waiting troopers. '*A bientôt!*' they called to Cigarette, who nodded farewell to them with a cigar in her mouth, and busied herself pouring some brandy into the old copper caldron in which some black coffee and muddy water, three parts sand, was boiling. A few moments later, and they were out of sight amongst the confusion, the crowds, and the flickering shadows of the camp. When they were quite gone, she came softly to him; she could not see him well in the gloom, but she touched his hand.

'Dieu! how cold you are. He is gone.'

He could not answer her to thank her, but he crushed in his the little warm brown palm. She felt a shiver shake his limbs.

'Is he your enemy?' she asked.

'No.'

'What then?'

'The man I love best on earth.'

'Ah!' She had felt a surprise she had not spoken that he should flee thus from any foe. 'He thinks you dead, then?'

'Yes.'

'And must always think so?'

'Yes.' He held her hand still, and his own wrung it hard—the grasp of comrade to comrade, not of man to woman. 'Child, you are bold, generous, pitiful; for God's sake, get me sent out of this camp tonight. I am powerless.'

There was that in the accent which struck his listener to the heart. He was powerless, fettered hand and foot as though he were a prisoner; a night's absence, and he would be shot as a deserter. He had grown accustomed to this rendering up of all his life to the rules of others; but now and then the galled spirit chafed, the netted stag strained at the bonds.

'I will try,' said Cigarette, simply, without anything of her audacity or of her vanity in the answer. 'Go you to the fire; you are cold.'

'Are you sure he will not return?'

'Not he. They are gone to eat and drink; I go with them. What is it you fear?'

'My own weakness.'

She was silent. She could just watch his features by the dim light, and she saw his mouth quiver under the fulness of his beard. He felt that if he looked again on the face of the man he loved he might be broken into self-pity, and unloose his silence, and shatter all the work of so many years. He had been strong where men of harder fibre and less ductile temper might have been feeble; but he never thought that he had been so; he only thought that he had acted on impulse, and had remained true to his act through the mere instinct of honour—an instinct inborn in his blood and his order—an instinct natural and unconscious with him as the instinct by which he drew his breath.

'You are a fine soldier,' said Cigarette, musingly; 'such men are not weak.'

'Why? We are only strong as tigers are strong—just the strength of the talon and fang. I do not know. I was weak as water once; I may be again, if—if——'

He scarcely knew that he was speaking aloud; he had forgotten her! His whole heart seemed burnt as with fire by the memory of that one face so familiar, so well loved, yet from which he must shrink as though some cowardly sin were between them. The wretchedness on him seemed more than he could bear; to know that this man was so near that the sound of his voice raised could summon him, yet that he must remain as dead to him—remain as one dead after a craven and treacherous guilt.

He turned suddenly, almost violently, upon Cigarette:

'You have surprised my folly from me; you know my secret so far; but you are too brave to betray me, you are too generous to tell of this? I can trust you to be silent?'

Her face flushed scarlet with astonished anger; her little childlike form grew instinct with haughty and fiery dignity.

'Monsieur, that question from one soldier of France to another is insult. We are not dastards!'

There was a certain grave reproach that mingled with the indignant scorn of the answer, and showed that her own heart was wounded by the doubt, as well as her military pride by the aspersion. Even amidst the conflict of pain at war in him he felt that, and hastened to soothe it.

'Forgive me, my child; I should not have wronged you with the question. It is needless, I know. Men can trust you to the death, they say.'

'To the death—yes.'

The answer was thoughtful, dreamy, almost sad, for Cigarette. His thoughts were too far from her in their tumult of awakened memories to note the tone as he went rapidly on:

'You have ingenuity, compassion, tact; you have power here, too, in your way; for the love of Heaven get me sent out on some duty before dawn! There is Biribi's murder to be avenged—would they give the errand to me?'

She thought a moment.

'We will see,' she said, curtly. 'I think I can do it. But go back, or you will be missed. I will come to you soon.'

She left him then, rapidly, drawing her hand quickly out of the clasp of his.

'*Que je suis bête! Que je suis bête!*' said Cigarette to herself; for she felt her heart aching to its core for the sorrow of this man who was nothing to her. He did not know what she had done for him in his suffering and delirium; he did not know how she had watched him all that night through, when she was weary, and bruised, and thirsting for sleep; he did not know; he held her hand as one comrade another's, and never looked to see if her eyes were blue or were black, were laughing or tear-laden. And yet she felt pain in his pain; she was always giving her life to his service. '*Que je suis bête! Que je suis bête!*' she murmured again. Many beside the little Friend of the Flag beat back as folly the noblest and purest thing in them.

Cecil mechanically returned to the fire at which the men of his *tribu* were cooking their welcome supper, and sat down near them, rejecting, with a gesture, the most savoury portion which, with their customary love and care for him, they were careful to select and bring to him. There had never been a time when they had found him fail to prefer them to himself, or fail to do them kindly service, if of such he had a chance; and they returned it with all that rough and silent attachment that can be so strong and so staunch in lives that may be black with crime or red with slaughter.

He sat like a man in a dream, whilst the loosened tongues of the men ran noisily on a hundred themes as they chaffed each other, exchanged a fire of bivouac jokes more racy than decorous, and gave themselves to the enjoyment of their rude meal, that had to them that savour which long hunger alone can give. Their voices came dull on his ear; the ruddy warmth of the fire was obscured to his sight; the din, the laughter, the stir all over the great camp, at the hour of dinner were lost on him. He was insensible to everything except the innumerable memories that thronged upon him, and the aching longing that filled his heart with the sight of the friend of his youth.

'He said once that he would take my hand before all the world always, come what would,' he thought. 'Would he take it now, I wonder? Yes; *he* never believed against me.'

And, as he thought, the same anguish of desire that had before smitten him to stand once more guiltless in the presence of men, and once more bear, untarnished, the name of his race and the honour of his fathers, shook him now as strong winds shake a tree that yet is fast rooted at its base, though it sway awhile beneath the storm.

'How weak I am!' he thought, bitterly. 'What does it matter? Life is so short, one is a coward indeed to fret over it. I cannot undo what I did. I cannot, if I would. To betray him *now*! God! not for a kingdom, if I had the chance! Besides, she may live still; and, even were she dead, to tarnish her name to clear my own would be a scoundrel's baseness— baseness that would fail as it merited; for who could be brought to believe me now?'

The thoughts unformed drifted through his mind, half dulled, half sharpened by the deadly pain, and the rush of old brotherly love that had arisen in him as he had seen the face of his friend beside the watch-fire of the French bivouac. It was hard; it was cruelly hard; he had, after a long and severe conflict, brought himself into contentment with his lot, and taught himself oblivion of the past, and interest in the present, by active duties and firm resolve; he had vanquished all the habits, controlled most of the weakness, and banished nearly all the frailties and indulgences of his temperament in the long ordeal of African warfare. It was cruelly hard that now when he had obtained serenity, and more than half attained forgetfulness, these two—her face and his—must come before him, one to recall the past, the other to embitter the future!

As he sat with his head bent down and his forehead leaning on his arm, while the hard biscuit that served for a plate stood unnoticed beside him, with the food that the soldiers had placed on it, he did not hear Cigarette's step till she touched him on the arm. Then he looked up; her eyes were looking on him with a tender, earnest pity.

'Hark! I have done it,' she said, gently. 'But it will be an errand very close to death, that you must go on——'

He raised himself erect, eagerly.

'No matter that! Ah, Mademoiselle, how I thank you!'

'Chut! I am no Paris demoiselle!' said Cigarette, with a dash of her old acrimony. 'Ceremony in a camp—pouf! You must have been a Court Chamberlain once, weren't you? Well, I have done it. Your officers were talking yonder of a delicate business; they were uncertain who best to employ. I put in my speech—it was dead against military etiquette, but I did it—I said to M. le Général: "You want the best rider, the most silent tongue, and the surest steel in the squadrons? Take Bel-à-faire-peur, then?" "Who is that?" asked the General; he would have sent out of camp anybody but Cigarette for the interruption. "Mon Général," said I, "the Arabs asked that, too, the other day at Zarâila." "What!" he cried, "the man Victor—who held the ground with his Chasseurs? I know—a fine soldier. M. le Colonel, shall we send him?" The Black Hawk had scowled thunder on you; he hates you more still since that affair of Zarâila, specially, because the General has reported your conduct with such praise that they cannot help but promote you. Well, he had looked thunder, but now he laughed. "Yes, mon Général," he answered him, "take him, if you like. It is fifty to one whoever goes on that business will not come back alive, and you will rid me of the most insolent fine gentleman in my squadrons." The General hardly heard him; he was deep in thought; but he asked a good deal about you from the Hawk, and Châteauroy spoke for your fitness for the errand they are going to send you on, very truthfully, for a wonder. I don't know why; but he wants you to be sent, I think; most likely that you may be cut to pieces. And so they will send for you in a minute. I have done it as you wished, "*le diable prends le fruit!*" '

There was something of her old brusquerie and recklessness in the closing sentences; but it had not her customary débonnaire lightness. She knew too well that the chances were as a hundred to one that he would never return alive from this service on which he had entreated to be dispatched. Cecil grasped both her hands in his with warm gratitude, that was still, like the touch of his hands, the gratitude of comrade to comrade, not of man to woman.

'God bless you, Cigarette! You are a true friend, my child. You have done me immeasurable benefits——'

'Oh-hé. I am a true friend,' said the Little One, something pettishly. She would have preferred another epithet. 'If a man wants to get shot as a very great favour, I always let him pleasure himself. Give a man his own way, if you wish to be kind to him. You are children, all of you, nothing but children, and if the toy that pleases you best is death, why—you must have it. Nothing else would content you. I know you. You always want what flies from you, and are tired of what lies to your hand. That is always a man.'

'And a woman too, is it not?'

Cigarette shrugged her shoulders.

'Oh, I dare say. We love what is new—what is strange. We are hum-ming-tops; we will only spin when we are fresh wound up with a string to our liking.'

'Make an exception of yourself, my child. You are always ready to do a good action, and never tire of that. From my heart, I thank you. I wish to Heaven I could prove it better.'

She drew her hands away from him.

'A great thing I have done, certainly! Got you permission to go and throw a cartel at old King Death; that is all! There! Loup-à-griffes-de-fer is coming to you. That is your summons.'

The orderly so nicknamed approached, and brought the bidding of the General in command of the Cavalry for Cecil to render himself at once to his presence. These things brook no second's delay in obedience; he went with a quick adieu to Cigarette, and the little Friend of the Flag was left in his vacant place beside the fire.

And there was a pang at her heart.

'Ten to one he goes to his death,' she thought. But Cigarette, *volage* little mischief though she was, could reach very high in one thing; she could reach a love that was unselfish, and one that was heroic.

A few moments, and Cecil returned.

'Rake,' he said rapidly, in the French he habitually used, 'saddle my horse and your own. I am allowed to choose one of you to accompany me.'

Rake, in paradise, and the envied of every man in the squadrons, turned to his work—with him a task of scarce more than a second; and Cecil approached his little Friend of the Flag.

'My child, I cannot attempt to thank you. But for you, I should have been tempted to send my lance through my own heart.'

'Keep its lunge for the Arbicos, *mon ami*,' said Cigarette brusquely— the more brusquely because that new and bitter pang was on her. 'As for me, I want no thanks.'

'No; you are too generous. But not the less do I wish I could render them more worthily than by words. If I live, I will try; if not, keep this in my memory. It is the only thing I have.'

He put into her hand the ring she had seen in the little *bonbonnière*; a ring of his mother's that he had saved when he had parted with all else, and that he had put off his hand and into the box of Petite Reine's gift the day he had entered the Algerian army.

Cigarette flushed scarlet with passions he could not understand, and she could not have disentangled.

'The ring of your mistress! Not for me, if I know it! Do you think I want to be paid?'

'The ring was my mother's,' he answered her simply. 'And I offer it only *en souvenir*.'

She lost all her hot colour, and all her fiery wrath; his grave and gentle courtesy always strangely stilled and rebuked her; but she raised the ring off the ground where she had flung it, and placed it back in his hand.

'If so, still less should you part with it. Keep it; it will bring you happiness one day. As for me, I have done nothing, pardieu!'

'You have done what I value the more for that noble disclaimer. May I thank you thus, Little One?'

He stooped and kissed her; a kiss that the lips of a man will always give to the bright youthful lips of a woman, but a kiss, as she knew well, without passion, even without tenderness in it.

With a sudden impetuous movement, with a shyness and a refusal that had never been in her before, she wrested herself from him, her face burning, her heart panting, and plunged away from him into the depth of the shadow; and he never sought to follow her, but threw himself into saddle as his grey was brought up: another instant, and, armed to the teeth, he rode out of the camp into the darkness of the silent, melancholy, lonely Arab night.

CHAPTER V

Seul au Monde

The errand on which he went was one, as he was well aware, from which it were a thousand chances to one that he ever issued alive.

It was to reach a distant branch of the Army of Occupation with dispatches for the chief in command there, and to do this he had to pass through a fiercely hostile region, occupied by Arabs with whom no sort of peace had ever been made, the most savage as well as the most predatory of the wandering tribes. His knowledge of their tongue, and his friendship with some men of their nation, would avail him nothing here; for their fury against the Franks was intense, and it was said that all prisoners who had fallen into their hands had been put to death with merciless barbarities. This might be true or untrue; wild tales were common among Algerian campaigners; whichever it were, he thought little of it as he rode out on to the lonely plains. Every kind of hazardous adventure and every variety of peril had been familiar with him in this African life; and now there were thoughts and memories on him which deadened every recollection of merely physical risk.

'We must ride as hard and as fast as we can, and *as silently*,' were the only words he exchanged with Rake, as he loosened his grey to a hand-gallop.

'All right, sir,' answered the trooper, whose warm blood was dancing, and whose blue eyes were alive like fire with delight. That he had been absent on a far-away foraging raid on the day of Zarâila had been nothing short of agony to Rake, and the choice made of him for this duty was to him a gift of paradise. He loved fighting for fighting's sake; and to be beside Cecil was the greatest happiness life held for him.

They had two hundred miles to traverse, and had received only the command he had passed on Rake, to ride 'hard, fast, and silently'. To the hero of Zarâila the General had felt too much soldierly sympathy to add the superfluous injunction to do his uttermost to carry safely and success-fully to their destination the papers that were placed in his sabre-tasche. They knew well that the errand would be done, or the Chasseur's *main de femme, mais main de fer*, would be stiffened and nerveless in death.

It was just nightfall; the afterglow had faded only a few moments before. Giving their horses, which they were to change once ten hours for the distance, and two for bait and for rest, he reckoned that they would reach the camp before the noon of the coming day, as the beasts, fresh and fast in the camp, flew like greyhounds beneath them.

Another night-ride that they had ridden together came to the minds of both; but they spoke not a word as they swept on, their sabres shaken loose in their sheaths, their lances well gripped, and the pistols with which they had been supplied sprung in their belts ready for instant action if a call should come for it. Every rood of the way was as full of unseen danger as if laid over mines. They might pass in safety; they might any moment be cut down by ten score against two. From every hanging scarp of rugged rock a storm of musket-balls might pour; from every screen of wild-fig foliage a shower of lances might whistle through the air; from every darkling grove of fir-trees an Arab band might spring and swoop on them;—but the knowledge scarcely recurred to the one save to make him shake his sword more at loose for quick disengagement, and only made the sunny blue eyes of the other sparkle with a vivid and longing zest.

The night grew very chill as it wore on; the north wind rose, rushing against them with a force and icy touch that seemed to freeze their bones to the marrow after the heat of the day and the sun that had scorched them so long. There was no regular road; they went across the country, their way sometimes leading over level land, over which they swept like lightning, great plains succeeding one another with wearisome monotony; sometimes, on the contrary, lying through ravines, and defiles, and gloomy woods, and broken hilly spaces, where rent bare rocks were thrown on one another in gigantic confusion, and the fantastic shapes of the wild fig and the dwarf palm gathered a hideous grotesqueness in the darkness. For there was no moon. and the stars were often hidden by the storm-rack of leaden clouds that drifted over the sky; and the only sound they heard was the cry of the jackal, or the shriek of the night-bird, and now and then the sound of shallow watercourses, where the parched beds of hidden brooks had been filled by the autumnal rain.

The first five-and-twenty miles passed without interruption, and the horses laid well and warmly to their work. They halted to rest and bait the beasts in a rocky hollow, sheltered from the blasts of the *bise*, and green with short sweet grass, sprung up afresh after the summer drought.

'Do you ever think of *him*, sir?' said Rake, softly, with a lingering love in his voice, as he stroked the greys and tethered them.

'Of whom?'

'Of the King, sir. If he's alive, he's gettin' a rare old horse now.'

'Think of him! I wish I did not, Rake.'

'Wouldn't you like to see him agen, sir?'

'What folly to ask! You know——'

'Yes, sir, I know,' said Rake, slowly. 'And I know—leastways I picked it out of a old paper—that your elder brother died, sir, like the old Lord, and Mr Berk's got the title.'

Rake had longed and pined for an opportunity to dare say this thing which he had learned, and which he could not tell whether or no Cecil knew likewise. His eyes looked with straining eagerness through the gloom into his master's; he was uncertain how his words would be taken. To his bitter disappointment, Cecil's face showed no change, no wonder.

'I have heard that,' he said, calmly—as calmly as though the news had no bearing on his fortunes, but was some stranger's history.

'Well, sir, but he *ain't* the Lord?' pleaded Rake, passionately. 'He won't never be while you're living, sir?'

'Oh yes he is! I am dead, you know.'

'But he *won't*, sir!' reiterated Rake. 'You're Lord Royallieu if ever there was a Lord Royallieu, and if ever there will be one.'

'You mistake. An outlaw has no civil rights, and can claim none.'

The man looked very wistfully at him; all these years through he had never learned why his master was thus 'dead' in Africa, and he had too loyal a love and faith ever to ask, or ever to doubt but that Cecil was the wronged and not the wrongdoer.

'You ain't a outlaw, sir,' he muttered. 'You could take the title if you would.'

'Oh no! I left England under a criminal charge; I should have to disprove that before I could inherit.'

Rake crushed bitter oaths into muttered words as he heard. 'You could disprove it, sir, of course, right and away, if you chose.'

'No; or I should not have come here. Let us leave the subject. It was settled long ago. My brother is Lord Royallieu. I would not disturb him if I had the power, and I have not it. Look, the horses are taking well to their feed.'

Rake asked him no more: he had never had a harsh word from Cecil in their lives; but he knew him too well for all that to venture to press on him a question thus firmly put aside. But his heart ached sorely for his master; he would so gladly have seen 'the king among his own again', and would have striven for the restoration as strenuously as ever a Cavalier strove for the White Rose; and he sat in silence, perplexed and ill-satisfied, under the shelter of the rock, with the great, dim, desolate African landscape stretching before him, with here and there a gleam of light upon it when the wind swept the clouds apart. His volatile speech was chilled, and his buoyant spirits were checked. That Cecil was justly outlawed he would have thought it the foulest treason to believe for one instant; yet he felt that he might as soon seek to wrench up the great stones above him from their base as seek to change the resolution of this man, whom he had once known pliant as a reed and careless as a child.

They were before long in saddle again and off, the country growing wilder at each stride the horses took.

'It is all alive with Arabs for the next ten leagues,' said Cecil, as he settled himself in his saddle. 'They have come northward and been sweeping the country like a locust-swarm, and we shall blunder on some of them sooner or later. If they cut me down, don't wait, but slash my sabre-tasche loose and ride off with it.'

'All right, sir,' said Rake, obediently; but he thought to himself, 'Leave you alone with them demons? Damn me if I will.'

And away they went once more, in speed and in silence, the darkness of full night closing in on them, the skies being black with the heavy drift of rising storm-clouds.

Meantime Cigarette was feasting with the officers of his regiment. The dinner was the best that the camp-scullions could furnish in honour of the two or three illustrious tourists who were on a visit to the head-quarters of the Algerian Army; and the Little One, the heroine of Zarâila, and the toast of every mess throughout Algeria, was as indispensable as the champagnes. Not that she was altogether herself tonight; she was feverish, she was bitter, she was full of stinging ironies; but that delicious gaiety, like a kitten's play, was gone from her, and its place, for the first time in her life, was supplied by unreal and hectic excitement. In truth, whilst she laughed, and coquetted, and fenced with the bright two-edged blade of her wit, and tossed down the wines into her little throat like a trooper, she was thinking nothing at all of what was around her, and very little of what she said or she did. She was thinking of the starless night out yonder, of the bleak arid country, of the great, dim, measureless plains; of one who was passing through them all, and one who might never return.

It was the first time that the absent had ever troubled her present; it was the first time that ever this foolish, senseless, haunting, unconquerable fear for another had approached her: fear!—she had never known it for herself, why should she feel it now for him?—a man whose lips had touched her own as lightly, as indifferently, as they might have touched the leaves of a rose or the curls of a dog!

She felt her face burn with the flush of a keen, unbearable, passionate shame. Men by the score had wooed her love, to be flouted with the insouciant mischief of her coquetry, and forgotten tomorrow if they were shot today; and now he—he whose careless, calm caress would make her heart vibrate and her limbs tremble with an emotion she had never known—he valued her love so little that he never even knew that he had roused it! To the proud young warrior of France a greater degradation, a deadlier humiliation, than this could not have come to her.

Yet she was true as steel to him; true with the strong and loyal fealty that is inborn with such natures as hers. To have betrayed what he had trusted to her, because she was neglected and wounded by him, would have been a feminine baseness of which the soldier-like soul of Cigarette would have been totally incapable. Her revenge might be fierce, and rapid, and sure, like the revenge of a soldier; but it could never be stealing and traitorous, and never like the revenge of a woman.

Not a word escaped her that could have given a clue to the secret with which he had involuntarily weighted her; she only studied with interest and keenness the face and the words of this man whom he had loved, and from whom he had fled as criminals flee from their accusers.

'What is your name?' she asked him, curtly, in one of the pauses of the amorous and witty nonsense that circulated in the tent in which the officers of Chasseurs were entertaining him.

'Well—some call me Seraph.'

'Ah! you have *petits noms* then in Albion? I should have thought she was too sombre and too stiff for them. Besides?'

'Lyonnesse.'

'What a droll name! What are you?'

'A soldier.'

'Good! What grade?'

'A Colonel of Guards.'

Cigarette gave a little whistle to herself; she remembered that a Marshal of France had once said of a certain Chasseur, 'He has the seat of the English Guards.'

'My pretty catechist, M. le Duc does not tell you his title,' cried one of the officers.

Cigarette interrupted him with a toss of her head.

'Ouf! Titles are nothing to me. I am a child of the People. So you are a Duke, are you, M. le Seraph? Well, that is not much, to my thinking. Bah! there is Fialin made a Duke in Paris, and there are aristocrats here wearing privates' uniforms, and littering down their own horses. Bah! Have you that sort of thing in Albion?'

'Attorneys throned on high, and gentlemen glad to sweep crossings? Oh yes!' laughed her interlocutor. 'But you speak of aristocrats in your ranks—that reminds me. Have you not in this corps a soldier called Louis Victor?'

He had turned as he spoke to one of the officers, who answered him in the affirmative; while Cigarette listened with all her curiosity and all her interest, that needed a deeper name, heightened and tight-strung.

'A fine fellow,' continued the Chef d'Escadron to whom he had appealed. 'He behaved magnificently the other day at Zarâila; he must be distinguished for it. He is just sent on a perilous errand, but though so quiet he is a *croc-mitaine*, and woe to the Arabs who slay him! Are you acquainted with him?'

'Not in the least. But I wished to hear all I could of him. I have been told he seems above his present position. Is it so?'

'Likely enough, monsieur; he seems a gentleman. But then we have many gentlemen in the ranks, and we can make no difference for that. Cigarette can tell you more of him; she used to complain that he bowed like a Court chamberlain.'

'Oh-hé—I did!' cried Cigarette, stung into instant irony because pained and irritated by being appealed to on the subject. 'And of course, when so many of his officers have the manners of Pyrenean bears, it is a little awkward for him to bring us the manners of a Palace!'

Which effectually chastised the Chef d'Escadron, who was one of those who had a *ton de garnison* of the roughest, and piqued himself on his powers of fence much more than on his habits of delicacy.

'Has this Victor any history?' asked the English Duke.

'He has written one with his sword; a fine one,' said Cigarette, curtly. 'We are not given here to care much about any other.'

'Quite right; I asked because a friend of mine who had seen his carvings wished to serve him if it were possible; and——'

'Ho! That is Miladi, I suppose?' Cigarette's eyes flashed fire instantly, in wrath and suspicion. 'What did she tell you about him, *la belle dédaigneuse*?'

'I am ignorant of whom you speak?' he answered, with something of surprise and of annoyance.

'Are you?' said Cigarette, in derision. 'I doubt that. Of whom should I speak but of *her*? Bah! She insulted him, she offered him gold, she sent my men the spoils of her table, as if they were paupers, and he thinks it all divine because it is done by Madame la Princesse Corona d'Amägué! Faugh! when he was delirious, the other night, he could babble of nothing but of her—of her—of her!'

The jealous, fiery impatience in her vanquished every other thought; she was a child in much, she was untutored in all; she had no thought that by her scornful vituperation of 'Miladi' she could either harm Cecil or betray herself. But she was amazed to see the English guest change colour with a haughty anger that he strove to subdue as he half rose and answered her with an accent in his voice that reminded her—she knew not why—of Bel-à-faire-peur and of Marquise.

'Madame la Princesse Corona d'Amägué is my sister; why do you venture to couple the name of this Chasseur with hers?'

Cigarette sprang to her feet, vivacious, imperious, reckless, dared to anything by the mere fact of being publicly arraigned.

'Pardieu! Is it insult to couple the silver pheasant with the Eagles of France?—a pretty idea, truly! So she is your sister, is she? Miladi? Well, then, tell her from me to think twice before she outrages a soldier with "patronage"; and tell her, too, that had I been he I would have ground my ivory toys into powder before I would have let them become the playthings of a *grande dame* who tendered me gold for them!'

The Englishman looked at her with astonishment that was mingled with a vivid sense of intense annoyance and irritated pride, that the name he cherished closest should be thus brought in, at a camp dinner, on the lips of a vivandière and in connection with a trooper of Chasseurs.

'I do not understand your indignation, mademoiselle,' he said, with an impatient stroke to his beard. 'There is no occasion for it. Madame Corona d'Amägué, my sister,' he continued, to the officers present, 'became accidentally acquainted with the skill at sculpture of this Corporal of yours; he appeared to her a man of much refinement and good breeding. She chanced to name him to me, and feeling some pity——'

'M. le Duc!' cried the ringing voice of Cigarette, loud and startling as a bugle-note, while she stood like a little lioness, flushed with the draughts of champagne and with the warmth of wrath at once jealous and generous, 'keep your compassion until it is asked of you. No soldier of France needs it; that I promise you. I know this man that you talk of "pitying". Well, I saw him at Zarâila three weeks ago; he had drawn up his men to die with them rather than surrender and yield up the guidon; I dragged him half dead, when the field was won, from under his horse, and his first conscious act was to give the drink that I brought him to a wretch who had thieved from him. Our life here is hell upon earth to such as he, yet none ever heard a lament wrung out of him; he is gone to the chances of death tonight as most men go to their mistresses' kisses; he is a soldier Napoleon would have honoured. Such a one is not to have the patronage of a Miladi Corona, nor the pity of a stranger of England. Let the first respect him; let the last imitate him!'

And Cigarette, having pronounced her defence and her eulogy with the vibrating eloquence of some orator from a tribune, threw her champagne goblet down with a crash, and, breaking through the arms outstretched to detain her, forced her way out despite them, and left her hosts alone in their lighted tent.

'*C'est Cigarette!*' said the Chef d'Escadron, with a shrug of his shoulders, as of one who explained, by that sentence, a whole world of irreclaimable eccentricities.

'A strange little Amazon!' said their guest. 'Is she in love with this Victor, that I have offended her so much with his name?'

The Major shrugged his shoulders.

'I don't know that, monsieur,' answered one. 'She will defend a man in his absence, and rate him to his face most soundly. Cigarette whirls about like a little paper windmill, just as the breeze blows; but, as the windmill never leaves its stick, so she is always constant to the Tricolour.'

Their guest said little more on the subject; in his own thoughts he was bitterly resentful that, by the mention of this Chasseur's fortunes, he should have brought in the name he loved so well—the purest, fairest, haughtiest name in Europe—into a discussion with a vivandière at a camp dinner.

Châteauroy, throughout, had said nothing; he had listened in silence, the darkness lowering still more heavily upon his swarthy features; only now he opened his lips for a few brief words:

'Mon cher Duc, tell Madame not to waste the rare balm of her pity. The fellow you enquire for was an outcast and an outlaw when he came to us. He fights well—it is often a blackguard's virtue!'

His guest nodded, and changed the subject; his impatience and aversion at the introduction of his sister's name into the discussion made him drop the theme unpursued, and let it die out forgotten.

Venetia Corona associated with an Algerian trooper! If Cigarette had been of his own sex, he could have dashed the white teeth down her throat for having spoken of the two in one breath.

And as, later on, he stretched his gallant limbs out on his narrow camp palliasse, tired with a long day in saddle under the hot African sun, the Seraph fell asleep with his right arm under his handsome golden head, and thought no more of this unknown French sabreur.

But Cigarette remained wakeful.

She lay curled up in the straw against her pet horse, Etoile-Filante, with her head on the beast's glossy flank and her hand amongst his mane. She often slept thus in camp, and the horse would lie still and cramped for hours rather than awaken her, or, if he rose, would take the most watchful heed to leave unharmed the slender limbs, the flushed cheeks, the frank fair brow of the sleeper beneath him, that one stroke of his hoof could have stamped out into a bruised and shapeless mass.

Tonight Etoile-Filante slept, and his mistress was awake—wide awake, with her eyes looking out into the darkness beyond, with a passionate mist of unshed tears in them, and her mouth quivering with pain and with wrath. The vehement excitation had not died away in her, but there had come with it a dull, spiritless, aching depression. It had roused her to fury to hear the reference to her rival spoken—of that aristocrat whose name had been on Cecil's lips when he had been delirious. She had kept his secret loyally, she had defended him vehemently; there was something that touched her to the core in the thought of the love with which he had recognized this friend who, in ignorance, spoke of him as of some unknown French soldier. She could not tell what the history was, but she could divine nearly enough to feel its pathos and its pain. She had known, in her short life, more of men and of their passions and of their fortunes than many lives of half a century in length can ever do; she could guess, nearly enough to be wounded with its sorrow, the past which had exiled the man who had kept by him his lost mother's ring as the sole relic of years to which he was dead as utterly as though he were lying in his coffin. No matter what the precise reason was—women, or debt, or accident, or ruin—these two, who had been familiar comrades, were now as strangers to each other; the one slumbered in ignorance near her, the other had gone out to the close peril of death, lest the eyes of his friend should recognise his face and read his secret. It troubled her, it weighed on her, it smote her with a pang. It might be that now, even now—this very moment, whilst her gaze watched the dusky shadows of the night chase one another along the dreary plains—a shot might have struck down this life that had been stripped of name and fame and country; even now all might be over!

And Cigarette felt a cold sickly shudder seize her that never before, at death or danger, had chilled the warm swift current of her bright French blood. In bitter scorn at herself, she muttered hot oaths between her pretty teeth.

Mère de Dieu! he had touched her lips as carelessly as her own kiss would have touched the rose-hued waxen petals of a cluster of oleander-blossoms; and she cared for him still!

Whilst the Seraph slept dreamlessly, with the tents of the French camp around him, and the sleepless eyes of Cigarette watched afar off the dim distant forms of the videttes as they circled slowly round at their outpost duty, eight leagues off, through a vast desert of shadow and silence the two horsemen swept swiftly on. Not a word had passed between them; they rode close together in unbroken stillness; they were scarcely visible to each other, for there was no moon, and storm-clouds obscured the skies. Now and then their horses' hoofs struck fire from a flint stone, and the flash sparkled through the darkness; often not even the sound of their gallop was audible on the grey, dry, loose soil.

Every rood of the road was sown thick with peril; no frowning ledge of rock, with pine-roots in its clefts, but might serve as the barricade behind which some foe lurked; no knot of cypress-shrubs, black even on that black sheet of shadow, but might be pierced with the steel tubes of levelled waiting muskets.

Pillaging, burning, devastating wherever they could, in what was to them a holy war of resistance to the infidel and the invader, the predatory tribes had broken out into a revolt which the rout of Zaraila, heavy blow though it had been to them, had by no means ended. They were still in arms, infesting the country everywhere southward, defying regular pursuit, impervious to regular attacks, carrying on the harassing guerilla warfare at which they were such adepts, and causing thus to their Frankish foe more irritation and more loss than decisive engagements would have produced. They feared nothing, had nothing to lose, and could subsist almost upon nothing. They might be driven into the desert, they might even be exterminated after long pursuit; but they would never be vanquished. And they were scattered now far and wide over the country; every cave might shelter, every ravine might enclose them; they appeared here, they appeared there; they swooped down on a convoy, they carried sword and flame into a settlement, they darted like a flight of hawks upon a foraging-party, they picked off any vidette as he wheeled his horse round in the moonlight; and every yard of the miles which the two grey chargers of the Chasseurs d'Afrique must cover ere their service was done was as rife with death as though its course lay over the volcanic line of an earthquake.

They had reached the centre of the plain when the sound they had long looked for rang on their ears, piercing the heavy breathless stillness of the night. It was the Allah-il-Allah of their foes, the war-cry of the Moslem. Out of the gloom—whether from long pursuit or some near hiding-place they could not tell—there broke suddenly upon them the fury of an Arab onslaught. In the darkness all they could see were the flash of steel, the flame of fierce eyes against their own, the white steam of smoking horses, the spray of froth flung off the snorting nostrils, the rapid glitter of the curved flissas—whether two, or twenty, or twice a hundred were upon them they could not know—they never did know. All of which they were conscious was that in an instant, from the tranquil melancholy around them of the great, dim, naked space, they were plunged into the din, the fury, the heat, the close, crushing, horrible entanglement of conflict, without the power to perceive or to number their foes, and only able to follow the sheer simple instincts of attack and of defence. All they were sensible of was one of those confused moments, deafening, blinding, filled with violence and rage and din——an eternity in semblance, a second in duration—that can never be traced, never be recalled, yet in whose feverish excitement men do that which, in their calmer hours, would look to them a fable of some Amadis of Gaul.

How they were attacked, how they resisted, how they struck, how they were encompassed, how they thrust back those who were hurled on them in the black night, with the north sea-wind like ice upon their faces, and the loose African soil drifting up in clouds of sand around them, they could never have told. Nor how they strained free from the armed ring that circled them, and beat aside the shafts of lances and the blades of swords, and forced their chargers breast to breast against the fence of steel, and through the tempest of rage, and blows, and shouts, and wind, and driven sand, cut their way through the foe whose very face they scarce could see, and plunged away into the shadows across the desolation of the plain, pursued, whether by one or by a thousand they could not guess, for the gallop was noiseless on the powdered soil, and the Arab yell of baffled passion and slaughterous lust was half drowned in the rising of the wind-storm. Had it been day, they would have seen their passage across the level table-land traced by a crimson stream upon the sand, in which the blood of Frank and Arab blended equally.

As it was, they dashed headlong down through the darkness that grew yet denser and blacker as the storm rose. For miles the ground was level before them, and they had only to let the half-maddened horses, that had as by a miracle escaped all injury, rush on at their own will through the whirl of the wind that drove the dust upward in spiral columns and brought icy breaths of the north over the sear, sunburnt, southern wastes.

For a long space they had no sense but that of rapid ceaseless motion through the thick gloom and against the pressure of the violent blasts. The speed of their gallop and the strength of the currents of air were like some narcotic that drowned and that dizzied perception. In the intense darkness neither could see, neither hear, the other; the instinct of the beasts kept them together, but no word could be heard above the roar of the storm, and no light broke the sombre veil of shadow through which they passed as fast as leopards course through the night. The first faint streak of dawn grew grey in the east when Cecil felt his charger stagger and sway beneath him, and halt, worn out and quivering in every sinew with fatigue. He threw himself off the animal in time to save himself from falling with it as it reeled and sank to the ground.

'Massena cannot stir another yard,' he said. 'Do you think they follow us still?'

There was no reply.

He strained his sight to pierce the darkness, but he could distinguish nothing; the gloom was still too deep. He spoke more loudly; still there was no reply. Then he raised his voice in a shout; it rang through the silence, and, when it ceased, the silence reigned again.

A deadly chill came on him. How had he missed his comrade? They must be far apart, he knew, since no response was given to his summons; or—the alternative rose before him with a terrible foreboding.

That intense quiet had a repose as of death in it, a ghastly loneliness that seemed filled with desolation. His horse was stretched before him on the sand, powerless to rise and drag itself a rood onward, and fast expiring. From the plains around him not a sound came, either of friend or foe. The consciousness that he was alone, that he had lost for ever the only friend left to him, struck on him with that conviction which so often foreruns the assurance of calamity. Without a moment's pause, he plunged back in the direction he had come, leaving the charger on the ground to pant its life out as it must, and sought to feel his way along, so as to seek as best he could the companion he had deserted. He still could not see a rood before him, but he went on slowly, with some vague hope that he should ere long reach the man whom he knew death or the fatality of accident alone would keep from his side. He could not feel or hear anything that gave him the slightest sign or clue to aid his search; he only wandered farther from his horse, and risked falling afresh into the hands of his pursuers; he shouted again with all his strength, but his own voice alone echoed over the plains, while his heart stood still with the same frozen dread that a man feels when, wrecked on some barren shore, his cry for rescue rings back on his own ear over the waste of waters.

The flicker of the dawn was growing lighter in the sky, and he could see dimly now, as in some winter day's dark twilight, though all around him hung the leaden mist, with the wild winds driving furiously. It was with difficulty almost that he kept his feet against their force; but he was blown onward by their current, though beaten from side to side, and he still made his way forward. He had repassed the ground already traversed by some hundred yards or more, which seemed the length of many miles in the hurricane that was driving over the earth and sky, when some outline still duskier than the dusky shadow caught his sight; it was the body of a horse, standing on guard over the fallen body of a man.

Another moment and he was beside them.

'My God! Are you hurt?'

He could see nothing but an indistinct and shapeless mass, without form or colour, to mark it out from the brooding gloom and from the leaden earth. But the voice he knew so well answered him with the old love and fealty in it; eager with fear for him.

'When did you miss me, sir? I didn't mean you to know; I held on as long as I could; and when I couldn't no longer, I thought you was safe not to see I'd knocked over, so dark as it was.'

'Great Heavens! You are wounded, then?'

'Just finished, sir. Lord! it don't matter. Only you ride on, Mr Cecil; ride on, I say. Don't mind *me*.'

'What is it? When were you struck? O Heaven! I never dreamt——'

Cecil hung over him, striving in vain through the shadows to read the truth from the face on which he felt by instinct the seal of death was set.

'I never meant you should know, sir. I meant just to drop behind, and die on the quiet. You see, sir, it was just this way; they hit me as we forced through them. There's the lance-head in my loins now. I pressed it in hard, and kept the blood from flowing, and thought I should hold out so till the sun rose. But I couldn't do it so long; I got sick and faint after a while, and I knew well enough it was death. So I dropped down while I'd sense left to check the horse and get out of saddle in silence. I hoped you wouldn't miss me, in the darkness and the noise the wind was making; and you didn't hear me then, sir; I was glad.'

His voice was checked in a quick gasping breath; his only thought had been to lie down and die in solitude so that his master might be saved.

A great sob shook Cecil as he heard; no false hope came to him; he felt that this man was lost to him for ever, that this was the sole recompense which the cruelty of Africa would give to a fidelity passing the fidelity of woman; these throes of dissolution the only payment with which fate would ever requite a loyalty that had held no travail weary, no exile drear, and no danger worthy counting, so long as they were encountered and endured in his own service.

'Don't take on about it, sir,' whispered Rake, striving to raise his head that he might strain his eyes better through the gloom to see his master's face. 'It was sure to come some time; and I ain't in no pain—to speak of. Do leave me, Mr Cecil—leave me, for God's sake, and save yourself!'

'Did you leave me?'

The answer was very low, and his voice shook as he uttered it; but through the roar of the hurricane Rake heard it.

'That was different, sir,' he said, simply. 'Let me lie here, and go you on. It'll soon be over, and there's nought to be done.'

'O God! is no help possible?'

'Don't take on, sir; it's no odds. I allays was a scamp, and scamps die game, you know. My life's been a rare spree, count it all and all; and it's a great good thing, you see, sir, to go off quick like this. I might have been laid in hospital. If you'd only take the beast and ride on, sir——'

'Hush! hush! Would you make me coward, or brute, or both?'

The words broke in an agony from him. The time had been when he had been himself stretched in what he had thought was death, in just such silence, in just such solitude, upon the bare baked earth, far from men's aid, and near only to the hungry eyes of watching beasts of prey. Then he had been very calm, and waited with indifference for the end; now his eyes swept over the remorseless wastes, that were growing faintly visible under the coming dawn, with all the impatience, the terror, of despair. Death had smitten down many beside him; buoyant youth and dauntless manhood he had seen a thousand times swept under the great waves of war and lost for ever; but it had an anguish for him here that he would never have known had he felt his own life-blood well out over the sand. The whole existence of this man had been sacrificed for him, and its only reward was a thrust of a lance in a midnight fray—a grave in an alien soil.

His grief fell dully on ears half deafened already to the sounds of the living world. The exhaustion that follows on great loss of blood was upon the soldier who for the last half-hour had lain there in the darkness and the stillness, quietly waiting death, and not once seeking even to raise his voice for succour lest the cry should reach and should imperil his master.

The morning had broken now, but the storm had not lulled. The northern winds were sweeping over the plains in tenfold violence, and the rains burst and poured, with the fury of waterspouts, on the crust of the parched, cracked earth. Around them there was nothing heard or seen except the leaden angry mists, tossed to and fro under the hurricane, and the white light of the coming day breaking lividly through the clouds. The world held no place of more utter desolation, more unspeakable loneliness; and in its misery, Cecil, flung down upon the sands beside him, could do nothing except—helpless to aid, and powerless to save— watch the last breath grow feebler and feebler, until it faded out from the only life that had been faithful to him.

By the fitful gleams of day he could see the blood slowly ebbing from the great gap where the lance-head was still bedded with its wooden shaft snapped in two; he could see the drooped head that he had raised upon his knee, with the yellow northern curls that no desert suns had darkened; and Rake's eyes, smiling so brightly and so bravely still, looked up from under their weary lids to his.

'I'd never let you take my hand before, sir; just take it once now—will you?—while I can see you still.'

Their hands met as he asked it, and held each other close and long; all the loyal service of the one life, and all the speechless gratitude of the other, told better than by all words in that one farewell.

A light that was not from the stormy dusky morning shone over the soldier's face.

'Time was, sir,' he said, with a smile, 'when I used to think as how, some day or another, when I should have done something great and grand, and you was back amongst your own again, and they here had given me the Cross, I'd have asked you to have done that before all the Army, and just to have said to 'em, if so you'd liked, "He was a scamp, and he wasn't thought good for nought; but he kep' true to me, and you see it made him go straight, and I aren't ashamed to call him my friend." I used to think that, sir, though 'twas silly, perhaps. But it's best as it is—a deal best, no doubt. If you was only back safe in camp——'

'O God! cease! I am not worthy one thought of love like yours.'

'Yes you are, sir—leastways you was to me. When you took pity on me, it was just a toss up if I didn't go right to the gallows. Don't grieve that way, Mr Cecil. If I could just have seen you home again in your place, I should have been glad—that's all. You'll go back one day, sir; when you do, tell the King I ain't never forgot him.'

His voice grew faint as the last sentence stole from his lips; he lay quite still, his head leant back against his master; and the day came, with the north winds driving over the plains, and the grey mists tossed by them to and fro like smoke.

There was a long silence, a pause in which the wind-storm ceased and the clouds of the loosed sands sunk. Alone, with the wastes stretching around them, were the living and the dying man, with the horse standing motionless beside them, and, above, the gloom of the sullen sky. No aid was possible: they could but wait, in the stupefaction of despair, for the end of all to come.

In that awful stillness, in that sudden lull in the madness of the hurricane, death had a horror which it never wore in the riot of the battlefield, in the intoxication of the slaughter. There was no pity in earth or heaven; the hard hot ground sucked down its fill of blood; the icy air enwrapped them like a shroud.

The faithfulness of love, the strength of gratitude, were of no avail; the one perished, the other was powerless to save.

In that momentary hush, as the winds sank low, the heavy eyes, half sightless now, sought with their old wistful dog-like loyalty the face to which so soon they would be blind for ever.

'Would you tell me once, sir—now? I never asked—I never would have done—but maybe I might know in this last minute; you never did that sin you bear the charge on?'

'God is my witness, no.'

The light, that was like sunlight, shone once more in the aching, wandering eyes.

'I knew, I knew! It was——'

Cecil bowed his head over him, lower and lower.

'Hush! He was but a child; and I——'

With a sudden and swift motion, as though new life were thrilling in him, Rake raised himself erect, his arms stretched outward to the east, where the young day was breaking.

'I knew, I knew! *I* never doubted. You will go back to your own some day, and men shall learn the truth—thank God, thank God!'

Then, with that light still on his face, his head fell backward; and with one quick, brief sigh his life fled out for ever.

*   *   *

The time passed on; the storm had risen afresh; the violence of the gusts blew yellow sheets of sand whirling over the plains. Alone, with the corpse across his knees, Cecil sat motionless as though turned to stone. His eyes were dry and fixed; but ever and again a great tearless sob shook him from head to foot. The only life that linked him with the past, the only love that had suffered all things for his sake, were gone, crushed out as though they never had been, like some insect trodden in the soil.

He had lost all consciousness, all memory, save of that lifeless thing which lay across his knees, like a felled tree, like a broken log, with the glimmer of the tempestuous day so chill and white upon the upturned face.

He was alone on earth; and the solitudes around him were not more desolate than his own fate.

He was like a man numbed and stupified by intense cold; his veins seemed stagnant, and his sight could only see those features that became so terribly serene, so fearfully unmoved with the dread calm of death. Yet the old mechanical instincts of a soldier guided him still; he vaguely knew that his errand had to be done, must be done, let his heart ache as it would, let him long as he might to lie down by the side of his only friend, and leave the torture of life to grow still in him also for evermore.

Instinctively, he moved to carry out the duty trusted to him. He looked east and west, north and south; there was nothing in sight that could bring him aid; there were only the dust-clouds hurled in billows hither and thither by the bitter winds still blowing from the sea. All that could be done had to be done by himself alone. His own safety hung on the swiftness of his flight: for aught he knew, at every moment, out of the mist and the driven sheets of sand there might rush the desert-horses of his foes. But this memory was not with him: all he thought of was that burden stretched across his limbs, which, laid down one hour here unwatched, would be the prey of the jackal and the vulture. He raised it reverently in his arms, and with long laborious effort drew its weight up across the saddle of the charger which stood patiently waiting by, turning its docile eyes with a plaintive wondering sadness on the body of the rider it had loved. Then he mounted, himself; and with the head of his lost comrade borne up upon his arm, and rested gently on his breast, he rode westward over the great plain to where his mission lay.

The horse paced slowly beneath the double load of dead and living; he would not urge the creature faster on; every movement that shook the drooping limbs, or jarred the repose of that last sleep, seemed desecration. He passed the place where his own horse was stretched: the vultures were already there. He shuddered; and then pressed faster on, as though the beasts and birds of prey would rob him of his burden ere he could give it sanctuary. And so he rode, mile after mile, over the barren land, with no companion save the dead.

The winds blew fiercely in his teeth; the sand was in his eyes and hair; the way was long, and weary, and sown thick with danger; but he knew of nothing, felt and saw nothing, save that one familiar face so strangely changed and transfigured by that glory with which death had touched it.

CHAPTER VI

'*Je vous Achète votre Vie*'

Thus burdened, he made his way for over two leagues. The hurricane never abated, and the blinding dust rose around him in great waves. The horse fell lame; he had to dismount, and move slowly and painfully over the loose heavy soil on foot, raising the drooping head of the lifeless rider. It was bitter, weary, cruel travail, of an intolerable labour, of an intolerable pain.

Once or twice he grew sick and giddy, and lost for a moment all consciousness; but he pressed onwards, resolute not to yield and leave the vultures, hovering aloft, their prey. He was still somewhat weakened by the wounds of Zarâila; he had been bruised and exhausted by the skirmish of the past night; he was weary and heart-broken; but he did not yield to his longing to sink down on the sands, and let his life ebb out: he held patiently onward through the infinite misery of the passage. At last he drew near the caravanserai where he had been directed to obtain a change of horses. It stood mid-way in the distance that he had to traverse, and almost alone where the face of the country changed, and was more full of colour, and more broken into rocky and irregular surface.

As a man walks in a dream, he led the sinking beast towards its shelter, as its irregular corner towers became dimly perceptible to him through the dizzy mists that had obscured his sight. By sheer instinct he found his route straight towards the open arch of its entrance-way, and into the square courtyard thronged with mules, and camels, and horses; for the caravanserai stood on the only road that led through that district to the south, and was the only house of call for drovers, or shelter for travellers and artists of Europe who might pass that way. The groups in the court paused in their converse and in their occupations, and looked in awe at the grey charger with its strange burden, and the French Chasseur who came so blindly forward like a man feeling his passage through the dark. There was something in the sight that had a vague terror for them before they clearly saw what this thing was which was thus brought into their presence. Cecil moved slowly on into their midst, his hand on the horse's rein; then a great darkness covered his sight; he swayed to and fro, and fell senseless on the grey stone of the paved court, while the muleteers and the camel-drivers, the Kabyls and the French, who were mingled there, crowded around him in fear and in wonder. When consciousness returned to him, he was lying on a stone bench in the shadow of the wall, with the coolness of the fountain water bubbling near, and a throng of lean, bronzed, eager faces about him in the mid-day sunlight which had broken through the wind-storm.

Instantly he remembered all.

'Where is *he*?' he asked.

They knew that he meant the dead man, and answered him in a hushed murmur of many voices. They had placed the body gently down within, in a darkened chamber.

A shiver passed over him; he stretched his hand out for water that they held to him:

'Saddle me a fresh horse; I have my work to do.'

He knew that for no friendship, or grief, or suffering, or self-pity might a soldier pause by the wayside whilst his errand was still undone, his duty unfulfilled.

He drank the water thirstily; then reeling slightly still from the weakness that was still upon him, he rose rejecting their offers of aid. 'Take me to him,' he said, simply. They understood him; there were French soldiers amongst them, and they took him, without question or comment, across the court to the little square stone cell within one of the towers, where they had laid the corpse, with nothing to break the quiet and the solitude except the low soft cooing of some doves that had their home in its dark corners, and flew in and out at pleasure through the oval aperture that served as window.

He motioned them all back with his hand, and went into the gloom of the chamber alone. Not one amongst them followed.

When he came forth again the reckless and riotous *fantassins* of France turned silently and reverentially away, so that they should not look upon his face. For it was well known throughout the Army that no common tie had bound together the exiles of England; and the fealty of comrade to comrade was sacred in their sight.

The fresh animal, saddled, was held ready outside the gates. He crossed the court, moving still like a man without sense of what he did; he had the instinct to carry out the mission trusted to him, instantly and accurately, but he had no distinct perception or memory of aught else, save of those long-familiar features of which, ere he could return, the cruel sun of Africa would not have spared one trace.

He passed under the shadow of the gateway arch; a shadow black and intense against the golden light which, with the ceasing of the storm, flooded the land in the full morning. There were movement, noise, change, haste in the entrance. Besides the arrival of the detachment of the line and of a string of northward-bound camels, the retinue of some travellers of rank was preparing for departure, and the resources of the humble caravanserai were taxed beyond their powers. The name that some of the hurrying grooms shouted loudly in their impatience broke through his stupor and reached him. It was that of the woman whom, however madly, he loved with all the strength of a passion born out of utter hopelessness. He turned to the outrider nearest him:

'You are of the Princess Corona's suite? What does she do here?'

'Madame travels to see the country and the war.'

'The war? This is no place for her. The land is alive with danger—rife with death.'

'Miladi travels with M. le Duc, her brother. Miladi does not know what fear is.'

'But——'

The remonstrance died on his lips; he stood gazing out from the gloom of the arch at a face close to him, on which the sun shone full, a face unseen for twelve long years, and which, a moment before laughing and careless in the light, changed, and grew set, and rigid, and pale with the pallor of an unutterable horror. His own flushed, and moved, and altered with a wholly different emotion—emotion that was above all of an intense and yearning tenderness. For a moment both stood motionless and speechless; then, with a marvellous self-command and self-restraint, Cecil brought his hand to his brow in military salute, passed with the impassiveness of a soldier who passed a gentleman, reached his charger, and rode away upon his errand over the brown and level ground.

He had known his brother in that fleeting glance; but he hoped that his brother would see no more in him than a French trooper who bore resemblance by a strange hazard to one long believed to be dead and gone. The instinct of generosity, the instinct of self-sacrifice, moved him now as, long ago one fatal night, they had moved him to bear the sin of his mother's darling as his own.

Full remembrance, full consideration of what he had done, never came to him as he dashed on across the many leagues that still lay between him and his goal. His one impulse had been to spare the other from the knowledge that he lived; his one longing was to have the hardness and the bitterness of his own life buried in the oblivion of a soldier's grave.

* * *

Within six-and-thirty hours the instructions he bore were in the tent of the Chef du Bataillon whom they were to direct, and he himself returned to the caravanserai to fulfil with his own hand to the dead those last offices which he would delegate to none. It was night when he arrived; all was still and deserted. He enquired if the party of tourists was gone; they answered him in the affirmative; there only remained the detachment of the French Infantry, which were billeted there for a while.

It was in the coolness and the hush of the night, with the great stars shining clearly over the darkness of the plains, that they made the single grave, under a leaning shelf of rock, with the sombre fans of a pine spread above it, and nothing near but the sleeping herds of goats. The sullen echo of the soldiers' muskets gave its only funeral requiem; and the young lambs and kids in many a future springtime would come and play, and browse, and stretch their little tired limbs upon its sod, its sole watchers in the desolation of the plains.

When all was over, and the startled flocks had settled once again to rest and slumber, Cecil still remained there alone. Thrown down upon the grave, he never moved as hour after hour went by. To others that lonely and unnoticed tomb would be as nothing; only one amongst the thousand marks left on the bosom of the violated earth by the ravenous and savage lusts of war. But to him it held all that had bound him to his lost youth, his lost country, his lost peace; all that had remained of the years that were gone, and were now as a dream of the night. This man had followed him, cleaved to him, endured misery and rejected honour for his sake; and all the recompense such a life received was to be stilled for ever by a spear-thrust of an unknown foe, unthanked, undistinguished, unavenged! It seemed to him like murder—murder with which his own hand was stained.

The slow night hours passed; in the stillness that had succeeded to the storm of the past day, there was not a sound except the bleating of the young goats straying from the herd. He lay prostrate under the black boughs of the pine; the exhaustion of great fatigue was on him, a grief, acute as remorse, consumed him for the man who, following his fate, had only found at the end a nameless and lonely grave in the land of his exile.

He started with a thrill of almost superstitious fear as through the silence he heard a name whispered—the name of his childhood, of his past.

He sprang to his feet, and as he turned in the moonlight he saw once more his brother's face, pale as the face of the dead, and strained with an agonising dread. Concealment was no longer possible: the younger man knew that the elder lived; knew it by a strange and irresistible certainty that needed no proof, that left no place for hope or fear in its chill, leaden, merciless conviction.

For some moments neither spoke. A flood of innumerable memories choked thought or word in both. They knew each other—all was said in that.

Cecil was the first to break the silence. He moved nearer with a rapid movement, and his hand fell heavily on the other's shoulder.

'Have you lived stainlessly *since*?'

The question was stern as the demand of a judge. His brother shuddered beneath this touch, and covered his face with his hands.

'God is my witness, yes! But you—you—they said that you were dead! . . .'

Cecil's hand fell from his shoulder. There was that in the words which smote him more cruelly than any Arab steel could have done; there was the accent of *regret*.

'I am dead,' he said simply, 'dead to the world and you.'

He who bore the title of Royallieu covered his face.

'How have you lived?' he whispered hoarsely.

'Honourably. Let that suffice. And you?'

The other looked up at him with a piteous appeal—the old timorous terrified appeal that had been so often seen on the boy's face, strangely returning on the gracious and mature beauty of the man's.

'In honour too, I swear! That was my first disgrace, and my last. You bore the weight of my shame? Good God, what can I say? Such nobility, such sacrifice——'

He would have said enough, more than enough, to satisfy the one who had lost all for his sake, had there but been once in his voice no *fear*, but only love. As it was, that which he still thought of was himself alone. Whilst crushed with the weight of his brother's surpassing generosity, he still was filled with only one thought that burned through the darkness of his bewildered horror, and that thought was his own jeopardy. Even in the very first hours of his knowledge that the man whom he had believed dead was living, living and bearing the burden of the guilt he should have borne, what he was filled with was the imminence of his own peril.

Cecil stood in silence looking at him. He saw the boyish loveliness he remembered so well altered into the stronger and fuller beauty of the man. He saw that life had gone softly, smoothly, joyously, with this weak and feminine nature; and that, in the absence of temptation to evil, its career had been fair and straight in the sight of the world. He saw that his brother had been, in one word, happy. He saw that happiness had done for this character what adversity had done for his own. He saw that by it had been saved a temperament that calamity would have wrecked. He stood and looked at him, but he spoke not one word; whatever he felt, he restrained from all expression.

The younger man still hid his face upon his hands, as if, even in those pale grey moonbeams, he shunned the light that was about him.

'We believed you were dead,' he murmured, wildly. 'They said so; there seemed every proof. But when I saw you yesterday, I knew you— I knew you, though you passed me as a stranger. I stayed on here; they told me you would return. God! what agony this day and night have been!'

Cecil was silent still; he knew that this agony had been *the dread* lest he should be living.

There were many emotions at war in him—scorn, and pity, and wounded love, and pride too proud to sue for a gratitude denied, or quote a sacrifice that was almost without parallel in generosity, all held him speechless. To overwhelm the sinner before him with reproaches, to count and claim the immeasurable debts due to him, to upbraid and to revile the wretched weakness that had left the soil of a guilt not his own to rest upon him—to do aught of this was not in him. Long ago he had accepted the weight of an alien crime, and borne it as his own; to undo now all that he had done in the past, to fling out to ruin now the one whom he had saved at such a cost, to turn, after twelve years, and forsake the man, all coward though he was, whom he had shielded for so long—this was not possible to him. Though it would be but his own birthright that he would demand, his own justification that he would establish, it would have seemed to him like a treacherous and craven thing. No matter that the one for whom the sacrifice had been made was unworthy of it, he held that every law of honour and of justice forbade him now to abandon his brother, and yield him up to the retribution of his early fault. It might have been a folly in the first instance, it might even have been a madness, that choice of standing in his brother's place to receive the shame of his brother's action; but it had been done so long before—done on the spur of generous affection, and actuated by the strange hazard that made the keeping of a woman's secret demand the same reticence which also saved the young lad's name; to draw back from it now would have been a cowardice impossible to his nature.

All seemed uttered, without words, by their gaze at one another. He could not speak with tenderness to this craven who had been false to the fair repute of their name:—and he would not speak with harshness. He felt too sick at heart, too weary, too filled with pain, to ask aught of his brother's life; it had been saved from temptation, and therefore saved from evil; that knowledge sufficed to him.

The younger man stood half stupified, half maddened. In the many years that had passed by, although his character had not changed, his position had altered greatly; and in the last few months he had enjoyed all the power that wealth and independence and the accession to his title could bestow. He felt some dull, hot, angered sense of wrong done to him by the fact that the rightful heir of them still lived; some chafing, ingrate, and unreasoning impatience with the saviour of his whole existence; some bitter pangs of conscience that he would be baser yet, base beyond all baseness, to remain in his elder's place, and accept this sacrifice still, whilst knowing now the truth.

'Bertie—Bertie!' he stammered, in hurried appeal—and the name of his youth touched the hearer of it strangely, making him for the moment forget all save that he looked once more upon one of his own race—'on my soul, I never doubted that the story of your death was true. No one did. All the world believed it. If I had known you lived, I would have said that you were innocent; I would—I would have told them how I forged your friend's name and your own when I was so desperate that I scarce knew what I did. But they said that you were killed, and I thought then—then—it was not worth while; it would have broken my father's heart. God help me! I was a coward!'

He spoke the truth; he was a coward; he had ever been one. Herein lay the whole story of his fall, his weakness, his sin, and his ingratitude. Cecil knew that never will gratitude exist where craven selfishness holds reign; yet there was an infinite pity mingled with the scorn that moved him. After the years of bitter endurance he had passed, the heroic endurance he had witnessed, the hard and unending miseries that he had learned to take as his daily portion, this feebleness and fear roused his wondering compassion almost as a woman's weakness would have done. Still he never answered; the hatred of the stain that had been brought upon their name by his brother's deed (stain none the less dark, in his sight, because hidden from the world), his revulsion from this man, who was the only creature of their race who ever had turned poltroon, the thousand remembrances of childhood that uprose before him, the irresistible yearning for some word from the other's lips that should tell of some lingering trace in him of the old love strong enough to kill, for the moment at least, the selfish horror of personal peril—all these kept him silent.

His brother misinterpreted that silence.

'I am in your power—utterly in your power,' he moaned in his fear. 'I stand in your place; I bear your title; you know that our father and our brother are dead? All that I have inherited is yours—do you know that, since you have never claimed it?'

'I know it.'

'And you have never come forward to take your rights?'

'What I did not do to clear my own honour, I was not likely to do merely to hold a title.'

The meaning of his answer drifted beyond the ear on which his words fell; it was too high to be comprehended by the lower nature. The man who lived in prosperity and peace, and in the smile of the world, and the purple of power, looked bewildered at the man who led the simple, necessitous, perilous, semi-barbaric existence of an Arab-Franco soldier.

'But—great Heaven!—this life of yours? It must be wretchedness?'

'Perhaps. It has at least no disgrace in it.'

The reply had the only sternness of contempt that he had suffered himself to show. It stung down to his listener's soul.

'No—no!' he murmured. 'You are happier than I. You have no remorse to bear! And yet—to tell the world that I am guilty!——'

'You need never tell it: I shall not.'

He spoke quite quietly, quite patiently. Yet he well knew, and had well weighed, all he surrendered in that promise—the promise to condemn himself to a barren and hopeless fate for ever.

'You will not?'

The question died almost inaudible on his dry, parched tongue. The one passion of fear upon him was for himself; even in that moment of supplication his disordered thoughts hovered wildly over the chances of whether, if his elder brother even now asserted his innocence and claimed his birthright, the world and its judges would ever believe him.

Cecil for a while again was silent, standing there by the newly made grave of the soldier who had been faithful as those of his own race and of his own order never had been. His heart was full. The ingratitude and the self-absorption of this life for which his own had been destroyed smote him with a fearful suffering. And only a few hours before he had looked once more on the face of the beloved friend of his youth;—a deadlier sacrifice than to lay down wealth, and name, and heritage, and the world's love, was to live on leaving that one comrade of his early days to believe him dead after a deed of shame.

His brother sank down on the mound of freshly flung earth, sinking his head upon his arms with a low moan. Time had not changed him greatly; it had merely made him more intensely desirous of the pleasures and the powers of life, more intensely abhorrent of pain, of censure, of the contempt of the world. As, to escape these in his boyhood, he had stooped to any degradation, so, to escape them in his manhood, he was capable of descending to any falsehood or any weakness. His was one of those natures which, having no love of evil for evil's sake, still embrace any form of evil which may save them from the penalty of their own weakness. Now, thus meeting one whom for twelve years he had believed must rise from the tomb itself to reproach or to accuse him, unstrung his every nerve, and left him with only one consciousness—the desire at all costs to be saved.

Cecil's eyes rested on him with a strange melancholy pity: he had loved his brother as a youth—loved him well enough to take and bear a heavy burden of disgrace in his stead. The old love was not dead; but stronger than itself was his hatred of the shame that had touched their race by the wretched crime that had driven him into exile, and his wondering scorn for the feeble and self-engrossed character that had lived contentedly under false colours, and with a hidden blot screened by a fictitious semblance of honour. He could not linger with him; he did not know how to support the intolerable pain that oppressed him in the presence of the only living creature of his race; he could not answer for himself what passionate and withering words might not escape him; every instant of their interview was a horrible temptation to him—the temptation to demand from this coward his own justification before the world—the temptation to seize out of these unworthy hands his birthright and his due.

But the temptation, sweet, insidious, intense, strengthened by the strength of right, and well-nigh overwhelming with all its fair delicious promise for the future, did not conquer him. What resisted it was his own simple instinct of justice; an instinct too straight and true either to yield to self-pity or to passionate desire—justice which made him feel that, since he had chosen to save this weakling once for their lost mother's sake, he was bound for ever not to repent nor to retract. He gazed awhile longer, silently, at the younger man, who sat still rocking himself wearily to and fro on the loose earth of the freshly filled grave. Then he went and laid his hand on his brother's shoulder: the other started and trembled; he remembered that touch in days of old.

'Do not fear me,' he said, gently and very gravely. 'I have kept your secret twelve years; I will keep it still. Be happy—be as happy as you can. All I bid of you in return is so to live that in your future your past shall be redeemed.'

The words of the saint to the thief, '*Je vous achète votre vie*', were not more merciful, not more noble, than the words with which he purchased, at the sacrifice of his own life, the redemption of his brother's. The other looked at him with a look that was half of terror—terror at the magnitude of this ransom that was given to save him from the bondage of evil.

'My God! You cannot mean it! And you——?'

'I shall lead the life fittest for me: I am content in it. It is enough.'

The answer was very calm, but it choked him in its utterance. Before his memory rose one fair, proud face. 'Content!' Ah, Heaven!—it was the only lie that had ever passed his lips.

His hand lay still upon his brother's shoulder, leaning more heavily there, in the silence that brooded over the hushed plains.

'Let us part now, and for ever. Leave Algeria at once. That is all I ask.'

Then, without another word that could add reproach or seek for gratitude, he turned and went away over the great dim level of the African waste, whilst the man whom he had saved sat as in stupor, gazing at the brown shadows, and the sleeping herds, and the falling stars that ran across the sky, and doubting whether the voice he had heard and the face upon which he had looked were not the visions of a waking dream.

CHAPTER VII

'Venetia'

How that night was spent Cecil could never recall in full. Vague memories remained with him of wandering over the shadowy country, of seeking by bodily fatigue to kill the thoughts rising in him, of drinking at a little water-channel in the rocks as thirstily as some driven deer, of flinging himself down at length, worn out, to sleep under the hanging brow of a mighty wall of rock; of waking when the dawn was reddening the east with the brown plains around him, and far away under a knot of palms was a goatherd with his flock like an idyl from the old pastoral life of Syria. He stood looking at the light which heralded the sun, with some indefinite sense of heavy loss, of fresh calamity, upon him. It was only slowly that he remembered all. Years seemed to have been pressed into the three nights and days since he had sat by the bivouac-fire, listening to the fiery words of the little Friend of the Flag.

The full consciousness of all that he had surrendered in yielding up afresh his heritage rolled in on his memory like the wave of some heavy sea that sweeps down all before it.

When that tear-blotted and miserable letter had reached him in the green alleys of the Stephanien, and confessed to him that his brother had relied on the personal likeness between them and the similarity of their handwriting to pass off as his the bill in which his own name and that of his friend was forged, no thought had crossed him to take upon himself the lad's sin. It had only been when, brought under the charge, he must, to clear himself, have at once accused the boy, and have betrayed the woman whose reputation was in his keeping, that rather by generous impulse than by studied intention he had taken up the burden that he had now carried for so long. Whether or no the money-lenders had been themselves in reality deceived he could never tell; but it had been certain that, having avowed themselves confident of his guilt, they could never shift the charge on to his brother in the face of his own acceptance of it. So he had saved the youth without premeditation or reckoning of the cost. And now that the full cost was known to him, he had not shrunk back from its payment. Yet that payment was one that gave him a greater anguish than if he had laid down his life in physical martyrdom.

To go back to the old luxury, and ease, and careless peace, to go back to the old fresh fair English woodlands, to go back to the power of command and the delight of free gifts, to go back to men's honour, and reverence, and high esteem—these would have been sweet enough— sweet as food after long famine. But far more than these would it have been to go back and take the hand of his friend once more in the old unclouded trust of their youth; to go back, and stand free and blameless amongst his peers, and know that all that man could do to win the heart and the soul of a woman he could at his will do to win hers whose mere glance of careless pity had sufficed to light his life to passion. And he had renounced all this. This was the cost: and he had paid it—paid it because the simple, natural, inflexible law of justice had demanded it.

One whom he had once chosen to save he could not now have deserted, except by what would have been, in his sight, dishonour. Therefore, when the day broke, and the memories of the night came with his awakening, he knew that his future was without hope—without it as utterly as was ever that of any captive shut in darkness, and silence, and loneliness, in a prison, whose only issue was the oubliettes. There is infinite misery in the world, but this one misery is rare; or men would perish from the face of the earth as though the sun withdrew its light.

Alone in that dreary scene, beautiful from its vastness and its solemnity, but unutterably melancholy, unutterably oppressive, he also wondered whether he lived or dreamed.

From amongst the reeds the plovers were rising; over the barren rocks the dazzling lizards glided; afar off strayed the goats: that was the only sign of animal existence. He had wandered a long way from the caravanserai, and he began to retrace his steps, for his horse was there, and although he had received licence to take leisure in returning, he had no home but the camp, no friends but those wild-eyed, leopard-like, ferocious sons of the razzia and the slaughter, who would throng around him like a pack of dogs, each eager for the first glance, the first word; these companions of his adversity and of his perils, whom he had learned to love, with all their vices and all their crimes, for sake of the rough, courageous love that they could give in answer.

He moved slowly back over the desolate tracks of land stretched between him and the Algerian halting-place. He had no fear that he would find his brother there. He knew too well the nature with which he had to deal to hope that old affection would so have outweighed present fear that his debtor would have stayed to meet him yet once more. On the impulse of the ungovernable pain which the other's presence had been, he had bidden him leave Africa at once; now he almost wished that he had bid him stay. There was a weary unsatisfied longing for some touch of love or of gratitude from this usurper, whom he had raised in his place. He would have been rewarded enough if *one* sign of gladness that he lived had broken through the egotism and the stricken fear of the man whom he remembered as a little golden-headed child, with the hand of their dying mother lying in benediction on the fair silken curls.

He had asked no questions. He had gone back to no recriminations. He guessed all it needed him to know; and he recoiled from the recital of the existence whose happiness was purchased by his own misery, and whose dignity was built on sand. His sacrifice had not been in vain. Placed out of the reach of temptation, the plastic, feminine, unstable character had been without a stain in the sight of men; but it was little better at the core; and he wondered, in his suffering, as he went onward through the beauty of the young day, whether it had been worth the bitter price he had paid to raise this bending reed from out the waters which would have broken and swamped it at the outset. It grew fair, and free, and flower-crowned now, in the midst of a tranquil and sunlit lake; but was it of more value than a drifted weed bearing the snake-egg hidden at its root?

He had come so far out of the ordinary route across the plains that it was two hours or more before he saw the dark grey square of the caravanserai walls, and to its left that single leaning pine growing out of a cleft within the rock that overhung the spot where the keenest anguish of all his life had known had been encountered and endured—the spot which yet, for sake of the one laid to rest there, beneath the sombre branches, would be for ever dearer to him than any other place in the soil of Africa.

Whilst yet the caravanserai was distant, the piteous cries of a mother-goat caught his ear; she was bleating beside a watercourse, into which her kid of that spring had fallen, and whose rapid swell, filled by the recent storm, was too strong for the young creature. Absorbed as he was in his own thoughts, the cry reached him and drew him to the spot; it was not in him willingly to let any living thing suffer, and he was always gentle to all animals. He stooped and, with some little difficulty, rescued the little goat for its delighted dam.

As he bent over the water, he saw something glitter beneath it; he caught it in his hand and brought it up; it was the broken half of a chain of gold, with a jewel in each link. He changed colour as he saw it; he remembered it as one that Venetia Corona had worn on the morning that he had been admitted to her. It was of peculiar workmanship, and he recognized it at once. He stood with the toy in his hand, looking long at the shining links, with their flashes of precious stones. They seemed to have voices that spoke to him of her about whose beautiful white throat they had been woven—voices that whispered incessantly in his ear, 'Take up your birthright, and you will be free to sue to her at least, if not to win her.' No golden and jewelled plaything ever tempted a starving man to theft as this tempted him now to break the pledge he had just given.

His birthright! He longed for it for this woman's sake—for the sake, at least, of the right to stand before her as an equal, and to risk his chance with others who sought her smile—as he had never done for any other thing which, with that heritage, would have become his. Yet he knew that, even were he to be false to his word, and go forward and claim his right, he would never be able to prove his innocence; he could never hope to make the world believe him unless the real criminal made that confession which he held himself forbidden, by his own past action, ever to extort.

He gazed long at the broken costly toy, while his heart ached with a cruel pang; then he placed it in safety in the little blue enamel box, beside the ring which Cigarette had flung back to him, and went onward to the caravanserai. She was no longer there, in all probability; but the lost bagatelle would give him, some time or another, a plea on which to enter her presence. It was a pleasure to him to know that; though he knew also that every added moment spent under the sweet sovereignty of her glance was so much added pain, so much added folly, to the dream-like and baseless passion with which she had inspired him.

The trifling incident of the goat's rescue and the chain's trouvaille, slight as they were, still were of service to him. They called him back from the past to the present; they broke the stupor of suffering that had fastened on him; they recalled him to the actual world about him in which he had to fulfil his duties as a trooper of France.

It was almost noon when, under the sun-scorched branches of the pine that stretched its sombre fans up against the glittering azure of the morning skies, he approached the gates of the Algerine house-of-call—a study for the colour of Gérôme, with the pearly grey of its stone tints, and the pigeons wheeling above its corner towers, while under the arch of its entrance a string of mules, maize-laden, were guided; and on its bench sat a French fantassin, singing gaily songs of Paris whilst he cut open a yellow gourd.

Cecil went within, and bathed, and dressed, and drank some of the thin cool wine that found its way hither in the wake of the French army. Then he sat down for a while at one of the square cabin-like holes which served for casements in the tower he occupied, and, looking out into the court, tried to shape his thoughts and plan his course. As a soldier he had no freedom, no will of his own, save for this extra twelve or twenty-four hours which they had allowed him for leisure in his return journey. He was obliged to go back to his camp, and there, he knew, he might again encounter one whose tender memories would be as quick to recognize him as the craven dread of his brother had been. He had always feared this ordeal, although the arduous service in which his chief years in Africa had been spent, and the remote expeditions on which he had always been employed, had partially removed him from the ever-present danger of such recognition until now. And now he felt that if once the brave kind eyes of his old friend should meet his own, concealment would be no longer possible, yet, for sake of that promise he had sworn in the past night, it must be maintained at every hazard, every cost. Vacantly he sat and watched the play of the sunshine in the prismatic water of the courtyard fountain, and the splashing, and the pluming, and the murmuring of the doves and pigeons on its edge. He felt meshed in a net from which there was no escape—none—unless, on his homeward passage, a thrust of Arab steel should give him liberty.

The trampling of horses on the pavement below roused his attention. A thrill of hope went through him that his brother might have lingering conscience, latent love enough, to have made him refuse to obey the bidding to leave Africa. He rose and leaned out. Amidst the little throng of riding-horses, grooms, and attendants who made an open way through the polyglot crowd of an Algerian caravanserai at noon, he saw the one dazzling face of which he had so lately dreamed by the water-freshet in the plains. It was but a moment's glance, for she had already dismounted from her mare, and was passing within with two other ladies of her party; but in that one glance he knew her. His discovery of the chain gave him a plea to seek her; should he avail himself of it? He hesitated awhile; it would be safest, wisest, best, to deliver up the trinket to her courier, and pass on his way without another look at that beauty which could never be his, which could never lighten for him even with the smile that a woman may give her equal or her friend. She could never be aught to him save one more memory of pain, save one remembrance the more to embitter the career which not even hope would ever illumine. He knew that it was only madness to go into her presence, and feed, with the cadence of her voice, the gold light of her hair, the grace and graciousness of her every movement, the love which she would deem such intolerable insult, that, did he ever speak it, she would order her people to drive him from her like a chidden hound. He knew that; but he longed to indulge the madness despite it, and he did so. He went down into the court below, and found her suite.

'Tell your mistress that I, Louis Victor, have some jewels which belong to her, and ask her permission to restore them to her hands,' he said to one of her equerries.

'Give them to me, if you have picked them up,' said the man, putting out his hand for them.

Cecil closed his own upon them:

'Go and do as I bid you.'

The equerry paused, doubtful whether or no to resist the tone and the words. A Frenchman's respect for the military uniform prevailed; he went within.

In the best chamber of the caravanserai, Venetia Corona was sitting, listless in the heat, when her attendant entered. The *grandes dames* who were her companions in their tour through the seat of war, were gone to their siesta. She was alone, with a scarlet burnous thrown about her, and upon her all the languor and idleness common to the noontide, which was still very warm, though, in the autumn, the nights were so icily cold on the exposed level of the plains. She was lost in thought, moreover. She had heard, the day before, a story that had touched her—of a soldier who had been slain crossing the plains, and had been brought, through the hurricane and the sandstorm, at every risk, by his comrade, who had chosen to endure all peril and wretchedness rather than leave the dead body to the vultures and the kites. It was a nameless story to her—the story of two obscure troopers, who, for aught she knew, might have been two of the riotous and savage brigands that were common in the Army of Africa. But the loyalty and the love shown in it had moved her; and to the woman whose life had been cloudless and cradled in ease from her birth, there was that in the suffering and the sacrifice which the anecdote suggested, that had at once the fascination of the unknown, and the pathos of a life so far removed from her, so little dreamed of by her, that all its coarser cruelty was hidden, whilst only its unutterable sadness and courage remained before her sight.

Had she, could she, ever have seen it in its realities, watched and read and understood it, she would have been too intensely revolted to have perceived the actual latent nobility possible in such an existence; as it was, she heard but of it in such words as alone could meet the ear of a great lady, she gazed at it only in pity from a far distant height, and its terrible tragedy had solemnity and beauty for her.

When her servant approached her now with Cecil's message, she hesitated some few moments in surprise. She had not known that he was in her vicinity; the story she had heard had been simply of two unnamed Chasseurs d'Afrique, and he himself might have fallen on the field weeks before, for aught that she had heard of him. Some stray rumours of his defence of the encampment of Zarâila, and of the fine prowess shown in his last charge, alone had drifted to her. He was but a trooper; and he fought in Africa. The world had no concern with him, save the miniature world of his own regiment.

She hesitated some moments; then gave the required permission. 'He has once been a gentleman: it would be cruel to wound him,' thought the imperial beauty, who would have refused a Prince or neglected a Duke with chill indifference, but who was too generous to risk the semblance of humiliation to the man who could never approach her save upon such sufferance as was in itself mortification to one whose pride survived his fallen fortunes.

Moreover, the interest he had succeeded in awakening in her, the mingling of pity and of respect that his words and his bearing had aroused, was not extinct; had, indeed, only been strengthened by the vague stories that had of late floated to her of the day of Zarâila; of the day of smoke and steel and carnage, of war in its grandest yet its most frightful shape, of the darkness of death which the courage of human souls had power to illumine as the rays of the sun the tempest-cloud. Something more like quickened and pleasured expectation than any one amongst her many lovers had ever had power to rouse, moved her as she heard of the presence of the man who, in that day, had saved the honour of his Flag. She came of an heroic race; she had heroic blood in her; and heroism, physical and moral, won her regard as no other quality could ever do. A man capable of daring greatly, and of suffering silently, was the only man who could ever hope to hold her thoughts.

The room was darkened from the piercing light without; and in its gloom, as he was ushered in, the scarlet of her cashmere and the gleam of her fair hair was all that, for the moment, he could see. He bowed very low that he might get his calmness back before he looked at her; and her voice in its lingering music came on his ear.

'You have found my chain, I think? I lost it in riding yesterday. I am greatly indebted to you for taking care of it.'

She felt that she could only thank, as she would have thanked an equal who should have done her this sort of slight service, the man who had brought to her the gold pieces with which his Colonel had insulted him.

'It is I, Madame, who am the debtor of so happy an accident.'

His words were very low, and his voice shook a little over them; he was thinking not of the jewelled toy that he came here to restore, but of the inheritance that had passed away from him for ever, and which, possessed, would have given him the title to seek what his own efforts could do to wake a look of tenderness in those proud eyes which men ever called so cold, but which he felt might still soften, and change, and grow dark with the thoughts and the passions of love, if the soul that gazed through them were but once stirred from its repose.

'Your chain is here, Madame, though broken, I regret to see,' he continued, as he took the little box from his coat and handed it to her. She took it, and thanked him, without, for the moment, opening the enamel case as she motioned him to a seat at a little distance from her own.

'You have been in terrible scenes since I saw you last,' she continued. 'The story of Zaraîla reached us. Surely they cannot refuse you the reward of your service now?'

'It will make little difference, Madame, whether they do or not.'

'Little difference! How is that?'

'To my own fate, I meant. Whether I be a Captain or a Corporal cannot alter——'

He paused; he dreaded lest the words should escape him which should reveal to her that which she would regard as such intolerable offence, such insolent indignity, when felt for her by a soldier in the grade he held.

'No? Yet such recognition is usually the ambition of every military life.'

A very weary smile passed over his face.

'I have no ambition, Madame. Or, if I have, it is not a pair of epaulettes that will content it.'

She understood him; she comprehended the bitter mockery that the tawdry, meretricious rewards of regimental decoration seemed to the man who had waited to die at Zaraîla as patiently and as grandly as the Old Guard at Waterloo.

'I understand! The rewards are pitifully disproportionate to the services in any army. Yet how magnificently you and your men, as I have been told, held your ground all through that fearful day!'

'We did our duty—nothing more.'

'Well! is not that the rarest thing amongst men?'

'Not amongst soldiers, Madame.'

'Then you think that every trooper in a regiment is actuated by the finest and most impersonal sentiment that can actuate human beings!'

'I will not say that. Poor wretches! they are degraded enough, too often. But I believe that more or less in every good soldier, even when he is utterly unconscious of it, is an impersonal love for the honour of his Flag, an uncalculating instinct to do his best for the reputation of his corps. We are called human machines; we are so, since we move by no will of our own; but the lowest among us will at times be propelled by one single impulse—a desire to die greatly. It is all that is left to most of us to do.'

She looked at him with that old look which he had seen once or twice before in her, of pity, respect, sympathy, and wonder, all in one. He spoke to her as he had never spoken to any living being. The grave, quiet, listless impassiveness that still was habitual with him—relic of the old habits of his former life—was very rarely broken, for his real nature or his real thoughts to be seen beneath it. But she, so far removed from him by position and by circumstance, and distant with him as a great lady could not but be with a soldier of whose antecedents and whose character she knew nothing, gave him sympathy, a sympathy that was sweet and rather felt than uttered; and it was like balm to a wound, like sweet melodies on a weary ear, to the man who had carried his secret so silently and so long, without one to know his burden or to soothe his pain.

'Yes,' she said, thoughtfully, while over the brilliancy of her face there passed a shadow. 'There must be infinite nobility amongst these men, who live without hope—live only to die. That soldier, a day or two ago, who brought his dead comrade through the hurricane, risking his own death rather than leave the body to the carrion-birds—you have heard of him? What tenderness, what greatness, there must have been in that poor fellow's heart.'

'Oh no. That was nothing.'

'Nothing! They have told me he came every inch of the way in danger of the Arabs' shot and steel. He had suffered so much to bring the body safe across the plains; he fell down insensible on his entrance here.'

'You set too much store on it. I owed him a debt far greater than any act like that could ever repay.'

'*You!* Was it you?'

'Yes, Madame. He who perished had a thousand-fold more of such nobility as you have praised than I.'

'Ah? Tell me of him,' she said, simply; but he saw that the lustrous eyes bent on him had a grave sweet sadness in them, that was more precious and more pitiful than a million utterances of regret could ever have been.

Those belied her much, who said that she was heartless; though grief
had never touched her, she could feel keenly the grief of other lives. He
obeyed her bidding now, and told her, in brief words, the story, which
had a profound pathos spoken there, where without through the oval
unglazed casement in the distance there was seen the tall dark leaning
pine that overhung the grave of yesternight; the story, over which his
voice oftentimes fell with the hush of a cruel pain in it, and which he
could have related to no other save herself. It had an intense melancholy
and a strange beauty in its brevity and its simplicity, told in that gaunt,
still, darkened chamber of the caravanserai, with the grey gloom of its
stone walls around, and the rays of the golden sunlight from without
straying in to touch the glistening hair of the proud head that bent
forward to listen to the recital. Her face grew paler as she heard; and
a mist was over the radiance of her azure eyes: that death in the loneliness of the plains moved her deeply with the grand simplicity of its
unconscious heroism. And, though he spoke little of himself, she felt,
with all the divination of a woman's sympathies, how he who told her
this thing had suffered by it—suffered far more than the comrade whom
he had laid down in the grave where, far off in the noonday warmth,
the young goats were at rest on the sod. When he ceased, there was a
long silence; he had lost even the memory of her in the memory of the
death that he had painted to her; and she was moved with that wondering
pain, that emotion, half dread and half regret, with which the contemplation of calamities that have never touched and that can never touch
them will move women far more callous, far more world-chilled, than
herself.

In the silence her hands toyed listlessly with the enamel bonbonnière,
whose silver had lost all its bright enamelling, and was dinted and dulled
till it looked no more than lead. The lid came off at her touch as she
musingly moved it round and round; the chain and the ring fell into her
lap; the lid remained in her hand, its interior unspoiled and studded in its
centre with one name in turquoise letters—VENETIA.

She started as the word caught her eye and broke her reverie; the
colour came warmer into her cheek; she looked closer and closer at the
box, then with a rapid movement turned her head and gazed at her
companion.

'How did you obtain this?'

'The chain, Madame? It had fallen in the water.'

'The chain! No! the box!'

He looked at her in surprise.

'It was given me very long ago.'

'And by whom?'

'By a young child, Madame.'

Her lips parted slightly, the flush on her cheeks deepened; the beautiful face, which the Roman sculptor had said only wanted tenderness to make it perfect, changed, moved, was quickened with a thousand shadows of thought.

'The box is mine! I gave it! And you?'

He rose to his feet and stood entranced before her, breathless and mute.

'And you?' she repeated.

He was silent still; gazing at her. He knew her now—how had he been so blind as never to guess the truth before, as never to know that those imperial eyes and that diadem of golden hair could belong alone but to the women of one race?

'And you?' she cried once more, while she stretched her hand out to him. 'And you—you are Philip's friend? you are Bertie Cecil?'

Silently he bowed his head: not even for his brother's sake, or for sake of his pledged word, could he have lied to *her*.

But her outstretched hands he would not see; he would not take. The shadow of an imputed crime was stretched between them.

'Petite Reine!' he murmured! 'Ah, God! how could I be so blind?'

She grew very pale as she sank back again upon the couch from which she had risen. It seemed to her as though a thousand years had drifted by since she had stood beside this man under the summer leaves of the Stephanien, and he had kissed her childish lips, and thanked her for her loving gift. And now—they had met thus!

He said nothing. He stood paralysed, gazing at her. There had been no added bitterness needed in the cup which he drank for his brother's sake, yet this bitterness surpassed all other: it seemed beyond his strength to leave *her* in the belief that he was guilty. She in whom all fair and gracious things were met, she who was linked by her race to his past and his youth; she whose clear eyes in her childhood had looked upon him in that first hour of the agony that he had suffered then, and still suffered on, in the cause of a coward and an ingrate.

She was pale still; and her eyes were fixed on him with a gaze that recalled to him the look with which 'Petite Reine' had promised that summer day to keep his secret, and tell none of that misery of which she had been witness.

'They thought that you were dead,' she said at length, whilst her voice sank very low. 'Why have you lived like this?'

He made no answer.

'It was cruel to Philip,' she went on, whilst her voice still shook. 'Child though I was, I remember his passion of grief when the news came that you had lost your life. He has never forgotten you. So often now he will still speak of you! He is in your camp. We are travelling together. He will be here this evening. What delight it will give him to know his dearest friend is living! But why—why—have kept *him* ignorant, if you were lost to all the world beside?'

Still he answered her nothing. The truth he could not tell; the lie he would not. She paused, waiting reply. Receiving none, she spoke once more, her words full of that exquisite softness which was far more beautiful in her than in women less tranquil, less chill, and less negligent in ordinary moments.

'Mr Cecil, I divined rightly! I knew that you were far higher than your grade in Africa; I felt that in all things, save in some accident of position, we were equals. But why have you condemned yourself to this misery? Your life is brave, is noble, but it must be a constant torture to such as you? I remember well what you were—so well, that I wonder we have never recognized each other before now. The existence you lead in Algeria must be very terrible to you, though it is greater, in truth, than your old years of indolence?'

He sank down beside her on a low seat, and bowed his head on his hands for some moments. He knew that he must leave this woman whom he loved, and who knew him now as one whom in her childhood she had seen caressed and welcomed by all her race, to hold him guilty of this wretched, mean, and fraudulent thing, under whose charge he had quitted her country. Great dews of intense pain gathered on his forehead; his whole mind, and heart, and soul revolted against this brand of a guilt not his own that was stamped on him; he could have cried out to her the truth in all the eloquence of a breaking heart.

But he knew that his lips had been sealed by his own choice for ever; and the old habits of his early life were strong upon him still. He lifted his head and spoke gently, and very quietly, though she caught the tremour that shook through the words:

'Do not let us speak of myself. You see what my life is; there is no more to be said. Tell me rather of your own story—you are no longer the Lady Venetia? You have been wedded and widowed, they say?'

'The wife of an hour—yes! But it is of yourself that I would hear. Why have left the world, and, above all, why have left *us*, to think you dead? I was not so young when we last saw you, but that I remember well how all my people loved you.'

Had she been kept in ignorance of the accusation beneath which his flight had been made? He began to think so. It was possible. She had been so young a child when he had left for Africa; then the story was probably withheld from reaching her, and now, what memory had the world to give a man whose requiem it had said twelve long years before? In all likelihood she had never heard his name, save from her brother's lips, that had been silent on the shame of his old comrade.

'Leave my life alone, for God's sake!' he said, passionately. 'Tell me of your own—tell me, above all, of *his*. He loved me, you say?—O Heaven! he did. Better than any creature that ever breathed; save the man whose grave lies yonder.'

'He does so still,' she answered, eagerly; 'Philip's is not a heart that forgets. It is a heart of gold, and the name of his earliest friend is graven on it as deeply now as ever. He thinks you dead; tonight will be the happiest hour he has ever known when he shall meet you here.'

He rose hastily, and moved thrice to and fro the narrow floor whose rugged earth had been covered with furs and rugs lest it should strike a chill to her as she passed over it: the torture grew unsupportable to him. And yet, it had so much of sweetness that he was powerless to end it— sweetness in the knowledge that she knew him now her equal, at least by birth; in the change that it had made in her voice and her glance, whilst the first grew tender with olden memories, and the last had the smile of friendship; in the closeness of the remembrances that seemed to draw and bind them together; in the swift sense that in an instant, by the utterance of a name, the ex-barrier of caste which had been between them had fallen now and for ever.

She watched him with grave musing eyes. She was moved, startled, softened to a profound pity for him, and filled with a wondering of regret; yet a strong emotion of relief, of pleasure, rose above these. She had never forgotten the man to whom, in her childish innocence, she had brought the gifts of her golden store; she was glad that he lived, though he lived thus; glad with a quicker, warmer, more vivid emotion than any that had ever occupied her for any man living or dead except her brother. The interest she had vaguely felt in a stranger's fortunes, and which she had driven contemptuously away as unworthy of her harbouring was justified for one whom her people had known and valued whilst she had been in her infancy, and of whom she had never heard from her brother's lips aught except constant regret and imperishable attachment. For it was true, as Cecil divined, that the dark cloud under which his memory had passed to all in England had never been seen by her eyes, from which, in childhood, it had been screened, and, in womanhood, withheld, because his name had been absolutely forgotten by all save the Seraph, to whom it had been fraught with too much pain for its utterance to be ever voluntary.

'What is it you fear from Philip?' she asked him, at last, when she had waited vainly for him to break the silence. 'You can remember him but ill if you think that there will be anything in his heart save joy when he shall know that you are living. You little dream how dear your memory is to him——'

He paused before her abruptly.

'Hush, hush! or you will kill me! Why!—three nights ago I fled the camp as men flee pestilence, because I saw his face in the light of the bivouac-fire and dreaded that he should so see mine!'

She gazed at him in troubled amaze; there was that in the passionate agitation of this man who had been serene through so much danger, and unmoved beneath so much disaster, that startled and bewildered her.

'You fled from Philip? Ah! how you must wrong him! What will it matter to him whether you be prince or trooper, wear a peer's robes or a soldier's uniform? His friendship never yet was given to externals. But—stay!—that reminds me of your inheritance. Do you know that Lord Royallieu is dead? that your younger brother bears the title, thinking you perished at Marseilles? He was here with me yesterday; he has come to Algeria for the autumn. Whatever your motive may have been to remain thus hidden from us all, you must claim your own rights now. You must go back to all that is so justly yours. Whatever your reason be to have borne with all the suffering and the indignity that have been your portion here, they will be ended now.'

Her beauty had never struck him so intensely as at this moment, when, in urging him to the demand of his rights, she so unconsciously tempted him to betray his brother and to forsake his word. The indifference and the careless coldness that had to so many seemed impenetrable and unalterable in her were broken and had changed to the warmth of sympathy, of interest, of excitation. There was a world of feeling in her face, of eloquence in her eyes, as she stooped slightly forward with the rich glow of the cashmeres about her, and the sun-gleam falling across her brow. Pure, and proud, and noble in every thought, and pressing on him only what was the due of his birth and his heritage, she yet unwittingly tempted him with as deadly a power as though she were the vilest of her sex, seducing him downward to some infamous dishonour.

To do what she said would be but his actual right, and would open to him a future so fair that his heart grew sick with longing for it; and yet to yield, and to claim justice for himself, was forbidden him as utterly as though it were some murderous guilt. He had promised never to sacrifice his brother; the promise held him like the fetters of a galley.

'Why do you not answer me?' she pursued, whilst she leaned nearer with wonder, and doubt, and a certain awakening dread shadowing the blue lustre of her eyes that were bent so thoughtfully, so searchingly, upon him. 'Is it possible that you have heard of your inheritance, of your title and estates, and that you voluntarily remain a soldier here? Lord Royallieu must yield them the instant you prove your identity, and in that there could be no difficulty. I remember you well now, and Philip, I am certain, will only need to see you once to——'

'Hush, for pity's sake! Have you never heard—have none ever told you——'

'What?'

Her face grew paler with a vague sense of fear; she knew that he had been equable and resolute under the severest tests that could try the strength and the patience of man, and she knew, therefore, that no slender thing could agitate and could unman him thus.

'What is it I should have heard?' she asked him, as he kept his silence.

He turned from her so that she could not see his face.

'That, when I became dead to the world, I died with the taint of crime on me!'

'Of crime?'

An intense horror thrilled through the echo of the word; but she rose, and moved, and faced him with the fearless resolve of a woman whom no half-truth would blind, and no shadowy terror appal.

'Of crime? What crime?'

Then, and then only, he looked at her, a strange, fixed, hopeless, yet serene look, that she knew no criminal ever would or could have given.

'I was accused of having forged your brother's name.'

A faint cry escaped her; her lips grew white, and her eyes darkened and dilated.

'Accused! But wrongfully?'

His breath came and went in quick sharp spasms.

'I could not prove that.'

'Not prove it? Why?'

'I could not.'

'But he—Philip—never believed you guilty?'

'I cannot tell. He may; he must.'

'But you *are* not!'

It was not an interrogation, but an affirmation that rang out in the silver clearness of her voice. There was not a single intonation of doubt in it;

there was rather a haughty authority that forbade even himself to say that one of his race and that one of his order could have been capable of such ignoble and craven sin.

His mouth quivered, a bitter sigh broke from him; he turned his eyes on her with a look that pierced her to the heart.

'Think me guilty or guiltless, as you will; I cannot answer you.'

His last words were suffocated with the supreme anguish of their utterance. As she heard it, the generosity, the faith, the inherent justice, and the intrinsic sweetness that were latent in her beneath the negligence and the chillness of external semblance rose at once to reject the baser, to accept the nobler, belief offered to her choice. She had lived much in the world, but it had not corroded her; she had acquired keen discernment from it, but she had preserved all the courageous and the chivalrous instincts of her superb nature. She looked at him now, and stretched her hands out towards him with a royal and gracious gesture of infinite eloquence.

'You are guiltless, whatever circumstance may have arrayed against you, whatever shadow of evil may have fallen falsely on you. Is it not so?'

He bowed his head low over her hands as he took them. In that moment half the bitterness of his doom passed from him; he had at least her faith. But his face was bloodless as that of a corpse, and the loud beatings of his heart were audible on the stillness. This faith must live on without one thing to show that he deserved it; if, in time to come, it should waver and fall, and leave him in the darkness of the foul suspicion under which he dwelt, what wonder would there be?

He lifted his head and looked her full in the eyes; her own closed involuntarily, and filled with tears. She felt that the despair and the patience of that look would haunt her until her dying day.

'I *was* guiltless; but none could credit it then; none would do so now; nor can I seek to make them. Ask me no more; give me your belief, if you can—God knows what precious mercy it is to me; but leave me to fulfil my fate, and tell no living creature what I have told you now.'

The great tears stood in her eyes, and blinded her as she heard. Even in the amaze and the vagueness of this first knowledge of the cause of his exile she felt instinctively, as the Little One also had done, that some great sacrifice, some great fortitude and generosity, lay within this sealed secret of his sufferance of wrong. She knew, too, that it would be useless to seek to learn that which he had chosen to conceal; that for no slender cause could he have come out to lead this life of whose sufferings she could gauge the measure; that nothing save some absolute and imperative reason could have driven him to accept such living death as was his doom in Africa.

'Tell no one!' she echoed. 'What! not Philip even? not your oldest friend? Ah! be sure, whatever the evidence might be against you, his heart never condemned you for one instant.'

'I believe it. Yet all you can do for me, all I implore you to do for me, is to keep silence for ever on my name. Today, accident has made me break a vow I never thought but to keep sacred. When you recognized me, I could not deny myself, I could not lie to you; but, for God's sake, tell none of what has passed between us!'

'But why?' she pursued—'why? You lie under this charge still—you cannot disprove it, you say; but why not come out before the world, and state to all what you swear now to me, and claim your right to bear your father's honours? If you were falsely accused, there must have been some guilty in your stead; and if——'

'Cease, for pity's sake! Forget I ever told you I was guiltless! Blot my memory out; think of me as dead, as I have been, till your eyes called me back to life. Think that I am branded with the theft of your brother's name; think that I am vile, and shameless, and fallen as the lowest wretch that pollutes this army; think of me as what you will, but *not* as innocent!'

The words broke out in a torrent from him, bearing down with them all his self-control, as the rush of waters bears away all barriers that have long dammed their course. They were wild, passionate, incoherent; unlike any that had ever passed his lips, or been poured out in her presence. He felt mad with the struggle that tore him asunder, the longing to tell the truth to her, though he should never after look upon her face again, and the honour which bound silence on him for sake of the man whom he had sworn under no temptation to dispossess and to betray.

She heard him silently, with her grand meditative eyes, in which the slow tears still floated, fixed upon him. Most women would have thought that conscious guilt spoke in the violence of his self-accusation: she did not. Her intuition was too fine, her sympathies too true. She felt that he feared, not that she should unjustly think him guilty, but that she should justly think him guiltless. She knew that this, whatever its root might be, was the fear of the stainless, not of the criminal life.

'I hear you,' she answered him, gently; 'but I do not believe you, even against yourself. The man whom Philip loved and honoured never sank to the base fraud of a thief.'

Her glorious eyes were still on him as she spoke, seeming to read his very soul. Under that glance all the manhood, all the race, all the pride, and the love, and the courage within him refused to bear in her sight the shame of an alien crime, and rose in revolt to fling off the bondage that forced him to stand as a criminal before the noble gaze of this woman. His eyes met hers full, and rested on them without wavering; his head was raised, and his carriage had a fearless dignity.

'No. I was innocent. But in honour I must bear the yoke that I took on me long ago: in honour I can never give you or any living soul the *proof* that this crime was not mine. I thought that I should go to my grave without any ever hearing of the years that I have passed in Africa, without any ever learning the name I used to bear. As it is, all I can ask is now— to be forgotten.'

His voice fell before the last words, and faltered over them. It was bitter to ask only for oblivion from the woman whom he loved with all the strength of a sudden passion born in utter hopelessness; the woman whose smile, whose beauty, whose love might even possibly have been won as his own in the future, if he could have claimed his birthright. So bitter, that rather than have spoken those words of resignation he would have been led out by a platoon of his own soldiery and shot in the autumn sunlight beside Rake's grave.

'You ask what will not be mine to give,' she answered him, while a great weariness stole through her own words, for she was bewildered, and pained, and oppressed with a new strange sense of helplessness before this man's nameless suffering. 'Remember—I knew you so well in my earliest years, and you are so dear to the one dearest to me. It will not be possible to forget such a meeting as this. Silence, of course, you can command from me, if you insist on it; but——'

'I command nothing from you; but I implore it. It is the sole mercy you can show. Never, for God's sake! speak of me to your brother or to mine.'

'Do you so mistrust Philip's affection?'

'No. It is because I trust it too entirely.'

'Too entirely to do what?'

'To deal it fruitless pain. As you love him—as you pity me—pray that he and I never meet!'

'But why? if all this could be cleared——'

'It never can be.'

The baffled sense of impotence against the granite wall of some immovable calamity which she had felt before came on her. She had been always used to be obeyed, followed, and caressed; to see obstacles crumble, difficulties disappear, before her wish; she had not been tried by any sorrow, save when, a mere child still, she felt the pain of her father's death; she had been lapped in softest luxury, crowned with easiest victory. The sense that here there was a tragedy whose meaning she could not reach, that there was here a fate that she could not change or soften, brought a strange unfamiliar feeling of weakness before a hopeless and cruel doom that was no more to be altered by her will than the huge bare rocks of Africa out yonder in the glare of noon were to be lifted by her hand. For she knew that this man, who made so light of perils that would have chilled many to the soul in terror, and who bore so quiet and serene a habit beneath the sharpest stings and hardest blows of his adversities, would not speak thus without full warrant, would not consign himself to this renunciation of every hope, unless he were compelled to it by a destiny from which there was no escape.

She was silent some moments, her eyes resting on him with that grave and luminous regard which no man had ever charged to one more tender or less calmly contemplative. He had risen again, and paced to and fro the narrow chamber, his head bent down, his chest rising and falling with the laboured, quickened breath. He had thought that the hour in which his brother's ingratitude had pierced his heart had been the greatest suffering he had ever known, or ever could know; but a greater had waited on him here, in the fate to which the jewelled toy that he had lifted from the water had accidentally led him, not dreaming to what he came.

'Lord Royallieu,' she said, softly, at length, while she rose and moved towards him, the scarlet of the trailing cashmeres gathering dark ruby lights in them as they caught sun and shadow; and at the old name, uttered in her voice, he started, and turned, and looked at her as though he saw some ghost of his past life rise from its grave. 'Why look at me so?' she pursued ere he could speak. 'Act how you will, you cannot change the fact that you are the bearer of your father's title. So long as you live, your brother Berkeley can never take it legally. You may be a Chasseur of the African Army, but none the less are you a Peer of England.'

'What matter that?' he muttered. 'Why tell me that? I have said I am dead. Leave me buried here, and let him enjoy what he may—what he can.'

'But this is folly—madness——'

'No; it is neither. I have told you I should stand as a felon in the eyes of the English law; I should have no civil rights; the greatest mercy fate can show me is to let me remain forgotten here. It will not be long, most likely, before I am thrust into the African sand, to rot like that brave soul out yonder. Berkeley will be the lawful holder of the title then; leave him in peace and possession now.'

He spoke the words out to the end—calmly, and with unfaltering resolve. But she saw the great dews gather on his temples, where silver threads were just glistening among the bright richness of his hair, and she heard the short, low, convulsive breathing with which his chest heaved as he spoke. She stood close beside him, and gazed once more full in his eyes, while the sweet imperious cadence of her voice answered him:

'There is more than I know of here. Either you are the greatest madman, or the most generous man that ever lived. You choose to guard your own secret; I will not seek to persuade it from you. But tell me one thing—*why* do you thus abjure your rights, permit a false charge to rest on you, and consign yourself for ever to this cruel agony?'

His lips shook under his beard as he answered her:

'Because I can do no less in honour. For God's sake do not *you* tempt me!'

A quick deep sigh escaped her as she heard, her face grew very pale as it had done before, and she moved slightly from him.

'Forgive me,' she said, after a long pause. 'I will never ask you that again.'

She could honour honour too well, and too well divine all that he suffered for its sake, ever to become his temptress in bidding him forsake it; yet, with a certain weariness, a certain dread, wholly unfamiliar to him, she realised that what he had chosen was the choice not of his present or of his future. It could have no concern for her—save that long years ago he had been the best-beloved friend of her best-beloved relative— whether or no he remained lost to all the world under the unknown name of a French Chasseur. And yet it smote her with a certain dull unanalysed pain; it gave her a certain emotion of powerlessness and of hopelessness to realize that he would remain all his years through, until an Arab's shot should set him free, under this bondage of renunciation, beneath this yoke of service. She stood silent long, leaning against the oval of the casement, with the sun shed over the glowing cashmeres that swept round her. He stood apart in silence also. What could he say to her? His whole heart longed with an unutterable longing to tell her the truth, and bid her be his judge between him and his duty; but his promise hung on him like a leaden weight. He must remain speechless; and leave her, for doubt to assail her, and for scorn to follow it in her thoughts of him, if so they would.

Heavy as had been the curse to him of that one hour in which honour had forbade him to compromise a woman's reputation, and old tenderness had forbade him to betray a brother's sin, he had never paid so heavy a price for his act as that which he paid now.

Through the yellow sunlight without, over the barren dust-strewn plains, in the distance there approached three riders, accompanied by a small escort of Spahis, with their crimson burnous floating in the autumnal wind. She started, and turned to him:

'It is Philip! He is coming for me from your camp today.'

His eyes strained through the sun-glare:

'Ah, God! I cannot meet him—I have not strength. You do not know——'

'I know how well he loved you.'

'Not better than I him! But I cannot—I dare not—unless I could meet him as we never shall meet upon earth, we must be apart for ever. For Heaven's sake promise me never to speak my name!'

'I promise until you release me.'

'And you can believe me innocent still, in face of all?'

She stretched her hands to him once more: 'I believe. For I know what you once were.'

Great burning tears fell from his eyes upon her hands as he bent over them:

'God bless you! You were an angel of pity to me in your childhood; in your womanhood you give me the only mercy I have known since the last day you looked upon my face! We shall be far sundered for ever—may I come to you once more?'

She paused in hesitation and in thought awhile, while for the first time in all her years a tremulous tenderness passed over her face; she felt an unutterable pity for this man, and for his doom. Then she drew her hands gently away from him:

'Yes, I will see you again.'

So much concession to such a prayer Venetia Corona had never before given. He could not command his voice to answer, but he bowed low before her as before an empress: another moment, and she was alone.

She stood looking out at the wide level country beyond, with the glare of the white strong light and the red burnous of the Franco-Arabs glowing against the blue but cloudless sky; she thought that she must be dreaming some fantastic story born of these desert solitudes.

Yet her eyes were dim with tears, and her heart ached with another's woe. Doubt of him never came to her; but there was a vague, terrible pathos in the mystery of his fate that oppressed her with a weight of future evil, unknown, and unmeasured.

'Is he a madman?' she mused. 'If not, he is a martyr; one of the greatest that ever suffered unknown to other men.'

*   *   *

In the coolness of the late evening in the court of the caravanserai her brother and his friends lounged with her and the two ladies of their touring and sketching party, while they drank their sherbet, and talked of the Gérôme colours of the place, and watched the flame of the after-glow burn out, and threw millet to the doves and pigeons straying at their feet.

'My dear Venetia!' cried the Seraph, carelessly, tossing handfuls of grain to the eager birds, 'I enquired for your Sculptor-Chasseur—that fellow Victor—but I failed to see him, for he had been sent on an expedition shortly after I reached the camp. They tell me he is a fine soldier; but by what the Marquis said, I fear he is but a handsome blackguard, and Africa, after all, may be his fittest place.'

She gave a bend of her head to show she heard him, stroking the soft throat of a little dove that had settled on the bench beside her.

'There is a charming little creature there, a little fire-eater—Cigarette they call her—who is in love with him, I fancy. Such a picturesque child!—swears like a trooper, too,' continued he who was now Duke of Lyonnesse. 'By the way, is Berkeley gone?'

'Left yesterday.'

'What for?—where to?'

'I was not interested to enquire.'

'Ah! you never liked him! Odd enough to leave without reason or apology?'

'He had his reasons, doubtless.'

'And made his apology to you?'

'Oh yes.'

Her brother looked at her earnestly; there was a care upon her face new to him.

'Are you well, my darling?' he asked her. 'Has the sun been too hot, or *la bise* too cold for you?'

She rose, and gathered her cashmeres about her, and smiled somewhat wearily her adieu to him.

'Both, perhaps. I am tired. Good night.'

CHAPTER VIII

The Gift of the Cross

One of the most brilliant of Algerian autumnal days shone over the great camp in the south. The war was almost at an end for a time; the Arabs were defeated and driven desert-wards; hostilities irksome, harassing, and annoying, like all guerrilla warfare, would long continue, but peace was virtually established, and Zarâila had been the chief glory that had been added by the campaign to the flag of Imperial France. The kites and the vultures had left the bare bones by thousands to bleach upon the sands, and the hillocks of brown earth rose in crowds where those, more cared for in death, had been hastily thrust beneath the brown crust of the earth. The dead had received their portion of reward—in the jackal's teeth, in the crow's beak, in the worm's caress. And the living received theirs in this glorious rose-flecked glittering autumn morning, when the breath of winter made the air crisp and cool, but the ardent noon still lighted with its furnace-glow the hillside and the plain.

The whole of the Army of the South was drawn up on the immense level of the plateau to witness the presentation of the Cross of the Legion of Honour.

It was full noon. The sun shone without a single cloud on the deep sparkling azure of the skies. The troops stretched east and west, north and south, formed up in three sides of one vast massive square. The battalions of Zouaves and of Zéphyrs; the brigade of Chasseurs d'Afrique; the squadrons of Spahis; the regiments of Tirailleurs and Turcos; the batteries of Flying Artillery, were all massed there, reassembled from the various camps and stations of the southern provinces to do honour to the day: to do honour in especial to one by whom the glory of the Tricolour had been saved unstained.

The red, white and blue of the standards, the brass of the eagle guidons, the grey tossed manes of the chargers, the fierce swarthy faces of the soldiery, the scarlet of the Spahis' cloaks, and the snowy folds of the Demi-Cavalerie turbans, the shine of the sloped lances, and the glisten of the carbine barrels, fused together in one sea of blended colour, flashed into a million of prismatic hues against the sombre bistre shadow of the sunburnt plains and the clear blue of the skies.

It had been a sanguinary, fruitless, cruel campaign; it had availed nothing except to drive the Arabs away from some hundred leagues of useless and profitless soil; hundreds of French soldiers had fallen by disease, and drought, and dysentery, as well as by shot and sabre, and were unrecorded save on the books of the bureaux, unlamented save, perhaps, in some little nestling hamlet amongst the great green woods of Normandy, or some wooden hut amongst the olives and the vines of Provence, where some woman toiling till sunset among the fields, or praying before some wayside saint's stone niche, would give a thought to the far-off and devouring desert that had drawn down beneath its sands the head that used to lie upon her bosom, cradled as a child's, or caressed as a lover.

But the drums rolled out their long deep thunder over the wastes; and the shot-torn standards fluttered gaily in the breeze blowing from the west; and the clear full music of the French bands echoed away to the dim distant terrible south, where the desert-scorch and the desert-thirst had murdered their bravest and best—and the Army was *en fête*. *En fête*, for it did honour to its darling. Cigarette received the Cross.

Mounted on her own little bright bay, Etoile-Filante, with tricolour ribbons flying from his bridle and amongst the glossy fringes of his mane, the Little One rode among her Spahis. A scarlet kepi was set on her thick silken curls, a tricolour sash was knotted round her waist, her wine-barrel was slung on her left hip, her pistols thrust in her *ceinturon*, and a light carbine held in her hand with the butt-end resting on her foot. With the sun on her childlike brunette face, her eyes flashing like brown diamonds in the light, and her marvellous horsemanship, showing its skill in a hundred désinvoltures and daring tricks, the little Friend of the Flag had come hither amongst her half-savage warriors, whose red robes surrounded her like a sea of blood.

And on a sea of blood she, the Child of War, had floated, never sinking in that awful flood, but buoyant ever above its darkest waves, catching ever some ray of sunlight upon her fair young head, and being oftentimes like a star of hope to those over whom its dreaded waters closed. Therefore they loved her, these grim, slaughterous, and lustful warriors, to whom no other thing of womanhood was sacred, by whom in their wrath or their crime no friend and no brother was spared, whose law was licence, and whose mercy was murder. They loved her, these brutes whose greed was like the tiger's, whose hate was like the devouring flame; and any who should have harmed a single lock of her curling hair would have had the spears of the African Mussulmans buried by the score in his body. They loved her, with the one fond proud triumphant love these vultures of the army ever knew; and today they gloried in her with fierce passionate delight. Today she was to her wild wolves of Africa what Jeanne of Vaucouleurs was to her brethren of France. And today was the crown of her young life. It is given to most, if the desire of their soul ever become theirs, to possess it only when long and weary and fainting toil has brought them to its goal; when beholding the golden fruit so far off, through so dreary a pilgrimage, dulls its bloom as they approach; when having so long centred all their thoughts and hopes in the denied possession of that one fair thing, they find but little beauty in it when that possession is granted to satiate their love. But thrice happy, and few as happy, are they to whom the dream of their youth is fulfilled *in* their youth, to whom their ambition comes in full sweet fruitage, whilst yet the colours of glory have not faded to the young, eager, longing eyes that watch its advent. And of these was Cigarette.

In the fair, slight, girlish body of the child-soldier there lived a courage as daring as Danton's, a patriotism as pure as Vergniaud's, a soul as aspiring as Napoléon's. Untaught, untutored, uninspired by poet's words or patriot's bidding, spontaneous as the rising and the blossoming of some wind-sown, sun-fed flower, there was, in this child of the battle and the razzia, the spirit of genius, the desire to live and to die greatly. It was unreasoned on, it was felt not thought, it was often drowned in the gaiety of young laughter and the ribaldry of military jest, it was often obscured by noxious influence and stifled beneath the fumes of lawless pleasure; but there, ever, in the soul and the heart of Cigarette, dwelt the germ of a pure ambition—the ambition to do some noble thing for France, and leave her name upon her soldiers' lips, a watchword and a rallying-cry for evermore. To be for ever a beloved tradition in the army of her country, to have her name remembered in the roll-call as '*Mort sur le champ d'honneur*'; to be once shrined in the love and honour of France, Cigarette—full of the boundless joys of life that knew no weakness and no pain, strong as the young goat, happy as the young lamb, careless as the young flower tossing on the summer breeze—Cigarette would have died contentedly. And now, living, some measure of this desire had been fulfilled to her, some breath of this imperishable glory had passed over her. France had heard the story of Zarâila; from the Throne a message had been passed to her; what was far beyond all else to her, her own Army of Africa had crowned her, and thanked her, and adored her as with one voice, and wheresoever she passed the wild cheers rang through the roar of musketry, as through the silence of sunny air, and throughout the regiments every sword would have sprung from its scabbard in her defence if she had but lifted her hand and said one word—'Zarâila!'

The Army looked on her with delight now. In all that mute, still, immovable mass that stretched out so far, in such gorgeous array, there was not one man whose eyes did not turn on her, whose pride did not centre in her—their Little One, who was so wholly theirs, and who had been under the shadow of their Flag ever since the curls, so dark now, had been yellow as wheat in her infancy. The Flag had been her shelter, her guardian, her plaything, her idol; the flutter of the striped folds had been the first thing at which her childish eyes had laughed; the preservation of its colours from the sacrilege of an enemy's touch had been her religion, a religion whose true following was, in her sight, salvation of the worst and the most worthless life; and that Flag she had saved, and borne aloft in victory at Zarâila. There was not one in all those hosts whose eyes did not turn on her with gratitude, and reverence, and delight in her as their own.

Not one: except where her own keen, rapid glance, far-seeing as the hawk's, lighted on the squadrons of the Chasseurs d'Afrique, and found amongst their ranks one face, grave, weary, meditative, with a gaze that seemed looking far away from the glittering scene to a grave that lay unseen leagues beyond, behind the rocky ridge.

'He is thinking of the dead man, not of me,' thought Cigarette; and the first taint of bitterness entered into her cup of joy and triumph, as such bitterness enters into most cups that are drunk by human lips. A whole Army was thinking of her, and of her alone; and there was a void in her heart, a thorn in her crown, because one among that mighty mass—one only—gave her presence little heed, but thought rather of a lonely tomb amongst the desolation of the plains.

But she had scarce time even for that flash of pain to quiver in impotent impatience through her. The trumpets sounded, the salvoes of artillery pealed out, the lances and the swords were carried up in salute; on to the ground rode the Marshal of France, who represented the imperial will and presence, surrounded by his staff, by generals of division and brigade, by officers of rank, and by some few civilian riders. An aide galloped up to her where she stood with the corps of her Spahis, and gave her his orders. The Little One nodded carelessly, and touched Etoile-Filante with the prick of the spur. Like lightning the animal bounded forth from the ranks, rearing and plunging, and swerving from side to side, while his rider, with exquisite grace and address, kept her seat like the little semi-Arab that she was, and with a thousand curves and bounds cantered down the line of the gathered troops, with the west wind blowing from the far-distant sea, and fanning her bright cheeks till they wore the soft scarlet flush of the glowing japonica flower. And all down the ranks a low, hoarse, strange, longing murmur went—the buzz of the voices which, but that discipline suppressed them, would have broken out in worshipping acclamations.

As carelessly as though she reined up before the Café door of the *As de Pique*, she arrested her horse before the great Marshal who was the impersonation of Authority, and put her hand up in the salute, with her saucy wayward laugh. He was the impersonation of that vast, silent, awful, irresponsible power which, under the name of the Second Empire, stretched its hand of iron across the sea, and forced the soldiers of France down into nameless graves, with the desert sand choking their mouths; but he was no more to Cigarette than any drummer-boy that might be present. She had all the contempt for the laws of rank of your thorough inborn democrat, all the gay insouciant indifference to station of the really free and untrammelled nature; and, in her sight, a dying soldier, lying quietly in a ditch to perish of shot-wounds without a word or a moan, was greater than all Messieurs les Maréchaux glittering in their stars and orders. As for impressing her, or hoping to impress her, with rank—pooh! You might as well have bid the sailing clouds pause in their floating passage because they came between royalty and the sun. All the sovereigns of Europe would have awed Cigarette not one whit more than a gathering of muleteers. 'Allied sovereigns—bah!' she would have said, 'what did that mean in '15? A chorus of magpies chattering over one stricken eagle!'

So she reined up before the Marshal and his staff, and the few great personages whom Algeria could bring around them, as indifferently as she had many a time reined up before a knot of grim Turcos, smoking under a barrack-gate. *He* was nothing to her: it was her Army that crowned her. 'The Generalissimo is the poppy-head, the men are the wheat; lay every ear of the wheat low, and of what use is the towering poppy that blazed so grand in the sun?' Cigarette would say with metaphorical unction, forgetful, like most allegorists, that her fable was one-sided and unjust in figure and deduction.

Nevertheless, despite her gay contempt for rank, her heart beat fast under its gold-laced jacket as she reined up Etoile and saluted. In that hot clear sun all the eyes of that immense host were fastened on her, and the hour of her longing desire was come at last. France had recognized that she had done greatly, and France, through the voice of this, its chief, spoke to her—France, her beloved, and her guiding-star, for whose sake the young brave soul within her would have dared and have endured all things. There was a group before her, large and brilliant, but at them Cigarette never looked; what she saw were the sunburnt faces of her 'children', of men who, in the majority, were old enough to be her grandsires, who had been with her through so many darksome hours, and whose black and rugged features lightened and grew tender whenever they looked upon their Little One. For the moment she felt giddy with sweet fiery joy; they were here to behold her thanked in the name of France.

The Marshal, in advance of all his staff, doffed his plumed hat and bowed to his saddle-bow as he faced her. He knew her well by sight, this pretty child of his Army of Africa, who had, before then, suppressed mutiny like a veteran, and led the charge like a Murat—this kitten with a lion's heart, this humming-bird with an eagle's swoop.

'Mademoiselle,' he commenced, while his voice, well skilled to such work, echoed to the furthest end of the long lines of troops, 'I have the honour to discharge today the happiest duty of my life. In conveying to you the expression of the Emperor's approval of your noble conduct in the present campaign, I express the sentiments of the whole Army. Your action on the day of Zarâila was as brilliant in conception as it was great in execution; and the courage you displayed was only equalled by your patriotism. May the soldiers of many wars remember you and emulate you. In the name of France, I thank you. In the name of the Emperor, I bring to you the Cross of the Legion of Honour.'

As the brief and soldierly words rolled down the ranks of the listening regiments, he stooped forward from his saddle and fastened the red ribbon on her breast; while from the whole gathered mass, watching, hearing, waiting breathlessly to give their tribute of applause to their darling also, a great shout rose as with one voice, strong, full, echoing over and over again across the plains in thunder that joined her name with the name of France and of Napoléon, and hurled it upward in fierce tumultuous idolatrous love to those cruel cloudless skies that shone above the dead. She was their child, their treasure, their idol, their young leader in war, their young angel in suffering; she was all their own, knowing with them one common mother—France. Honour to her was honour to them; they gloried with heart and soul in this bright young fearless life that had been amongst them ever since her infant feet had waded through the blood of slaughter-fields, and her infant lips had laughed to see the tricolour float in the sun above the smoke of battle.

And as she heard, her face became very pale, her large eyes grew dim and very soft, her mirthful mouth trembled with the pain of a too intense joy. She lifted her head, and all the unutterable love she bore her country and her people thrilled through the music of her voice:

'*Français!—ce n'était rien!*'

That was all she said; in that one first word of their common nationality, she spoke alike to the Marshal of the Empire and to the conscript of the ranks. 'Français!' that one title made them all equal in her sight; whoever claimed it was honoured in her eyes, and was precious to her heart, and when she answered them that it was nothing, this thing which they gloried in her, she answered but what seemed the simple truth in her code. She would have thought it 'nothing' to have perished by shot, or steel, or flame, in day-long torture of that one fair sake of France.

Vain in all else, and to all else wayward, here she was docile and submissive as the most patient child; here she deemed the greatest and the hardest thing that she could ever do far less than all that she would willingly have done. And as she looked upon the host whose thousand and ten thousand voices rang up to the noonday sun in her homage, and in hers alone, a light like a glory beamed upon her face that for once was white and still and very grave—none who saw her face then, ever forgot that look.

In that moment she touched the full sweetness of a proud and pure ambition, attained and possessed in all its intensity, in all its perfect splendour. In that moment she knew that divine hour which, born of a people's love and of the impossible desires of genius in its youth, comes to so few human lives—knew that which was known to the young Napoléon when, in the hot hush of the nights of July, France welcomed the Conqueror of Italy. And in that moment there was an intense stillness; the Army crowned as its bravest and its best a woman-child in the springtime of her girlhood.

Then Cigarette laid her hand on the Cross that had been the dream of her years since she had first seen the brazen glisten of the eagles above her wondering eyes of infancy, and loosened it from above her heart, and stretched her hand out with it to the great Chief.

'Monsieur le Maréchal, this is not for me.'

'Not for you! The Emperor bestows it——'

Cigarette saluted with her left hand, still stretching to him the decoration with the other.

'It is not for me—not whilst I wear it unjustly.'

'Unjustly! What is your meaning? My child, you talk strangely. The gifts of the Empire are not given lightly.'

'No; and they shall not be given unfairly. Listen.' The colour had flushed back, bright and radiant, to her cheeks; her eyes glanced with their old daring; her contemptuous, careless eloquence returned, and her voice echoed, every note distinct as the notes of a trumpet-call, down the ranks of the listening soldiery. 'Hark you! The Emperor sends me this Cross; France thanks me; the Army applauds me. Well, I thank them, one and all. Cigarette was never yet ungrateful; it is the sin of the coward. But I say I will not take what is unjustly mine, and this preference to me is unjust. I saved the day at Zarâila?—oh-hé! *grande chose ça!* And how?—by scampering fast on my mare, and asking for a squadron or two of my Spahis—that was all. If I had not done so much—I, a soldier of Africa— why, I should have deserved to have been shot like a cat—bah!—should I not? It was not I who saved the battle. Who was it? It was a Chasseur d'Afrique, I tell you. What did he do? Why, this. When his officers were all gone down, he rallied, and gathered his handful of men, and held the ground with them all through the day—two—four—six—eight—ten hours in the scorch of the sun. The Arbicos, even, were forced to see that was grand; they offered him life if he would yield. All his answer was to form his few horsemen into line as well as he could for the slain, and charge—a last charge in which he knew not one of his troop could live through the swarms of the Arbis around them. That I saw with my own eyes. I and my Spahis just reached him in time. Then who is it that saved the day, I pray you?—I, who just ran a race for fun and came in at the fag-end of the thing, or this man who lived the whole day through in the carnage, and never let go of the guidon, but only thought how to die greatly? I tell you, the Cross is his, and not mine. Take it back, and give it where it is due.'

The Marshal listened, half amazed, half amused—half prepared to resent the insult to the Empire and to discipline, half disposed to award that submission to her caprice which all Algeria gave to Cigarette.

'Mademoiselle,' he said, with a grave smile, 'the honours of the Empire are not to be treated thus. But who is this man for whom you claim so much?'

'Who is he?' echoed Cigarette, with all her fiery disdain for authority ablaze once more like brandy in a flame. 'Oh-hé! Napoléon Premier would not have left his Marshals to ask that! He is the finest soldier in Africa, if it be possible for one to be finer than another where all are so great. They know that; they pick him out for all the dangerous missions. But the Black Hawk hates him, and so France never hears the truth of all that he does. I tell you, if the Emperor had seen him as I saw him on the field of Zaraila, his would have been the Cross, and not mine.'

'You are generous, my Little One.'

'No; I am just.'

Her brave eyes glowed in the sun, her voice rang as clear as a bell. She raised her head proudly and glanced down the line of her army. She was just—that was the one virtue in Cigarette's creed without which you were poltroon, or liar, or both.

She alone knew what neglect, what indifference, what unintentional but none the less piercing insults she had to avenge; she alone knew of that pain with which she had heard the name of her patrician rival murmured in delirious slumber after Zaraila; she alone knew of that negligent caress of farewell with which her lips had been touched as lightly as his hand caressed a horse's neck or a bird's wing. But these did not weigh with her one instant to make her withhold the words that she deemed deserved; these did not balance against him one instant the pique and the pain of her own heart in opposition to the due of his courage and his fortitude.

Cigarette was rightly proud of her immunity from the weaknesses of her sex; she had neither meanness nor selfishness.

The Marshal listened gravely, the groups around him smilingly. If it had been any other than the Little One, it would have been very different; as it was, all France and all Algeria knew Cigarette.

'What may be the name of this man whom you praise so greatly, my pretty one?' he asked her.

'That I cannot tell, Monsieur le Maréchal. All I know is he calls himself here Louis Victor.'

'Ah! I have heard much of him. A fine soldier, but——'

'A fine soldier without a "but",' interrupted Cigarette, with rebellious indifference to the rank of the great man she corrected, 'unless you add, "but never done justice by his Chief".'

As she spoke, her eyes for the first time glanced over the various personages who were mingled amongst the staff of the Marshal, his invited guests for the review upon the plains. The colour burned more duskily in her cheek, her eyes glittered with hate; she could have bitten her little, frank, witty tongue through and through for having spoken the name of that Chasseur who was yonder, out of earshot, where the lance-heads of his squadrons glistened against the blue skies. She saw a face which, though seen but once before, she knew instantly again—the face of 'Miladi'. And she saw it change colour, and lose its beautiful hue, and grow grave and troubled as the last words passed between herself and the French Marshal.

'Ah! can she *feel*!' wondered Cigarette, who, with a common error of such vehement young democrats as herself, always thought that hearts never ached in the Patrician Order, and thought so still when she saw the listless proud tranquillity return, not again to be altered, over the perfect features that she watched with so much violent instinctive hate. 'Did she heed his name, or did she not? What are their faces in that Order? Only alabaster masks!' mused the child. And her heart sank, and bitterness mingled with her joy, and the soul that had a moment before been so full of all pure and noble emotion, all high and patriotic and idealistic thought, was dulled and soiled and clogged with baser passions. So ever do unworthy things drag the loftier nature earthward.

She scarcely heard the Marshal's voice as it addressed her with a kindly indulgence, as to a valued soldier and a spoilt pet in one.

'Have no fear, Little One. Victor's claims are not forgotten, though we may await our own time to investigate and reward them. No one ever served the Empire and remained unrewarded. For yourself, wear your Cross proudly. It glitters above not only the bravest but the most generous heart in the service.'

None had ever won such warm words from the redoubted chief, whose speech was commonly rapid and stern as his conduct of war, and who usually recompensed his men for fine service rather with a barrel of brandy to season their rations than with speeches of military eulogium. But it failed to give delight to Cigarette. She felt resting upon her the calm gaze of those brilliant azure eyes; and she felt, as she had done once in her rhododendron shelter, as though she were some very worthless, rough, rude, untaught, and coarse little barbarian, who was, at best, but fit for a soldier's jest and a soldier's riot in the wild licence of the barrack-room or the campaigning tent. It was only the eyes of this woman, whom he loved, which ever had power to awaken that humiliation, that impatience of herself, that consciousness of something lost and irrevocable, which moved her now.

Cigarette was proud with an intense pride of all her fiery liberty from every feminine trammel, of all her complete immunity from every scruple and every fastidiousness of her sex. But, for once, within sight of that noble and haughty beauty a poignant, cruel, wounding sense of utter inferiority, of utter debasement, possessed and weighed down her lawless and indomitable spirit. Some vague weary feeling that her youth was fair enough in the sight of men, but that her older years would be very dark, very terrible, came on her even in this hour of the supreme joy, the supreme triumph, of her life. Even her buoyant and cloudless nature did not escape that mortal doom which pursues and poisons every ambition in the very instant of its full fruition.

The doubt, the pain, the self-mistrust were still upon her as she saluted once again, and paced down the ranks of the assembled divisions; whilst every lance was carried, every sword lifted, every bayonet presented to the order, '*Portez vos armes!*' as she went; greeted as though she were an Empress, for that Cross which glittered on her heart, for that courage wherewith she had saved the Tricolour.

The great shouts rent the air; the clash of the lowered arms saluted her; the drums rolled out upon the air; the bands of the regiments of Africa broke into the fiery rapture of a war-march; the folds of the battle-torn flags were flung out wider and wider on the breeze. Grey-bearded men gazed on her with tears of delight upon their grizzled lashes, and young boys looked at her as the children of France once gazed upon Jeanne d'Arc, where Cigarette, with the red ribbon on her breast, rode slowly in the noonday light along the line of troops.

It was the paradise of which she had dreamed; it was the homage of the army she adored; it was one of those hours in which life is transfigured, exalted, sublimated into a divine glory by the pure love of a people; and yet in that instant, so long, so passionately desired, the doom of all genius was hers. There was the stealing pain of a weary unrest amidst the sunlit and intoxicating joy of satisfied aspiration.

The eyes of Venetia Corona followed her with something of ineffable pity. 'Poor little unsexed child!' she thought. 'How pretty and how brave she is! and—how true to him!'

The Seraph, beside her in the group around the flagstaff, smiled and turned to her.

'I said that little Amazon was in love with this fellow Victor; how loyally she stood up for him. But I dare say she would be as quick to send a bullet through him, if he should ever displease her.'

'Why? Where there is so much courage, there must be much nobility, even in the abandonment of such a life as hers.'

'Ah, you do not know what half-French, half-African natures are. She would die for him just now very likely; but if he ever forsake her, she will be quite as likely to run her dirk through him.'

'Forsake her! what is he to her?'

There was a certain impatience in the tone, and something of contemptuous disbelief, that made her brother look at her in wonder.

'What on earth can the loves of a camp concern *her*?' he thought, as he answered: 'Nothing that I know of; but this charming little tigress is very fond of him. By the way, can you point the man out to me? I am curious to see him.'

'Impossible! There are ten thousand faces, and the cavalry squadrons are so far off.'

She spoke with indifference, but she grew a little pale as she did so, and the eyes that had always met his so frankly, so proudly, were turned from him. He saw it, and it troubled him with a trouble the more perplexed that he could assign to himself no reason for it. That it could be caused by any interest felt for a Chasseur d'Afrique by the haughtiest lady in all Europe would have been too preposterous and too insulting a supposition for it ever to occur to him. And he did not dream the truth—the truth that it was her withholding, for the first time in all her life, any secret from him which caused her pain; that it was the fear lest he should learn that his lost friend was living thus which haunted her with that unspoken anxiety.

They were travelling here with the avowed purpose of seeing the military operations of the south; she could not have prevented him from accepting the Marshal's invitation to the review of the African Army without exciting comment and interrogation; she was forced to let events take their own course, and shape themselves as they would; yet an apprehension, a dread, that she could hardly form into distinct shape, pursued her. It weighed on her with an infinite oppression—this story which she alone had had revealed to her, this life whose martyrdom she alone had seen, and whose secret even she could not divine. It affected her more powerfully, it grieved her more keenly, than she herself knew. It brought her close, for the only time in her experience, to a life absolutely without a hope, and one that accepted the despair of such a destiny with silent resignation; it moved her as nothing less, as nothing feebler or of more common type, could ever have found power to do. There were a simplicity and a greatness in the mute, unpretentious, almost unconscious, heroism, of this man, who, for the sheer sake of that which he deemed the need of 'honour', accepted the desolation of his entire future, which attracted her as nothing else had ever done, which made her heart ache when she looked at the glitter of the Franco-Arab squadrons, where their sloped lances glistened in the sun, with a pang that she had never felt before. Moreover, as the untutored, half-barbaric, impulsive young heart of Cigarette had felt, so felt the high-bred, cultured, world-wise mind of Venetia Corona—that this man's exile was no shame, but some great sacrifice; a sacrifice whose bitterness smote her with its own suffering, whose mystery wearied her with its own perplexity, as she gazed down the line of the regiments to where the shot-bruised Eagle of Zaraïla gleamed above the squadrons of the Chasseurs d'Afrique.

He, in his place amongst those squadrons, knew her, though so far distant, and endured the deadliest trial of patience which had come to him whilst beneath the yoke of African discipline. To leave his place was to incur the heaviest punishment; yet he could almost have risked that sentence rather than wait there. Only seven days had gone by since he had been with her under the roof of the caravanserai; but it seemed to him as if these days had aged him more than all the twelve years that he had passed upon the Algerian soil. He was thankful that the enmity of his relentless chief had placed such shadow of evil report between his name and the rewards due to his service, that even the promised recognition of his brilliant actions at Zaraïla and elsewhere was postponed awhile on the plea of investigation. He was thankful that the honours which the whole Army expected for him, and which the antagonism of Châteauroy would soon be powerless to avert any longer from their meet bestowal, did not force him to go up there in the scorching light of the noon, and take those honours as a soldier of France, under the eyes of the man he loved, of the woman he adored.

As it was, he sat motionless as a statue in his saddle, and never looked westward to where the tricolours of the flagstaff drooped above the head of Venetia Corona.

Thus, he never heard the gallant words spoken in his behalf by the loyal lips that he had not cared to caress. As she passed down the ranks, indeed, he saw and smiled on his little champion; but the smile had only a weary kindness of recognition in it, and it wounded Cigarette more than though he had struck her through the breast with his lance.

The moment that he dreaded came; the troops broke up and marched past the representative of their Empire, the cavalry at the head of the divisions. He passed amongst the rest; he raised his lance so that it hid his features as much as its slender shaft could do; the fair and noble face on which his glance flashed was very pale and very grave; the one beside her was sunny and frank, and unchanged by the years that had drifted by, and its azure eyes, so like her own, sweeping over the masses with all the swift, keen appreciation of a military glance, were so eagerly noting carriage, accoutrement, harness, horses, that they never once fell upon the single soldier whose heart so unutterably longed for, even whilst it dreaded, his recognition.

Venetia gave a low, quick breath of mingled pain and relief as the last of the Chasseurs paced by. The Seraph started, and turned his head:

'My darling! Are you not well?'

'Perfectly!'

'You do not look so?—and you forgot now to point me out this special trooper. I forgot him too.'

'He goes there—the tenth from here.'

Her brother looked; it was too late:

'He is taller than the others. That is all I can see now that his back is turned. I will seek him out when——'

'Do no such thing!'

'And why! It was your own request that I enquired——'

'Think me changeable as you will. Do nothing to seek him, to enquire for him——'

'But *why*? A man who at Zarâila——'

'Never mind! Do not let it be said you notice a Chasseur d'Afrique at *my* instance.'

The colour flushed her face as she spoke; it was with the scorn, the hatred, of this shadow of an untruth with which she for the sole time in life soiled her lips. He, noting it, shook himself restlessly in his saddle. If he had not known her to be the noblest and the haughtiest of all the imperial women who had crowned his house with their beauty and their honour, he could have believed that some interest, degrading as disgrace, moved her towards this foreign trooper, and caused her altered wishes and her silence. As it was, so much insult to her as would have existed in the mere thought was impossible to him; yet it left him annoyed and vaguely disquieted.

The subject did not wholly fade from his mind throughout the entertainments that succeeded to the military inspection, in the great white tent glistening with gilded bees and brightened with tricolour standards which the ingenuity of the soldiers of the administration had reared as though by magic amidst the barrenness of the country, and in which the skill of camp cooks served up a delicate banquet. The scene was very picturesque, and all the more so for the widespread changing panorama without of the canvas city of the camp. It was chiefly designed to pleasure the great lady who had come so far southward; all the resources which could be employed were exhausted to make the occasion memorable and worthy of the dignity of the guests whom the Viceroy of the Empire delighted to honour. Yet she, seated there on his right hand, where the rich skins and cashmeres and carpets were strewn on a dais, saw in reality little save a confused blending of hues, and metals, and orders, and weapons, and snowy beards, and olive faces, and French elegance and glitter fused with the grave majesty of Arab pomp. For her thoughts were not with the scene around her, but with the soldier who was without in that teeming crowd of tents, who lived in poverty, and danger, and the hard slavery of unquestioning obedience, and asked only to be as one dead to all who had known and loved him in his youth. It was in vain that she repelled the memory; it usurped her, and would not be displaced.

Meantime, in another part of the camp, the heroine of Zarâila was feasted, not less distinctively, if more noisily and more familiarly, by the younger officers of the various regiments. La Cigarette, many a time before the reigning spirit of suppers and carouses, was banqueted with all the éclat that befitted that Cross which sparkled on her blue and scarlet vest. High throned on a pyramid of knapsacks, canteens, and rugs, toasted a thousand times in all the brandies and red wines that the stores would yield, sung of in improvised odes that were chanted by voices which might have won European fame as tenor or as basso, caressed and sued with all the rapid, fiery, lightly-come and lightly-go love of the camp, with twice a hundred flashing, darkling eyes bent on her in the hot admiration that her vain coquette spirit found delight in, ruling as she would with jest, and caprice, and command, and bravado all these men who were terrible as tigers to their foes, the Little One reigned alone;

and—like many who have reigned before her—found lead in her sceptre, dross in her diadem, satiety in her kingdom.

When it was over, this banquet that was all in her honour, and that three months before would have been a paradise to her, she shook herself free of the scores of arms outstretched to keep her captive, and went out into the night alone. She did not know what she ailed, but she was restless, oppressed, weighed down with a sense of dissatisfied weariness that had never before touched the joyous and elastic nature of the child of France.

And this, too, in the moment when the very sweetest and loftiest of her ambitions was attained! when her hand wandered to that decoration on her heart which had been ever in her sight what the crown of wild olive and the wreath of summer grasses were to the youths and to the victors of the old dead classic years! As she stood in solitude under the brilliancy of the stars, tears, unfamiliar and unbidden, rose in her eyes as they gazed over the hosts around her.

'How they live only for the slaughter! how they perish like the beasts of the field!' she thought. Upon her, as on the poet or the patriot who could translate and could utter the thought as she could not, there weighed the burden of that heart-sick consciousness of the vanity of the highest hope, the futility of the noblest effort, to bring light into the darkness of the suffering, toiling, blind throngs of human life.

'There is only one thing worth doing—to die greatly!' thought the aching heart of the child-soldier, unconsciously returning to the only end that the genius and the greatness of Greece could find as issue to the terrible jest, the mysterious despair, of all existence.

CHAPTER IX

The Desert Hawk and the Paradise-bird

Some way distant, parted by a broad strip of unoccupied ground from the camp, were the grand marquees set aside for the Marshal and for his guests. They were twelve in number, gaily decorated as far as decoration could be obtained in the southern provinces of Algeria, and had, Arab-like, in front of each the standard of the Tricolour. Before one were two

other standards also—the flags of England and of Spain. Cigarette, looking on from afar, saw the alien colours wave in the torchlight flickering on
them. 'That is *hers*,' thought the Little One, with the mournful and noble
emotions of the previous moments swiftly changing into the violent,
reasonless, tumultuous hatred at once of a rival and of an Order.

Cigarette was a thorough democrat; when she was two years old she
had sat on the topmost pile of a Parisian barricade, with the red bonnet on
her curls, and had clapped her tiny hands for delight when the bullets
flew, and the 'Marseillaise' rose above the cannonading; and the spirit
of the musketry and of the 'Marseillaise' had together passed into her
and made her what she was. She was a genuine democrat; and nothing
short of the pure isonomy of the Greeks was tolerated in her political
philosophy, though she could not have told what such a word had meant
for her life. She had all the furious prejudices and all the instinctive truths
in her of an uncompromising *Rouge*; and the sight alone of those lofty
standards, signalizing the place of rest of the 'aristocrats', whilst her
'children's' lowly tents wore in her sight all the dignity and all the
distinction of the true field, would have aroused her ire at any time. But
now a hate tenfold keener moved her; she had a jealousy of the one in
whose honour those two foreign ensigns floated, that was the most bitter
thing which had ever entered her short and sunny life—a hate the hotter
because tinged with that sickening sense of self-humiliation, because
mingled with that wondering emotion at beholding something so utterly
unlike to all that she had known or dreamt.

She had it in her, could she have had the power, to mercilessly and
brutally destroy this woman's beauty, which was so far above her reach, as
she had once destroyed the ivory wreath; yet, as that of the snow-white
carving had done, so did this fair and regal beauty touch her, even in the
midst of her fury, with a certain reverent awe, with a certain dim sense of
something her own life had missed. She had trodden the ivory in pieces
with all the violence of childish, savage, uncalculating hate, and she
had been chidden, as by a rebuking voice, by the wreck which her action
had made at her feet: so could she now, had it been possible, have ruined
and annihilated the loveliness that filled his heart and his soul; but so
would she also, the moment her instinct to avenge herself had been
sated, have felt the remorse and the shame of having struck down a
delicate and gracious thing that even in its destruction had a glory that
was above her.

Even her very hate attracted her to the sight, to the study, to the
presence of this woman, who was as dissimilar to all of womanhood that
had ever crossed her path, in camp and barrack, as the pure, white,

gleaming lily of the hothouse is unlike the wind-tossed, sand-stained, yellow leaf downtrodden in the mud. An irresistible fascination drew her towards the self-same pain which had so wounded her a few hours before—an impulse more intense than curiosity, and more vital than caprice, urged her to the vicinity of the only human being who had ever awakened in her the pangs of humiliation, the throbs of envy.

And she went to that vicinity, now that the daylight had just changed to evening, and the ruddy torch-glare was glowing everywhere from great pine-boughs thrust in the ground, with their resinous branches steeped in oil and flaring alight. There was not a man that night in camp who would have dared oppose the steps of the young heroine of the Cross wherever they might choose, in their fantastic flight, to wander. The sentinels passing up and down the great space before the marquees challenged her, indeed, but she was quick to give the answering password, and they let her go by them, their eyes turning after the little picturesque form that every soldier of the Corps of Africa loved almost like the flag beneath which he fought. Once in the magic circle, she paused awhile, the desire that urged her on, and the hate that impelled her backward, keeping her rooted there in the dusky shadow which the flapping standards threw.

To creep covertly into her rival's presence, to hide herself like a spy to see what she wished, to show fear, or hesitation, or deference, were not in the least what she contemplated. What she intended was to confront this fair, strange, cold, cruel thing, and see if she were of flesh and blood like other living beings, and do the best that could be done to outrage, to scourge, to challenge, to deride her with all the insolent artillery of camp ribaldry, and show her how a child of the people could laugh at her rank, and affront her purity, and scorn her power. Definite idea there was none in her; she had come on impulse; but a vague longing in some way to break down that proud serenity which galled her so sharply, and bring hot blood of shame into that delicate face, and cast indignity on that imperious and unassailable pride, consumed her.

She longed to do as some girl of whom she had once been told by an old Invalide had done in the '89—a girl of the people, a fisher-girl of the Cannébière, who had loved one above her rank, a noble who deserted her for a woman of his own order, a beautiful, soft-skinned, lily-like scornful aristocrat, with the silver ring of merciless laughter and the languid lustre of sweet contemptuous eyes. The Marseillaise bore her wrong in silence—she was a daughter of the south and of the populace, with a dark, brooding, burning beauty, strong and fierce, and braced with the salt lashing of the sea and with the keen breath of the stormy mistral. She held her peace while the great lady was wooed and won, while the marriage joys came with the purple vintage-time, while the people were made drunk at the bridal of their châtelaine in those hot, ruddy, luscious autumn days.

She held her peace; and the Terror came, and the streets of the city by the sea ran blood, and the scorch of the sun blazed, every noon, on the scaffold. Then she had her vengeance. She stood and saw the axe fall down on the proud snow-white neck that never had bent till it bent there, and she drew the severed head into her own bronzed hands and smote the lips his lips had kissed, a cruel blow that blurred their beauty out, and twined a fish-hook in the long and glistening hair and drew it, laughing as she went, through dust, and mire, and gore, and over the rough stones of the town, and through the shouting crowds of the multitudes, and tossed it out on to the sea, laughing still as the waves flung it out from billow to billow, and the fish sucked it down to make their feast. '*Voilà tes secondes noces!*' she cried, where she stood and laughed by the side of the grey angry water watching the tresses of the floating hair sink downward like a heap of sea-tossed weed.

That horrible story came to the memory of Cigarette now as it had been told her by the old soldier who, in his boyhood, had seen the entry of the Marseillaise to Paris. She knew what the woman of the people had felt when she had bruised and mocked and thrown out to the devouring waters that fair and fallen head.

'I could do it—I could do it,' she thought, with the savage instinct of her many-sided nature dominant, leaving uppermost only its ferocity— the same ferocity as had moved the southern woman to wreak her hatred on the senseless head of her rival. The school in which the child-soldier had been reared had been one to foster all those barbaric impulses, to leave in their inborn uncontrolled force all those native desires which the human shares with the animal nature. There had been no more to teach her that these were criminal or forbidden than there is to teach the young tigress that it is cruel to tear the antelope for food. What Cigarette was, that nature had made her; she was no more trained to self-control, or to the knowledge of good, than is the tiger's cub as it wantons in its play under the great broad tropic leaves.

Now, she acted on her impulse; her impulse of open scorn of rank, of reckless vindication of her right to do just whatsoever pleasured her; and she went boldly forward and dashed aside, with no gentle hand, the folds that hung before the entrance of the tent, and stood there with the gleam of the starry night and the glow of the torches behind her, so that her picturesque and brightly coloured form looked painted on a dusky lurid background of shadow and of flame.

The action startled the occupants of the tent, and made them both look up: they were Venetia Corona and a Levantine woman, who was her favourite and most devoted attendant, and had been about her from her birth. The tent was the first of three set aside for her occupance, and had been adorned, with as much luxury as was procurable, and with many of the rich and curious things of Algerian art and workmanship so far as they could be hastily collected by the skill and quickness of the French intendance. Cigarette stood silently looking at the scene on which she had thus broken without leave or question; she saw nothing of it except one head lifted in surprise at her entrance—just such a head, just so proudly carried, just so crowned with gleaming hair, as that which the Marseillaise had dragged through the dust of the streets and cast out to the lust of the sharks. Venetia hesitated a moment in astonished wonder; then, with the grace and the courtesy of her race, rose and approached the entrance of her tent in which that figure, half a soldier, half a child, was standing with the fitful reddened light behind. She recognized whose it was.

'Is it you, *ma petite?*' she said, kindly. 'Come within. Do not be afraid——'

She spoke with the gentle consideration of a great lady to one whom she admired for her heroism, compassionated for her position, and thought naturally in need of such encouragement. She had liked the frank, fearless, ardent brunette face of the Little Friend of the Flag; she had liked her fiery and indomitable defence of the soldier of Zaraïla; she felt an interest in her as deep as her pity, and she was above the scruples which many women of her rank might have had as to the fitness of entering into conversation with this child of the Army. She was gentle to her as to a young bird, a young kitten, a young colt; what her brother had said of the Vivandière's love for one whom the girl only knew as a trooper of Chasseurs filled with an indefinable compassion the woman who knew him as her own equal and of her own Order.

Cigarette, for once, answered nothing; her eyes very lowering, burning, savage.

'You wish to see me?' Venetia asked once more. 'Come nearer. Have no fear——'

The one word unloosed the spell which had kept Cigarette speechless; the one word was an insult beyond endurance, that lashed all the worst spirit in her into flame.

'Fear!' she cried, with a camp oath, whose blasphemy was happily unintelligible to her listener. 'Fear! You think *I* fear *you*!—the darling of the Army, who saved the squadron at Zaraïla, who has seen a thousand days of bloodshed, who has killed as many men with her own hand as any Lascar among them all—fear *you*, you hothouse flower, you paradise-bird, you silver pheasant, who never did aught but spread your dainty colours in the sun, and never earned so much as the right to eat a piece of black bread, if you had your deserts! Fear *you*——I! Why! Do you not know that I could kill you where you stand as easily as I could wring the neck of any one of those gold-winged orioles that flew above your head today, and who have more right to live than you, for they do at least labour in their own fashion for their food, and their drink, and their dwelling? Dieu de Dieu! Why, I have killed Arabs, I tell you—great gaunt grim men—and made them bite the dust under my fire. Do you think I would check for a moment at dealing *you* death, you beautiful, useless, honeyed, poisoned, painted exotic, that has every wind tempered to you, and thinks the world only made to bear the fall of your foot!'

The fury of words was poured out without pause, and with an intense passion vibrating through them; the wine was hot in her veins, the hate was hot in her heart; her eyes glittered with murderous meaning, and she darted with one swift bound to the side of the rival she loathed, with the pistol half out of her belt; she expected to see the one she threatened recoil, quail, hear the threat in terror; she mistook the nature with which she dealt. Venetia Corona never moved, never gave a sign of the amazement that awoke in her; but she put her hand out and clasped the barrel of the weapon, while her eyes looked down into the flashing, looming, ferocious ones that menaced her, with calm contemptuous rebuke in which something of infinite pity was mingled.

'Child, are you mad?' she said, gravely. 'Brave natures do not stoop to assassination, which you seem to deify. If you have any reason to feel evil against me, tell me what it is: I always repair wrong if I can; but as for those threats—they are most absurd if you do not mean them, they are most wicked if you do.'

The tranquil, unmoved, serious words stilled the vehement passion she rebuked with a strange and irresistible power; under her gaze the savage lust in Cigarette's eyes died out, and their lids drooped over them; the dusky scarlet colour faded from her cheeks; for the first time in her life she felt humiliated, vanquished, awed. If this 'aristocrat' had shown one sign of fear, one trace of apprehension, all her violent and reckless hatred would have reigned on, and, it might have been, have rushed from threat to execution; but showing the only quality, that of courage, for which she had respect, her great rival confused and disarmed her. She was only sensible, with a vivid agonizing sense of shame, that her only cause of hatred against this woman was that he loved her. And this she would have died a thousand deaths rather than have acknowledged.

She let the pistol pass into Venetia's grasp; and stood, irresolute and ashamed, her fluent tongue stricken dumb, her intent to wound, and sting, and outrage with every vile coarse jest she knew, rendered impossible to execute. The purity and the dignity of her opponent's presence had their irresistible influence, an influence too strong for even her débonnaire and dangerous insolence. She hated herself in that moment more than she hated her rival.

Venetia laid the loaded pistol down, away from both, and seated herself on the cushions from which she had risen. Then she looked once more long and quietly at her unknown antagonist.

'Well?' she said, at length. 'Why do you venture to come here? And why do you feel this malignity towards a stranger who never saw you until this morning?'

Under the challenge the fiery spirit of Cigarette rallied, though a rare and galling sense of intense inferiority, of intense mortification, was upon her; though she would almost have given the Cross which was on her breast that she had never come into this woman's sight.

'Oh-hé!' she answered, recklessly, with the red blood flushing her face again at the only evasion of truth of which the little desperado, with all her sins, had ever been guilty. 'I hate you, Miladi, because of your Order—because of your nation—because of your fine dainty ways— because of your aristocrat's insolence—because you treat my soldiers like paupers—because you are one of those who do no more to have the right to live than the purple butterfly that flies in the sun, and who oust the people out of their dues as the cuckoo kicks the poor birds that have reared it, out of the nest of down to which it never has carried a twig or a moss!'

Her listener heard with a slight smile of amusement and of surprise that bitterly discomfited the speaker. To Venetia Corona the girl-soldier seemed mad; but it was a madness that interested her, and she knew at a glance that this child of the Army was of no common nature and no common mind.

'I do not wish to discuss democracy with you,' she answered, with a tone that sounded strangely tranquil to Cigarette after the scathing acrimony of her own. 'I should probably convince you, as little as you would convince me; and I never waste words. But I heard you today claim a certain virtue—justice. How do you reconcile with that, your very hasty condemnation of a stranger of whose motives, actions, and modes of life it is impossible you can have any accurate knowledge?'

Cigarette once again was silenced; her face burned, her heart was hot with rage. She had come prepared to upbraid and to outrage this patrician with every jibe and grossness camp usage could supply her with, and—she stood dumb before her! She could only feel an all-absorbing sense of being ridiculous, and contemptible, and puerile in her sight.

'You bring two charges against me,' said Venetia, when she had vainly awaited answer. 'That I treat your comrades like paupers, and that I rob the people—my own people I imagine you to mean—of their dues. In the first, how will you prove it?—in the second, how can you know it?'

'Pardieu, Miladi!' swore Cigarette, recklessly, seeking only to hold her own against the new sense of inferiority and of inability that oppressed her. 'I was in the hospital when your fruits and your wines came; and as for your people, I don't speak of them—they are all slaves, they say, in Albion, and will bear to be yoked like oxen if they think they can turn any gold in the furrows!—I speak of *the* people. Of the toiling, weary, agonized, joyless, hapless multitudes who labour on, and on, and on, ever in darkness, that such as you may bask in sunlight and take your pleasures wrung out of the death-sweat of millions of work-murdered poor! What right have you to have your path strewn with roses, and every pain spared from you; to only lift your voice and say, "Let that be done," to see it done?—to find life one long sweet summer day of gladness and abundance, while they die out in agony by thousands, ague-stricken, famine-stricken, crime-stricken, age-stricken, for want only of one ray of the light of happiness that falls from dawn to dawn like gold upon *your* head?'

Vehement and exaggerated as the upbraiding was, her hearer's face grew very grave, very thoughtful, as she spoke; those luminous earnest eyes, whose power even the young democrat felt, gazed wearily down into hers.

'Ah, child! do you think *we* never think of that? You wrong me—you wrong my Order. There are many besides myself who turn over that terrible problem as despairingly as you can ever do. As far as in us lies, we strive to remedy its evil; the uttermost effort can do little, but that little is only lessened—fearfully lessened—whenever Class is arrayed against Class by that blind antagonism which animates yourself.'

Cigarette's intelligence was too rapid not to grasp the truths conveyed by those words; but she was in no mood to acknowledge them.

'Nom de Dieu, Miladi!' she swore in her teeth. 'If you do turn over the problem—you aristocrats—it is pretty work, no doubt! just putting the bits of a puzzle-ball together so long as the game pleases you, and leaving the puzzle in chaos when you are tired! Oh-hé! I know how fine ladies and fine gentlemen play at philanthropies! But I am a child of the People, mark you; and I only see how birth is an angel that gives such as you eternal sunlight and eternal summer, and how birth is a devil that drives down the millions into a pit of darkness, of crime, of ignorance, of misery, of suffering, where they are condemned before they have opened their eyes to existence, where they are sentenced before they have left their mothers' bosoms in infancy. You do not know what that darkness is. It is night—it is ice—it is hell!'

Venetia Corona sighed wearily as she heard; pain had been so far from her own life, and there was an intense eloquence in the low deep words that seemed to thrill through the stillness.

'Nor do you know how many shadows chequer that light which you envy! But I have said; it is useless for me to argue these questions with you. You commence with a hatred of a class; all justice is over wherever that element enters. If I were what you think, I should bid you leave my presence which you have entered so rudely. I do not desire to do that. I am sure that the heroine of Zarâila has something nobler in her than mere malignity against a person who can never have injured her; and I would endure her insolence for the sake of awakening her justice. A virtue, that was so great in her at noon, cannot be utterly dead at nightfall?'

Cigarette's fearless eyes drooped under the gaze of those bent so searchingly, yet so gently, upon her; but only for a moment: she raised them afresh with their old dauntless frankness.

'Dieu! you shall never say you wanted justice and truth from a French soldier, and failed to get them! I hate you, never mind why—I *do*, though you never harmed me. I came here for two reasons; one, because I wanted to look at you close—you are not like anything that I ever saw; the other, because I wanted to wound you, to hurt you, to outrage you, if I could find a way how. And you will not let me do it. I do not know what it is in you.'

In all her courted life, the great lady had had no truer homage than lay in that irate reluctant wonder of this fiery foe.

She smiled slightly.

'My poor child, it is rather something in yourself—a native nobility that will not allow you to be as unjust and as insolent as your soul desires——'

Cigarette gave a movement of intolerable impatience:

'Pardieu! do not pity *me*, or I shall give you a taste of my "insolence" in earnest! You may be a sovereign *grande dame* everywhere else, but you can carry no terror with you for me, I promise you!'

'I do not seek to do so. If I did not feel interest in you, do you suppose I should suffer for a moment the ignorant rudeness of an ill-bred child? You fail in the tact, as in the courtesy, that belong to your nation.'

The rebuke was gentle, but it was all the more severe for its very serenity. It cut Cigarette to the quick; it covered her with an overwhelming sense of mortification and of failure. She was too keen and too just, despite all her vanity, not to feel that she deserved the condemnation, and not to know that her opponent had all the advantage and all the justice on her side. She had done nothing by coming here; nothing except to appear as an insolent and wayward child before her superb rival, and to feel a very anguish of inferiority before the grace, the calm, the beauty, the nameless potent charm of this woman, whom she had intended to humiliate and injure!

The inborn truth within her, the native generosity and candour that soon or late always overruled every other element in the Little One, conquered her now. She dashed down her Cross on the ground and trod passionately on the decoration she adored:

'I disgrace it the first day I wear it! You are right, though I hate you, and you are as beautiful as a sorceress! There is no wonder he loves you!'

'He! Who?'

There was a colder and more utterly amazed hauteur in the interrogation than had come into her voice throughout the interview, yet on her fair face a faint warmth rose.

The words were out, and Cigarette was reckless what she said, almost unconscious, indeed, in the violence of the many emotions in her.

'The man who carves the toys you give your dog to break!' she answered bitterly. 'Dieu de Dieu! he loves you. When he was down with his wounds after Zaraïla, he said so; but he never knew what he said, and he never knew that I heard him. You are like the women of his old world; though through you he got treated like a dog, he loves you!'

'Of whom do you venture to speak?'

The cold calm dignity of the answer, whose very tone was a rebuke, came strangely after the violent audacity of Cigarette's speech.

'Sacre bleu! of him, I tell you, who was made to bring his wares to you like a hawker. And you think it insult, I will warrant!—insult for a soldier who has nothing but his courage, and his endurance, and his heroism under suffering to ennoble him, to dare to love Madame la Princesse Corona! I think otherwise. I think that Madame la Princesse Corona never had a love of so much honour, though she has had princes and nobles and all the men of her rank, no doubt, at her feet, through that beauty that is like a spell!'

Hurried headlong by her own vehemence, and her own hatred for her rival which drove her to magnify the worth of the passion of which she was so jealous, that she might lessen, if she could, the pride of her on whom it was lavished, she never paused to care what she said, or heed what its consequences might become. She felt incensed, amazed, irritated, to see no trace of any emotion come on her hearer's face: the hot, impetuous, expansive, untrained nature underrated the power for self-command of the Order she so blindly hated.

'You speak idly and at random, like the child you are,' the *grande dame* answered her with chill contemptuous rebuke. 'I do not imagine that the person you allude to made you his confidante in such a matter?'

'He!' retorted Cigarette. 'He belongs to your class, Miladi. He is as silent as the grave. You might kill him, and he would never show it hurt. I only know what he muttered in his fever.'

'When you attended him?'

'Not I!' cried Cigarette, who saw for the first time that she was betraying herself. 'He lay in the scullion's tent where I was; that was all; and he was delirious with the shot-wounds. Men often are——'

'Wait! Hear me a little while, before you rush on in this headlong and foolish speech,' interrupted her auditor, who had in a moment's rapid thought decided on her course with this strange wayward nature. 'You err in the construction you have placed on the words, whatever they were, which you heard. The gentleman—he is a gentleman—whom you speak of bears me no love. We are almost strangers. But, by a strange chain of circumstances he is connected with my family; he once had great friendship with my brother; for reasons that I do not know, but which are imperative with him, he desires to keep his identity unsuspected by every one; an accident alone revealed it to me, and I have promised him not to divulge it. You understand?'

Cigarette gave an affirmative gesture. Her eyes were fastened sullenly, yet with a deep bright glow in them, upon her companion; she was beginning to see her way through his secret—a secret she was too intrinsically loyal even now to dream of betraying.

'You spoke very nobly for him today. You have the fealty of one brave character to another, I am sure?' pursued Venetia Corona, purposely avoiding all hints of any warmer feeling on her listener's part, since she saw how tenacious the girl was of any confession of it. 'You would do him service if you could, I fancy; am I right?'

'Oh yes!' answered Cigarette, with an over-assumption of carelessness. 'He is *bon-zig*; we always help each other. Besides, he is very good to my men. What is it you want of me?'

'To preserve secrecy on what I have told you for his sake; and to give him a message from me.'

Cigarette laughed scornfully; she was furious with herself for standing obediently like a chidden child to hear this patrician's bidding, and to do her will. And yet, try how she would, she could not shake off the spell under which those grave, sweet, lustrous eyes of command held her.

'Pardieu, Miladi! Do you think I babble like any young *bleu* drunk with his first measure of wine? As for your message, you had better let him come and hear what you have to say; I cannot promise to remember it!'

'Your answer is reckless; I want a serious one. You spoke like a brave and a just friend to him today; are you willing to act as such tonight? You have come here strangely, rudely, without pretext or apology; but I think better of you than you would allow me to do if I judged only from the surface. I believe that you have loyalty, as I know that you have courage.'

Cigarette set her teeth hard.

'What of that? I have them *en militaire*, that is all.'

'This of it. That one who has them will never cherish malice unjustifiably, or fail to fulfil a trust.'

Cigarette's clear brown skin grew very red.

'That is true,' she muttered, reluctantly. Her better nature was growing uppermost, though she strove hard to keep the evil one predominant.

'Then you will cease to feel hatred towards me for so senseless a reason as that I belong to an aristocracy that offends you; and you will remain silent on what I tell you concerning the one whom you know as Louis Victor?'

Cigarette nodded assent; the sullen fire-glow still burnt in her eyes, but she succumbed to the resistless influence which the serenity, the patience, and the dignity of this woman had over her. She was studying Venetia Corona all this while with the keen rapid perceptions of envy and of jealousy, studying her features, her form, her dress, her attitude, all the many various and intangible marks of birth and breeding which were so new to her, and which made her rival seem so strange, so dazzling, so marvellous a sorceress to her; and all the while the sense of her own inferiority, her own worthlessness, her own boldness, her own debasement was growing upon her, eating sharply as aquafortis into brass, into the metal of her vanity and her pride, humiliating her unbearably, yet making her heart ache with a sad pathetic pity for herself.

'He *is* of your Order, then?' she asked, abruptly.

'He was—yes.'

'Oh-hé!' cried Cigarette, with her old irony. 'Then he must be always, mustn't he? You think too much of your blue blood, you patricians, to fancy it can lose its royalty, whether it run under a King's purple or a Roumi's canvas shirt. Blood tells, they say! Well, perhaps it does. Some say *my* father was a Prince of France—may be! So, he is of your Order? Bah! I knew that the first day I saw his hands. Do you want me to tell you why he lives amongst us, buried like this?'

'Not if you violate any confidence to do so.'

'Pardieu! he makes no confidence, I promise you. Not ten words will Monseigneur say, if he can help it, about anything. He is as silent as a lama; it is *populacier* to talk! But we learn things without being told in camp; and I know well enough he is here to save some one else, in some one's place; it is a *sacrifice*, look you, that nails him down to this martyrdom.'

Her auditor was silent; she thought as the vivandière thought, but the pride in her, the natural reticence and reserve of her class, made her shrink from discussing the history of one whom she knew—shrink from having any argument on his past or future with a saucy, rough, fiery young camp-follower, who had broken thus unceremoniously on her privacy. Yet she needed greatly to be able to trust Cigarette; the child was the only means through which she could send him a warning that must be sent; and there were a bravery and a truth in her which attracted the 'aristocrat', to whom she was as singular and novel a rarity as though she were some young savage of desert western isles.

'Look you, Miladi,' said Cigarette, half sullenly, half passionately, for the words were wrenched out of her generosity, and choked her in their utterance, 'that man suffers: his life here is a hell upon earth—I don't mean for the danger, he is *bon soldat*; but for the indignity, the subordination, the licence, the brutality, the tyranny. He is as if he were chained to the galleys. He never says anything, oh no! he is of your kind, you know! But he suffers. Mort de Dieu! he suffers. Now, if you be his friend, can you do nothing for him? Can you ransom him in no way? Can you go away out of Africa and leave him in this living death to get killed and thrust into the sand, like his comrade the other day?'

Her hearer did not answer; the words made her heart ache; they cut her to the soul. It was not for the first time that the awful desolation of his future had been present before her; but it was the first time that the fate to which she would pass away and leave him had been so directly in words before her. Cigarette, obeying the generous impulses of her better nature, and abandoning self with the same reckless impetuosity with which a moment before she would, if she could, have sacrificed her rival, saw the advantage gained, and pursued it with rapid skill. She was pleading against herself: no matter, in that instant she was capable of crucifying herself, and only remembering mercy to the absent.

'I have heard,' she went on, vehemently, for the utterance to which she forced herself was very cruel to her, 'that you of the Noblesse are staunch as steel to your own people. It is the best virtue that you have. Well, he is of your people. Will you go away in your negligent indifference, and leave him to eat his heart out in bitterness and misery? He was your brother's friend; he was known to you in his early time; you have said so. And you are cold enough and cruel enough, Miladi, not to make one effort to redeem him out of bondage?—to go back to your palaces, and your pleasures, and your luxuries, and your flatteries, and be happy, while this man is left on bearing his yoke here?—and it is a yoke that galls, that kills!—bearing it until, in some day of desperation, he rebels, and is shot like a dog; or, in some day of mercy, a naked blade cuts its way to his heart, and makes its pulse cease for ever? If you do, you patricians are worse still than I thought you!'

Venetia heard her without interruption; a great sadness came over her face as the vivid phrases followed each other. She was too absorbed in the subject of them to heed the challenge and the insolence of their manner. She knew that the Little One who spoke them loved him, though so tenacious to conceal her love; and she was touched, not less by the magnanimity which, for his sake, sought to release him from the African service, than by the hopelessness of his coming years as thus prefigured before her.

'Your reproaches are unneeded,' she replied, slowly and wearily. 'I could not abandon one who was once the friend of my family to such a fate as you picture without very great pain. But I do not see how to alter this fate, as you think I could do with so much ease. I am not in its secret; I do not know the reason of its seeming suicide; I have no more connection with its intricacies than you have. This gentleman has chosen his own path; it is not for me to change his choice or spy into his motives.'

Cigarette's flashing searching eyes bent all their brown light on her.

'Madame Corona, you are courageous; to those who are so, all things are possible.'

'A great fallacy! You must have seen many courageous men vanquished. But what would you imply by it?'

'That you can help this man if you will.'

'Would that I could; but I can discern no means——'

'Make them.'

Even in that moment her listener smiled involuntarily at the curt imperious tones, decisive as Napoléon's '*Partons!*' before the Passage of the Alps.

'Be certain, if I can, I will. Meantime, there is one pressing danger of which you must be my medium to warn him. He and my brother must not meet. Tell him that the latter, knowing him only as Louis Victor, and interested in the incidents of his military career, will seek him out early tomorrow morning before we quit the camp. I must leave it to him to avoid the meeting as best he may be able.'

Cigarette smiled grimly.

'You do not know much of the camp. Victor is only a *bas-officier*; if his officers call him up, he must come, or be thrashed like a slave for contumacy. He has no will of his own.'

Venetia gave an irrepressible gesture of pain.

'True; I forgot. Well, go and send him to me. My brother must be taken into his confidence, whatever that confidence reveals. I will tell him so. Go and send him to me; it is the last chance.'

Cigarette gave no movement of assent; all the jealous rage in her flared up afresh to stifle the noble and unselfish instincts under which she had been led during the later moments. A coarse and impudent scoff rose to her tongue, but it remained unuttered; she could not speak it under that glance, which held the evil in her in subjection, and compelled her reluctant reverence against her will.

'Tell him to come here to me,' repeated Venetia, with the calm decision of one to whom any possibility of false interpretation of her motives never occurred, and who was habituated to the free action that accompanied an unassailable rank. 'My brother must know what I know. I shall be alone, and he can make his way hither, without doubt, unobserved. Go and say this to him. You are his loyal little friend and comrade.'

'If I be, I do not see why I am to turn *your* lackey, Madame,' said Cigarette, bitterly. 'If you want him, you can send for him by other messengers!'

Venetia Corona looked at her steadfastly, with a certain contempt in the look.

'Then your pleading for him was all insincere? Let the matter drop, and be good enough to leave my presence, which, you will remember, you entered unsummoned and undesired.'

The undeviating gentleness of the tone made the rebuke cut deeper, as her first rebuke had cut, than any sterner censure or more peremptory dismissal could have done. Cigarette stood irresolute, ashamed, filled with rage, torn by contrition, impatient, wounded, swayed by jealous rage and by the purer impulses she strove to stifle.

The Cross she had tossed down caught her sight as it glittered on the carpet strewn over the hard earth; she stooped and raised it; the action sufficed to turn the tide with her impressionable, ardent, capricious nature: she would not disgrace *that*.

'I will go,' she muttered in her throat; 'and you—you——O God! no wonder men love you when even I cannot hate you!'

Almost ere the words were uttered she had dashed aside the hangings before the tent-entrance, and had darted out into the night air. Venetia Corona gazed after the swiftly flying figure as it passed over the starlit ground, lost in amazement, in pity, and in regret, wondering afresh if she had only dreamt of this strange interview in the Algerian camp, which seemed to have come and gone with the blinding rapidity of lightning.

'A little tigress!' she thought; 'and yet with infinite nobility, with wonderful germs of good in her. Of such a nature what a rare life might have been made! As it is, her childhood we smile at and forgive; but, great Heaven! what will be her maturity, her old age! Yet how she loves him! And she is so brave she will not show it.'

With the recollection came the remembrance of Cigarette's words as to his own passion for herself, and she grew paler as it did so. 'God forbid he should have *that* pain too!' she murmured. 'What could it be save misery for us both?'

Yet she did not thrust the fancy from her with contemptuous nonchalance as she had done every other of the many passions she had excited and disdained; it had a great sadness and a greater terror for her. She dreaded it unspeakably for him; also perhaps, unconsciously, she dreaded it slightly for herself.

She wished now that she had not sent for him. But it was done; it was for sake of their old friendship; and she was not one to vainly regret what was unalterable, or to desert what she deemed generous and right for the considerations of prudence or of egotism.

CHAPTER X

Ordeal by Fire

Amidst the mirth, the noise, the festivity, which reigned throughout the camp as the men surrendered themselves to the enjoyment of the largesses of food and of wine allotted to them by their Marshal's command in commemoration of Zaráila, one alone remained apart, silent and powerless to rouse himself even to the forced semblance, the forced endurance, of their mischief and their pleasure. They knew him well, and they also loved him too well to press such participation on him. They knew that it was no lack of sympathy with them that made him so grave amidst their mirth, so mute amidst their volubility. Some thought that he was sorely wounded by the delay of the honours promised him. Others, who knew him better, thought that it was the loss of his brother-exile which weighed on him, and made all the scene around him full of pain. None approached him; but whilst they feasted in their tents, making the celebration of Zaraîla equal to the Jour de Mazagran, he sat along over a picket-fire on the far outskirts of the camp.

His heart was sick within him. To remain here was to risk with every moment that ordeal of recognition which he so unutterably dreaded; and to flee was to leave his name to the men, with whom he had served so long, covered with obloquy and odium, buried under all the burning shame and degradation of a traitor's and deserter's memory. The latter course was impossible to him; the only alternative was to trust that the vastness of that great concrete body, of which he was one unit, would suffice to hide him from the discovery of the friend whose love he feared as he feared the hatred of no foe. He had not been seen as he had passed the flagstaff; there was little fear that in the few remaining hours any chance could bring the illustrious guest of a Marshal to the outposts of the scattered camp.

Yet he shuddered as he sat in the glow of the fire of pine-wood; she was so near, and he could not behold her!—though he might never see her face again; though they must pass out of Africa, home to the land that he desired as only exiles can desire, whilst he still remained silent, knowing that, until death should release him, there could be no other fate for him, save only this one, hard, bitter, desolate, uncompanioned, unpitied, unrewarded life. But to break his word as the price of his freedom was not possible to his nature or in his creed. This fate was, in chief, of his own making: he accepted it without rebellion, because rebellion would have been in this case both cowardice and self-pity.

He was not conscious of any heroism in this; it seemed to him the only course left to a man who, in losing the position, had not abandoned the instincts of a gentleman.

The evening wore away, unmeasured by him; the echoes of the soldiers' mirth came dimly on his ear; the laughter, and the songs, and the music were subdued into one confused murmur by distance; there was nothing near him except a few tethered horses, and far away the mounted figure of the vidette who kept watch beyond the boundaries of the encampment. The fire burned on, for it had been piled high before it was abandoned; the little white dog of his regiment was curled at his feet; he sat motionless, sunk in thought, with his head drooped upon his breast. The voice of Cigarette broke on his musing.

'*Beau sire*, you are wanted yonder.'

He looked up wearily: could he never be at peace! He did not notice that the tone of the greeting was rough and curt; he did not notice that there was a stormy darkness, a repressed bitterness, stern and scornful, on the Little One's face; he only thought that the very dogs were left sometimes at rest and unchained, but a soldier never.

'You are wanted!' repeated Cigarette, with imperious contempt.

He rose, on the old instinct of obedience.

'For what?'

She stood looking at him without replying; her mouth was tightly shut in a hard line that pressed inward all its soft and rosy prettiness. She was seeing how haggard his face was, how heavy his eyes, how full of fatigue his movements. Her silence recalled him to the memory of the past day.

'Forgive me, my dear child, if I have seemed without sympathy in all your honours,' he said, gently, as he laid his hand on her shoulder. 'Believe me, it was unintentional. No one knows better than I how richly you deserved them; no one rejoices more that you should have received them.'

The very gentleness of the apology stung her like a scorpion; she shook herself roughly out of his hold.

'*Point de phrases!* All the army is at my back; do you think I cannot do without *you*? Sympathy too! Bah! We don't know those fine words in camp. You are wanted, I tell you—go!'

'But where?'

'To your Silver Pheasant yonder—go!'

'Who? I do not——'

'Dame! Can you not understand? Miladi wants to see you; I told her I would send you to her. You can use your dainty sentences with her; she is of your Order!'

'What! *she* wishes——'

'Go!' reiterated the Little One with a stamp of her boot. 'You know the great tent where she is throned in honour—Morbleu!—as if the oldest and ugliest hag that washes out my soldiers' linen were not of more use and more deserved such lodgement than Madame la Princesse, who has never done aught in her life, not even brushed out her own hair of gold! She waits for you. Where are your palace manners? Go to her, I tell you. She is of your own people: *we* are not!'

The vehement imperious phrases coursed in disorder one after another, rapid and harsh, and vibrating with a hundred repressed emotions. He paused one moment, doubting whether she did not play some trick upon him; then, without a word, left her, and went rapidly through the evening shadows.

Cigarette stood looking after him with a gaze that was very evil, almost savage, in its wrath, in its pain, in its fiery jealousy, that ached so hotly in her, and was chained down by that pride, which was as intense in the Vivandière of Algeria as ever it could be in any Duchess of a Court. Reckless, unfeminine, hardened, vitiated in much, as all her sex would have deemed, and capable of the utmost abandonment to her passion had it been returned, the haughty young soul of the child of the People was as sensitively delicate in this one thing as the purest and chastest amongst women could have been; she dreaded above every other thing that he should ever suspect that she loved him, or that she desired his love.

Her honour, her generosity, her pity for him, her natural instinct to do the thing that was right, even to her foes, any one of the unstudied and unanalysed qualities in her had made her serve him even at her rival's bidding. But it had cost her none the less hardly because so manfully done; none the less did all the violent ruthless hate, the vivid childlike fury, the burning, intolerable jealousy of her nature combat in her with the cruel sense of her own unlikeness with that beauty which had subdued even herself, and with that nobler impulse of self-sacrifice which grew side by side with the baser impulses of passion.

As she crouched down by the side of the fire, all the gracious spiritual light that had been upon her face was gone; there was something of the goaded, dangerous, sullen ferocity of a brave animal hard pressed and over driven.

Her native generosity, the loyal disinterestedness of her love for him, had overborne the jealousy, the wounded vanity, and the desire of vengeance that reigned in her. Carried away by the first, she had, for the hour, risen above the last, and allowed the nobler wish to serve and rescue him prevail over the baser egotism. Nothing with her was ever premeditated; all was the offspring of the caprice or the impulse of the immediate moment. And now the reaction followed: she was only sensible of the burning envy that consumed her of this woman who seemed to her more than mortal in her wonderful fair loveliness, in her marvellous difference from everything of their sex that the camp and the barrack ever showed.

'And I have sent him to her when I should have fired my pistol into her breast!' she thought, as she sat by the dying embers. And she remembered once more the story of the Marseilles fisherwoman. She understood that terrible vengeance under the hot southern sun, beside the ruthless southern seas.

Meanwhile, he, who so little knew or heeded how he occupied her heart, passed unnoticed through the movements of the military crowds, crossed the breadth that parted the encampment from the marquees of the Generals and their guests, gave the countersign and approached unarrested, and so far unseen save by the sentinels, the tents of the Corona suite. The Marshal and his male visitors were still over their banquet wines; she had withdrawn early, on the plea of fatigue; there was no one to notice his visit except the men on guard, who concluded that he went by command. In the dusky light, for the moon was very young, and the flare of the torches made the shadows black and uncertain, no one recognised him; the few soldiers stationed about saw one of their own troopers, and offered him no opposition, made him no question. He knew the password; that was sufficient. The Levantine waiting near the entrance drew the tent-folds aside and signed to him to enter: another moment, and he was in the presence of her mistress, in that dim amber light from the standing candelabra, in that heavy soft-scented air perfumed from the aloe-wood burning in a brazier, through which he saw, half blinded at first coming from the darkness without, that face which subdued and dazzled even the antagonism and the lawlessness of Cigarette.

He bowed low before her, preserving that distant ceremonial due from the rank he ostensibly held to hers.

'Madame, this is very merciful! I know not how to thank you.'

She motioned to him to take a seat near to her, while the Levantine, who knew nothing of the English tongue, retired to the farther end of the tent.

'I only kept my word,' she answered, 'for we leave the camp tomorrow; Africa next week.'

'So soon!'

She saw the blood forsake the bronzed fairness of his face, and leave a dusky pallor there. It wounded her as if she suffered herself. For the first time she believed what the Little One had said—that this man loved her.

'I sent for you,' she continued, hurriedly, her graceful languor and tranquillity, for the first time, stirred and quickened by emotion, almost by embarrassment. 'It was very strange, it was very painful, for me to trust that child with such a message. But you know us of old; you know we do not forsake our friends for considerations of self-interest or outward semblance. We act as we deem right; we do not heed untrue constructions. There are many things I desire to say to you——'

She paused; he merely bent his head; he could not trust the calmness of his voice in answer.

'First,' she continued, 'I must entreat you to allow me to tell Philip what I know. You cannot conceive how intensely oppressive it becomes to me to have any secret from him. I never concealed so much as a thought from my brother in all my life, and to evade even a mute question from his brave frank eyes makes me feel a traitress to him.'

'Anything else,' he muttered. 'Ask me anything else. For God's sake, do not let him dream that I live!'

'But why? You still speak to me in enigmas. Tomorrow, moreover, before we leave, he intends to seek you out as what he thinks you—a soldier of France. He is interested by all he hears of your career; he was first interested by what I told him of you when he saw the ivory carvings at my villa. I asked the little vivandière to tell you this, but, on second thoughts, it seemed best to see you myself once more, as I had promised.'

There was a slow weariness in the utterance of the words. She had said that she could not reflect on leaving him to such a fate as this of his in Africa without personal suffering, or without an effort to induce him to reconsider his decision to condemn himself to it for evermore.

'That French child,' she went on, rapidly, to cover both the pain that she felt and that she dealt, 'forced her entrance here in a strange fashion; she wished to see me, I suppose, and to try my courage too. She is a little brigand, but she has a true and generous nature, and she loves you very loyally.'

'Cigarette?' he asked, wearily; his thoughts could not stay for either pity or interest for her in this moment. 'Oh no!—I trust not. I have done nothing to win her love; and she is a fierce little *condottiera* who disdains all such weakness. She forced her way in here? That was unpardonable; but she seems to bear a singular dislike to you.'

'Singular indeed! I never saw her until today.'

He answered nothing; the conviction stole on him that Cigarette hated her because he loved her.

'And yet she brought you my message?' pursued his companion. 'That seems her nature—violent passions, yet thorough loyalty. But time is precious. I must urge on you what I bade you come to hear. It is to implore you to put your trust, your confidence, in Philip. You have acknowledged to me that you are guiltless—no one who knows what you once were could ever doubt it for an instant—then let him hear this, let him be your judge as to what course is right and what wrong for you to pursue. It is impossible for me to return to Europe knowing you are living thus and leaving you to such a fate. What motive you have to sentence yourself to such eternal banishment I am ignorant; but all I ask of you is, confide in him. Let him learn that you live; let him decide whether or not this sacrifice of yourself be needed. His honour is as punctilious as that of any man on earth; his friendship you can never doubt. Why conceal anything from him?'

His eyes turned on her with that dumb agony which once before had chilled her to the soul.

'Do you think, if I *could* speak in honour, I should not tell *you* all?'

A flush passed over her face, the first that the gaze of any man had ever brought there. She understood him.

'But,' she said, gently and hurriedly, 'may it not be that you overrate the obligations of honour? I know that many a noble-hearted man has inexorably condemned himself to a severity of rule that a dispassionate judge of his life might deem very exaggerated, very unnecessary. It is so natural for an honourable man to so dread that he should do a dishonourable thing through self-interest or self-pity, that he may very well overestimate the sacrifice required of him through what he deems justice or generosity. May it not be so with you? I can conceive no reason that can be strong enough to require of you such fearful surrender of every hope, such utter abandonment of your own existence.'

Her voice failed slightly over the last words; she could not think with calmness of the destiny that he accepted. Involuntarily some prescience of pain that would for ever pursue her own life unless his were rescued lent an intense earnestness, almost entreaty, to her argument. She did not bear him love as yet; she had seen too little of him, too lately only known him as her equal; but there were in her, stronger than she knew, a pity, a tenderness, a regret, an honour for him that drew her towards him with an indefinable attraction, and would sooner or later warm and deepen into love. Already it was sufficient, though she deemed it but compassion and friendship, to make her feel that an intolerable weight would lie heavy on her future if his should remain condemned to this awful isolation and oblivion while she alone of all the world should know and hold his secret.

He started from her side as he heard, and paced to and fro the narrow limits of the tent like a caged animal. For the first time it grew a belief to him, in his thoughts, that were he free, were he owner of his heritage, he could rouse her heart from its long repose and make her love him with the soft and passionate warmth of his dead Arab mistress—a thing that had been as distant from her negligence and her pride as warmth from the diamond or the crystal. He felt as if the struggle would kill him. He had but to betray his brother, and he would be unchained from his torture; he had but to break his word, and he would be at liberty. All the temptation that had before beset him paled and grew as nought beside this possibility of the possession of her love which dawned upon him now.

She, knowing nothing of this which moved him, believed only that he weighed her words in hesitation, and strove to turn the balance.

'Hear me,' she said, softly. 'I do not bid you decide; I only bid you confide in Philip—in one who, as you must well remember, would sooner cut off his own hand than counsel a base thing, or do an unfaithful act. You are guiltless of this charge under which you left England; you endure it rather than do what you deem dishonourable to clear yourself. That is noble—that is great. But it is possible, as I say, that you may exaggerate the abnegation required of you. Whoever was the criminal, should suffer. Yours is magnificent magnanimity; but it may surely be also false justice alike to yourself and the world.'

He turned on her almost fiercely in the suffering she dealt him:

'It *is*! I was a madness—a Quixotism—the wild, unconsidered act of a fool. What you will! But it is done; it was done for ever—so long ago—when your young eyes looked on me in the pity of your innocent childhood. I cannot redeem its folly now by adding to its baseness. I cannot change the choice of a madman by repenting of it with a coward's caprice. Ah, God! you do not know what you do—how you tempt. For pity's sake, urge me no more. Help me—strengthen me—to be true to my word. Do not bid me do evil that I may enter paradise through my sin!'

He threw himself down beside her as the incoherent words poured out, his arms flung across the pile of cushions on which he had been seated, his face hidden on them. His teeth clenched on his tongue till the blood flowed; he felt that if the power of speech remained with him he should forswear every law that had bound him to silence, and tell her all, whatever the cost.

She looked at him, she heard him, moved to a greater agitation than ever had had sway over her, for the first time the storm-winds that swept by her did not leave her passionless and calm; this man's whole future was in her hands. She could bid him seek happiness, dishonoured; or cleave to honour, and accept wretchedness for ever.

It was a fearful choice to hold.

'Answer me. Choose for me!' he said, vehemently. 'Be my law, and be my God!'

She gave a gesture almost of fear.

'Hush, hush! The woman does not live who should be that to any man.'

'You shall be it to me! Choose for me!'

'I cannot! You leave so much in darkness and untold——'

'Nothing that you need know to decide your choice for me, save one thing only—that I love you.'

She shuddered.

'This is madness! What have you seen of me?'

'Enough to love you while my life shall last, and love no other woman. Ah! I was but an African trooper in your sight, but in my own I was your equal. You only saw a man to whom your gracious alms and your gentle charity were to be given, as a queen may stoop in mercy to a beggar; but I saw one who had the light of my old days in her smile, the sweetness of my old joys in her eyes, the memories of my old world in her every grace and gesture. You forget! I was nothing to you; but you were so much to me. I loved you the first moment that your voice fell on my ear. It is madness! Oh yes! I should have said so too in those old years. A madness I would have sworn never to feel. But I have lived a hard life since then, and no men ever love like those who suffer. Now you know all; know the worst that tempts me. No famine, no humiliation, no obloquy, no loss I have known, ever drove me so cruelly to buy back my happiness with the price of dishonour as this one desire, to stand in my rightful place before men, and be free to strive with you for what they have not won!'

As she heard, all the warmth, all the life, faded out of her face; it grew as white as his own, and her lips parted slightly, as though to draw her breath was oppressive. The wild words overwhelmed her with their surprise not less than they shocked her with their despair. An intense truth vibrated through them, a truth that pierced her and reached her heart as no other supplication ever had done. She had no love for him yet, or she thought not; she was very proud, and resisted such passions; but in that moment the thought swept by her that such love might be possible. It was the nearest submission to it she had ever given. She heard him in unbroken silence; she kept silence long after he had spoken. So far as her courage and her dignity could be touched with it, she felt something akin to terror at the magnitude of the choice left to her.

'You give me great pain, great surprise,' she murmured. 'All I can trust is, that your love is of such sudden birth that it will die as rapidly——'

He interrupted her:

'You mean that, under no circumstances—not even were I to possess my inheritance—could you give me any hope that I might wake your tenderness?'

She looked at him full in the eyes with the old, fearless, haughty instinct of refusal to all such entreaty, which had made her so indifferent—and many said so pitiless—to all. At his gaze, however, her own changed and softened, grew shadowed, and then wandered from him.

'I do not say that. I cannot tell——'

The words were very low: she was too truthful to conceal from him what half dawned on herself—the possibility that, more in his presence and under different circumstances, she might feel her heart go to him with a warmer and a softer impulse than that of friendship. The heroism of his life had moved her greatly.

His head dropped down again upon his arms.

'It is *possible* at least! I am blind—mad. Make my choice for me! I know not what I do.'

The tears that had gathered in her eyes fell slowly down over her colourless cheeks; she looked at him with a pity that made her heart ache with a sorrow only less than his own. The grief was for him chiefly: yet something of it for herself. Some sense of present bitterness that fell on her from his fate, some foreboding of future regret that would inevitably and for ever follow her when she left him to his loneliness and his misery, smote on her with a weightier pang than any her caressed and cloudless existence had encountered. Love was dimly before her as the possibility he called it; remote, unrealized, still unacknowledged, but possible under certain conditions, only known as such when it was also impossible through circumstance.

He had suffered silently; endured strongly; fought greatly: these were the only means through which any man could have ever reached her sympathy, her respect, her tenderness. Yet though a very noble and a very generous woman, she was also a woman of the world. She knew that it was not for her to say even thus much to a man who was in one sense well-nigh a stranger, and who stood under the accusation of a crime whose shadow he allowed to rest on him unmoved. She felt sick at heart: she longed unutterably, with a warmer longing than had moved her previously, to bid him, at all cost, lay bare his past, and throw off the imputed shame that hung on him. Yet all the grand traditions of her race forbade her to counsel the acceptance of an escape whose way led through a forfeiture of honour.

'Choose for me, Venetia!' he muttered at last once more.

She rose with what was almost a gesture of despair, and thrust the gold hair off her temples.

'Heaven help me, I cannot—I dare not! And—I am no longer capable of being just!'

There was an accent almost of passion in her voice; she felt that so greatly did she desire his deliverance, his justification, his return to all which was his own, desired even his presence amongst them in her own world, that she could no longer give him calm and unbiased judgement. He heard, and the burning tide of a new joy rushed on him, checked almost ere it was known, by the dread lest for her sake she should ever give him so much pity that such pity became love.

He started to his feet and looked down imploringly into her eyes—a look under which her own never quailed or drooped, but which they answered with that same regard which she had given him when she had declared her faith in his innocence.

'If I thought it possible you could ever care——'

She moved slightly from him; her face was very white still, and her voice, though serenely sustained, shook as it answered him:

'If I could—believe me, I am not a woman who would bid you forsake your honour to spare yourself or me. Let us speak no more of this! What can it avail, except to make you suffer greater things? Follow the counsels of your own conscience. You have been true to them hitherto; it is not for me, or through me, that you shall ever be turned aside from them.'

A bitter sigh broke from him as he heard.

'They are noble words. And yet it is so easy to utter, so hard to follow them. If you had one thought of tenderness for me, you could not speak them.'

A flush passed over her face.

'Do not think me without feeling—without sympathy—pity——'

'These are not love.'

She was silent; they were, in a sense, nearer to love than any emotion she had ever known.

'If you loved me,' he pursued passionately—'ah, God! the very word from me to you sounds insult; and yet there is not one thought in me that does not honour you—if you loved me, could you stand there and bid me drag on this life for ever, nameless, friendless, hopeless, having all the bitterness but none of the torpor of death, wearing out the doom of a galley-slave, though guiltless of all crime?'

'Why speak so? You are unreasoning; a moment ago you implored me not to tempt you to the violation of what you hold your honour; because I bid you be faithful to it, you deem me cruel!'

'Heaven help me! I scarce know what I say. I ask you, if you were a woman who cared for me, could you decide thus?'

'These are wild questions,' she murmured; 'what can they serve? I believe that I should—I am sure that I should. As it is—as your friend——'

'Ah, hush! Friendship is crueller than hate.'

'Cruel?'

'Yes; the worst cruelty when we seek love—a stone proffered us when we ask for bread in famine!'

There was desperation, almost ferocity, in the answer; she was moved and shaken by it—not to fear, for fear was not in her nature, but to something of awe, and something of the despairing hopelessness that was in him.

'Lord Royallieu,' she said, slowly, as if the familiar name were some tie between them, some cause of excuse for these, the only love-words she had ever heard without disdain and rejection—'Lord Royallieu, it is unworthy of you to take this advantage of an interview which I sought, and sought for your own sake. You pain me, you wound me. I cannot tell how to answer you. You speak strangely, and without warrant.'

He stood mute and motionless before her, his head sunk on his chest. He knew that she rebuked him justly; he knew that he had broken through every law he had proscribed himself, and that he had sinned against that code of chivalry which should have made her sacred from such words whilst they were those he could not utter, nor she hear, except in secrecy and shame. Unless he could stand justified in her sight and in that of all men, he had no right to seek to wring out tenderness from her regret and from her pity. Yet all his heart went out to her in one irrepressible entreaty.

'Forgive me, for pity's sake! After tonight I shall never look upon your face again.'

'I do forgive,' she said, gently, whilst her voice grew very sweet. 'You endure too much already for one needless pang to be added by me. All I wish is, that you had never met me, so that this last, worst, thing had not come unto you!'

A long silence fell between them; where she leaned back amongst her cushions, her face was turned from him. He stood motionless in the shadow, his head still dropped upon his breast, his breathing loud and slow and hard. To speak of love to her was forbidden to him, yet the insidious temptation wound closer and closer round his strength. He had only to betray the man he had sworn to protect, and she would know his innocence, she would hear his passion; he would be free, and she—he grew giddy as the thought rose before him—she might, with time, be brought to give him other tenderness than that of friendship. He seemed to touch the very supremacy of joy, to reach it almost with his hand, to have honours, and peace, and all the glory of her haughty loveliness, and all the sweetness of her subjugation, and all the soft delights of passion before him in their golden promise:—and he was held back in bands of iron, he was driven out from them desolate and accurst.

Unlike Cain, he had suffered in his brother's stead, yet, like Cain, he was branded, and could only wander out into the darkness and the wilderness.

She watched him many minutes, he unconscious of her gaze; and whilst she did so, many conflicting emotions passed over the colourless delicacy of her features; her eyes were filled and shadowed with many altering thoughts; her heart was waking from its rest, and the high, generous, unselfish nature in her strove with her pride of birth, her dignity of habit.

'Wait,' she said, softly, with the old imperial command of her voice subdued, though not wholly banished. 'I think you have mistaken me somewhat. You wrong me if you think that I could be so callous, so indifferent, as to leave you here without heed as to your fate. Believe in your innocence you know that I do, as firmly as though you substantiated it with a thousand proofs; reverence your devotion to your honour you are certain that I must, or all better things were dead in me.'

Her voice sank inaudible for the instant; she recovered her self-control with an effort.

'You reject my friendship—you term it cruel—but at least it will be faithful to you; too faithful for me to pass out of Africa and never give you one thought again. *I believe in you.* Do you not know that that is the highest trust, to my thinking, that one human life can show in another's? You decide that it is your duty not to free yourself from this bondage, not to expose the actual criminal, not to take up your rights of birth. I dare not seek to alter that decision. But I cannot leave you to such a future without infinite pain, and there must—there shall be—means through which you will let me hear of you—through which, at least, I can know that you are living.'

She stretched her hands towards him with that same gesture with which she had first declared her faith in his guiltlessness; the tears trembled in her voice and swam in her eyes. As she had said, she suffered for him exceedingly. He, hearing those words which breathed the only pity that had never humiliated him, and the loyal trust which was but the truer because the sincerity of faith in lieu of the insanity of love dictated it, made a blind, staggering, unconscious movement of passionate dumb agony. He seized her hands in his and held them close against his breast one instant, against the loud hard panting of his aching heart.

'God reward you! God keep you! If I stay, I shall tell you all; let me go, and forget that we ever met! I am dead—let me be dead to you!'

With another instant he had left the tent and passed out into the red glow of the torchlit evening. And Venetia Corona dropped her proud head down upon the silken cushions where his own had rested, and wept as women weep over their dead—in such emotion as had never come to her in all the course of her radiant, victorious, and imperious life.

It seemed to her as if she had seen him slain in cold blood, and had never lifted her hand or her voice against his murder.

His voice rang in her ear; his face was before her with its white still rigid anguish; the scorching accents of his avowal of love seemed to search her very heart. If this man perished in any of the thousand perils of war she would for ever feel herself his assassin. She had his secret, she had his soul, she had his honour in her hands; and she could do nothing better for them both than to send him from her to eternal silence, to eternal solitude!

Her thoughts grew unbearable; she rose impetuously from her couch and paced to and fro the narrow confines of her tent. Her tranquillity was broken down; her pride was abandoned; her heart, at length, was reached and sorely wounded. The only man she had ever found, whom it would have been possible to her to have loved, was one already severed from her by a fate almost more hideous than death.

And yet, in her loneliness, the colour flushed back into her face; her eyes gathered some of their old light; one dreaming shapeless fancy floated vaguely through her mind.

If, in the years to come, she knew him in all ways worthy, and learned to give him back this love he bore her, it was in her to prove that love at no matter what cost to her pride and her lineage. If his perfect innocence were made clear in her own sight, there was greatness and there was unselfishness enough in her nature to make her capable of regarding alone his martyrdom and his heroism, and disregarding the opinion of the world. If hereafter she grew to find his presence the necessity of her life, and his sacrifice of that nobility, and of that purity, she now believed it, she—proud as she was with the twin pride of lineage and of character— would be capable of incurring the odium and the marvel of all who knew her by uniting her fate to his own, by making manifest her honour and her tenderness for him, though men saw in him only a soldier of the Empire, only a base-born trooper, beneath her, as Riom beneath the daughter of D'Orléans. She was of a brave nature, of a great nature, of a daring courage, and of a superb generosity. Abhorring dishonour, full of glory in the stainless history of her race, and tenacious of the dignity and of the magnitude of her House, she yet was too courageous and too haughty a woman not to be capable of braving calumny if conscious of her own pure rectitude beneath it, not to be capable of incurring false censure, if encountered in the path of justice and of magnanimity. It was possible, even on herself it dawned as possible, that so great might become her compassion and her tenderness for this man, that she would, in some distant future, when the might of his love and the severity of his suffering should prevail with her, say to him:

'Keep your secret from the world as you will. Prove your innocence only to me; let me and the friend of your youth alone know your name and your rights. And knowing all, knowing you myself to be hero and martyr in one, I shall not care what the world thinks of you, what the world says of me. I will be your wife: I have lands and riches, and honours, and greatness enough to suffice for us both.'

If ever she loved him exceedingly, she would become capable of this sacrifice from the strength, and the graciousness, and the fearlessness of her nature, and such love was not so distant from her as she thought.

*   *   *

Outside her tent there was a peculiar mingling of light and shadow; of darkness from the moonless and now cloud-covered sky, of reddened warmth from the tall burning pine-boughs thrust into the soil in lieu of other illumination. The atmosphere was hot from the flames, and chilly with the breath of the night winds; it was oppressively still, though from afar off the sounds of laughter in the camp still echoed, and near at hand the dull and steady tramp of the sentinels fell on the hard parched soil. Into that blended heat and cold, dead blackness and crimson glare, he reeled out from her presence; drunk with pain as deliriously as men grow drunk with raki. The challenge rang on the air:

'Who goes there?'

He never heard it. Even the old long-accustomed habits of a soldier's obedience were killed in him.

'Who goes there?' the challenge rang again.

Still he never heard, but went on blindly. From where the tents stood, there was a stronger breadth of light through which he had passed, and was passing still—a light strong enough for it to be seen whence he came, but not strong enough to show his features.

'Halt, or I fire!' The sentinel brought the weapon to his shoulder and took a calm, close, sure aim. Cecil did not speak; the password he had forgotten as though he had never heard or never given it.

Another figure than that of the soldier on guard came out of the shadow, and stood between him and the sentinel. It was that of Châteauroy; he was mounted on his grey horse and wrapped in his military cloak, about to go the round of the cavalry camp. Their eyes met in the wavering light like the glow from a furnace-mouth: in a glance they knew each other.

'It is one of my men,' said the chief carelessly to the sentinel. 'Leave me to deal with him.'

The guard saluted, and resumed his beat.

'Why did you refuse the word, sir?'

'I did not hear.'

'And why did you not hear?'

There was no reply.

'Why are you absent from your squadron?'

There was no reply still.

'Have you no tongue, sir? The matraque shall soon make you speak! Why are you here?'

There was again no answer.

Châteauroy's teeth ground out a furious oath; yet a flash of brutal delight glittered in his eyes. At last he had hounded down this man, so long out of his reach, into disobedience and contumacy.

'Why are you here, and where have you been?' he demanded once more.

'I will not say.'

The answer, given at length, was tranquil, low, slowly and distinctly uttered, in a deliberate refusal, in a deliberate defiance.

The dark and evil countenance above him grew livid with fury.

'I can have you thrashed like a dog for that answer, and I will. But first listen here, *beau sire*! I know as well as though you had confessed to me. Your silence cannot shelter your great mistress's shame. Ah-ha! la Faustine! So Madame votre Princesse is so cold to her equals, only to choose her lovers out of my blackguards, and take her midnight intrigues like a camp courtesan!'

Cecil's face changed terribly as the vile words were spoken. With the light and rapid spring of a leopard, he reached the side of his commander, one hand on the horse's mane, the other on the wrist of his chief, that it gripped like an iron vice.

'*You lie!* And you know that you lie. Breathe her name once more, and, by God, as we are both living men, I will have your life for your outrage!'

And, as he spoke, with his left hand he smote the lips that had blasphemed against her.

It was broken asunder at last—all the long and bitter patience, all the calm and resolute endurance, all the undeviating serenity beneath provocation, which had never yielded through twelve long years, but which had borne with infamy and with tyranny in such absolute submission for sake of those around him, who would revolt at his sign, and be slaughtered for his cause. The promise he had given to endure all things for their sakes— the sakes of his soldiery, of his comrades—was at last forgotten. All he remembered was the villany that dared touch her name, the shame that through him was breathed on her. Rank, duty, bondage, consequence, all were forgotten in that one instant of insult that mocked in its odious lie at her purity. He was no longer the soldier bound in obedience to submit to the indignities that his chief chose to heap on him; he was a gentleman who defended a woman's honour, a man who avenged a slur on the life that he loved.

Châteauroy wrenched his wrist out of the hold that crushed it, and drew his pistol. Cecil knew that the laws of active service would hold him but justly dealt with if the shot laid him dead in that instant for his act and his words.

'You can kill me—I know it. Well, use your prerogative; it will be the sole good you have ever done to me.'

And he stood erect, patient, motionless, looking into his chief's eyes with a calm disdain, with an unuttered challenge, that, for the first moment, wrung something of savage respect and of sullen admiration out from the soul of his great foe.

He did not fire; it was the only time in which any trait of abstinence from cruelty had ever been seen in him. He signed to the soldiers of the guard with one hand, whilst with the other he still covered with his pistol the man whom martial law would have allowed him to have shot down, or have cut down, at his horse's feet.

'Arrest him,' he said, simply.

Cecil offered no resistance; he let them seize and disarm him without an effort at the opposition which could have been but a futile, unavailing, trial of brute force. He dreaded lest there should be one sound that should reach *her* in that tent where the triad of standards drooped in the dusky distance. He had been, moreover, too long beneath the yoke of this despotic and irresponsible authority to waste breath or to waste dignity in vain contest with the absolute and the immutable. He was content with what he had done—content to have met once, not as soldier to chief, but as man to man, the tyrant who held his fate.

For once, beneath the spur of that foul outrage to the dignity and the innocence of the woman he had quitted, he had allowed a passionate truth to force its way through the barriers of rank and the bonds of subservience. Insult to himself he had borne as the base prerogative of his superior, but insult to her he had avenged with the vengeance of equal to equal, of the one who loved, upon the one who calumniated, her.

And as he sat in the darkness of the night with the heavy tramp of his guards for ever on his ear, there was peace rather than rebellion in his heart—the peace of one heart sick with strife and with temptation, who beholds in death a merciful ending to the ordeal of existence. 'I shall die in her cause at least,' he thought. 'I could be content if I were only sure that she would never know.'

For this was the chief dread which hung on him, that she should ever know, and in knowing, suffer for his sake.

The night rolled on, the army around him knew nothing of what had happened. Châteauroy, conscious of his own coarse guilt against the guest of his Marshal, kept the matter untold and undiscovered, under the plea that he desired not to destroy the harmony of the general rejoicing. The one or two field-officers with whom he took counsel agreed to the wisdom of letting the night pass away undisturbed. The accused was the idol of his own squadron: there was no gauge what might not be done by troops heated with excitement and drunk with wine, if they knew that their favourite comrade had set the example of insubordination, and would be sentenced to suffer for it. Beyond these, and the men employed in his arrest and guard, none knew what had chanced; not the soldiery beneath that vast sea of canvas, many of whom would have rushed headlong to mutiny and to destruction at his word; not the woman who in the solitude of her wakeful hours was haunted by the memory of his love-words, and felt steal on her the unacknowledged sense that, if his future were left to misery, happiness could never more touch her own; not the friend of his early days, laughing and drinking gaily with the officers of the staff.

None knew; not even Cigarette. She sat alone, so far away that none sought her out, beside the picket-fire that had long died out, with the little white dog of Zarâila curled on the scarlet folds of her skirt. Her arms rested on her knees, and her temples were leant on her hands, tightly twisted amongst the dark silken curls of her boyish hair. Her face had the same dusky savage intensity upon it; and she never once moved from that rigid attitude.

She had the Cross on her heart—the idol of her long desire, the star to which her longing eyes had looked up, ever since her childhood, through the reek of carnage and the smoke of battle: and she would have flung it away like dross to have had his lips touch hers once with *love*.

'*Que je suis folle!*' she muttered in her throat, '*que je suis folle!*'

And she knew herself mad; for the desires and the delights of love die swiftly, but the knowledge of honour abides always. Love would have made her youth sweet with an unutterable gladness, to glide from her and leave her weary, dissatisfied, forsaken. But that Cross, the gift of her country, the symbol of her heroism, would be with her always, and light her for ever with the honour of which it was the emblem; and if her life should last until youth should pass away, and age come, and with age death, her hand would wander to it on her dying bed, and she would smile as she died to hear the living watchers murmur: 'That life had glory—that life was lived for France.'

She knew this: but she was young; she was a woman-child, she had the ardour of voluptuous youth in her veins, she had the desolation of abandoned youth in her heart. And honour looked so cold beside love!

She rose impetuously; the night was far spent, the camp was very still, the torches had long died out, and a streak of dawn was visible in the east. She stood awhile looking very earnestly across the wide black city of tents.

'I shall be best away for a time. I grow mad, treacherous, wicked here,' she thought. 'I will go and see Blanc-Bec.'

Blanc-Bec was the old soldier of the Army of Italy.

In a brief while she had saddled and bridled Etoile-Filante, and ridden out of the camp without warning or farewell to any: she was as free to come and to go as though she were a bird on the wing.

Thus Cigarette went, knowing nothing of his fate. And with the sunrise went also the woman whom he loved—in ignorance.

CHAPTER XI

The Vengeance of the Little One

The warm transparent light of an African autumnal noon shone down through the white canvas roof of a great tent in the heart of the encamped divisions at the headquarters of the Army of the South. Within the tent there was a densely packed throng—an immense, close, hushed, listening crowd, of which every man wore the uniform of France, and of which the mute undeviating attention, forbidden by discipline alike to be broken by sound of approval or of dissent, had in it something that was almost terrible, contrasted with the vivid eagerness in their eyes and the strained absorption of their countenances; for they were in court, and that court was the Conseil de Guerre of their own southern camp.

The prisoner was arraigned on the heaviest charge that can be laid against the soldier of any army; and yet, as the many eyes of the military crowd turned on him where he stood surrounded by his guard, his crime against his chief was forgotten, and they only remembered—Zarâila.

Many of those present had seen him throughout that day of blood, at the head of his decimated squadron, with the guidon held aloft above every foe; to them, that tall slender form standing there, with a calm weary dignity that had nothing of the passion of the mutinous, or the consciousness of the criminal, in its serene repose, had shed upon it the lustre of a heroism that made them ready almost to weep like women that the death of a mutineer should be the sole answer given by France to the saviour of her honour.

He preserved entire reticence in court. The instant the *acte d'accusation* had been read to him, he had seen that his chief would not dare to couple with it the proud pure name he had dared to outrage; his most bitter anxiety was thus at an end. For all the rest, he was tranquil.

No case could be clearer, briefer, less complex, more entirely incapable of defence. The soldiers of the guard gave evidence as to the violence and fury of the assault. The sentinel bore witness to having heard the refusal to reply; a moment after he had seen the attack made, and the blow given. The accuser merely stated that, meeting his *sous-officier* out of the bounds of the cavalry camp, he had asked him where he had been, and why he was there, and, on his commanding an answer, had been assaulted in the manner described, with violence sufficient to have cost his life had not the guard been so near at hand. When questioned as to what motive he could assign for the act, he replied that he considered his corporal had always incited evil feeling and mutinous conduct in the squadrons, and had, he believed, that day attributed to himself his failure to receive the Cross. The statement passed without contradiction by the prisoner, who, to the interrogations and entreaties of his legal *défenseur*, only replied that the facts were stated accurately as they occurred, and that his reasons for the deed he declined to assert.

When once more questioned as to his country and his past by the president, he briefly declined to give answer. When asked if the names by which he was enrolled were his own, he replied that they were two of his baptismal names, which had served his purpose on entering the army. When asked if he accepted as true the charge of exciting sedition among the troops, he replied that it was so little true that over and over again the men would have mutinied if he had given them a sign, and that he had continually induced them to submit to discipline sheerly by force of his own example. When interrogated as to the cause of the language he had used to his commanding officer, he said briefly that the language deserved the strongest censure as from a soldier to his colonel, but that it was justified as he had used it, which was as man to man, though he was aware the plea availed nothing in military law, and was impermissible for the safety of the service. When it was enquired of him if he had not repeatedly inveighed against his commanding officer for severity, he briefly denied it; no man had ever heard him say a syllable that could have been construed into complaint; at the same time he observed that all the squadrons knew perfectly well that personal enmity and oppression had been shown him by his chief throughout the whole time of his association with the regiment. When pressed as to the cause that he assigned for this, he gave, in a few comprehensive outlines, the story of the capture and the deliverance of the Emir's bride. This was all that could be elicited from him; and even this was answered only out of deference to the authority of the court, and from his unwillingness to set a bad example before the men with whom he had served so long. When it was finally demanded of him if he had aught to urge in his own extenuation, he paused a moment, with a gaze under which even the hard eagle eyes grew restless, looked across to Châteauroy, and addressed his antagonist rather than the president:

'Only this, that a tyrant, a liar, and a traducer cannot wonder if men prefer death to submission beneath insult. But I am well aware this is no vindication of my act as a soldier; and I have no desire to say words which, whatever their truth, might become hereafter dangerous legacies and dangerous precedents to the army.'

That was all which he answered; and neither his counsel nor his accusers could extort another syllable from him.

He knew that what he had done was justified to his own conscience; but he did not seek to dispute that it was unjustifiable in military law. True, had all been told, it was possible enough that his judges would exonerate him morally, even if they condemned him legally; his act would be seen blameless as a man's, even whilst still punishable as a soldier's: but to purchase immunity for himself at cost of bringing the fairness of her fame into the coarse babble of men's tongues was an alternative, craven and shameful, which never even once glanced across his thoughts.

He had kept faith to a woman whom he had known heartless and well-nigh worthless; it was not to the woman whom he loved with all the might of an intense passion, and whom he knew pure and glorious as the morning sun, that he would break his faith now.

All through the three days that the Conseil sat, his look and his manner never changed; the first was quite calm, though very weary, the latter courteous but resolute with the unchanged firmness of one who knew his own past action justified; for the rest, many noticed that, during the chief of the long exhausting hours of his examination and his trial, his thoughts seemed far away, and he appeared to recall them to the present with difficulty, and with nothing of the vivid suspense of an accused, whose life and death swung in the judgement-balance.

In truth, he had no dread as he had no hope left; he knew well enough that by the blow which had vindicated her honour he had forfeited his own existence. All he wished was that his sentence had been dealt without this formula of debate and of delay, which could have issue but in one end. There was not one man in court who was not more moved than he, more quick to terror and regret for his doom. To many amongst his comrades who had learned to love the gentle, silent 'aristocrat', who bore every hardship so patiently, and humanized them so imperceptibly by the simple force of an unvaunted example, those three days were torture. Wild, brutal brigands, whose year was one long razzia of plunder, rapine, and slaughter, felt their lips tremble like young girls' when they asked how the issue went for him; and blood-stained marauders, who thought as little of assassination for a hidden pot of gold as butchers of drawing a knife across a sheep's throat, grew still and fear-stricken with a great awe when the muttering passed through the camp that they would see no more amongst their ranks that 'woman's face' which they had beheld so often foremost in the fight, with a look on it that thrilled their hearts like their forbidden chant of the Marseillaise.

For when the third day closed, they knew that he must die.

And there were men, hard as steel, ravenous of blood as vultures, who, when they heard that sentence given, choked great sobs down into the cavernous depths of their sinewy breasts; but he never gave sigh or sign. He never moved once while the decree of death was read to him; and there was no change in the weary calmness of his eyes. He bent his head in acquiescence.

'*C'est bien!*' he said, simply.

It seemed well to him: dead, his secret would lie in the grave with him, and the long martyrdom of his life be ended.

*   *   *

In the brightness of the noon Cigarette leaned out of her little oval casement that framed her head like an old black oak carving—a head with the mellow bloom on its cheeks, and the flash of scarlet above its dark curls, and the robin-like grace of poise and balance in it as it hung out there in the sun.

Cigarette had been there a whole hour in thought; she!—who never had wasted a moment in meditation or reverie, and who found the long African day all too short for her busy, abundant, joyous life, that was always full of haste and work, just as a bird's will seem so, though the bird have no more to do than to fly at its will through summer air, and feed at its will from brook and from berry, from a ripe ear of the corn or from a deep cup of the lily.

For the first time she was letting time drift away in the fruitless labour of vain purposeless thought, because, for the first time also, happiness was not with her.

They were gone for ever—all the elastic joyaunce, all the free fair hours, all the dauntless gaiety of childhood, all the sweet harmonious laughter of a heart without a care. They were gone for ever; for the touch of love and of pain had been laid on her; and never again would her radiant eyes smile cloudlessly like the young eagle's at a sun that rose but to be greeted as only youth can greet another dawn of a life that is without a shadow.

And she leaned wearily here with her cheek lying on the cold grey Moorish stone; the colour and the brightness were in the rays of the noon, in the rich hues of her hair and her mouth, in the scarlet glow of her dress: there was no brightness in her face. The eyes were vacant as they watched the green lizard glide over the wall beyond, and the lips were parted with a look of unspeakable fatigue; the tire, not of the limbs, but of the heart. She had come thither hoping to leave behind her on the desert wind that alien care, that new strange passion, which sapped her strength, and stung her pride, and made her evil with such murderous lust of vengeance; but they were with her still. Only something of the deadly biting ferocity of jealousy had changed into a passionate longing to be as that woman was who had his love; into a certain hopeless sickening sense of having for ever lost that which alone could have given her such beauty and such honour in the sight of men as those this woman had.

To her it seemed impossible that this patrician who had his passion should not return it. To the child of the camp, though she often mocked at caste, all the inexorable rules, all the reticent instincts of caste, were things unknown. She would have thought love could have bridged over any gulf; she would have failed to comprehend all the thousand reasons which would have forbidden any bond between the great aristocrat and a man of low grade and of dubious name. She only thought that the one she envied was free, was powerful, was high enough to hold and exercise an irresponsible will; she only thought of love as she had always seen it, quickly born, hotly cherished, wildly indulged, and without tie or restraint. And she had left them together! Her heart ached, her lips grew white, her eyes through their new sadness had something of the sullen, savage fire with which she had levelled her pistol at Venetia

Corona, where she leaned out from the oval attic-lattice above the deserted Moorish court below—the only thing of life and youth in the grim desolation of the God-forgotten place.

'And I came without my vengeance!' she mused. To the nature that felt the ferocity of the vendetta a right and a due, and to the jealous fever that still flamed after some such awful retribution as that which the fisher-girl of Marseilles had dealt by the southern sea, there was wounding humiliation in her knowledge that she had left her rival unharmed, and had come hither, out from his sight and his presence, lest he should see in her one glimpse of that folly which she would have killed herself under her own steel rather than have betrayed either for his contempt or his compassion.

'And I came without my vengeance!' she mused: in that oppressive noon, in that grey and lonely place, in that lofty tower-solitude, where there was nothing between her and the hot, hard, cruel blue of the heavens, vengeance looked the only thing that was left her, the only means whereby that void in her heart could be filled, that shame in her life be washed out. To love! and to love a man who had no love for her, whose eyes only beheld another's face, whose ears only thirsted for another's voice! Its degradation stamped her a traitress in her own sight—traitress to her code, to her pride, to her country, to her flag!

And yet at the core of her heart so tired a pang was aching! She who had gloried in being the child of the whole people, the daughter of the whole army, felt lonely and abandoned, as though she were some bird which an hour ago had been flying in all its joy amongst its brethren, and now, maimed with one shot, had fallen with broken pinion and torn plumage to lie alone upon the sand and die.

The touch of a bird's wing brushing her hair brought the dreamy comparison to her wandering thoughts. She started and lifted her head; it was a blue carrier-pigeon, one of the many she fed at that casement, and the swiftest and surest of several she sent with messages for the soldiers between the various stations and corps. She had forgotten she had left the bird at the encampment.

She caressed it absently, while the tired creature sank down on her bosom; then only she saw that there was a letter beneath one wing. She unloosed it, and looked at it without being able to tell its meaning; she could not read a word, printed or written. Military habits were too strong with her for the arrival not to change her reverie into action; whoever it was for, it must be seen. She gave the pigeon water and grain, then wound her way down the dark, narrow stairs, through the height of the tower, out into the passage below.

She found an old French cobbler sitting at a stall in a casement stitching leather; he was her customary reader and scribe in this quarter. She touched him with the paper. 'Bon Mathieu! wilt thou read this to me?'

He took it, and looked first at the superscription.

'It is for thee, Little One, and signed "Petit Pot-de-terre".'

Cigarette nodded listlessly.

'"Tis a good lad, and a scholar,' she answered, absently. 'Read on!'

And he read aloud:

'There is ill news. I send the bird on a chance to find thee. Bel-à-faire-peur struck the Black Hawk—a light blow, but with threat to kill following it. He had been tried, and is to be shot. There is no appeal to the Conseil de Révision. The case is clear; the Colonel could have cut him down, were that all. I thought you should know. We are all sorry. It was done on the night of the great fête. I am thy humble lover and slave.'

So the boy-Zouave's scrawl, crushed, and blotted, and written with great difficulty, ran in its brief phrases that the slow muttering of the old shoemaker drew out in tedious length.

Cigarette heard; she never made a movement or gave a sound, but all the blood fled out of her brilliant face, leaving it horribly blanched beneath its brown sun-scorch; and her eyes distended, senseless, sightless, were fastened on the old man's slowly moving mouth.

'Read it again!' she said, simply, when all was ended. He started, and looked up at her face: the voice had not one accent of its own tones left.

He obeyed, and read it once more to the end. Then a loud shuddering sigh escaped her, like the breath of one stifling under flames.

'Shot!' she said, vacantly. 'Shot!'

The old man rose hurriedly.

'Child! art thou ill?'

Her eyes turned on him without any consciousness in them.

'The blow was struck for *her*!' she muttered. 'It was that night, you hear—that night?'

'What night? Thou lookest so strangely! Dost thou love this doomed soldier?'

Cigarette laughed—a laugh whose echo thrilled horribly through the lonely Moresco courtway.

'Love? love? I hated him, look you! So I said. And I longed for my vengeance. It is come!'

She was still a moment; her mouth quivering as though she were under physical torture, her strained eyes fastened on the empty air, the veins in her throat swelling and throbbing till they glowed to purple. Then she crushed the letter in one hand, and flew, fleet as any antelope, through the streets of the Moorish quarter, and across the city to the quay.

The people ever gave way before her; but now they scattered like frightened sheep from her path. There was something that terrified them in that bloodless horror set upon her face, and in that fury of resistless speed with which she rushed upon her way.

Once only in her headlong career through the throngs she paused; it was as one face, on which the strong light of the noontide poured, came before her. The senseless look changed in her eyes; she wheeled out of her route, and stopped before the man who had thus arrested her. He was leaning idly over the stall of a Turkish bazaar, and her hand grasped his arm before he saw her.

'You have his face?' she muttered. 'What are you to him?'

He made no answer; he was too amazed.

'You are of his race,' she persisted. 'You are brethren by your look. What are you to him?'

'To whom?'

'To the man who calls himself Louis Victor? a Chasseur of my Army?'

Her eyes were fastened entirely on him; keen, ruthless, fierce, in this moment, as a hawk's. He grew pale, and murmured an incoherent denial. He sought to shake her off, first gently, then more rudely; he called her mad, and tried to fling her from him; but the lithe fingers only wound themselves closer on his arm.

'Be still—fool!' she muttered; and there was that in the accent that lent a strange force and dignity in that moment to the careless and mischievous plaything of the soldiery—force that overcame him, dignity that overawed him. 'You are of his people; you have his eyes, and his look, and his features. He disowns you, or you him. No matter which. He is of your blood; and he lies under sentence of death, do you know that?'

With a stifled cry, the other recoiled from her; he never doubted that she spoke the truth, none could who had looked upon her face.

Cigarette smiled—a smile that had the same terrible meaning in it as the laugh that had curdled the old Frenchman's veins in the stillness of the Moorish passage.

'Do not lie to *me*,' she said, curtly. 'It avails you nothing. Read that.'

She thrust before him the paper the pigeon had brought; his hand trembled sorely as he held it: he believed in that moment that this strange creature, half soldier, half woman, half brigand, half child, knew all his story and all his shame from his brother.

'Shot!' he echoed hoarsely as she had done, when he had read on to the end. 'Shot! O Heaven! and I——'

She drew him out of the thoroughfare into a dark recess within the bazaar; he submitted unresistingly; he was filled with the horror, the remorse, the overwhelming shock of his brother's doom.

'He will be shot,' she said, with a strange calmness. 'We shoot down many men in our Army. I know him well. He was justified in his act, I do not doubt; but discipline will not stay for that——'

'Silence, for mercy's sake! Is there no hope—no possibility?'

Her lips were parched like the desert sand as her dry hard words came through them. 'None. There is no appeal but the *Révision*; and none to that here. His chief could have cut him down on the instant. It took place in camp. You feel this thing; you are of his race, then?'

'I am his brother!'

She was silent; looking at him fixedly, it did not seem to her strange that she should thus have met one of his blood in the crowds of Algiers. She was absorbed in the one catastrophe whose hideousness seemed to eat her very life away, even whilst her nerve, and her brain, and her courage remained at their keenest and strongest.

'You are his brother,' she said, slowly, so much as an affirmation that his belief was confirmed that she had learned both their relationship and their history from Cecil. 'You must go to him, then.'

He shook from head to foot.

'Yes, yes! But it will be too late!'

She did not know that the words were cried out in all the contrition of an unavailing remorse; she gave them only their literal significance, and shuddered as she answered him:

'That you must risk. You must go to him. But, first, I must know more. Tell me his name, his rank.'

He was silent: coward and egotist though he was, both cowardice and egotism were killed in him under the overwhelming horror with which he felt himself as truly by moral guilt a fratricide as though he had stabbed his elder through the heart.

'Speak!' hissed Cigarette through her clenched teeth. 'If you have any kindness, any pity, any love for the man of your blood, who will be shot there like a dog, do not waste a second—answer me, tell me all.'

He turned his wild terrified glance upon her: he had in that moment no sense but to seize some means of reparation, to declare his brother's rights, to cry out to the very stones of the streets his own wrong and his victim's sacrifice.

'He is the head of my House!' he answered her, scarce knowing what he answered. 'He should bear the title that I bear now. He is here, in this misery, because he is the most merciful, the most generous, the most long-suffering of living souls! If he die, it is not they who have killed him; it is I!'

She listened, with her face set in that stern, fixed, resolute command which never varied: she neglected all that wonder, or curiosity, or interest would have made her ask at any other time, she only heeded the few great facts that bore upon the fate of the condemned.

'Settle with yourself for that sin,' she said, bitterly. '*Your* remorse will not save him. But do the thing that I bid you, if that remorse be sincere. Write me out here that title you say he should bear, and your statement that he is your brother, and should be the chief of your race; then sign it, and give it to me.'

He seized her hands, and gazed with imploring eyes into her face.

'Who are you? What are you? If you have the power to do it, for the love of God rescue him! It is I who have murdered him—I—who have let him live on in this hell for my sake!'

'For your sake!'

She flung his hands off her and looked him full in the face; that glance of the speechless scorn, the unutterable rebuke, of the woman-child who would herself have died a thousand deaths rather than have purchased a whole existence by a single falsehood or a single cowardice, smote him like a blow, and avenged his sin more absolutely than any public chastisement. The courage and the truth of a girl scorned his timorous fear and his living lie. His head sank, he seemed to shrink under her gaze; his act had never looked so vile to him as it looked now.

She gazed a moment longer at him in mute and wondering disdain that there should be on earth a male life capable of such fear and of such ignominy as these. Then the strong and rapid power in her took its instant ascendancy over the weaker nature.

'Monsieur, I do not know your story; I do not want. I am not used to men who let others suffer for them. What I require is your written statement of your brother's name and station; give it me.'

He made a gesture of consent; he would have signed away his soul, if he could, in the stupor of remorse which had seized him. She brought him pens and paper from the Turk's store, and dictated what he wrote:

I hereby affirm that the person serving in the Chasseurs d'Afrique under the name of Louis Victor is my elder brother, Bertie Cecil, lawfully, by inheritance, the Viscount Royallieu, Peer of England. I hereby also acknowledge that I have succeeded to and borne the title illegally, under the supposition of his death.

(Signed) BERKELEY CECIL.

He wrote it mechanically, the force of her will and the torture of his own conscience driving him, on an impulse, to undo in an instant the whole web of falsehood that he had let circumstance weave on and on to shelter him through twelve long years. He allowed her to draw the paper from him and fold it away in her belt. He watched her with a curious dreamy sense of his own impotence against the fierce and fiery torrent of her bidding.

'What is it you will do?' he asked her, as she took from him the acknowledgment to the world of his brother's life and name.

'The best that shall lie in my power; do you the same.'

'Can his life yet be saved?'

'His honour may—his honour shall.'

Her face had an exceeding beauty as she spoke; though it was stern and rigid still, a look that was sublime gleamed over it. She the waif and stray of a dissolute camp, knew better than the scion of his own race how the doomed man would choose the vindication of his honour before the rescue of his life. He laid his hand on her as she moved:

'Stay!—stay! One word——'

She flung him off her again:

'This is no time for words. Go to him—*coward!*—and let the balls that kill him reach you too, if you have one trait of manhood left in you!'

Then, swiftly as a swallow darts, she quitted him and flew on her headlong way, down through the pressure of the people, and the throngs of the marts, and the noise, and the colour, and the movement of the streets.

The sun was scarce declined from its noon before she rode out of the city, on a black half-bred horse of the Spahis, swift as the antelope and as wild, with her only equipment, some pistols in her holsters, and a bag of rice and a skin of water slung at her saddle-bow.

They asked her where she went; she never answered. Some noticed that her eyes looked at them as if she had no sense or comprehension of their presence. The hoofs struck sharp echoes out of the rugged stones of the tortuous streets, and the people were scattered like chaff before a breeze as she went at full gallop down through Algiers with the sun-ray glittering like fire on her Cross. Her comrades, used to see her ever with some song in the air and some laugh on the lips as she went, looked after her with wonder as she passed them, silent, and with her face white and stern as though the bright brown loveliness of it had been changed to alabaster.

'What is it with the Cigarette?' they asked each other. None could tell; the desert horse and his rider flew by them as a swallow flies. The gleam of her croix d'honneur and the colourless calm of the childlike face that wore the resolve of a Napoléon's on it, were the last they ever saw of Cigarette.

The words with which she had asked for the loan of a fresh young horse had been the only ones she had spoken. All her fluent untiring speech was gone—gone with the rose-hue from her cheek, with the laugh from her mouth, with the child's joyaunce from her heart; but the brave, staunch, dauntless spirit lived with a soldier's courage, with a martyr's patience.

And she rode straight through the scorch of the mid-day sun, along the sea-coast eastward. The dizzy swiftness would have blinded most who should have been carried through the dry air and under the burning skies at that breathless and pauseless speed; but she had ridden half-maddened colts with the skill of Arabs themselves; she had been tossed on a holster from her earliest years, and had clung with an infant's hands in fearless glee to the mane of rough-riders' chargers. She never swerved, she never sickened; she was borne on and on against the hard hot currents of the cleft air with only one sense—that she went so slowly, so slowly, when with every beat of the ringing hoofs one of the few moments of a doomed life fled away!

She had a long route before her; she had many leagues to travel, and there were but four-and-twenty hours, she knew well, left to the man who was condemned to death. Four-and-twenty hours left open for appeal—no more—betwixt the delivery and execution of the sentence. That delay was always interpreted by the French Code as a delay extending from the evening of one day to the dawn of the second day following; and some slight interval might then ensue, according as the General in command ordained. But the twenty-four hours was all of which she could be certain; and even of them some must have flown by since the carrier-pigeon had been loosed to her. She could not tell how long he had to live; and the agony in her made the headlong electric speed of the free-born horse of the plains seem tardy, and slackened, and weighed with lead, as they flew eastward by the line of the sea.

There were fifty miles between her and her goal; Abd-el-Kader's horse had once covered that space in three hours, so men of the Army of D'Aumale had told her: she knew what they had done she could do with this brave swift beast, who asked nothing of her but to be left to dash onward at his own will as fleetly as his wild sires of the desert herds had galloped ere ever the curb or the spur of a rider had touched them. Once only she paused, to let him lie a brief while, and cool his foam-flaked sides, and crop some short sweet grass that grew where a cleft of water ran and made the bare earth green. She sat quite motionless while he rested; she was keenly alive to all that could best save his strength and further her travel; but she watched him during those few minutes of rest and inaction with a fearful look of hunger in her eyes—the worst hunger, that which craves Time and cannot seize it fast enough. Then she mounted again, and again went on, in her flight.

She swept by cantonments, villages, soldiers on the march, douairs of peaceful Arabs, strings of mules and camels, caravans of merchandise: nothing arrested her; she saw nothing that she passed, as she rode over the hard dust-covered shadowless roads, over the weary sun-scorched monotonous country, over the land without verdure and without foliage, the land that yet has so weird a beauty, so irresistible a fascination; the land to which men, knowing that death waits for them in it, yet return with so mad an infatuation as her lovers went back across the waters to Circe.

The horse was reeking with smoke and foam, and the blood was coursing from his flanks, as she reached her destination at last, and threw herself off his saddle as he sank faint and quivering to the ground. Whither she had come was to a fortress where the Marshal of France, who was the Viceroy of Africa, had arrived that day in his progress of inspection throughout the provinces. Soldiers clustered round her eagerly beneath the gates, a thousand questions pouring from their curious tongues. She pointed to the animal with one hand, to the gaunt pile of stone that bristled with cannon with the other.

'Have a care of him; and lead me to the Chief!'

She spoke quietly; but a certain sensation of awe and fear moved those who heard. She was not the Child of the Army whom they knew so well. She was a creature, desperate, hard pressed, mute as death, strong as steel; above all, hunted by despair.

They hesitated to take her message, to do her bidding. The one whom she sought was great and supreme here as a king; they dreaded to approach his staff, to ask his audience.

Cigarette looked at them a moment, then loosened her Cross and held it out to an Adjutant standing beneath the gates.

'Take that to the man who gave it me. Tell him Cigarette waits; and that with each moment that she waits a soldier's life is lost. Go!'

The Adjutant took it, and went. Over and over again she had brought intelligence of an Arab movement, news of a contemplated razzia, warning of an internal revolt, or tidings of an encounter on the plains, that had been of priceless value to the army which she served. It was not lightly that Cigarette's words were ever received when she spoke as she spoke now; nor was it impossible that she now brought to them that which would brook neither delay nor trifling.

She waited patiently; all the iron discipline of military life had never bound her gay and lawless spirit down, but now she was singularly still and mute. Only there gleamed thirstily in her eyes that fearful avarice which begrudges every moment in its flight as never the miser grudges his hoarded gold into the robber's grasp.

A few minutes, and the decoration was brought back to her, and her demand granted. She was summoned to the Marshal's presence. She was taken within the casemate of the fortress, to where the mighty Chief stood amongst his officers. It was the ordnance-room, a long vast silent chamber filled with stands of arms, with all the arts and appliances of war brought to their uttermost perfection, and massed in all the resource of a great empire against the sons of the desert, who had nothing to oppose to them save the despair of a perishing nationality and a stifled freedom.

The Marshal, leaning against a brass field-piece, turned to her with a smile in his keen stern eyes.

'You, my young *décorée*! What brings you here?'

She came up to him with her rapid leopard-like grace, and he started as he saw the change upon her features, that had lost all the sunlight of their childhood and their mirth, and were so dark, so colourless, so still. She was covered with sand and dust, and with the animal's blood-flecked foam. The beating of her heart from the fury of the gallop had drained every hue from her face; her voice was scarcely articulate in its breathless haste as she saluted him:

'Monseigneur, I have come from Algiers since noon——'

'From Algiers!'

He and his officers echoed the name of the city in incredulous amaze; they knew how far from them down along the sea-line the white town lay.

'Since noon, to rescue a life—the life of a great soldier, of a guiltless man. He who saved the honour of France at Zaraïla is to die the death of a mutineer at dawn!'

'What!—your Chasseur?'

A dusky scarlet fire burned through the pallor of her face; but her eyes never quailed, and the torrent of her eloquence returned under the pangs of shame that were beaten back under the noble instincts of her love.

'Mine!—since he is a soldier of France; yours, too, by that title. I am come here, from Algiers, to speak the truth in his name, and to save him for his own honour and the honour of my Empire. See here! At noon, I have this paper, sent by a swift pigeon—read it! You see how he is to die, and why. Well, by my Cross, by my Flag, by my France, I swear that not

a hair of his head shall be touched, not a drop of blood in his veins shall
be shed!'

He looked at her, astonished at the grandeur and the courage which
could come on this child of razzias and revelries, and give to her all the
fearless command of some young empress. But his face darkened and set
sternly as he read the paper; it was the vilest crime in the sight of a proud
soldier, this crime against discipline, of the man for whom she pleaded.

'You speak madly,' he said, with cold brevity. 'The offence merits the
chastisement. I shall not attempt to interfere.'

A convulsion went over her face; but her spirit only rose the stronger
for the despair that strove to crush it.

'Wait! you will hear, at least, Monseigneur?'

'I will hear you—yes; but I tell you, once for all, I never change
sentences that are pronounced by conseils de guerre; and this crime is the
last for which you should attempt to plead for mercy with me.'

'Hear me, at least!' she cried, with passionate ferocity—the ferocity of
a dumb animal wounded by a shot. 'You do not know what this man is—
how he has had to endure; I do. I have watched him; I have seen the
brutal tyranny of his chief, who hated him because the soldiers loved him.
I have seen his patience, his obedience, his long-suffering beneath insults
that would have driven any other to revolt and murder. I have seen him—
I have told you how—at Zaraîla, thinking never of death or of life, only of
our Flag, that he has made his own, and under which he has been forced
to lead the life of a galley-slave, of a dog——'

'The finer soldier he be, the less pardonable his offence.'

'That I deny! If he were a dolt, a brute, a thing of wood, as many are,
he would have no right to vengeance; as it is, he is a gentleman, a hero, a
martyr; may he not forget for one hour that he is a slave? Look you! I have
seen him so tried, that I told him—I, who love my Army better than any
living thing under the sun—that I would forgive him if he forgot duty and
dealt with his tyrant as man to man. And he always held his soul in
patience. Why? Not because he feared death—he desired it; but because
he loved his comrades, and suffered in peace and in silence lest, through
him, they should be led into evil——'

His eyes softened as he heard her; but the inflexibility of his voice
never altered.

'It is useless to argue with me,' he said, briefly; 'I never change a
sentence.'

'But I say that you *shall*!' As the audacious words were flung forth, she
looked him full in the eyes, while her voice rang with its old imperious
oratory. 'You are a great chief; you are as a monarch here; you hold the

gifts and the grandeur of the Empire; but, *because* of that—because you are as France in my eyes—I swear, by the name of France, that you shall see justice done to him—after death, if you cannot in life. Do you know who he is—this man whom his comrades will shoot down at sunrise as they shoot down the murderer and the ravisher in their crimes?'

'He is a rebellious soldier; it is sufficient.'

'He is *not*! He is a man who vindicated a woman's honour; he is a man who suffers in a brother's place; he is an aristocrat exiled to a martyrdom; he is a hero who has never been greater than he will be great in his last hour. Read that! What you refuse to justice, and mercy, and courage, and guiltlessness, you will grant, maybe, to your Order.'

She forced into his hand the written statement of Cecil's name and station. All the hot blood was back in her cheek, all the fiery passion back in her eyes. She lashed this potent ruler with the scourge of her scorn as she had lashed a drunken horde of plunderers with her whip. She was reckless of what she said; she was conscious only of one thing—the despair that consumed her.

The French Marshal glanced his eye on the fragment, carelessly and coldly. As he saw the words, he started, and read on with wondering eagerness.

'Royallieu!' he muttered—'Royallieu!'

The name was familiar to him; he it was who, when he had murmured, 'That man has the seat of the English Guards,' as a Chasseur d'Afrique had passed him, had been ignorant that in that Chasseur he saw one whom he had known in many a scene of Court splendour and Parisian pleasure. The years had been many since Cecil and he had met, but not so many but that the name brought memories of friendship with it, and moved him with emotion.

He turned with grave anxiety to Cigarette.

'You speak strangely; how came this in your hands?'

'Thus: the day that you gave me the Cross, I saw Madame la Princesse Corona. I hated her, and I went—no matter! From her I learned that he whom we call Louis Victor was of her rank, was of old friendship with her house, was exiled and nameless, but for some reason unknown to her. She needed to see him; to bid him farewell, so she said. I took the message for her; I sent him to her.' Her voice grew husky and savage, but she forced her words on with the reckless sacrifice of self that moved her. 'He went to her tent, alone, at night; that was, of course, whence he came when Châteauroy met him. I doubt not the Black Hawk had some foul thing to hint of his visit, and that the blow was struck for her—for her! Well; in the streets of Algiers I saw a man with a face like his own; different, but the same race, look you. I spoke to him; I taxed him. When he found that the one whom I spoke of was under sentence of death, he grew mad—he cried out that he was his brother, and had murdered him; that it was for his sake that the cruelty of this exile had been borne, and that if his brother perished, he would be his destroyer. Then I bade him write down that paper, since these English names were unknown to me, and I brought it hither to you that you might see under his hand and with your own eyes that I have uttered the truth. And now is that man to be killed like a mad beast whom you fear? Is that death the reward France will give for Zaraïla?'

Her eyes were fixed with a fearful intensity of appeal upon the stern face bent above her; her last arrow was sped; if this failed, all was over. As he heard, he was visibly moved; he remembered the felon's shame that in years gone by had fallen across the banished name of Bertie Cecil; the history seemed clear as crystal to him seen beneath the light shed on it from other days.

His hand fell heavily on the gun-carriage.

'Mort de Dieu! it was his brother's sin, not his!'

There was a long silence; those present who knew nothing of all that was in his memory felt instinctively that some dead weight of alien guilt was lifted off a blameless life for ever.

She drew a deep, long, sighing breath; she knew that he was safe. Her hands unconsciously locked on the great Chief's arms; her eyes looked up, senselessly in their rapture and their dread, to his.

'Quick, quick!' she gasped. 'The hours go so fast; while we speak here, he——'

The words died in her throat. The Marshal swung round with a rapid sign to a staff officer.

'Pens and ink! instantly! My brave child, what can we say to you? I will send an aide-de-camp to arrest the execution of the sentence. It must be deferred till we know the whole truth of this; if it be as it looks now, he shall be saved if the Empire can save him!'

She looked up in his eyes with a look that froze his very heart.

'His honour!' she muttered; 'his honour—if not his life!'

He understood her; he bowed his haughty head low down to hers.

'True. We will cleanse that, if all other justice be too late.'

The answer was infinitely gentle, infinitely solemn. Then he turned and wrote his hurried order, and bade his aide-de-camp go with it without a second's loss; but Cigarette caught it from his hand.

'To me! to me! No other will go so fast!'

'But, my child, you are worn out already.'

She turned on him her beautiful wild eyes, in which the blinding passionate tears were floating.

'Do you think I would tarry for *that*? Ah! I wish that I had let them tell me of God, that I might ask Him now to bless you! Quick, quick! Lend me your swiftest horse, one that will not tire. And send a second order by your officer; the Arabs may kill me as I go, and then they will not know!'

He stooped and touched her little brown, scorched, feverish hand with reverence.

'My child, Africa has shown me much heroism, but none like yours. If you fall, *he* shall be safe, and France will know how to avenge its darling's loss.'

She turned and gave him one look, infinitely sweet, infinitely eloquent.

'Ah!—France!' she said, so softly that the word was but a sigh of unutterable tenderness. The old imperishable early love was not dethroned; it was there still before all else. France was without rival with her.

Then, without another second's pause, she flew from them, and vaulting into the saddle of a young horse which stood without in the courtyard, rode once more, at full speed, out into the pitiless blaze of the sun, out to the wasted desolation of the plains.

The order of release, indeed, was in her bosom; but the chances were as a million to one that she would reach him with it in time, ere with the rising of the sun his life would have set for ever.

All the horror of remorse was on her; the bitter jealousy in which she had desired vengeance on him seemed to have rendered her a murderess. She loved him with an exceeding passion; and only in this extremity, when it was confronted with the imminence of death, did the fullness and the greatness of that love make their way out of the petulant pride and the wounded vanity which had obscured them. She had been ere now a child and a hero; beneath this blow which struck at him she changed— she became a woman and a martyr.

And she rode at full speed through the night, as she had done through the daylight, her eyes glancing all around in the keen instinct of a trooper, her hand always on the butt of her belt pistol. For she knew well what the danger was of these lonely, unguarded, untravelled leagues that yawned in such distance between her and her goal.

The Arabs, beaten, but only rendered furious by defeat, swept down on to those plains with the old guerilla skill, the old marvellous rapidity. She knew that with every second shot or steel might send her reeling from her saddle, that with every moment she might be surrounded by some desperate band who would spare neither her sex nor her youth. But this intoxication of peril, the wine-draught she had drunk from her infancy, was all which sustained her in that race with death. It filled her veins with their old heat, her heart with its old daring, her nerves with their old matchless courage: but for it she would have dropped, heart-sick with terror and despair, ere her errand could have been done; under it she had the coolness, the keenness, the sagacity, the sustained force, and the supernatural strength of some young hunted animal. They might slay her so that she left perforce her mission unaccomplished; but no dread of such a fate had power to appal her or arrest her. While there should be breath in her, she would go on to the end.

There were many hours' hard riding before her, at the swiftest pace her horse could make; and she was already worn by the leagues already traversed. Although this was nothing new that she did now, yet as time flew on and she flew with it, ceaselessly, through the dim solitary barren moonlit land, her brain now and then grew giddy, her heart now and then stood still with a sudden numbing faintness. She shook the weakness of her with the resolute scorn for it of her nature, and succeeded in its banishment. They had put in her hand, as she had passed through the fortress gates, a lance with a lantern muffled in Arab fashion, so that the light was unseen from before, while it streamed over her herself, to enable her to guide her way if the moon should be veiled by clouds. With that single starry gleam, aslant on a level with her eyes, she rode through the ghastly twilight of the half-lit plains, now flooded with lustre as the moon emerged, now engulfed in darkness as the stormy western winds drove the cirri over it. But neither darkness nor light differed to her; she noted neither; she was like one drunk with strong wine, and she had but one dread—that her horse would give way under the unnatural strain, and that she would reach too late, when the life she went to save would have fallen for ever, silent unto death, as she had seen the life of Marquise fall.

Hour on hour, league on league, passed away; she felt the animal quiver under the spur, and she heard the catch in his panting breath as he strained to give his fleetest and best that told her how, ere long, the racing speed, the extended gallop, at which she kept him, would tell, and beat him down despite his desert strain. She had no pity; she would have killed twenty horses under her to reach her goal; she was giving her own life, she was willing to lose it if by its loss she did this thing, to save the man condemned to die with the rising of the sun. She did not spare herself; and she would have spared no living thing, to fulfil the mission that she undertook. She loved with the passionate blindness of her sex, with the absolute abandonment of the southern blood. If to spare him she must have bidden thousands fall, she would have given the word for their destruction without a moment's pause.

Once from some screen of gaunt and barren rock a shot was fired at her, and flew within a hair's breadth of her brain; she never even looked around to see whence it had come; she knew it was from some Arab prowler of the plains. Her single spark of light through the half-veiled lantern passed as swiftly as a shooting-star across the plateau. And as she felt the hours steal on—so fast, so hideously fast—with that horrible relentlessness, 'ohne hast, ohne rast', which tarries for no despair, as it hastens for no desire, her lips grew dry as dust, her tongue clove to the roof of her mouth, the blood beat like a thousand hammers on her brain.

What she dreaded came.

Midway in her course, when, by the stars, she knew midnight was past, the horse vainly strove with hard-drawn gasps to answer the demand made on him by the spur and by the lance-shaft with which he was goaded onward. In the lantern-light she saw his head stretched out in the racing agony, his distended eyeballs, his neck covered with foam and blood, his heaving flanks that seem bursting with every throb that his heart gave; she knew that, half a league more forced from him, and he would drop like a dead thing never to rise again. Her eyes swept over the dusky plains with all the horror in them of one who sees murder done and cannot raise his hand to ward it off: the dawn was near at hand, and the leagues betwixt her and the camp were so many! She let the bridle drop upon the poor beast's neck, and threw her arms above her head with a shrill wailing cry, whose despair echoed over the noiseless plains like the cry of a shot-stricken animal. She saw it all: the rising of the rosy golden day; the stillness of the hushed camp; the tread of the few picked men; the open coffin by the open grave; the levelled carbines gleaming in the first rays of the sun. . . . She had seen it so many times—seen it, to the awful end, when the living man fell down in the morning light a shattered, senseless, soulless, crushed-out mass.

That single moment was all the soldier's nature in her gave to the abandonment of despair, to the paralysis that seized her. With that one cry from the depths of her breaking heart, the weakness spent itself: she knew that action alone could aid him. She looked across, southward and northward, east and west, to see if there were aught near from which she could get aid. If there were none, the horse must drop down to die, and with his life the other life would perish as surely as the sun would rise.

Her eyes caught sight in the distance of some dark thing moving rapidly—a large cloud skimming the earth. She kept her gaze fixed on the advancing cloud, till, with the marvellous surety of her desert-trained vision, she disentangled it from the floating mists and wavering shadows, and recognised it, as it was—a band of Arabs.

If she turned eastward out of her route, the failing strength of her horse would be fully enough to take her into safety from their pursuit, or even from their perception, for they were coming straightly and swiftly across the plain. If she were seen by them, she was certain of her fate. They could only be the desperate remnant of the decimated tribes—the foraging raiders of starving and desperate men, hunted from refuge to refuge, and carrying fire and sword in their vengeance wherever an unprotected caravan or a defenceless settlement gave them the power of plunder and of slaughter, that spared neither age nor sex. She was known throughout the length and the breadth of the land to the Arabs: she was neither child nor woman to them; she was but the soldier who had brought up the French reserve at Zarâila; she was but the foe who had seen them defeated, and ridden down with her comrades in their pursuit in twice a score of vanquished, bitter, intolerably shameful days. Some amongst them had sworn by their God to put her to a fearful death if ever they made her captive, for they held her in superstitious awe, and thought the spell of the Frankish successes would be broken if she were slain. She knew that; yet, knowing it, she looked at their advancing band one moment, then turned her horse's head and rode straight towards them.

'They will kill me, but that may save him,' she thought. 'Any other way he is lost.'

So she rode directly towards them; rode so that she crossed their front, and placed herself in their path, standing quite still, with the cloth torn from the lantern, so that its light fell full about her, as she held it above her head. In an instant they knew her. They were some two score of the remnant who had escaped from the carnage of Zarâila; they recognized her with all the rapid unerring surety of hate. They gave the shrill wild war-shout of their tribe, and the whole mass of gaunt, dark, mounted figures with their weapons whirling round their heads enclosed her; a cloud of kites settled down with their black wings and cruel beaks upon one young silvery-plumed gerfalcon.

She sat unmoved, and looked up at the naked blades that flashed above her: there was no fear upon her face, only a calm resolute proud beauty, very pale, very still in the light that gleamed on it from the lantern-rays.

'I surrender,' she said, briefly: she had never thought to say these words of submission to her scorned foes; she would not have been brought to utter them to spare her own existence. Their answer was a yell of furious delight, and their bare blades smote each other with a clash of brutal joy: they had her, the Frankish child who had brought shame and destruction on them at Zarâila, and they longed to draw their steel through the fair young throat, to plunge their lances into the bright bare bosom, to twine her hair round their spear handles, to rend her delicate limbs apart, as a tiger rends the antelope, to torture, to outrage, to wreak their vengeance on her.

Their chief, only, motioned their violence back from her, and bade them leave her untouched. At him she looked, still with the same fixed, serene, scornful resolve: she had encountered these men often in battle, she knew well how rich a prize she was to him. But she had one thought alone with her; and for it she subdued contempt, and hate, and pride, and every passion in her.

'I surrender,' she said, with the same tranquillity. 'I have heard that you have sworn by your God and your Prophet to tear me limb from limb because that I—a child, and a woman-child—brought you to shame and to grief on the day of Zarâila. Well, I am here; do it. You can slake your will on me. But inasmuch as you are brave men, and as I have ever met you in fair fight, let me speak one word with you first.'

Through the menaces and the rage around her, fierce as the yelling of starving wolves around a frozen corpse, her clear brave tones reached the ear of the chief in the lingua-sabir that she used. He was a young man, and his ear was caught by that tuneful voice, his eyes by that youthful face with the glow of the lantern-light upon its colourless, scornful, earnest features. He signed upward the swords of his followers, and motioned them back as their arms were stretched to seize her, and their shouts clamoured for her slaughter.

'Speak on,' he said briefly to her.

'You have sworn to take my body, sawn in two, to Ben-Ihreddin?' she pursued, naming the Arab leader whom her Spahis had driven off the field of Zarâila. 'Well, here it is; you can take it to him; and you will receive the piasters, and the horse, and the arms that he has promised to whosoever shall slay me. I have surrendered; I am yours. But you are bold men, and the bold are never mean; therefore I will ask one thing of you. There is a man yonder, in my camp, condemned to death with the dawn. He is innocent. I have ridden from Algiers today with the order of his release. If it is not there by sunrise, he will be shot; and he is guiltless as a child unborn. My horse is worn out; he could not go another half league. I knew that, since he had failed, my comrade would perish, unless I found a fresh beast or a messenger to go in my stead. I saw your band come across the plain. I knew that you would kill me, because of your oath and of your Emir's bribe; but I thought that you would have greatness enough in you to save this man who is condemned, without crime, who must perish unless you, his foes, have pity on him. Therefore I came. Take the paper that frees him; send your fleetest and surest with it, under a flag of truce, into our camp by the dawn; let him tell them there that I, Cigarette, gave it him—he must say no word of what you have done to me, or his white flag will not protect him from the vengeance of my Army—and then receive your reward from your Emir when you lay my head down for his horse's hoofs to trample into the dust. Answer me—is the pact fair? Ride on with this paper northward, and then kill me with what torments you choose.'

She spoke with calm unwavering resolve, meaning that which she uttered to its very uttermost letter. She knew that these men had thirsted for her blood; she offered it to be shed to gain for him that messenger on whose speed his life was hanging; she knew that a price was set upon her head, but she delivered herself over to the hands of her tormentors so that thereby she might purchase his redemption.

As they heard, silence fell upon the brutal clamorous herd around—the silence of amaze and of respect. The young chief listened gravely; by the glistening of his keen black eyes, he was surprised and moved, though, true to his teaching, he showed neither emotion as he answered her:

'Who is this Frank for whom you do this thing?'

'He is the warrior to whom you offered life on the field of Zarâila because his courage was as the courage of gods.'

She knew the qualities of the desert character; knew how to appeal to its reverence and to its chivalry.

'And for what does he perish?' he asked.

'Because he forgot for once that he was a slave; and because he has borne the burden of a guilt that was not his own.'

They were quite still now, closed around her; these ferocious plunderers, who had been thirsty a moment before to sheathe their weapons in her body, were spellbound by the sympathy of courageous souls; some vague perception that there was a greatness in this little tigress of France, whom they had sworn to hunt down and slaughter, which surpassed all they had known or dreamed.

'And you have given yourself up to us, that by your death you may purchase a messenger from us for this errand?' pursued their leader. He had been reared as a boy in the high tenets and the pure chivalries of the school of Abd-el-Kader; and they were not lost in him, despite the crimes and the desperation of his life.

She held the paper out to him with a passionate entreaty breaking through the enforced calm of despair with which she had hitherto spoken.

'Cut me in ten thousand pieces with your swords, but save *him*, as you are brave men, as you are generous foes!'

Then, with a single sign of his hand, their leader waved them back where they crowded around her, and leaped down from his saddle, and led the horse he had dismounted to her.

'Maiden,' he said, gently, 'we are Arabs, but we are not brutes. We swore to avenge ourselves on an enemy; we are not vile enough to accept a martyrdom. Take my horse—he is the swiftest of my troop—and go you on your errand; you are safe from me.'

She looked at him in stupor; the sense of his words was not tangible to her; she had had no hope, no thought, that they would ever deal thus with her; all she had ever dreamed of was so to touch their hearts and their generosity that they would spare one from amongst their troop to do the errand of mercy she had begged of them.

'You play with me!' she murmured, while her lips grew whiter and her great eyes larger in the intensity of her emotion. 'Ah! for pity's sake make haste and kill me, so that only this may reach him!'

The chief, standing by her, lifted her in his sinewy arms, up on to the saddle of his charger. His voice was very solemn, his glance was very gentle; all the nobility of the highest Arab nature was aroused in him at the heroism of a child, a girl, an infidel—one, in his sight, abandoned and shameful amongst her sex.

'Go in peace,' he said, simply; 'it is not with such as thee that we war.'

Then, and then only, as she felt the fresh reins placed in her hand, and saw the ruthless horde around her fall back and leave her free, did she understand his meaning, did she comprehend that he gave her back both liberty and life, and, with the surrender of the horse he loved, the noblest and most precious gift that the Arab ever bestows or ever receives. The unutterable joy seemed to blind her, and gleam upon her face like the blazing light of noon, as she turned her burning eyes full on him.

'Ah! now I believe that thine Allah is a greater god than the god of the Christians! If I live, thou shalt see me back ere another night; if I die, France will know how to thank thee!'

'We do not do the thing that is right for the sake that men may recompense us,' he answered her, gently. 'Fly to thy friend, and, here-after, do not judge that those who are in arms against thee must needs be as the brutes that seek out whom they shall devour.'

Then, with one word in his own tongue, he bade the horse bear her southward, and, as swiftly as a spear launched from his hand, the animal obeyed him and flew across the plains. He looked after her awhile, through the dim tremulous darkness that seemed cleft by the rush of the gallop as the clouds are cleft by lightning, while his tribe sat silent in their saddles in moody unwilling consent, savage in that they had been deprived of prey, moved in that they were sensible of this martyrdom which had been offered to them.

'Verily the courage of a woman has put the best amongst us unto shame,' he said, rather to himself than them, as he mounted the stallion brought him from the rear and rode slowly northward, unconscious that the thing he had done was great, because conscious only that it was just.

And, borne by the fleetness of the desert-bred beast, she went away through the heavy bronze-hued dullness of the night. Her brain had no sense, her hands had no feeling, her eyes had no sight; the rushing as of waters was loud on her ears, the giddiness of fasting and of fatigue sent the gloom eddying round and round like a whirlpool of shadow. Yet she had remembrance enough left to ride on, and on, and on without once flinching from the agonies that racked her cramped limbs and throbbed in her beating temples; she had remembrance enough to strain her blind eyes towards the east and murmur, in her terror of that white dawn which must soon break, the only prayer that had been ever uttered by the lips no mother's kiss had ever touched:

'*O God! keep the day back!*'

CHAPTER XII

In the Midst of her Army

There was a line of light in the eastern sky. The camp was very still. It was the hour for the mounting of the guard, and, as the light spread higher and higher, whiter and whiter, as the morning came, a score of men advanced slowly and in silence to a broad strip of land screened from the great encampment by the rise and fall of the ground, and stretching far and even, with only here and there a single palm to break its surface, over which the immense arc of the sky bent, grey and serene, with only the one colourless gleam eastward that was changing imperceptibly into the warm red flush of opening day.

Sunrise and solitude: they were alike chosen lest the army that hon-oured, the comrades that loved him, should rise to his rescue, casting off the yoke of discipline, and remembering only that tyranny and that wretchedness under which they had seen him patient and unmoved throughout so many years of servitude.

He stood tranquil beside the coffin within which his broken limbs and shot-pierced corpse would so soon be laid for ever. There was a deep sadness on his face, but it was perfectly serene. To the words of the priest who approached him he listened with respect, though he gently declined the services of the Church. He had spoken but very little since his arrest; he was led out of the camp in silence, and waited in silence now, looking across the plains to where the dawn was growing richer and brighter with every moment that the numbered seconds of his life drifted slowly and surely away.

When they came near to bind the covering over his eyes, he motioned them away, taking the bandage from their hands and casting it far from him.

'Did I ever fear to look down the depths of my enemies' muskets?'

It was the single outbreak, the single reproach, that escaped from him—the single utterance by which he ever quoted his services to France. Not one who heard him dared again force on him that indignity which would have blinded his sight, as though he had ever dreaded to meet death.

That one protest having escaped from him, he was once more calm, as though the vacant grave yawning at his feet had been but a couch of down to rest his tired limbs. His eyes watched the daylight deepen, and widen, and grow into one sheet of glowing roseate warmth; but there was no regret in the gaze; there was a fixed fathomless resignation that moved with a vague sense of awe those who had come to slay him, and who had been so used to slaughter that they were wont to fire their volley into their comrade's breast as callously as into the ranks of their antagonists.

'It is best thus,' he thought, 'if only she never knows——'

Over the slope of brown and barren earth that screened the camp from view there came, at the very moment that the ramrods were drawn out with a shrill sharp ring from the carbine-barrels, a single figure, tall, stalwart, lithe, with the spring of the deerstalker in its rapid step, and the sinew of the northern races in its mould.

Cecil never saw it; he was looking at the east, at the deepening of the morning flush, and his head was turned away.

The newcomer went straight to the Adjutant in command, and addressed him with brief preface, hurriedly and low.

'Your prisoner is Victor of the Chasseurs?—he is to be shot this morning?'

The officer assented; he suffered the interruption, recognizing the rank of the speaker.

'I heard of it yesterday; I rode all night from Oran. I feel great pity for this man, though he is unknown to me,' the stranger pursued, in rapid whispered words. 'His crime was——?'

'A blow to his Colonel, Monseigneur.'

'And there is no possibility of a reprieve?'

'None.'

'May I speak with him an instant? I have heard it thought that he is of my country, and of a rank above his standing in his regiment here.'

'You may address him, M. le Duc; but be brief. Time presses.'

He thanked the officer for the unusual permission, and turned to approach the prisoner. At that moment Cecil turned also, and their eyes met. A great shuddering cry broke from them both; his head sank as though the bullets had already pierced his breast, and the friend who believed him dead stood gazing at him, paralysed.

For a moment there was an awful silence; then the Seraph's voice rang out with a horror in it that thrilled through the careless, callous hearts of the watching soldiery.

'Who is that man? *He* died—he died so long ago! And yet——'

Cecil's head was sunk on his chest; he never spoke, he never moved; he knew the helpless, hopeless misery that waited for the one who found him living only to find him also standing beside his open grave. He saw nothing; he only felt the crushing force of his friend's arms flung round him, as though seizing him to learn whether he were a living creature or a spectre dreamed of in delirium.

'*Who are you*? Answer me, for pity's sake!'

As the swift, hoarse, incredulous words poured on his ear, he, not seeking to unloose the other's hold, lifted his head and looked full in the eyes that had not met his own for twelve long years. In that one look all was uttered; the strained, eager, doubting gaze that read answer in it needed no other.

'You live still! Oh! thank God—thank God!'

And as the thanksgiving escaped him, he forgot all save the breathless joy of this resurrection; forgot that at their feet the yawning grave was open and unfilled. Then, and only then, under that recognition of the friendship that had never failed and never doubted, the courage of the condemned gave way, and his limbs shook with a great shiver of intolerable torture; and at the look that came upon his face, the man who loved him remembered all—remembered that he stood there in the morning light only to be shot down like a beast of prey.

Holding him there still, he swore a great oath that rolled like thunder down the air.

'You! and perishing here! If they send their shots through *you*, they shall reach me first in their passage! Why have you lived like this? Why have been lost to *me*, if you were dead to all the world beside?'

They were the words that his sister had spoken; Cecil's lips quivered as he heard, his voice was scarcely audible as it panted through them.

'I was accused——'

'Ay! But by whom? Not by me. Never by me!'

'God reward you. You have never doubted?'

'Doubted? Was your honour not as my own?'

'I can die at peace then; you know me guiltless——'

'Great Heaven! Death shall not touch you. As I stand here, not a hair of your head shall be harmed——'

'Hush! Justice must take its course. One thing only—has *she* heard?'

'Nothing. She is in Algiers. But you can be saved; you *shall* be saved; they do not know what they do!'

'Yes! They but follow the sentence of the law. Do not regret it. It is best thus.'

'Best!—that you should be slaughtered in cold blood!'

He knew what the demands of discipline exacted, he knew what the inexorable tyranny of the Army enforced, he knew that he had found the life lost to him for so long only to stand by and see it struck down like a shot stag's.

Cecil's eyes looked at him with a regard in which all the sacrifice, all the patience, all the martyrdom of his life were spoken.

'Best, because a lie I could never speak to you, and the truth I could never tell to you. Do not let your sister know; it might give her pain. I have loved her; that is useless, like all the rest. Give me your hand once more, and then—let them do their duty. Turn your head away; it will soon be over!'

Almost ere he asked it, his friend's hands closed upon both his own, keeping the promise made so long before in the old years gone: those gentle weary words rent his very soul, and he knew that he was powerless here; he knew that he could no more stay this doom of death than he could stay the rising of the sun up over the eastern heavens.

The clear voice of the officer in command rang shrilly through the stillness.

'Monseigneur, make your farewell. I can wait no longer.'

The Seraph started, and flung himself round with the grand challenge of a lion struck by a puny spear; his face flushed crimson; his words were choked in his throbbing throat.

'As I live, you shall not fire! I forbid you! I swear by my honour and the honour of England that he shall not perish. He is of my country; he is of my order. I will appeal to your Emperor; he will accord me his life the instant I ask it. Give me only an hour's reprieve—a few moments' space, to speak to your chiefs, to seek out your General——'

'It is impossible, Monseigneur.'

The curt, calm answer was inflexible: against the sentence and its execution there could be no appeal.

Cecil laid his hand upon his old friend's shoulders.

'It will be useless,' he murmured. 'Let them act; the quicker the better.'

'What! you think I can look on and see you die?'

'Would you had never known I lived——'

The officer made a gesture to the guard to separate them.

'Monseigneur, submit to the execution of the law, or I must arrest you.'

The Seraph flung off the detaining hand of the guard, and swung round, gazing close into the Adjutant's immovable face, which before that gaze lost its coldness and its rigour, and changed to a great pity for this stranger who had found the companion of his youth in the trooper who stood condemned to perish there.

'An hour's reprieve; for mercy's sake, grant that!'

'I have said, it is impossible.'

'But you do not dream who he is——'

'It matters not.'

'He is an English noble, I tell you——'

'He is a soldier who has broken the law; that suffices.'

'O Heaven! have you no humanity?'

'We have justice.'

'Justice! If you have justice, let your chiefs hear his story; let his name be made known; give me an hour's space to plead for him. Your Emperor would grant me his life, were he here; yield me an hour—a half-hour—anything that will give me time to save him——'

'It is out of the question; I must obey my orders. I regret you should have this pain; but if you do not cease to interfere, my soldiers must make you.'

Where the guards held him, Cecil saw and heard. His voice rose with all its old strength and sweetness:

'My friend, do not plead for me. For the sake of our common country and our old love, let us both meet this with silence and with courage.'

'You are a madman!' cried the man, whose heart felt breaking under this doom he could neither avert nor share. 'You think that they shall kill you before my eyes?—you think I shall stand by to see you murdered? What crime have you done? None, I dare swear, save being moved, under insult, to act as the men of your race ever acted! Why have lived as you have done? why not have trusted my faith and my love? If you had believed in my faith as I believed in your innocence, this misery never had come to us!'

'Hush! hush! or you will make me die like a coward.'

He dreaded lest he should do so; this ordeal was greater than his power to bear it. With the mere sound of this man's voice a longing, so intense in its despairing desire, came on him for this life which they were about to kill in him for ever.

The words stung his hearer well-nigh to madness; he turned on the soldiers with all the fury of his race that slumbered so long, but when it awoke was like the lion's rage. Invective, entreaty, conjuration, command, imploring prayer, and ungoverned passion poured in tumultuous words, in agonized eloquence, from his lips: all answer was a quick sign of the hand; and, ere he saw them, a dozen soldiers were round him, his arms were seized, his magnificent frame was held as powerless as a lassoed bull; for a moment there was a horrible struggle, then a score of ruthless hands locked him in as in iron gyves, and forced his mouth to silence and his eyes to blindness; this was all the mercy they could give—to spare him the sight of his friend's slaughter.

Cecil's eyes strained on him with one last longing look, then he raised his hand and gave the signal for his own death-shot:

'*Droit à mon cœur!*'

The levelled carbines covered him; he stood erect, with his face full towards the sun: ere they could fire, a shrill cry pierced the air:

'Wait! in the name of France.'

Dismounted, breathless, staggering, with her arms flung upward, and her face bloodless with fear, Cigarette came over the rising ground.

The cry of command pealed out upon the silence in the voice that the Army of Africa loved as the voice of their Little One. And the cry came too late; the volley was fired, the crash of sound thrilled across the words that bade them pause, the heavy smoke rolled out upon the air, the death that was doomed was dealt.

But beyond the smoke-cloud he staggered slightly, yet stood erect still, unharmed, grazed only by some few of the balls. The flash of fire had not been so fleet as the swiftness of her love; and on his breast she had thrown herself, and flung her arms about him, and tossed her head backward with her old dauntless sunlit smile as the balls pierced her bosom, and broke her limbs, and were turned away by that shield of warm young life from him.

She had saved him. She would perish in his stead.

Her arms were gliding from about his neck, and her failing limbs were sinking to the earth, as he caught her up ere she dropped to his feet.

'O God! my child! they have killed you!'

He suffered more, as the cry broke from him, than if the bullets had brought him that death which he saw at one glance had stricken down for ever all the glory of her childhood, all the gladness of her youth.

She laughed—all the clear, imperious, arch laughter of her sunniest hours unchanged.

'Chut! It is the powder and ball of France! *that* does not hurt. If it was an Arbico's bullet now! But wait! Here is the Marshal's order. He suspends your sentence; I have told him all. You are safe!—do you hear?—you are safe! How he looks! Is he grieved to live? *Mes Français!* tell him clearer than I can tell—here is the order. The General must have it. No—not out of my hand till the General sees it. Fetch him, some of you—fetch him to me.'

'Great Heaven! you have given your life for mine!'

The words broke from him in an agony as he held her upward against his heart, himself so blind, so stunned, with the sudden recall from death to life, and with the sacrifice whereby life was thus brought to him, that he could scarce see her face, scarce hear her voice, but only dimly, incredulously, terribly knew, in some vague sense, that she was dying, and dying thus for him.

She smiled up in his eyes, while even in that moment, when her limbs were broken down like a wounded bird's, and the shots had pierced through from her shoulder to her bosom, a hot scarlet flush came over her cheeks as she felt his touch and rested on his heart.

'A life! *Tiens!* what is it to give? We hold it in our hands every hour, we soldiers, and toss it in change for a draught of wine. Lay me down on the ground—at your feet—so! I shall live longest that way, and I have much to tell. How they crowd round! *Mes soldats*, do not make that grief and that rage over me. They are sorry they fired; that is foolish. They were only doing their duty, and they could not hear me in time.'

But the brave words could not console those who had killed the Child of the Tricolour; they flung their carbines away, they beat their breasts, they cursed themselves and the mother who had borne them; the silent, rigid, motionless phalanx that had stood there in the dawn to see death dealt in the inexorable penalty of the law was broken up into a tumultuous, breathless, heart-stricken, infuriated throng, maddened with remorse, convulsed with sorrow, turning wild eyes of hate on him as on the cause through which their darling had been stricken. He, laying her down with unspeakable gentleness as she had bidden him, hung over her, leaning her head against his arm, and watching in paralysed horror the helplessness of the fleeting spirit, the slow flowing of the blood beneath the Cross that shone where that young heroic heart so soon would beat no more.

'Oh, my child, my child!' he moaned, as the full might and meaning of this devotion which had saved him at such cost rushed on him. 'What am I worth that you should perish for me? Better a thousand times have left me to my fate! Such nobility, such sacrifice, such love——'

The hot colour flushed her face once more; she was strong to the last to conceal that passion for which she was still content to perish in her youth.

'Chut! We are comrades, and you are a brave man. I would do the same for any of my Spahis. Look you, I never heard of your arrest till I heard too of your sentence. Then I knew it was too late, unless I could get to the Chief. So I went——'

She paused a moment, and her features grew white, and quivered with the pain of the death-wounds, and with the oppression that seemed to lie like lead upon her chest. But she forced herself to be stronger than the anguish which assailed her strength; and she motioned them all to be silent as she spoke on while her voice still should serve her.

'They will tell you how I did it—I have not time. The Marshal gave his word you shall be saved; there is no fear. That is your friend who bends over me, is it not? A fair face—a brave face. You will go back to your land—you will live among your own people—and *she*, she will love you now—now she knows you are of her Order!'

Something of the old thrill of jealous dread and hate quivered through the words, but the purer, nobler nature vanquished it; she smiled up in his eyes, heedless of the tumult round them:

'You will be happy. That is well. Look you—it is nothing that I did. I would have done it for any one of my soldiers. And for this'—she touched the blood flowing from her side with the old, bright, brave smile—'it was an accident; they must not grieve for it. My men are good to me; they will feel such regret and remorse; but do not let them. I am glad to die.'

The words were unwavering and heroic, but for one moment a convulsion went over her face; the young life was so strong in her, the young spirit was so joyous in her, existence was so new, so fresh, so bright, so dauntless a thing to Cigarette. She loved life: the darkness, the loneliness, the annihilation of death were horrible to her as the blackness and the solitude of night to a young child. Death, like night, can be welcome only to the weary, and she was weary of nothing on the earth that bore her buoyant steps; the suns, the winds, the delights of the sight, the joys of the senses, the music of her own laughter, the mere pleasure of the air upon her cheeks, or of the blue sky above her head, were all so sweet to her. Her welcome of her death-shot was the only untruth that had ever soiled her fearless lips. Death was terrible; yet she was content—content to have come to it for his sake.

There was a ghastly stricken silence round her. The order she had brought had just been glanced at, but no other thought was with the most callous there than the heroism of her act, than the martyrdom of her death.

The colour was fast passing from her lips, and a mortal pallor settling there in the stead of that rich brght hue, once warm as the scarlet heart of the pomegranate. Her head leant back on Cecil's breast, and she felt the great burning tears fall one by one upon her brow as he hung speechless over her: she put her hand upward and touched his eyes softly:

'Chut! What is it to die—just to die? You have *lived* your martyrdom; I could not have done that. Listen, just one moment. You will be rich. Take care of the old man—he will not trouble you long—and of Vole-qui-veut and Etoile, and Boule Blanche, and the rat, and all the dogs, will you? They will show you the Château de Cigarette in Algiers. I should not like to think that they would starve.'

She felt his lips move with the promise he could not find voice to utter; and she thanked him with that old childlike smile which had lost nothing of its light:

'That is good; they will be happy with you. And see that the Arab has back his horse—it was he who saved you; they must never harm him. And make my grave somewhere where my Army passes; where I can hear the trumpets, and the arms, and the passage of the troops—O God! I forgot! I shall not wake when the bugles sound. It will all *end* now, will it not? That is horrible, horrible!'

A shudder shook her as, for the moment, the full sense that all her glowing, redundant, sunlit, passionate life was crushed out for ever from its place upon the earth forced itself on and overwhelmed her. But she was of too brave a mould to suffer any foe—even the foe that conquers kings—to have power to appal her. She raised herself, and looked at the soldiery around her, amongst them the men whose carbines had killed her, whose anguish was like the heart-rending anguish of women.

'Mes Français! That was a foolish word of mine. How many of my bravest have fallen in death; and shall I be afraid of what they welcomed?

Do not grieve like that. You could not help it; you were doing your duty. If the shots had not come to me, they would have gone to him; and he has been unhappy so long, and borne wrong so patiently, he has earned the right to live and to enjoy. Now I—I have been happy all my days, like a bird, like a kitten, like a foal, just from being young and taking no thought. I should have had to suffer if I had lived; it is much best as it is——'

Her voice failed her when she had spoken the heroic words; loss of blood was fast draining all strength from her, and her limbs quivered in a torture she could not wholly conceal: he for whom she perished hung over her in wretchedness greater far than hers; it seemed a hideous dream to him that this child lay dying in his stead.

'Can nothing save her?' he cried aloud. 'O Christ! that you had fired one moment sooner!'

She heard; and looked up at him with a look in which all the hopeless, imperishable love she had resisted and concealed so long spoke with an intensity she never dreamed.

'She is content,' she whispered softly. 'You did not understand her rightly; that was all.'

'*All!* My God, how I have wronged you!'

The full strength and nobility of this devotion he had disbelieved in and neglected rushed on him as he met her eyes; for the first time he saw her as she was, for the first time he saw all of which the splendid heroism of this untrained nature would have been capable under a different fate. And it struck him suddenly, heavily, as with a blow; it filled him with a passion of remorse.

'My darling! My darling! what have I done to be worthy of such love?' he murmured, while the tears fell faster from his blinded eyes, and his head drooped until his lips met hers. At the first utterance of that word between them, at the unconscious fervour of his kisses that had the anguish of a farewell in them, the colour suddenly flushed all over her blanched face; she trembled in his arms; and a great shivering sigh ran through her. It came too late, this warmth of love. She learned what its sweetness might have been only when her lips grew numb, and her eyes sightless, and her heart without pulse, and her senses without consciousness.

'Hush!' she said, with a look that pierced his heart. 'Keep those kisses for Miladi. She will have the right to love you; she is of your "aristocrates", she is not "unsexed". As for me—I am only a little trooper who has saved my comrade! My soldiers, come round me one moment; I shall not long find words.'

Her eyes closed as she spoke; a deadly faintness and coldness passed over her; and she gasped for breath. A moment, and the resolute courage in her conquered: her eyes opened and rested on the war-worn faces of her 'children'—rested in a long, last look of unspeakable wistfulness and softness.

'I cannot speak as I would,' she said at length, while her voice grew very faint. 'But I have loved you. All is said!'

All was uttered in those four brief words: 'She had loved them.' The whole story of her young life was told in the single phrase. And the gaunt, battle-scarred, murderous, ruthless veterans of Africa who heard her could have turned their weapons against their own breasts, and sheathed them there, rather than have looked on to see their darling die.

'I have been too quick in anger sometimes—forgive it,' she said, gently. 'And do not fight and curse amongst yourselves; it is bad amidst brethren. Bury my Cross with me, if they will let you; and let the colours be over my grave, if you can. Think of me when you go into battle; and tell them in France——'

For she first time her own eyes filled with great tears as the name of her beloved land paused upon her lips; she stretched her arms out with a gesture of infinite longing, like a lost child that vainly seeks its mother.

'If I could only see France once more! France——'

It was the last word upon her utterance; her eyes met Cecil's in one fleeting upward glance of unutterable tenderness, then, with her hands still stretched out westward to where her country was, and with the dauntless heroism of her smile upon her face like light, she gave a tired sigh as of a child that sinks to sleep, and in the midst of her Army of Africa the Little One lay dead.

*   *   *

In the shadow of his tent at midnight, he whom she had rescued stood looking down at a bowed stricken form before him with an exceeding yearning pity in his gaze.

The words had at length been spoken that had lifted from him the burden of another's guilt: the hour at last had come in which his eyes had met the eyes of his friend, without a hidden thought between them. The sacrifice was ended; the martyrdom was over: henceforth this doom of exile and of wretchedness would be but as a hideous dream; henceforth his name would be stainless amongst men, and the desire of his heart would be given him. And in this hour of release the strongest feeling in him was the sadness of an infinite compassion; and where his brother was stretched prostrate in shame before him, Cecil stooped and raised him tenderly.

'Say no more,' he murmured. 'It has been well for me that I have suffered these things. For yourself—if you do indeed repent and feel that you owe me any debt, atone for it, and pay it by letting your own life be strong in truth and fair in honour.'

And it seemed to him that he himself had done no great or righteous thing in that servitude for another's sake, whose yoke was now lifted off him for evermore: but, looking out over the sleepless camp where one young child alone lay in a slumber that never would be broken, his heart ached with the sense of some great priceless gift received, and undeserved, and cast aside, even whilst in the dreams of passion that now knew its fruition possible, and the sweetness of communion with the friend whose faith had never forsaken him, he retraced the years of his exile, and thanked God that it was thus with him at the end.

CHAPTER XIII

At Rest

Under the green spring-tide leafage of English woodlands, made musical with the movement and the song of innumerable birds that had their nests among the hawthorn-boughs and deep cool foliage of elm and beech, an old horse stood at pasture. Sleeping, with the sun on his grey silken skin, and the flies driven off with a dreamy switch of his tail, and the grasses odorous about his hoofs, from dog-violets, and cowslips, and wild thyme: sleeping, yet not so surely, but at one voice he started, and raised his head with all the eager grace of his youth, and gave a murmuring noise of welcome and delight. He had known that voice in an instant, though for so many years his ear had never thrilled to it: Forest King had never forgotten.

Now, scarce a day passed but what it spoke to him some word of greeting or of affection, and his black soft eyes would gleam with their old fire, because its tone brought back a thousand memories of bygone victory—only memories now, when Forest King, in the years of age, dreamed out his happy life under the fragrant shade of the forest wealth of Royallieu.

With his arm over the horse's neck, the exile, who had returned to his birthright, stood silent awhile, gazing out over the land on which his eyes never wearied of resting; the glad, cool, green, dew-freshened earth that was so sweet and full of peace, after the scorched and blood-stained plains, whose sun was as flame, and whose breath was as pestilence. Then his glance came back and dwelt upon the face beside him, the proud and splendid woman's face that had learned its softness and its passion from him alone.

'It was worth banishment to return,' he murmured to her. 'It was worth the trials that I bore to learn the love that I have known——'

She, looking upward at him with those deep, lustrous, imperial eyes that had first met his own in the glare of the African noon, passed her hand over his lips with a gesture of tenderness far more eloquent from her than from women less prone to weakness.

'Ah, hush! when I think of what *her* love was, how worthless looks my own! how little worthy of the fate it finds! What have I done that every joy should become mine, when she——'

Her mouth trembled, and the phrase died unfinished; strong as her own love had grown, it looked to her unproven and without desert, beside that which had chosen to perish for his sake. And where they stood with the future as fair beyond them as the light of the day around them, he bowed his head as before some sacred thing at the whisper of the child who had died for him. The thoughts of both went back to a place in a desert land where the folds of the tricolour drooped over one little grave turned westward towards the shores of France—a grave, made where the beat of drum, and the sound of moving squadrons, and the ring of the trumpet-call, and the noise of the assembling battalions could be heard by night and day; a grave, where the troops as they passed it by, saluted and lowered their arms in tender reverence, in faithful unasked homage, because beneath the Flag they honoured there was carved in the white stone one name that spoke to every heart within the Army she had loved, one name on which the Arab sun streamed as with a martyr's glory:

CIGARETTE,
ENFANT DE L'ARMÉE, SOLDAT DE LA FRANCE.

OXFORD

MORE OXFORD PAPERBACKS

This book is just one of nearly 1000 Oxford Paperbacks currently in print. If you would like details of other Oxford Paperbacks, including titles in the World's Classics, Oxford Reference, Oxford Books, OPUS, Past Masters, Oxford Authors, and Oxford Shakespeare series, please write to:

UK and Europe: Oxford Paperbacks Publicity Manager, Arts and Reference Publicity Department, Oxford University Press, Walton Street, Oxford OX2 6DP.

Customers in UK and Europe will find Oxford Paperbacks available in all good bookshops. But in case of difficulty please send orders to the Cash-with-Order Department, Oxford University Press Distribution Services, Saxon Way West, Corby, Northants NN18 9ES. Tel: 0536 741519; Fax: 0536 746337. Please send a cheque for the total cost of the books, plus £1.75 postage and packing for orders under £20; £2.75 for orders over £20. Customers outside the UK should add 10% of the cost of the books for postage and packing.

USA: Oxford Paperbacks Marketing Manager, Oxford University Press, Inc., 200 Madison Avenue, New York, N.Y. 10016.

Canada: Trade Department, Oxford University Press, 70 Wynford Drive, Don Mills, Ontario M3C 1J9.

Australia: Trade Marketing Manager, Oxford University Press, G.P.O. Box 2784Y, Melbourne 3001, Victoria.

South Africa: Oxford University Press, P.O. Box 1141, Cape Town 8000.

OXFORD BOOKS

THE OXFORD BOOK OF ENGLISH GHOST STORIES

Chosen by Michael Cox and R. A. Gilbert

This anthology includes some of the best and most frightening ghost stories ever written, including M. R. James's 'Oh Whistle, and I'll Come to You, My Lad', 'The Monkey's Paw' by W. W. Jacobs, and H. G. Wells's 'The Red Room'. The important contribution of women writers to the genre is represented by stories such as Amelia Edwards's 'The Phantom Coach', Edith Wharton's 'Mr Jones', and Elizabeth Bowen's 'Hand in Glove'.

As the editors stress in their informative introduction, a good ghost story, though it may raise many profound questions about life and death, entertains as much as it unsettles us, and the best writers are careful to satisfy what Virginia Woolf called 'the strange human craving for the pleasure of feeling afraid'. This anthology, the first to present the full range of classic English ghost fiction, similarly combines a serious literary purpose with the plain intention of arousing pleasing fear at the doings of the dead.

'an excellent cross-section of familiar and unfamiliar stories and guaranteed to delight' *New Statesman*

OXFORD BOOKS

THE NEW OXFORD BOOK OF IRISH VERSE

Edited, with Translations, by Thomas Kinsella

Verse in Irish, especially from the early and medieval periods, has long been felt to be the preserve of linguists and specialists, while Anglo-Irish poetry is usually seen as an adjunct to the English tradition. This original anthology approaches the Irish poetic tradition as a unity and presents a relationship between two major bodies of poetry that reflects a shared and painful history.

'the first coherent attempt to present the entire range of Irish poetry in both languages to an English-speaking readership' *Irish Times*

'a very satisfying and moving introduction to Irish poetry' *Listener*

THE OXFORD AUTHORS

General Editor: Frank Kermode

THE OXFORD AUTHORS is a series of authoritative editions of major English writers. Aimed at both students and general readers, each volume contains a generous selection of the best writings—poetry, prose, and letters—to give the essence of a writer's work and thinking. All the texts are complemented by essential notes, an introduction, chronology, and suggestions for further reading.

WORLD'S CLASSICS SHAKESPEARE

'not simply a better text but a new conception of Shakespeare. This is a major achievement of twentieth-century scholarship.' Times Literary Supplement

Hamlet
Macbeth
The Merchant of Venice
As You Like It
Henry IV Part I
Henry V
Measure for Measure
The Tempest
Much Ado About Nothing
All's Well that Ends Well
Love's Labours Lost
The Merry Wives of Windsor
The Taming of the Shrew
Titus Andronicus
Troilus & Cressida
The Two Noble Kinsmen
King John
Julius Caesar
Coriolanus
Anthony & Cleopatra

WORLD'S CLASSICS SHAKESPEARE
ALL'S WELL THAT ENDS WELL
Edited by Susan Snyder

Usually classified as a 'problem comedy', *All's Well That Ends Well* invites a fresh assessment. Its psychologically disturbing presentation of an aggressive, designing woman and a reluctant husband wooed by trickery won it little favour in earlier centuries, and both directors and critics have frequently tried to avoid or simplify its uncomfortable elements. More recently, several distinguished productions have revealed it as an exceptionally penetrating study of both personal and social issues.

In her introduction to *All's Well That Ends Well*, Susan Snyder makes the play's clashing ideologies of class and gender newly accessible. She explains how the very discords of style can be seen as a source of theatrical power and complexity, and offers a fully reconsidered, helpfully annotated text for both readers and actors.

WORLD'S CLASSICS SHAKESPEARE
HENRY V
Edited by Gary Taylor

Henry V, the climax of Shakespeare's sequence of English history plays, is an inspiring, often comic celebration of a young warrior-king. But it is also a study of the costly exhilarations of war, and of the penalties as well as the glories of human greatness.

Introducing this brilliantly innovative edition, Gary Taylor shows how Shakespeare shaped his historical material, examines controversial critical interpretations, discusses the play's fluctuating fortunes in performance, and analyses the range and variety of Shakespeare's characterization.

ILLUSTRATED HISTORIES IN OXFORD PAPERBACKS

THE OXFORD ILLUSTRATED HISTORY OF ENGLISH LITERATURE

Edited by Pat Rogers

Britain possesses a literary heritage which is almost unrivalled in the Western world. In this volume, the richness, diversity, and continuity of that tradition are explored by a group of Britain's foremost literary scholars.

Chapter by chapter the authors trace the history of English literature, from its first stirrings in Anglo-Saxon poetry to the present day. At its heart towers the figure of Shakespeare, who is accorded a special chapter to himself. Other major figures such as Chaucer, Milton, Donne, Wordsworth, Dickens, Eliot, and Auden are treated in depth, and the story is brought up to date with discussion of living authors such as Seamus Heaney and Edward Bond.

'[a] lovely volume . . . put in your thumb and pull out plums' Michael Foot

'scholarly and enthusiastic people have written inspiring essays that induce an eagerness in their readers to return to the writers they admire' *Economist*

OXFORD REFERENCE

THE CONCISE OXFORD COMPANION TO ENGLISH LITERATURE

Edited by Margaret Drabble and Jenny Stringer

Based on the immensely popular fifth edition of the *Oxford Companion to English Literature* this is an indispensable, compact guide to the central matter of English literature.

There are more than 5,000 entries on the lives and works of authors, poets, playwrights, essayists, philosophers, and historians; plot summaries of novels and plays; literary movements; fictional characters; legends; theatres; periodicals; and much more.

The book's sharpened focus on the English literature of the British Isles makes it especially convenient to use, but there is still generous coverage of the literature of other countries and of other disciplines which have influenced or been influenced by English literature.

From reviews of *The Oxford Companion to English Literature*:

'a book which one turns to with constant pleasure . . . a book with much style and little prejudice' Iain Gilchrist, *TLS*

'it is quite difficult to imagine, in this genre, a more useful publication' Frank Kermode, *London Review of Books*

'incarnates a living sense of tradition . . . sensitive not to fashion merely but to the spirit of the age' Christopher Ricks, *Sunday Times*

OXFORD POPULAR FICTION
THE ORIGINAL MILLION SELLERS!

This series boasts some of the most talked-about works of British and US fiction of the last 150 years—books that helped define the literary styles and genres of crime, historical fiction, romance, adventure, and social comedy, which modern readers enjoy.

Riders of the Purple Sage	Zane Grey
The Four Just Men	Edgar Wallace
Trilby	George Du Maurier
Trent's Last Case	E C Bentley
The Riddle of the Sands	Erskine Childers
Under Two Flags	Ouida
The Lost World	Arthur Conan Doyle
The Woman Who Did	Grant Allen

Forthcoming in October:

Olive	Dinah Craik
The Diary of a Nobody	George and Weedon Grossmith
The Lodger	Belloc Lowndes
The Wrong Box	Robert Louis Stevenson